SCHAUM'S SOLVED PROBLEMS SERIES

2500 SOLVED PROBLEMS IN

COLLEGE ALGEBRA AND TRIGONOMETRY

by

Philip Schmidt, Ph.D.
State University of New York
at New Paltz

McGRAW-HILL, INC.
New York St. Louis San Francisco Auckland Bogotá Caracas
Hamburg Lisbon London Madrid Mexico Milan Montreal
New Delhi Paris San Juan São Paulo Singapore
Sydney Tokyo Toronto

Philip Schmidt, Ph.D., *Associate Professor of Secondary Education, SUNY at New Paltz, New York*

Dr. Schmidt has a B.S. from Brooklyn College (with a major in mathematics) and an M.A. in mathematics and a Ph.D. in mathematics education from Syracuse University. He is currently professor of secondary education at SUNY New Paltz, where he coordinates the secondary mathematics education program. He is the author of *3000 Solved Problems in Precalculus* as well as numerous journal articles. He recently completed a revision of the late Barnett Rich's *Geometry*.

The following figures were reproduced with permission from Ayres, *Schaum's Outline Series: Theory and Problems of First Year College Mathematics*, McGraw-Hill, 1958: figures 5.11–5.13, 9.15–9.21, 9.32–9.45, 10.49–10.57, and 11.29–11.32.

Project supervision was done by The Total Book.

Library of Congress Cataloging-in-Publication Data

Schmidt, Philip A.
  2500 solved problems in college algebra and trigonometry / by
Philip Schmidt.
  p. cm.—(Schaum's solved problems series)
ISBN 0-07-055373-4
  1. Algebra—Problems, exercises, etc. 2. Trigonometry—Problems,
exercises, etc. I. Title. II. Title: Twenty-five hundred solved problems in college
algebra and trigonometry. III. Series.
QA157.S247 1991
512′.13′076—dc20   89-78522
                                                                CIP

1 2 3 4 5 6 7 8 9 0 SHP/SHP 9 5 4 3 2 1 0

ISBN 0-07-055373-4

CONTENTS

TO THE STUDENT

There are many ways in which you can use this book. If you are currently enrolled in a college algebra/trigonometry course, use the table of contents and index to locate problems for the topic that you are currently studying. Notice that for the most important (and most common) problem types, multiple examples are given. Read through several problems carefully, try to do the rest without looking at the given solutions, and then check yourself with the solutions. Keep in mind that for any topic, you should do as many problems as possible. You may not be able to solve them all; they range from very straightforward to moderate to difficult and very difficult. You will find that problem-solving techniques are scattered throughout the book: these include techniques for solving particular kinds of problems, as well as general mathematical techniques.

If you are using this book as a review (or study aid) prior to enrolling in a precalculus or calculus course, be aware that it can also help you review (or learn or relearn) topics in algebra and trigonometry *while you're in* the precalculus course.

I owe thanks to many people for their assistance with this book: John Aliano, Executive Editor at McGraw-Hill, for his great confidence; Cathy Decker-Coffey, for her superior typing of the manuscript; Gary Onderdonk, for his proofreading of the manuscript; my son, Reed Schmidt, for showing such excitement at a mathematics book; and my wife, Jan Zlotnik Schmidt, herself a writer, for providing me with loving support and large blocks of time so that this book could be completed.

CHAPTER 1

Review of Basic Algebra

1.1 SETS AND THE NATURAL NUMBERS

1.1 Use the listing method to describe the set $\{x \mid x$ is a consonant in the word "Mississippi"$\}$.

▌ The word "Mississippi" contains the consonants m, s, p. (*Recall*: We do not list the element s four times in the set even though it occurs four times in "Mississippi.") Thus, the set we are looking for is $\{m, s, p\}$.

1.2 Use set-builder notation to designate the set $\{1, 3, 5, 7, 9\}$.

▌ The numbers 1, 3, 5, 7, 9 can easily be described as the odd integers that are greater than or equal to 1 and less than or equal to 9. In set-builder notation, the set is $\{x \mid 1 \le x \le 9$ and x is an odd integer$\}$.

For Probs. 1.3, 1.4, and 1.5, let $A = \{1, 2, 7, 9, 14\}$ and $B = \varnothing$.

1.3 Find $A \cap B$.

▌ $A \cap B = \{x \mid x \in A$ and $x \in B\}$. In this case, 1, 2, 7, 9, and 14 belong to A, and there is nothing in B, so no element can belong to both sets. $A \cap B = \varnothing$.

1.4 Find $A \cup B$.

▌ Remember that, by definition, $A \cup B = \{x \mid x \in A$ or $x \in B\}$. In this case, 1, 2, 7, 9, and 14 belong to A, and B contains no elements. So $A \cup B = \{1, 2, 7, 9, 14\}$.

1.5 Find $A - B$.

▌ $A - B = \{x \mid x \in A$ and $x \notin B\}$. Since in this example there are no elements in B, $A - B = \{x \mid x \in A\} = \{1, 2, 7, 9, 14\}$. (See Fig. 1.1.)

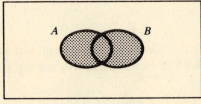

$A \cup B$ is shaded

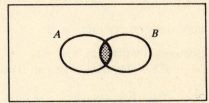

$A \cap B$ is shaded

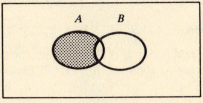

$A - B$ is shaded

Fig. 1.1

For Probs. 1.6, 1.7, and 1.8, let $A = \{1, 7, 11, 15\}$, $B = \{7, 14\}$, and $C = \{11, 26\}$.

1.6 Find $(A - B) \cup C$.

▎$A - B = \{x \mid x \in A \text{ and } x \notin B\} = \{1, 11, 15\}$ and $C = \{11, 26\}$. Thus, $(A - B) \cup C = \{1, 11, 15, 26\}$.

1.7 Find $B \times C$.

▎Recall that $B \times C = \{(b, c) \mid b \in B \text{ and } c \in C\}$. In this example, then, $B \times C = \{(7, 11), (7, 26), (14, 11), (14, 26)\}$.

1.8 Find $(A \cap C) \times C$.

▎$A \cap C = \{11\}$ (since 11 is the only element common to A and C). $(A \cap C) \times C = \{11\} \times \{11, 26\} = \{(11, 11), (11, 26)\}$.

1.9 Let $A = \{2, 6\}$ and $B = \{2, 6, 8\}$. Is $A \subseteq B$? Is $A \subset B$?

▎To determine whether $A \subseteq B$ ("A is a subset of B"), we need to find out whether every element of A is also an element of B. In this case, A contains 2 and 6, and both are elements of B, so $A \subseteq B$. To determine whether $A \subset B$ ("A is a proper subset of B"), we need to find out if $A \subseteq B$ and $A \neq B$. In this example, we already have determined that $A \subseteq B$ and since $A = \{2, 6\}$ and $B = \{2, 6, 8\}$, $A \neq B$. So $A \subset B$.

For Probs. 1.10 to 1.13, refer to Fig. 1.2.

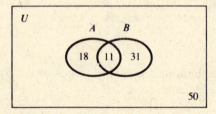

Fig. 1.2

1.10 How many elements are there in set B?

▎We notice that set B contains 31 elements that are not in set A and shares 11 other elements with A. We conclude that B contains 42 elements.

1.11 How many elements are there in B'?

▎B' (sometimes written as $\mathscr{C}B$) $= \{x \in U \mid x \notin B\}$, where U is the universal set. Here U contains $18 + 11 + 31 + 50 = 110$ elements, and B contains 42 elements. Thus, B' contains 68 elements.

1.12 How many elements are there in $A \cap B$? In $(A \cap B)'$?

▎From the Venn diagram, we can easily see that $A \cap B$ contains 11 elements. Thus $(A \cap B)' = \mathscr{C}(A \cap B) = \{x \in U \mid x \notin A \cap B\}$ contains $110 - 11 = 99$ elements.

1.13 How many elements are there in $A \cap B'$?

▎In Fig. 1.3, the crosshatched region is $A \cap B'$, that is, the region containing those elements common to A and B'. That region contains $18 + 11 = 29$ elements (refer to Fig. 1.2).

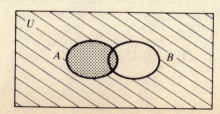

Fig. 1.3

1.14 How many subsets does $\{a, b\}$ contain? Exhibit these subsets.

❚ If set A contains n elements, where n is a whole number, then the number of subsets of $A = 2^n$. In this case, $n = 2$, so there are $2^2 = 4$ subsets. They are $\{a, b\}, \varnothing, \{a\}, \{b\}$.

1.15 Suppose that A contains m elements and B contains n elements, where m and n are natural numbers. Find the number of subsets of $A \times B$.

❚ $A \times B$ will contain $m \cdot n$ elements (since each element of A must be paired with each element of B). So, $A \times B$ will contain 2^{mn} subsets.

For Probs. 1.16 to 1.18, answer true or false, and explain your answer.

1.16 If $M \cap N = N$, then $N \subseteq M$.

❚ If $M \cap N = N$, then every element of N is also an element of $M \cap N$, which means it is an element of M. Thus, $N \subseteq M$. True.

1.17 The sets $\{ \ \}, \varnothing$, and $\{\varnothing\}$ all contain no elements.

❚ $\{ \ \}$ and $\varnothing$ both are empty, so they contain no elements. But $\{\varnothing\}$ contains the element $\varnothing$. So the answer is false.

1.18 If $M \cap N = \varnothing$, then at least one of M and N must be empty.

❚ Consider the set $M = \{1, 2\}$ and $N = \{e, f\}$. Now $M \cap N = \varnothing$ (they have no elements in common), but neither M nor N is empty. False.

1.19 Find a set containing only natural numbers that is equivalent to $\{a, e, \alpha\}$.

❚ Two sets are equivalent if they are in a one-to-one correspondence with each other. The set $\{1, 2, 3\}$ contains only natural numbers, and since $\{1, 2, 3\}, \ \{a, e, \alpha\}$, the sets are equivalent.

1.20 Suppose that m and n are natural numbers and $m \geq n$. Must $m - n$ be a natural number? Why?

❚ No. Suppose $m = F$ and $n = F$; then m is natural, and $m \geq n$ ($F \geq F$). But $m - n = 0$, and 0 is *not* a natural number. (0 is a whole number; 0 is an integer, but the smallest natural number is 1.)

1.2　REAL NUMBERS AND THE NUMBER LINE

For Probs. 1.21 to 1.27, all letters represent real numbers. For each statement, tell which of the following properties or definitions are used.

Commutative	Identity	Division
Associative	Inverse	Zero
Distributive	Subtraction	Negative

1.21 $(-1) + [-(-1)] = 0$

❚ For any real number a, $a + (-a) = 0$. We say $-a$ is the *additive identity* of a. Here replace a by -1, and $a + (-a) = 0 = (-1) + [-(-1)]$.

1.22 $7 \div 9 = 7(\frac{1}{9})$

❚ If t and u are real and $u \neq 0$, then $t \div u = t(1/u)$ by the definition of division. In this case, $t = 7$, $u = 9$ (and thus $u \neq 0$), and we have a direct application of the definition of division.

1.23 $1(-\frac{2}{3}) = -\frac{2}{3}$

❚ For any real number r, $1 \cdot r = r$. Here 1 is the *multiplicative identity*. In this case, r is simply replaced by $-\frac{2}{3}$.

1.24 $8 - 12 = 8 + (-12)$

▐ If a and b are real numbers, then $a - b$ is defined as $a + (-b)$. Thus, $8 - 12 = 8 + (-12)$ is an application of the definition of subtraction.

1.25 $7(s + t) = 7s + 7t$

▐ $a(b + c) = ab + ac$ is the *distributive law* for real numbers. Here $a = 7$, $b = s$, $c = t$.

1.26 $(3p + 9) + 3 = 3p + (9 + 3)$

▐ $(a + b) + c = a + (b + c)$ is the *associative law* for addition of real numbers. In this case, $a = 3p$, $b = 9$, $c = 3$.

1.27 $x + ym = x + my$

▐ Here ym is being replaced by my. And $ym = my$ is the *commutative law* for multiplication of real numbers.

For Probs. 1.28 to 1.30, arrange each list so that the items may be separated by $<$.

1.28 $\frac{2}{3}, -\frac{3}{4}, \frac{5}{6}, -1, \frac{4}{5}, -\frac{4}{3}, -\frac{1}{4}$

▐ Recall that $<$ means "less than." Its interpretation on the number line is "to the left of." Thus, $5 < 6$ means 5 is to the left of 6, or 5 is less than 6. In this problem, we arrange our list so that the numbers begin with the smallest and increase: $-\frac{4}{3} < -1 < -\frac{3}{4} < -\frac{1}{4} < \frac{2}{3} < \frac{4}{5} < \frac{5}{6}$.

1.29 $7, |14|, \sqrt{3}, 0.\overline{04}, \frac{3}{99}$

▐ Recall that $|14| = 14$, $1 < \sqrt{3} < 2$ and $0.\overline{04} = \frac{4}{99}$ (see 1.43 below). Thus, the solution here is $\frac{3}{99} < 0.\overline{04} < \sqrt{3} < 7 < |14|$.

1.30 $|-15|, 15, 0, 10$

▐ There is no solution here. The reason is that $|-15| = 15$. Thus, there is no way to place a $<$ between the first two (equal) elements in this list.

For Probs 1.31 through 1.37, perform the indicated operation.

1.31 $(-2)(3)(-5)$

▐ Recall that, for any two signed numbers of like sign, the product is nonnegative; for any two of unlike signs, the product is nonpositive. Thus

$$(-2)(3) = -6 \text{ (unlike signs)}$$

$$(6)(-5) = -30 \text{ (unlike signs)}$$

Therefore, $(-2)(3)(-5) = (-6)(-5)$

$$= +30$$

1.32 $40 + (-7)$

▐ To add two signed numbers of unlike sign, we subtract the smaller (in absolute value) from the larger and give the result the sign of the larger. Here 40 is larger than 7. Thus, we find $40 - 7 = 33$, and we use the sign of 40 (which is $+$). Thus, the answer is 33.

1.33 $14 + (-72)$

▐ See 1.32 above. Now $72 - 14 = 58$. Since $72 > 14$, we use the sign of 72 (which is $-$). Thus, $14 + (-72) = -58$.

1.34 $-8-(-6)+2$

$\blacksquare$ Remove parentheses first: $-(-6)=6$. Then

$$-8+6+2=(-8+6)+2$$
$$=-2+2$$
$$=0$$

1.35 $\dfrac{3-\frac{2}{3}}{5+\frac{5}{6}}$

$\blacksquare$ First simplify the numerator and denominator separately; then perform the division. (By the way, this is the first of several times you will be seeing complex fractions in this book.)

$$3-\tfrac{2}{3}=\tfrac{9}{3}-\tfrac{2}{3}=\tfrac{7}{3};$$
$$5+\tfrac{5}{6}=\tfrac{30}{6}+\tfrac{5}{6}=\tfrac{35}{6}.$$
$$\frac{\frac{7}{3}}{\frac{35}{6}}=\tfrac{7}{1}\cdot\tfrac{2}{35}=\tfrac{14}{35}=\tfrac{2}{5}.$$

1.36 $\dfrac{1\frac{1}{2}-2\frac{2}{3}}{3\frac{1}{5}-1\frac{1}{4}}$

$\blacksquare$ See Prob. 1.35.

$$1\tfrac{1}{2}-2\tfrac{2}{3}=\tfrac{3}{2}-\tfrac{8}{3}=\tfrac{9}{6}-\tfrac{16}{6}=-\tfrac{7}{6}$$
$$3\tfrac{1}{5}-1\tfrac{1}{4}=\tfrac{16}{5}-\tfrac{5}{4}=\tfrac{64}{20}-\tfrac{25}{20}=\tfrac{39}{20}$$
$$\frac{-\frac{7}{6}}{\frac{39}{20}}=\frac{-7}{6^{3}}\cdot\frac{20^{10}}{39}=\frac{-70}{117}.$$

[Remember that $-a/b=-(a/b)$.]

1.37 $\dfrac{4.9}{3-6.7}$

$\blacksquare$ $3-6.7=-3.7$ $\qquad$ (*Remember*: Use the sign of the larger number.)

$$\frac{4.9}{-3.7}=-1.32\overline{4324}\ldots$$
$$=-1.\overline{324}$$

(Do the division. Also the quotient is negative since the numerator's sign is different from the denominator's.)

For Probs. 1.38 to 1.42, simplify the given expression. All letters represent real numbers.

1.38 $10-(-7)$

$\blacksquare$ $10-(-7)=10+7=17$.

1.39 $[a+(-6)]+6$

$\blacksquare$ $[a+(-6)]+6=a+[(-6)+6]=a+0=a$.

1.40 $1-(1-b)$

$\blacksquare$ $1-(1-b)=1-1-(-b)=1-1+b=0+b=b$.

1.41 $2a-(b-2a)$

$\blacksquare$ $2a-(b-2a)=2a-b-(-2a)=2a-b+2a=2a+2a-b=4a-b$.

1.42 $(a - b)(-y)$

▮ $(a - b)(-y) = a(-y) - b(-y) = -ay + by$.

1.43 Convert the repeating decimal $0.04040\overline{4}$ to a fraction.

▮ Let $x = 0.040404\overline{04}$. Then $100x = 4.040404$, and $100x - x = 4.\overline{04} - 0.\overline{04} = 4$. So $99x = 4$ and $x = \frac{4}{99}$.

1.44 *Prove:* $(a + b) + (-a) = b$, for all real numbers a and b, justifying each step in your proof.

▮ $(a + b) + (-a) = (-a) + (a + b)$ (commutative law for addition) $= [(-a) + a] + b$ (associative law for addition) $= 0 + b$ (additive inverse) $= b$ (additive identity).

1.45 *Prove:* $a \cdot 0 = 0$ for every real number a.

▮ $a \cdot 0 = a(0 + 0)$ (additive identity) $= a \cdot 0 + a \cdot 0$ (distributive law). So $a \cdot 0 = a \cdot 0 + a \cdot 0$; but then $a \cdot 0$ must be the additive identity, since when you add it to $a \cdot 0$, you get $a \cdot 0$. Conclusion: $a \cdot 0 = 0$.

1.46 Evaluate each of the following: (*a*) $|-6|$, (*b*) $|1 - \sqrt{2}|$, (*c*) $7 - |-5|$, (*d*) $|-(-6)|$, (*e*) $|(-4) \cdot (6)|$.

▮ Recall that for any real number r, $|r| = \{r, r \geq 0; -r, r < 0\}$.

(*a*) Since $-6 < 0$, $|-6| = -(-6) = 6$.
(*b*) Since $1 - \sqrt{2} < 0$, $|1 - \sqrt{2}| = -(1 - \sqrt{2}) = \sqrt{2} - 1$.
(*c*) $7 - |-5| = 7 - [-(-5)] = 7 - 5 = 2$.
(*d*) $|-(-6)| = |6| = 6$.
(*e*) $|(-4) \cdot 6| = |-24| = 24$ or $|(-4) \cdot 6| = |-4| \cdot |6| = 4 \cdot 6 = 24$.

1.47 Find the distance between -4 and 17 on the number line.

▮ If a and b are points on the number line, the distance between a and b is denoted $d(a, b)$ or $d(b, a) = |a - b| = |b - a|$. In this case, $d(-4, 17) = |-4 - 17| = |-21| = 21$.

1.48 Tell whether each of the following is true or false, and why. All letters represent real numbers. (*a*) $a \leq a$. (*b*) If $a \leq 3$ and $3 \leq a$, then $a = 3$. (*c*) If $1 < a < 3$, then $a = 2$. (*d*) If $a < b$, then $-a < -b$. (*e*) If $a < b$, then $a^2 < b^2$.

▮ (*a*) $a \leq a$ is true for all real numbers since $\leq$ means "less than *or* equal to" and $a = a$ for all a.
(*b*) If $a \leq 3$, then a is 3 or lies to the left of 3 on the number line. If $3 \leq a$, then 3 is a or lies to the left of a on the number line. The only way this can happen is if $a = 3$. True.
(*c*) If $1 < a < 3$, then a could be 2, but it could also be 2.5 or $\pi - 1$, etc. False.
(*d*) Suppose $a < b$. Then $a - b < 0$. But if $a - b$ is negative, then $-(a - b)$ is positive. So $-(a - b) > 0$ or $-a + b > 0$, so $-a > -b$. So the given statement is false.
(*e*) This statement is false. Suppose $a = -2$ and $b = 1$. Then $a < b$. But $a^2 = 4$ and $b^2 = 1$, so $a^2 > b^2$.

1.49 *Prove:* $|x - y| = |y - x|$ for all real numbers x, y.

▮ Recall that for any real number a, $|a| = |-a|$. Thus, $|x - y| = |-(x - y)| = |-x + y| = |y - x|$.

1.3 INTEGRAL AND RATIONAL EXPONENTS

For Probs. 1.50 to 1.59, evaluate the given expression.

1.50 $7^6/7^4$

▮ $7^6/7^4 = 7^{6-4} = 7^2 = 49$. (*Recall:* $x^a/x^b = x^{a-b}$.)

1.51 $3^{41} \cdot 3^{-9}$

▮ Recall that $x^a \cdot x^b = x^{a+b}$. Thus, $3^{41} \cdot 3^{-9} = 3^{41+(-9)} = 3^{32}$.

1.52 $(2^6 \cdot 2^4)/(2^8 \cdot 2^9)$

$\blacksquare$ $(2^6 \cdot 2^4)/(2^8 \cdot 2^9) = 2^{10}/2^{17} = 2^{10-17} = 2^{-7} = 1/2^7 = \frac{1}{128}$.

1.53 $(2 + 2^{-1})/5 + (-8)^0 - 4^{3/2}$

$\blacksquare$ $(2 + 2^{-1})/5 + (-8)^0 - 4^{3/2} = (2 + \frac{1}{2})/5 + 1 - (4^{1/2})^3 = (\frac{5}{2})/5 + 1 - 2^3 = \frac{5}{10} + 1 - 8 = \frac{1}{2} - 7 = -6\frac{1}{2}$.

1.54 $(\frac{8}{27})^{2/3}$

$\blacksquare$ $(\frac{8}{27})^{2/3} = [(\frac{8}{27})^{1/3}]^2 = \frac{(8^{1/3})^2}{(27^{1/3})^2} = (\frac{2}{3})^2 = \frac{4}{9}$.

1.55 $(0.0001)^{3/4}$

$\blacksquare$ $(0.0001)^{3/4} = [(0.0001)^{1/4}]^3 = 0.001$.

1.56 $125^{-4/3}$

$\blacksquare$ $125^{-4/3} = (125^{1/3})^{-4} = 5^{-4} = 1/5^4 = \frac{1}{625}$.

1.57 $\sqrt[3]{\sqrt{\frac{1}{64}}}$

$\blacksquare$ $\sqrt[3]{\sqrt{\frac{1}{64}}} = \sqrt[3]{(\frac{1}{64})^{1/2}} = [(\frac{1}{64})^{1/2}]^{1/3} = (\frac{1}{64})^{1/6} = 1^{1/6}/64^{1/6} = 1/(2^6)^{1/6} = 1/2^1 = \frac{1}{2}$.

1.58 $\sqrt[5]{-32}$

$\blacksquare$ $\sqrt[5]{-32} = (-32)^{1/5} = [(-2)^5]^{1/5} = -2^{5(1/5)} = -2^1 = -2$.

1.59 $[4003(p + n)]^0$

$\blacksquare$ Recall that $x^0 = 1$ for any real number x. Thus, $[4003(p + n)]^0 = 1$ since $4003(p + n)$ is a real number.

For Probs. 1.60 to 1.72, simplify the given expression.

1.60 $x^2 \cdot x^{-9} \cdot x^{14}$

$\blacksquare$ $x^2 \cdot x^{-9} \cdot x^{14} = x^{2+(-9)+14} = x^7$.

1.61 $(xy^2)^{-6}$

$\blacksquare$ $(xy^2)^{-6} = x^{-6} \cdot (y^2)^{-6} = x^{-6} \cdot y^{-12} = x^{-6}/y^{12} = 1/x^6 y^{12}$.

1.62 $(2ab^2)^3(a^2c)^2$

$\blacksquare$ $(2ab^2)^3(a^2c)^2 = (2^3 a^3 b^6)(a^4 c^2) = (8a^3 b^6)(a^4 c^2) = 8a^7 b^6 c^2$.

1.63 $\dfrac{x^{-8} \cdot x^{-7}}{x^{-6}} \div \dfrac{x^{-5} \cdot x^{-4}}{x^{-3}}$

$\blacksquare$ $\dfrac{x^{-8} \cdot x^{-7}}{x^{-6}} \div \dfrac{x^{-5} \cdot x^{-4}}{x^{-3}} = \dfrac{x^{-15}}{x^{-6}} \cdot \dfrac{x^{-3}}{x^{-5} \cdot x^{-4}} = \dfrac{x^{-15}}{x^{-6}} \cdot \dfrac{x^{-3}}{x^{-9}} = x^{-9} \cdot x^6 = x^{-3} = \dfrac{1}{x^3}$.

1.64 $\dfrac{b^3 c^5}{a^7} \div \dfrac{b^2 c^7}{a^6}$

$\blacksquare$ $\dfrac{b^3 c^5}{a^7} \div \dfrac{b^2 c^7}{a^6} = \dfrac{b^3 c^5}{a^7} \cdot \dfrac{a^6}{b^2 c^7} = \dfrac{b^3 c^5}{b^2 c^7} \cdot \dfrac{a^6}{a^7} = bc^{-2} \cdot a^{-1} = \dfrac{b}{c^2 a}$.

1.65 $\dfrac{(2a^{-1})^2}{b^{-1}} \div \dfrac{(3a^{-2})^{-1}}{b^{-3}}$

▌ $\dfrac{(2a^{-1})^2}{b^{-1}} \div \dfrac{(3a^{-2})^{-1}}{b^{-3}} = \dfrac{4a^{-2}}{b^{-1}} \div \dfrac{3^{-1}a^2}{b^{-3}} = \dfrac{4a^{-2}}{b^{-1}} \cdot \dfrac{b^{-3}}{3^{-1}a^2} = \dfrac{4a^{-2}}{a^2} \cdot \dfrac{3b^{-3}}{b^{-1}} = 12a^{-4}b^{-2} = \dfrac{12}{a^4 b^2}.$

1.66 $[x^3/(4x^{-6})]^{-4/3}$

▌ $\left[\dfrac{x^3}{(4x^{-6})}\right]^{-4/3} = \left(\dfrac{x^9}{4}\right)^{-4/3} = \dfrac{x^{-12}}{4^{-4/3}} = \dfrac{4^{4/3}}{x^{12}}.$

1.67 $\dfrac{(p^{2/3}q^{1/12})^2}{(p^{3/5})^{1/3}}$

▌ $\dfrac{(p^{2/3}q^{1/12})^2}{(p^{3/5})^{1/3}} = \dfrac{p^{4/3}q^{1/6}}{p^{1/5}} = p^{4/3-1/5}q^{1/6} = p^{17/15}q^{1/6}.$

1.68 $\dfrac{\sqrt{a}\cdot a^{-2/3}}{\sqrt[6]{a^5}} + \dfrac{a^{-5/6}}{\sqrt[3]{a^2}\cdot a^{-1/2}}$

▌ $\dfrac{\sqrt{a}\cdot a^{-2/3}}{\sqrt[6]{a^5}} + \dfrac{a^{-5/6}}{\sqrt[3]{a^2}\cdot a^{-1/2}} = \dfrac{a^{1/2}a^{-2/3}}{a^{5/6}} + \dfrac{a^{-5/6}}{a^{2/3}a^{-1/2}}$

$$= \dfrac{a^{-1/6}}{a^{5/6}} + \dfrac{a^{-5/6}}{a^{1/6}} = \dfrac{a^{-1/6} + a^{4/6}a^{-5/6}}{a^{5/6}}$$

$$= \dfrac{a^{-1/6} + a^{-1/6}}{a^{5/6}} = \dfrac{2a^{-1/6}}{a^{5/6}}$$

$$= \dfrac{2}{a^{1/6}a^{5/6}} = \dfrac{2}{a^{6/6}} = \dfrac{2}{a}.$$

1.69 $\left(\frac{27}{8}x^{-3}y^{1/2}\right)^{-4/3}$

▌ $\left(\frac{27}{8}x^{-3}y^{1/2}\right)^{-4/3} = \left(\frac{27}{8}\right)^{-4/3} \cdot (x^{-3})^{-4/3} \cdot (y^{1/2})^{-4/3} = \left[\dfrac{1}{\left(\frac{27}{8}\right)^{4/3}}\right] \cdot x^{12/3} \cdot y^{-4/6}$

$$= \left(\dfrac{1}{\frac{81}{16}}\right) \cdot x^4 \cdot y^{-2/3} = \tfrac{16}{81} \cdot x^4 \cdot y^{-2/3} = \dfrac{16x^4}{81y^{2/3}}.$$

1.70 $\left(\dfrac{x^{-1/3}y^{1/2}}{x^{-1/4}y^{1/3}}\right)^6$

▌ $\left(\dfrac{x^{-1/3}y^{1/2}}{x^{-1/4}y^{1/3}}\right)^6 = \dfrac{(x^{-1/3})^6(y^{1/2})^6}{(x^{-1/4})^6(y^{1/3})^6} = \dfrac{x^{-2}y^3}{x^{-3/2}y^2} = x^{-1/2}y = \dfrac{y}{x^{1/2}}.$

1.71 $(9x^{1/3}y^{-1/2})^{3/2}(x^{-1/3}y^{1/4})$

▌ $(9x^{1/3}y^{-1/2})^{3/2}(x^{-1/3}y^{1/4}) = 9^{3/2}(x^{1/3})^{3/2}(y^{-1/2})^{3/2}(x^{-1/3}y^{1/4}) = 27x^{3/6}y^{-3/4}x^{-1/3}y^{1/4}$

$$= 27x^{1/6}y^{-2/4} = 27x^{1/6}y^{-1/2} = 27x^{1/6}/y^{1/2}.$$

1.72 $(a^{m/3}b^{n/2})^{-6}$, where $m, n \geq 0$

▌ $(a^{m/3}b^{n/2})^{-6} = a^{-6m/3}b^{-6n/2} = a^{-2m}b^{-3n} = 1/(a^{2m}b^{3n}).$

For Probs. 1.73 to 1.79, simplify and write in simplest radical form. Assume that all letters and radicands represent positive real numbers.

1.73 $\sqrt{16m^4y^8}$

▌ $\sqrt{16m^4y^8} = \sqrt{16}\sqrt{m^4}\sqrt{y^8} = \sqrt{4^2}\sqrt{(m^2)^2}\sqrt{(y^4)^2} = 4m^2y^4.$

1.74 $\sqrt[5]{32a^{15}b^{10}}$

▌ *(a)* $\sqrt[5]{32a^{15}b^{10}} = \sqrt[5]{32}\,\sqrt[5]{a^{15}}\,\sqrt[5]{b^{10}} = \sqrt[5]{2^5}\,\sqrt[5]{(a^3)^5}\,\sqrt[5]{(b^2)^5} = 2a^3b^2.$
(b) $\sqrt[5]{32a^{15}b^{10}} = 32^{1/5}(a^{15})^{1/5}(b^{10})^{1/5} = 2a^3b^2.$

1.75 $\sqrt{\sqrt[4]{5x}}$

▌ *(a)* Recall that $\sqrt[m]{\sqrt[n]{x}} = \sqrt[mn]{x}.$ So, $\sqrt{\sqrt[4]{5x}} = \sqrt[8]{5x}.$
(b) $\sqrt{\sqrt[4]{5x}} = (\sqrt[4]{5x})^{1/2} = [(5x)^{1/4}]^{1/2} = (5x)^{1/8} = \sqrt[8]{5x}.$

1.76 $2a\sqrt[3]{8a^8b^{13}}$

▌ $2a\sqrt[3]{8a^8b^{13}} = 2a\sqrt[3]{2^3(a^2)^3a^2(b^4)^3b} = 2a\sqrt[3]{2^3}\,\sqrt[3]{(a^2)^3a^2}\,\sqrt[3]{(b^4)^3b} = 2a \cdot 2 \cdot a^2 \cdot \sqrt[3]{a^2} \cdot b^4 \cdot \sqrt[3]{b} = 4a^3b^4 \cdot \sqrt[3]{a^2b}.$

1.77 $\sqrt[8]{3^6(u+v)^6}$

▌ $\sqrt[8]{3^6(u+v)^6} = \sqrt[8]{[3(u+v)]^6} = \sqrt[8]{\{[3(u+v)]^3\}^2} = \sqrt[4]{[3(u+v)]^3}.$

1.78 $\sqrt[3]{\dfrac{3y^5}{4x^4}}$

▌ $\sqrt[3]{\dfrac{3y^5}{4x^4}} = \sqrt[3]{\dfrac{3y^3y^2}{4x^3x}} = \dfrac{\sqrt[3]{3y^3y^2}}{\sqrt[3]{4x^3x}} = \dfrac{y\sqrt[3]{3y^2}}{x\sqrt[3]{4x}} \cdot \dfrac{\sqrt[3]{4^2x^2}}{\sqrt[3]{4^2x^2}}$ (rationalizing the denominator)

$= \dfrac{y\sqrt[3]{48x^2y^2}}{x\sqrt[3]{4^3x^3}} = \dfrac{y\sqrt[3]{48x^2y^2}}{x \cdot 4 \cdot x} = \dfrac{y\sqrt[3]{48x^2y^2}}{4x^2}.$

1.79 $\dfrac{2}{x^2 - \sqrt{x^4 + 2x^2 + 1}}$

▌ $\dfrac{2}{x^2 - \sqrt{x^4 + 2x^2 + 1}} = \dfrac{2}{x^2 - \sqrt{(x^2+1)^2}} = \dfrac{2}{x^2 - (x^2+1)} = \dfrac{2}{x^2 - x^2 - 1} = -2.$

1.4 ALGEBRAIC EXPRESSIONS

For Probs. 1.80 to 1.98, perform the indicated operations and simplify.

1.80 $13b + 7b - 6$

▌ $13b + 7b - 6 = (13b + 7b) - 6 = 20b - 6.$ Notice that no further simplification is possible.

1.81 $13b + 7 - 6$

▌ $13b + 7 - 6 = 13b + (7 - 6) = 13b + 1.$ Compare this to Prob. 1.80 above. Again, we can perform no further simplification.

1.82 $13b + 7b - 6b$

▌ $13b + 7b - 6b = (13 + 7 - 6)b = 14b.$ Alternatively, $13b + 7b - 6b = (13b + 7b) - 6b = 20b - 6b = 14b.$

1.83 $13b^2 + 7b^2 - 6b$

▌ $13b^2 + 7b^2 - 6b = (13b^2 + 7b^2) - 6b = 20b^2 - 6b.$

1.84 $3pq + 11pq - pq$

▌ $3pq + 11pq - pq = (3 + 11 - 1)pq = 13pq.$

1.85 $2(u-1)-(3u+2)-2(2u-3)$

▌ $2(u-1)-(3u+2)-2(2u-3)=2u-2-3u-2-4u+6=(2u-3u-4u)+(-2-2+6)=$ $-5u+2$.

1.86 $(x^2-2x+3)+(4x^2-x+6)$

▌ $(x^2-2x+3)+(4x^2-x+6)=x^2-2x+3+4x^2-x+6=(x^2+4x^2)+(-2x-x)+3+6=$ $5x^2-3x+9$.

1.87 $(x^2-2x+3)-(-5x^3-7x+1)$

▌ $x^2-2x+3-(-5x^3-7x+1)=x^2-2x+3+5x^3+7x-1=5x^3+x^2+(7x-2x)+3-1=$ $5x^3+x^2+5x+2$.

1.88 $4\sqrt[3]{y}-\sqrt[3]{y}+\sqrt{y}$

▌ $4\sqrt[3]{y}-\sqrt[3]{y}+\sqrt{y}=(4\sqrt[3]{y}-\sqrt[3]{y})+\sqrt{y}=3\sqrt[3]{y}+\sqrt{y}$.

1.89 $(x-2y)(x+3y)$

▌ $(x-2y)(x+3y)=x(x)+x(3y)+(-2y)(x)+(-2y)(3y)=x^2+3xy-2xy-6y^2=x^2+xy-6y^2$.

1.90 $(\sqrt{s}+\sqrt{t})^2$

▌ $(\sqrt{s}+\sqrt{t})^2=(\sqrt{s})^2+2\sqrt{s}\sqrt{t}+(\sqrt{t})^2=s+2\sqrt{st}+t$.

1.91 $(s+t)^3$

▌ $(s+t)^3=(s+t)^2(s+t)=(s^2+2st+t^2)(s+t)=s^2(s)+2st(s)+t^2(s)+s^2(t)+2st(t)+t^2(t)=$ $s^3+3s^2t+3st^2+t^3$.

1.92 $(3-x)(x^2-x-1)$

▌ $(3-x)(x^2-x-1)=3(x^2)+3(-x)+3(-1)+(-x)(x^2)+(-x)(-x)+(-x)(-1)=3x^2-3x-3-$ $x^3+x^2+x=x^3+4x^2-2x-3$.

1.93 $m-\{m-[m-(m-1)]\}$

▌ $m-\{m-[m-(m-1)]\}=m-\{m-[m-m+1]\}=m-\{m-1\}=m-m+1=1$.

1.94 $(5\sqrt{x}+2)(2\sqrt{x}-3)$

▌ $(5\sqrt{x}+2)(2\sqrt{x}-3)=5\sqrt{x}(2\sqrt{x})+5\sqrt{x}(-3)+2(2\sqrt{x})+2(-3)=10(\sqrt{x})^2-15\sqrt{x}+4\sqrt{x}-6=$ $10x-11\sqrt{x}-6$.

1.95 $(x-y)^2+2x(1+y)$

▌ $(x-y)^2+2x(1+y)=(x^2-2xy+y^2)+(2x+2xy)=x^2-2xy+2xy+y^2+2x=x^2+2x+y^2$.

1.96 $3m^{3/4}(4m^{1/4}-2m^8)$

▌ $3m^{3/4}(4m^{1/4}-2m^8)=3m^{3/4}4m^{1/4}+3m^{3/4}(-2m^8)=12m-6m^{35/4}$.

1.97 $\dfrac{a^2}{6}-\dfrac{a}{5}+\dfrac{1}{15}$

▌ $\dfrac{a^2}{6}-\dfrac{a}{5}+\dfrac{1}{15}=\dfrac{5\cdot a^2+6\cdot(-a)+2(1)}{30}=\dfrac{5a^2-6a+2}{30}$.

1.98 $\dfrac{x^2+1}{x}+\dfrac{x-2}{x^2}+\dfrac{x-3}{2x^2}$

▌ $$\frac{x^2+1}{x}+\frac{x-2}{x^2}+\frac{x-3}{2x^2}=\frac{2x(x^2+1)+2(x-2)+1(x-3)}{2x^2}$$

$$=\frac{(2x^3+2x)+(2x-4)+(x-3)}{2x^2}=\frac{2x^3+5x-7}{2x^2}.$$

1.99 Evaluate the polynomial $2x^2-3x-10$ when $x=-5$.

▌ Substituting -5 for x gives $2x^2-3x-10=2(-5)^2-3(-5)-10=2(25)+15-10=$ $50+15-10=55$.

1.100 Evaluate the polynomial p^2+2p+8 when $p=2v$.

▌ Substituting $2v$ for p, we get $p^2+2p+8=(2v)^2+2(2v)+8=4v^2+4v+8$.

1.101 Evaluate the polynomial y^2-3y-2 when $y=1-\sqrt{3}$.

▌ Substituting $1-\sqrt{3}$ for y, we get $y^2-3y-2=(1-\sqrt{3})^2-3(1-\sqrt{3})-2=(1-2\sqrt{3}+3)-3+$ $3\sqrt{3}-2=-1+(-2\sqrt{3}+3\sqrt{3})=\sqrt{3}-1$.

1.102 Given that $x+1/x=5$, find the value of x^2+1/x^2.

▌ Recall that $(x+1/x)^2=x^2+2x\cdot(1/x)+(1/x)^2=(x^2+1/x^2)+2$. Thus $(x+1/x)^2-2=x^2+1/x^2$. We know that $x+1/x=5$; thus, $5^2-2=x^2+1/x^2$, and $x^2+1/x^2=25-2=23$.

1.103 The length of a rectangle is 8 meters (m) more than its width. (***a***) If x represents the width of the rectangle, write an algebraic expression in terms of x that represents the area.
(***b***) Change the expression to a form without parentheses.

▌ (***a***) Consider Fig. 1.4. If x is the width and the length is 8 m more than the width, then the length must be $x+8$. Recall that $A=l\cdot w$ for a rectangle. In this case, $A=x(x+8)$.
(***b***) Without parentheses, $A=x^2+8x$.

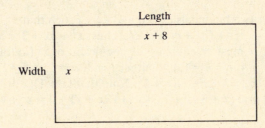

Length

$x+8$

Width x

Fig. 1.4

1.104 Establish that the following two formulas are true:

(***a***) $x^3+y^3=(x+y)(x^2-xy+y^2)$,   (***b***) $x^3-y^3=(x-y)(x^2+xy+y^2)$.

▌ (***a***) $(x+y)(x^2-xy+y^2)=x(x^2)+x(-xy)+x(y^2)+y(x^2)+y(-xy)+y(y^2)=$ $x^3-x^2y+xy^2+yx^2-xy^2+y^3=x^3+y^3$.

(***b***) $(x-y)(x^2+xy+y^2)=x(x^2)+x(xy)+x(y^2)-y(x^2)-y(xy)-y(y^2)=$ $x^3+x^2y+xy^2-x^2y-xy^2-y^3=x^3-y^3$. Thus, both formulas are established.

For Probs. 1.105 to 1.108, let n represent a number. Express each of the following algebraically.

1.105 27 more than the number

▌ Let $n=$ the number. Then "27 more than the number" means 27 more than n, which is 27 plus n. Thus, $27+n$ is the solution.

1.106   The number increased by 45

▌ The number is n. Thus, the number increased by 45 is the number plus 45, which is $n + 45$.

1.107   43 less than the number

▌ 43 less than 90 is $90 - 43$. Similarly, 43 less than the number is 43 less than $n = n - 43$.

1.108   43 less the number

▌ 43 less 10 means $43 - 10$; thus, 43 less the number means $43 - n$.

For Probs. 1.109 through 1.111, express each algebraically.

1.109 b decreased by one-half c

▌ 7 decreased by 4 means $7 - 4$; thus, b decreased by one-half c means $b - c/2$.

1.110   The average of m and 60

▌ To find the average of k items, we find their sum and divide by k. Thus, the average of m and 60 is $(m + 60)/2$.

1.111   The average of 60, x, y, and z

▌ See Prob. 1.110. Add 60, x, y, and z, and divide by the number of items. The solution is $\dfrac{60 + x + y + z}{4}$.

1.5 FACTORING AND FRACTIONAL EXPRESSIONS

For Probs. 1.112 to 1.138, factor the given polynomial completely (relative to the integers).

1.112 $x^2 + 4x + 3$

▌ We are looking for two binomials $cx + d$ and $ex + f$ such that $x^2 + 4x + 3 = (cx + d)(ex + f)$. Note that $(cx + d)(ex + f) = cex^2 + (de + cf)x + df$. Thus, $x^2 + 4x + 3 = cex^2 + (de + cf)x + df$. In order for this to take place, ce must be equal to the coefficient of x^2 (in this case, 1), and df must be 3. We conclude that $de + cf = 4$, $c = e = 1$, and either $f = 3$, $d = 1$ or $f = 1$, $d = 3$. Thus, $f = 3$, $d = 1$ or $f = 1$, $d = 3$. So $x^2 + 4x + 3 = (x + 3)(x + 1)$. Note that $(x + 1)(x + 3)$ is also correct. We can check our answer by noting that $(x + 3)(x + 1) = x^2 + (3x + x) + 3 = x^2 + 4x + 3$.

1.113 $x^2 - x - 6$

▌ We repeat the procedure from Prob. 1.112: $x^2 - x - 6 = (cx + d)(ex + f) = cex^2 + (de + cf)x + df$. We find that $ce = 1$, $de + cf = -1$, and $df = -6$. So $c = e = 1$, and then $d = -6$, $f = 1$ or $d = -1$, $f = 6$ or $d = -3$, $f = 2$ or $d = 3$, $f = -2$. Checking the quantity $de + cf$, we find that $d = -3$, $f = 2$ makes $de + cf = -1$. So $x^2 - x - 6 = (x - 3)(x + 2)$. [Note that $(x + 2)(x - 3)$ is also correct.]

1.114 $2x^2 + 7x + 6$

▌ Again we repeat the procedure given in Prob. 1.112. We want $ce = 2$, $de + cf = 7$, and $df = 6$. Then $c = 2$, $e = 1$ (or the reverse is fine also), $de + cf = 7$, and $d = 6$, $f = 1$ or $d = 3$, $f = 2$ or $d = 1$, $f = 6$ or $d = 2$, $f = 3$. Note that since $ce = 2$, $d = 6$, $f = 1$ is different from $d = 1$, $f = 6$. Checking these choices, we find that if $d = 3$ and $f = 2$, then $de + cf = 7$. So $2x^2 + 7x + 6 = (2x + 3)(x + 2)$.

1.115 $2x^2 - 2x - 4$

▌ (See Prob. 1.112 for a detailed outline of the method.) $2x^2 - 2x - 4 = (2x - 4)(x + 1)$. Check by noting that $(2x - 4)(x + 1) = 2x^2 - 2x - 4$. Note that, using the notation of Prob. 1.112, $ce = 2$, $df = -4$, and $cf + de = -2$. By trial and error, we conclude that $c = 2$, $e = 1$, $d = -4$, and $f = 1$.

1.116 $10x^2 - 14x - 12$

▮ Here we take a shortcut to the full technique described in Prob. 1.112. The reader may still apply that technique and will find that it works. Here we want factors whose product is $10x^2$. (Think! They are $10x$ and x, or $5x$ and $2x$.) We also want integral factors whose product is -12. (Think! They are ±12 and ∓1, or ±6 and ∓2, or ±4 and ∓3.) By trial and error, looking for a middle term of $-14x$, we find $10x^2 - 14x - 12 = (5x + 3)(2x - 4)$. *Note*: $3(2x) + (-4)(5x) = -14x$.

1.117 $6x^2 - 16x + 8$

▮ $6x^2 - 16x + 8 = (2x - 4)(3x - 2)$. Notice that $(2x)(3x) = 6x^2$, $(-4)(-2) = +8$, and $(-4)(3x) + (-2)(2x) = -16x$.

1.118 $4y^2 - 25$

▮ Recall that a polynomial of the form $r^2 - s^2$ is factored as follows: $r^2 - s^2 = (r + s)(r - s)$ (the "difference of two perfect squares" formula). See Fig. 1.5. In this case, we notice that $4y^2 = (2y)^2$ and $25 = 5^2$, and we conclude that $4y^2 - 25 = (2y + 5)(2y - 5)$. Check to see that $(2y + 5)(2y - 5) = 4y^2 - 25$ when you multiply.

> (a) Difference of two squares:
>
> $$r^2 - s^2 = (r + s)(r - s)$$
>
> (b) Difference of two cubes:
>
> $$r^3 - s^3 = (r - s)(r^2 + rs + s^2)$$
>
> (c) Sum of two cubes:
>
> $$(r^3 + s^3) = (r + s)(r^2 - rs + s^2)$$

Fig. 1.5

1.119 $m^3 - 6m - 3$

▮ We check the discriminant since we are having trouble finding factors. In this case, the discriminant $b^2 - 4ac$ is $(-6)^2 - 4(1)(-3) = 36 + 12 = 48$, which is not a perfect square. If the discriminant is not a perfect square, then the polynomial is not factorable relative to the integers.

1.120 $-x^2 - x + 12$

▮ Before we actually factor, ask yourself, What are the factors of 12 whose sum or difference is -1? (The middle term is $-x$.) The answer is 4 and 3. $-x^2 - x + 12 = -x^2 + 3x - 4x + 12 = (-x + 3)(x + 4)$.

1.121 $2t^2 + 75t - 200$

▮ $2t^2 + 75t - 200 = (2t - 5)(t + 40)$.

1.122 $2m^2 + 5mn + 3n^2$

▮ Notice that the solution must be of the form $(2m + 3n)(m + n)$ or $(2m + n)(m + 3n)$. In the first case, the middle term is the correct one $(5mn)$. Thus, $2m^2 + 5mn + 3n^2 = (2m + 3n)(m + n)$.

1.123 $x^2 + x + 1$

▮ Try to find factors here. You will discover that they are difficult to find. Check the discriminant $b^2 - 4ac$ (where $ax^2 + bx + c$ is the form of the polynomial). In this case, $b^2 - 4ac$ is $1^2 - 4(1)(1)$, which is negative. A negative discriminant signals that the polynomial is not factorable.

1.124 $27p^3 + 8q^3$

▮ This polynomial is of the form of a sum of two cubes: $27p^3 + 8q^3 = (3p)^3 + (2q)^3$. Look at Fig. 1.5 for the correct formula. Then $27p^3 + 8q^3 = (3p + 2q)[(3p)^2 - (3p)(2q) + (2q)^2] = (3p + 2q)(9p^2 - 6pq + 4q^2)$.

1.125 $t^2 + 25$

▮ Be careful! This is the *sum* of two squares. This is nonfactorable. Check the discriminant if you want proof: $b^2 - 4ac = 0^2 - 4(1)(25)$ which is negative.

1.126 $t^6 - a^{12}$

▮ Refer to Fig. 1.5. Here $t^6 - a^{12} = (t^2)^3 - (a^4)^3$ is of the form of a difference of two cubes. Using the proper formula gives $t^6 - a^{12} = (t^2 - a^4)[(t^2)^2 + t^2a^4 + (a^4)^2] = (t^2 - a^4)(t^4 + t^2a^4 + a^8)$. But $t^2 - a^4$ is of the form $(t)^2 - (a^2)^2$ which is the difference of two squares! $t^2 - a^4 = (t + a^2)(t - a^2)$. We conclude that $t^6 - a^{12} = (t + a^2)(t - a^2)(t^4 + t^2a^4 + a^8)$.

1.127 $2x^4 - 24x^3 + 40x^2$

▮ Notice that each term contains a factor of $2x^2$. Factor out that $2x^2$. $2x^4 - 24x^3 + 40x^2 = 2x^2(x^2 - 12x + 20)$. Now factor $x^2 - 12x + 20$. $x^2 - 12x + 20 = (x - 10)(x - 2)$. Thus, $2x^4 - 24x^3 + 40x^2 = 2x^2(x - 10)(x - 2)$.

1.128 $x^2y + 7xy$

▮ Notice that each term contains a factor of xy. So $x^2y + 7xy = xy(x + 7)$, and no further factoring is possible.

1.129 $4x^2 - 9y^2 + 4x + 1$

▮ We will use a regrouping technique here. $4x^2 - 9y^2 + 4x + 1 = 4x^2 + 4x + 1 - 9y^2 = (2x + 1)(2x + 1) - 9y^2 = (2x + 1)^2 - 9y^2$ (difference of two squares) $= (2x + 1 - 3y)(2x + 1 + 3y)$.

1.130 $x^3 + (y - z)^3$

▮ This is a "sum of two cubes" problem. $x^3 + (y - z)^3 = [x + (y - z)][x^2 - x(y - z) + (y - z)^2] = (x + y - z)[x^2 - xy + xz + (y - z)^2]$.

1.131 $a^4 - b^4$

▮ $a^4 - b^4 = (a^2)^2 - (b^2)^2 = (a^2 - b^2)(a^2 + b^2) = (a + b)(a - b)(a^2 + b^2)$. Notice that we are applying the "difference of two squares" formula twice here.

1.132 $s^6 - t^6$

▮ $s^6 - t^6 = (s^2)^3 - (t^2)^3 = (s^2 - t^2)(s^4 + s^2t^2 + t^4) = (s + t)(s - t)(s^4 + s^2t^2 + t^4)$.

1.133 $s^{12} - t^{12}$

▮ $s^{12} - t^{12} = (s^6)^2 - (t^6)^2 = (s^6 + t^6)(s^6 - t^6)$

$$= [(s^2)^3 + (t^2)^3][(s^2)^3 - (t^2)^3] \quad (\text{[sum of cubes][difference of cubes]})$$
$$= (s^2 + t^2)(s^4 - s^2t^2 + t^4)(s^2 - t^2)(s^4 + s^2t^2 + t^4)$$
$$= (s^2 + t^2)(s^4 - s^2t^2 + t^4)(s + t)(s - t)(s^4 + s^2t^2 + t^4).$$

1.134 $27 + 8/t^3$

▮ Although t^3 is in the denominator here, the techniques we have employed thus far still work. $27 + 8/t^3 = 3^3 + (2/t)^3 = (3 + 2/t)(9 - 6/t + 4/t^2)$.

1.135 $y^2 - 2xy + x^2 - y + x$

❚ $y^2 - 2xy + x^2 - y + x = (y - x)(y - x) - y + x = (y - x)^2 - (y - x) = (y - x)(y - x - 1)$.

1.136 $4a^4 + 8a^2b^2 + 9b^4$

❚ Direct attempts at factoring do not seem to work in this case. (You should try them!) Sometimes the following trick works: Notice that $4a^4 + 12a^2b^2 + 9b^4$ is factorable: $4a^4 + 12a^2b^2 + 9b^4 = (2a^2 + 3b^2)(2a^2 + 3b^2)$. We then write $8a^2b^2 + 9b^4 = (12a^2b^2 + 9b^4) - 4a^2b^2$. Then $4a^4 + 8a^2b^2 + 9b^4 = (4a^4 + 12a^2b^2 + 9b^4) - 4a^2b^2 = (2a^2 + 3b^2)^2 - (2ab)^2$ (difference of two squares) $= (2a^2 + 3b^2 - 2ab)(2a^2 + 3b^2 + 2ab)$.

1.137 $x^4 - 13x^2y^2 + 4y^4$

❚ We again use the technique described in Prob. 1.136:

$$
\begin{aligned}
x^4 - 13x^2y^2 + 4y^4 &= (x^4 - 4x^2y^2 + 4y^4) - 9x^2y^2 \\
&= (x^2 - 2y^2)^2 - 9x^2y^2 \\
&= (x^2 - 2y^2)^2 - (3xy)^2 \\
&= (x^2 - 2y^2 - 3xy)(x^2 - 2y^2 + 3xy) \\
&= (x^2 - 3xy - 2y^2)(x^2 + 3xy - 2y^2)
\end{aligned}
$$

1.138 $2y^{-2} - y^{-1} - 3$

❚ Don't be afraid of the negative exponents! $2y^{-2} - y^{-1} - 3 = (2y^{-1} - 3)(y^{-1} + 1)$. Remember that $2y^{-1}(y^{-1}) = 2y^{-2}$.

For Probs. 1.139 to 1.163, perform the indicated operations and reduce to lowest terms. Notice that factoring is important in many of these problems.

1.139 $\dfrac{3(x^2 - 1)}{x + 1}$

❚ $\dfrac{3(x^2 - 1)}{x + 1} = \dfrac{3(x + 1)(x - 1)}{x + 1} = \dfrac{3(x + 1)(x - 1)}{x + 1} = 3(x - 1)$.

1.140 $\dfrac{2y^2 - 2}{y^3 - y^2}$

❚ $\dfrac{2y^2 - 2}{y^3 - y^2} = \dfrac{2(y^2 - 1)}{y^2(y - 1)} = \dfrac{2(y + 1)(y - 1)}{y^2(y - 1)} = \dfrac{2(y + 1)}{y^2}$.

1.141 $\dfrac{x^2 - 2x - 15}{x^2 - 3x - 10}$

❚ $\dfrac{x^2 - 2x - 15}{x^2 - 3x - 10} = \dfrac{(x - 5)(x + 3)}{(x - 5)(x + 2)} = \dfrac{x + 3}{x + 2}$.

1.142 $\dfrac{x^2 - 4}{x^2 - 4x - 12}$

❚ $\dfrac{x^2 - 4}{x^2 - 4x - 12} = \dfrac{(x + 2)(x - 2)}{(x - 6)(x + 2)} = \dfrac{x - 2}{x - 6}$.

1.143 $\dfrac{4m-3}{18m^3} + \dfrac{3}{m} - \dfrac{2m-1}{6m^2}$

$\blacksquare$ $\dfrac{4m-3}{18m^3} + \dfrac{3}{m} - \dfrac{2m-1}{6m^2} = \dfrac{(4m-3) + 3(18m^2) - (2m-1)(3m)}{18m^3}$

$(18m^3$ is the least common denominator$)$

$= \dfrac{4m-3+54m^2-(6m^2-3m)}{18m^3} = \dfrac{4m-3+54m^2-6m^2+3m}{18m^3}$

$= \dfrac{48m^2+7m-3}{18m^3}.$

1.144 $\dfrac{x^2}{12} + \dfrac{x}{18} - \dfrac{1}{30}$

$\blacksquare$ $\dfrac{x^2}{12} + \dfrac{x}{18} - \dfrac{1}{30} = \dfrac{x^2(15)+x(10)-1(6)}{180}$ $(180$ is the least common denominator$) = \dfrac{15x^2+10x-6}{180}.$

1.145 $\dfrac{d^5}{3a} \div \left(\dfrac{d^2}{6a^2} \cdot \dfrac{a}{4d^3} \right)$

$\blacksquare$ $\dfrac{d^5}{3a} \div \left(\dfrac{d^2}{6a^2} \cdot \dfrac{a}{4d^3} \right) = \dfrac{d^5}{3a} \div \left(\dfrac{d^2}{4d^3} \cdot \dfrac{a}{6a^2} \right) = \dfrac{d^5}{3a} \div \dfrac{1}{(4d)(6a)}$

$= \dfrac{d^5}{3a} \div \dfrac{1}{24ad} = \dfrac{d^5}{3a} \cdot \dfrac{24ad}{1} = 8d^6.$

1.146 $\dfrac{x^2-2x-15}{x^3-3x^2-10x}$

$\blacksquare$ $\dfrac{x^2-2x-15}{x^3-3x^2-10x} = \dfrac{x^2-2x-15}{x(x^2-3x-10)} = \dfrac{\cancel{(x-5)}(x+3)}{x\cancel{(x-5)}(x+2)} = \dfrac{x+3}{x(x+2)}.$

1.147 $\dfrac{2x^2+x-1}{3x^2+2x-1} \cdot \dfrac{3x^2-2x-1}{2x^2-3x+1}$

$\blacksquare$ $\dfrac{2x^2+x-1}{3x^2+2x-1} \cdot \dfrac{3x^2-2x-1}{2x^2-3x+1} = \dfrac{\cancel{(2x-1)}\cancel{(x+1)}}{(3x-1)\cancel{(x+1)}} \cdot \dfrac{(3x+1)\cancel{(x-1)}}{\cancel{(2x-1)}\cancel{(x-1)}} = \dfrac{3x+1}{3x-1}.$

1.148 $\dfrac{t^2-t-6}{t^2+t-2} \cdot \dfrac{t^2+4t-5}{t^2+6t+5}$

$\blacksquare$ $\dfrac{t^2-t-6}{t^2+t-2} \cdot \dfrac{t^2+4t-5}{t^2+6t+5} = \dfrac{(t-3)\cancel{(t+2)}}{\cancel{(t+2)}\cancel{(t-1)}} \cdot \dfrac{\cancel{(t+5)}\cancel{(t-1)}}{\cancel{(t+5)}(t+1)} = \dfrac{t-3}{t+1}.$

1.149 $\dfrac{y^2-y-6}{y^2-2y+1} \cdot \dfrac{y^2+3y-4}{9y-y^3}$

$\blacksquare$ $\dfrac{y^2-y-6}{y^2-2y+1} \cdot \dfrac{y^2+3y-4}{9y-y^3} = \dfrac{(y-3)(y+2)}{(y-1)(y-1)} \cdot \dfrac{(y+4)(y-1)}{y(3-y)(3+y)}$

$= \dfrac{-\cancel{(3-y)}(y+2)}{\cancel{(y-1)}(y-1)} \cdot \dfrac{(y+4)\cancel{(y-1)}}{y\cancel{(3-y)}(3+y)}.$

Notice that we change $(y-3)$ to $-(3-y)$ so that we can cancel the $3-y$ in the denominator.

The above expression $= \dfrac{-(y+2)(y+4)}{y(y-1)(3+y)}.$

Philip Schmidt, Ph.D., *Associate Professor of Secondary Education, SUNY at New Paltz, New York*

Dr. Schmidt has a B.S. from Brooklyn College (with a major in mathematics) and an M.A. in mathematics and a Ph.D. in mathematics education from Syracuse University. He is currently professor of secondary education at SUNY New Paltz, where he coordinates the secondary mathematics education program. He is the author of *3000 Solved Problems in Precalculus* as well as numerous journal articles. He recently completed a revision of the late Barnett Rich's *Geometry*.

The following figures were reproduced with permission from Ayres, *Schaum's Outline Series: Theory and Problems of First Year College Mathematics*, McGraw-Hill, 1958: figures 5.11–5.13, 9.15–9.21, 9.32–9.45, 10.49–10.57, and 11.29–11.32.

Project supervision was done by The Total Book.

Library of Congress Cataloging-in-Publication Data

Schmidt, Philip A.
  2500 solved problems in college algebra and trigonometry / by
Philip Schmidt.
  p. cm.—(Schaum's solved problems series)
ISBN 0-07-055373-4
  1. Algebra—Problems, exercises, etc. 2. Trigonometry—Problems,
exercises, etc. I. Title. II. Title: Twenty-five hundred solved problems in college
algebra and trigonometry. III. Series.
QA157.S247 1991
512′.13′076—dc20   89-78522
                                                                      CIP

1 2 3 4 5 6 7 8 9 0 SHP/SHP 9 5 4 3 2 1 0

ISBN 0-07-055373-4

SCHAUM'S SOLVED PROBLEMS SERIES

2500 SOLVED PROBLEMS IN

COLLEGE ALGEBRA AND TRIGONOMETRY

by

Philip Schmidt, Ph.D.

State University of New York
at New Paltz

McGRAW-HILL, INC.
New York   St. Louis   San Francisco   Auckland   Bogotá   Caracas
Hamburg   Lisbon   London   Madrid   Mexico   Milan   Montreal
New Delhi   Paris   San Juan   São Paulo   Singapore
Sydney   Tokyo   Toronto

1.150 $\dfrac{2a^2 - 5a - 3}{a^2 + a - 2} \div \dfrac{3a^2 - 8a - 3}{a^2 - a - 6}$

$\blacksquare \quad \dfrac{2a^2 - 5a - 3}{a^2 + a - 2} \div \dfrac{3a^2 - 8a - 3}{a^2 - a - 6} = \dfrac{2a^2 - 5a - 3}{a^2 + a - 2} \cdot \dfrac{a^2 - a - 6}{3a^2 - 8a - 3}$

$\qquad\qquad = \dfrac{(2a + 1)\cancel{(a - 3)}}{\cancel{(a + 2)}(a - 1)} \cdot \dfrac{(a - 3)\cancel{(a + 2)}}{(3a + 1)\cancel{(a - 3)}}$

$\qquad\qquad = \dfrac{(2a + 1)(a - 3)}{(a - 1)(3a + 1)}.$

1.151 $\dfrac{2 - x}{2x + x^2} \cdot \dfrac{x^2 + 4x + 4}{x^2 - 4}$

$\blacksquare \quad \dfrac{2 - x}{2x + x^2} \cdot \dfrac{x^2 + 4x + 4}{x^2 - 4} = \dfrac{2 - x}{x(2 + x)} \cdot \dfrac{(x + 2)(x + 2)}{(x + 2)(x - 2)}$

$\qquad\qquad = \dfrac{-\cancel{(x - 2)}}{x\cancel{(2 + x)}} \cdot \dfrac{\cancel{(x + 2)}\cancel{(x + 2)}}{\cancel{(x + 2)}\cancel{(x - 2)}} = \dfrac{-1}{x}.$

[Notice the $2 - x = -(x - 2)$ trick here. Also remember that $2 + x$ and $x + 2$ are equal, and so they cancel!]

1.152 $\dfrac{2}{x^2 - 4} + \dfrac{6}{x - 2}$

$\blacksquare \quad \dfrac{2}{x^2 - 4} + \dfrac{6}{x - 2} = \dfrac{2}{(x + 2)(x - 2)} + \dfrac{6}{x - 2} = \dfrac{2(1) + 6(x + 2)}{(x + 2)(x - 2)} = \dfrac{2 + 6x + 12}{(x + 2)(x - 2)}$

$[(x + 2)(x - 2)$ is the least common denominator$] = \dfrac{6x + 14}{x^2 - 4}.$

1.153 $\dfrac{d + 5}{3d - 1} - \dfrac{2d - 1}{3d + 1}$

$\blacksquare \quad \dfrac{d + 5}{3d - 1} - \dfrac{2d - 1}{3d + 1} = \dfrac{(d + 5)(3d + 1) - (2d - 1)(3d - 1)}{(3d - 1)(3d + 1)}$

$\qquad\qquad = \dfrac{3d^2 + 16d + 5 - (6d^2 - 5d + 1)}{(3d - 1)(3d + 1)} = \dfrac{-3d^2 + 21d + 4}{(3d - 1)(3d + 1)}.$

1.154 $\dfrac{3x}{2x - 3} - \dfrac{7x}{2x + 1}$

$\blacksquare \quad \dfrac{3x}{2x - 3} - \dfrac{7x}{2x + 1} = \dfrac{3x(2x + 1) - 7x(2x - 3)}{(2x - 3)(2x + 1)} = \dfrac{6x^2 + 3x - (14x^2 + 21x)}{(2x - 3)(2x + 1)}$

$\qquad\qquad = \dfrac{-8x^2 + 24x}{(2x - 3)(2x + 1)} = \dfrac{2x(12 - 4x)}{(2x - 3)(2x + 1)}.$

1.155 $\dfrac{5x}{x^2 - y^2} + \dfrac{1}{x + y} - \dfrac{1}{x - y}$

$\blacksquare \quad$ Noting that $x^2 - y^2 = (x + y)(x - y)$, we have

$$\dfrac{5x}{x^2 - y^2} + \dfrac{1}{x + y} - \dfrac{1}{x - y} = \dfrac{5x(1) + 1(x - y) - 1(x + y)}{x^2 - y^2}$$

$$= \dfrac{5x + x - y - x - y}{x^2 - y^2} = \dfrac{5x - 2y}{x^2 - y^2}.$$

1.156 $\dfrac{1}{x-y} - \dfrac{2}{x+y} + \dfrac{3}{(2x-1)(x-y)}$

$\blacksquare$ $\dfrac{1}{x-y} - \dfrac{2}{x+y} + \dfrac{3}{(2x-1)(x-y)} = \dfrac{1(x+y)(2x-1) - 2(x-y)(2x-1) + 3(x+y)}{(x-y)(x+y)(2x-1)}$

$$= \frac{1(2x^2 + 2xy - x - y) - 2(2x^2 - 2xy - x + y) + (3x + 3y)}{(x-y)(x+y)(2x-1)}$$

$$= \frac{2x^2 + 2xy - x - y - 4x^2 + 4xy + 2x - 2y + 3x + 3y}{(x-y)(x+y)(2x-1)}$$

$$= \frac{-2x^2 + 6xy + 4x}{(x-y)(x+y)(2x-1)}.$$

1.157 $\dfrac{1}{x-1} + \dfrac{2x}{(x-1)^2} - \dfrac{4-x}{(x-1)^3}$

$\blacksquare$ $\dfrac{1}{x-1} + \dfrac{2x}{(x-1)^2} - \dfrac{4-x}{(x-1)^3} = \dfrac{1(x-1)^2 + 2x(x-1) - 1(4-x)}{(x-1)^3}$

$$= \frac{(x^2 - 2x + 1) + (2x^2 - 2x) - (4 - x)}{(x-1)^3}$$

$$= \frac{3x^2 - 3x - 3}{(x-1)^3} = \frac{3(x^2 - x - 1)}{(x-1)^3}.$$

1.158 $\dfrac{2 + 1/t}{2 - 1/t}$

$\blacksquare$ Notice that this expression involves a complex fraction. There are two basic ways of solving problems like this one. (1) Get rid of the complex fraction first, and then simplify.

$$\frac{2 + 1/t}{2 - 1/t} = \frac{2 + 1/t}{2 - 1/t} \cdot \frac{t}{t} = \frac{2t + 1}{2t - 1}$$

(t is the least common denominator of $1/t$ and $1/t$.)
   (2) Simplify the numerator and denominator first:

$$\frac{2 + 1/t}{2 - 1/t} = \frac{(2t+1)/t}{(2t-1)/t} = \frac{2t+1}{t} \div \frac{2t-1}{t} = \frac{2t+1}{t} \cdot \frac{t}{2t-1} = \frac{2t+1}{2t-1}.$$

1.159 $\dfrac{x+y}{x^{-1} - y^{-1}}$

$\blacksquare$ $\dfrac{x+y}{x^{-1} - y^{-1}} = \dfrac{x+y}{1/x - 1/y} = \dfrac{x+y}{1/x - 1/y} \cdot \dfrac{xy}{xy}$

(where xy is the least common denominator of $1/x$ and $1/y$)

$$= \frac{xy(x+y)}{xy/x - xy/y} = \frac{xy(x+y)}{y - x}.$$

1.160 $\dfrac{x+y}{x^{-1} + y^{-1}}$

$\blacksquare$ $\dfrac{x+y}{x^{-1} + y^{-1}} = \dfrac{x+y}{1/x + 1/y} = \dfrac{x+y}{1/x + 1/y} \cdot \dfrac{xy}{xy} = \dfrac{xy(x+y)}{y + x} = xy.$

1.161 $\dfrac{x/y - 2 + y/x}{x/y - y/x}$

▮ $\dfrac{x/y - 2 + y/x}{x/y - y/x} = \dfrac{x/y - 2 + y/x}{x/y - y/x} \cdot \dfrac{xy}{xy} = \dfrac{x^2 - 2xy + y^2}{x^2 - y^2} = \dfrac{(x-y)(x-y)}{(x+y)(x-y)} = \dfrac{x-y}{x+y}$.

1.162 $\dfrac{s^2/(s-t) - s}{t^2/(s-t) + t}$

▮ $\dfrac{s^2/(s-t) - s}{t^2/(s-t) + t} = \dfrac{s^2/(s-t) - s}{t^2/(s-t) + t} \cdot \dfrac{s-t}{s-t} = \dfrac{s^2 - s(s-t)}{t^2 + t(s-t)} = \dfrac{s^2 - s^2 + st}{t^2 + ts - t^2} = \dfrac{st}{ts} = 1$.

1.163 $1 + \dfrac{1}{1 + 1/(1 + 1/x)}$

▮ $\dfrac{1}{1 + 1/x} = \dfrac{1}{1 + 1/x} \cdot \dfrac{x}{x} = \dfrac{x}{x+1}$.

Thus, since $1 + \dfrac{1}{x} = \dfrac{1}{\dfrac{x}{x+1}}$,

$$1 + \cfrac{1}{1 + \cfrac{1}{1 + 1/x}} = 1 + \cfrac{1}{1 + \cfrac{x}{x+1}} = 1 + \cfrac{1}{1 + \cfrac{x}{x+1}} \cdot \dfrac{x+1}{x+1}$$

$$= 1 + \dfrac{x+1}{1(x+1) + x(1)} = 1 + \dfrac{x+1}{2x+1} = \dfrac{1(2x+1) + (x+1)}{2x+1} = \dfrac{3x+2}{2x+1}.$$

For Probs. 1.164 to 1.167, rationalize the denominators and reduce each fraction to lowest terms.

1.164 $\dfrac{\sqrt{x}}{1 + \sqrt{x}}$

▮ $\dfrac{\sqrt{x}}{1 + \sqrt{x}} = \dfrac{\sqrt{x}}{1 + \sqrt{x}} \cdot \dfrac{1 - \sqrt{x}}{1 - \sqrt{x}} = \dfrac{\sqrt{x} - x}{1 - x}$. ($1 - \sqrt{x}$ is the conjugate of $1 + \sqrt{x}$. When we multiply $1 + \sqrt{x}$ by $1 - \sqrt{x}$, we get $1 - x$, and the radical in the denominator is gone.)

1.165 $\dfrac{s\sqrt{3}}{\sqrt{3} - 1}$

▮ $\dfrac{s\sqrt{3}}{\sqrt{3} - 1} = \dfrac{s\sqrt{3}}{\sqrt{3} - 1} \cdot \dfrac{\sqrt{3} + 1}{\sqrt{3} + 1} = \dfrac{3s + s\sqrt{3}}{3 - 1} = \dfrac{3s + s\sqrt{3}}{2}$.

1.166 $\dfrac{1 - \sqrt{x+1}}{1 + \sqrt{x+1}}$

▮ $\dfrac{1 - \sqrt{x+1}}{1 + \sqrt{x+1}} = \dfrac{1 - \sqrt{x+1}}{1 + \sqrt{x+1}} \cdot \dfrac{1 - \sqrt{x+1}}{1 - \sqrt{x+1}}$

(Remember! $1 - \sqrt{x+1}$ is the conjugate of $1 + \sqrt{x+1}$. It is always the case that the conjugate of $\sqrt{m} + \sqrt{n}$ is $\sqrt{m} - \sqrt{n}$; the conjugate of $\sqrt{m} - \sqrt{n}$ is $\sqrt{m} + \sqrt{n}$.)

$$= \dfrac{1 - 2\sqrt{x+1} - (x+1)}{1 - (x+1)} = \dfrac{1 - 2\sqrt{x+1} - x - 1}{1 - x - 1} = \dfrac{-2\sqrt{x+1} - x}{-x}.$$

1.167 $\dfrac{x^2}{3 - \sqrt{x+3}}$

▮ $\dfrac{x^2}{3 - \sqrt{x+3}} = \dfrac{x^2}{3 - \sqrt{x+3}} \cdot \dfrac{3 + \sqrt{x+3}}{3 + \sqrt{x+3}} = \dfrac{x^2(3 + \sqrt{x+3})}{9 - (x+3)} = \dfrac{x^2(3 + \sqrt{x+3})}{6 - x}$.

For Probs. 1.168 and 1.169, rationalize the numerators.

1.168 $\dfrac{\sqrt{2+h} + \sqrt{2}}{h}$

▮ $\dfrac{\sqrt{2+h} + \sqrt{2}}{h} = \dfrac{\sqrt{2+h} + \sqrt{2}}{h} \cdot \dfrac{\sqrt{2+h} - \sqrt{2}}{\sqrt{2+h} - \sqrt{2}} = \dfrac{2 + h - 2}{h(\sqrt{2+h} - \sqrt{2})} = \dfrac{h}{h(\sqrt{2+h} - \sqrt{2})} = \dfrac{1}{\sqrt{2+h} - \sqrt{2}}$.

(Notice that we use the same procedure used for rationalizing the denominator.)

1.169 $\dfrac{2\sqrt{17} - 3x}{x+y}$

▮ $\dfrac{2\sqrt{17} - 3x}{x+y} = \dfrac{2\sqrt{17} - 3x}{x+y} \cdot \dfrac{2\sqrt{17} + 3x}{2\sqrt{17} + 3x} = \dfrac{4 \cdot 17 - 9x^2}{(x+y)(2\sqrt{17} + 3x)} = \dfrac{68 - 9x^2}{(x+y)(2\sqrt{17} + 3x)}$.

For Probs. 1.170 and 1.171, combine into single terms.

1.170 $\sqrt{\dfrac{5a}{8}} - \sqrt{\dfrac{2a}{5}}$

▮ $\sqrt{\dfrac{5a}{8}} - \sqrt{\dfrac{2a}{5}} = \dfrac{\sqrt{5a}}{\sqrt{8}} - \dfrac{\sqrt{2a}}{\sqrt{5}} = \dfrac{\sqrt{5a}}{2\sqrt{2}} - \dfrac{\sqrt{2a}}{\sqrt{5}}$

$= \dfrac{\sqrt{5}\sqrt{5a} - 2\sqrt{2}\sqrt{2a}}{2\sqrt{10}} = \dfrac{5\sqrt{a} - 4\sqrt{a}}{2\sqrt{10}} = \dfrac{\sqrt{a}}{2\sqrt{10}} = \dfrac{\sqrt{a}}{2\sqrt{10}} \cdot \dfrac{\sqrt{10}}{\sqrt{10}} = \dfrac{\sqrt{10a}}{20}$.

1.171 $\sqrt{\tfrac{1}{3}} - \dfrac{2}{\sqrt{3}} + \sqrt{12}$

▮ $\sqrt{\tfrac{1}{3}} - \dfrac{2}{\sqrt{3}} + \sqrt{12} = \dfrac{1}{\sqrt{3}} - \dfrac{2}{\sqrt{3}} + \sqrt{12} = \dfrac{1 - 2 + \sqrt{12}\sqrt{3}}{\sqrt{3}}$

$= \dfrac{-1 + \sqrt{36}}{\sqrt{3}} = \dfrac{-1 + 6}{\sqrt{3}} = \dfrac{5}{\sqrt{3}} = \dfrac{5}{\sqrt{3}} \cdot \dfrac{\sqrt{3}}{\sqrt{3}} = \dfrac{5\sqrt{3}}{3}$.

1.6 COMPLEX NUMBERS

(See Chap. 11 for a complete treatment of the complex numbers.) For Probs. 1.172 to 1.181, perform the indicated operations. Write the answer in the form $a + bi$.

1.172 $(1 + i) + (3 - 2i)$

▮ Recall that $(a + bi) \pm (c + di) = (a \pm c) + (b \pm d)i$. Then $(1 + i) + (3 - 2i) = 4 + (-1)i = 4 - i$.

1.173 $(-6 + 4i) + (2 - i)$

▮ $(-6 + 4i) + (2 - i) = -4 + 3i$.

1.174 $(2 - i) - (3 - 4i)$

▮ $(2 - i) - (3 - 4i) = (2 - 3) + (-1 + 4)i = -1 + 3i$.

1.175 $(-3-i)-(-2-3i)$

▮ $(-3-i)-(-2-3i)=(-3+2)+(-1+3)i=-1+2i$.

1.176 $(2-i)+(4i-3)$

▮ $(2-i)+(4i-3)=(2-i)+(-3+4i)=-1+3i$.

1.177 $3i+(4-2i)$

▮ $3i+(4-2i)=(0+3i)+(4-2i)=4+i$.

1.178 $(2+i)(3+2i)$

▮ Recall that $(a+bi)(c+di)=(ac-bd)+(bc+ad)i$. Then $(2+i)(3+2i)=2(3)-1(2)+(3+4)i=4+7i$.

1.179 $(3-i)(i-6)$

▮ $(3-i)(i-6)=(3-i)(-6+i)=(3)(-6)-(-1)(1)+(6+3)i=-17+9i$.

1.180 $\dfrac{2}{3+i}$

▮ We simplify this expression by multiplying the numerator and denominator by $3-i$, the conjugate of $3+i$:

$$\frac{2}{3+i}=\frac{2}{3+i}\cdot\frac{3-i}{3-i}=\frac{6-2i}{9+1}=\frac{6-2i}{10}=\frac{3}{5}-\frac{i}{5}.$$

1.181 $\dfrac{2+i}{1-i}$

▮ $\dfrac{2+i}{1-i}=\dfrac{2+i}{1-i}\cdot\dfrac{1+i}{1+i}=\dfrac{1+3i}{1+1}=\dfrac{1+3i}{2}=\frac{1}{2}+\frac{3}{2}i$.

For Probs. 1.182 to 1.187, find the indicated power of i.

1.182 i^3

▮ $i^3=i^2i=(-1)i=-i$.

1.183 i^4

▮ $i^4=(i^2)(i^2)=(-1)(-1)=1$.

1.184 i^7

▮ $i^7=i^4i^3=1(-i)=-i$.

1.185 i^8

▮ $i^8=i^4i^4=1(1)=1$.

1.186 i^{15}

▮ $i^{15}=(i^4)^3i^3=1^3(-i)=-i$.

1.187 i^{30}

▮ $i^{30}=(i^4)^7i^2=1^7(-1)=1(-1)=-1$.

For Probs. 1.188 to 1.191, perform the indicated operations and write all answers in the form $a + bi$.

1.188 $(4 - \sqrt{16}) + (2 + \sqrt{-25})$

∎ $\sqrt{-16} = 4i$; $\sqrt{-25} = 5i$. (*Recall*: $\sqrt{-16} = \sqrt{-1 \cdot 16} = \sqrt{-1}\sqrt{16} = i \cdot 4 = 4i$.)
$(4 - \sqrt{16}) + (2 + \sqrt{-25}) = (4 - 4i) + (2 + 5i) = 6 + i$.

1.189 $(2 + i) - (4 + \sqrt{-49})$

∎ $(2 + i) - (4 + \sqrt{-49}) = (2 + i) - (4 + 7i) = -2 - 6i$.

1.190 $(2 + \sqrt{-9})(1 + \sqrt{-4})$

∎ $(2 + \sqrt{-9})(1 + \sqrt{-4}) = (2 + 3i)(1 + 2i) = -4 + 7i$.

1.191 $\dfrac{1}{i}$

∎ $\dfrac{1}{i} = \dfrac{1}{0 + i} = \dfrac{1}{0 + i} \cdot \dfrac{0 - i}{0 - i} = \dfrac{-i}{-i^2} = \dfrac{-i}{1} = -i$.

1.192 Prove that $i^{4k} = 1$ for all natural numbers k.

∎ $i^{4k} = (i^4)^k = 1^k = 1$.

1.193 Solve for x and y: $3 - 2i = 4xi + 2y$.

∎ Two complex numbers $a + bi$ and $c + di$ are equal if and only if $a = c$ and $b = d$. So $3 - 2i = 4xi + 2y$ is rewritten as $3 - 2i = 2y + 4xi$; then $2y = 3$ and $4x = -2$, or $y = \frac{3}{2}$ and $x = -\frac{1}{2}$.

For Probs. 1.194 to 1.201, solve the given equation.

1.194 $x^2 - x + 1 = 0$

∎ $x = \dfrac{-b \pm \sqrt{b^2 - 4ac}}{2a} = \dfrac{1 \pm \sqrt{1^2 - 4(1)(1)}}{2(1)} = \dfrac{1 \pm \sqrt{-3}}{2} = \dfrac{1 \pm i\sqrt{3}}{2}$.

1.195 $2x^2 - 2x + 3 = 0$

∎ $x = \dfrac{2 \pm \sqrt{4 - 24}}{4} = \dfrac{2 \pm \sqrt{-20}}{4} = \dfrac{2 \pm 2\sqrt{-5}}{4} = \dfrac{1 \pm i\sqrt{5}}{2}$.

1.196 $x^2 + 5 = 0$

∎ $x^2 = -5$ or $x = \pm\sqrt{-5} = \pm i\sqrt{5}$.

1.197 $x^3 + 8 = 0$

∎ $x^3 = -8$ or $x = \sqrt[3]{-8} = -2$. (Compare this with Prob. 1.196 above.)

1.198 Prove that 0 is the additive identity for the complex numbers.

∎ Write 0 as $0 + 0i$. Then $(a + bi) + 0 = (a + bi) + (0 + 0i) = (a + 0) + (b + 0)i = a + bi$. Thus, $0 + 0i = 0 =$ additive identity.

1.199 Find the additive inverse of $c + di$, and prove that it is the additive inverse.

∎ We claim $-c - di$ is the additive inverse. Now $c + di + (-c - di) = c + (-c) + [d + (-d)]i = 0 + 0i =$ the additive identity. Thus, $-c - di$ is the additive inverse for $c + di$.

1.200 Prove that 1 is the multiplicative identity for the complex numbers.

▮ $1 = 1 + 0i$. $(a + bi)(1) = (a + bi)(1 + 0i) = a(1) - b(0) + (b + 0)i = a + bi$. Therefore, $1 = 1 + 0i$ is the multiplicative identity.

1.201 Find x and y such that $(3 + 2i)(x + yi) = 1$.

▮ $(3 + 2i)(x + yi) = (3x - 2y) + (2x + 3y)i$. Then $(3 + 2i)(x + yi) = (3x - 2y) + (2x + 3y)i = 1 = 1 + 0i$. Thus, $3x - 2y = 1$ and $2x + 3y = 0$. Solving these simultaneously (see Chap. 5 for the background on this if you need it), we get

$$
\begin{aligned}
6x - 4y &= 2 \\
\underline{6x + 9y} &= 0 \\
-13y &= 2 \\
y &= -\tfrac{2}{13}
\end{aligned}
$$

$6x - 4(-\tfrac{2}{13}) = 2$, $6x + \tfrac{8}{13} = 2 = \tfrac{26}{13}$, $6x = \tfrac{18}{13}$, or $x = \tfrac{18}{78} = \tfrac{3}{13}$.

CHAPTER 2
Equations and Inequalities

2.1 LINEAR EQUATIONS

For Probs. 2.1 to 2.18, solve the given equation, if possible. In each case, the replacement set for the variable is the set of real numbers.

2.1 $6x + 3 = 19x + 5$

$$
\begin{array}{rll}
6x + 3 = & 19x + 5 & \\
-3 & -3 & \text{(Add } -3 \text{ to both sides.)} \\
\hline
6x = & 19x + 2 & \\
-19x & -19x & \text{(Add } -19x \text{ to both sides.)} \\
\hline
(-\tfrac{1}{13})(-13x) = & 2 \cdot (-\tfrac{1}{13}) & \text{(Multiply both sides by } -\tfrac{1}{13}.) \\
x = & -\tfrac{2}{13} &
\end{array}
$$

Note that we added -3 to both sides so that constant terms could be found on only one side of the equation [in this case, the right-hand side (RHS)]. We then add $-19x$ to both sides to isolate the terms involving a variable. We multiplied by $-\tfrac{1}{13}$ to transform $-13x$ to x.

2.2 $3x - 10 = 17 - 3x$

$$
\begin{array}{rll}
3x - 10 = & 17 - 3x & \\
+10 & +10 & \text{(Add 10.)} \\
\hline
3x = & 27 - 3x & \\
3x & 3x & \text{(Add } 3x.) \\
\hline
\tfrac{1}{6} \cdot 6x = & 27 \cdot \tfrac{1}{6} & \text{(Multiply by } \tfrac{1}{6}.) \\
x = & \tfrac{27}{6} &
\end{array}
$$

2.3 $9 - 14t = 17t - 11$

$$
\begin{array}{rl}
9 - 14t = & 17t - 11 \\
+14t = & +14t \\
\hline
9 = & 31t - 11 \\
+11 & +11 \\
\hline
\tfrac{1}{31} \cdot 20 = & 31t \cdot \tfrac{1}{31} \\
\tfrac{20}{31} = & t
\end{array}
$$

Note that we first isolated the variable and then added the constant. This could have been done in reverse order. See Prob. 2.4.

2.4 $9 - 14t = 17t - 11$

$$
\begin{array}{rl}
9 - 14t = & 17t - 11 \\
-9 & -9 \\
\hline
-14t = & 17t - 20 \\
-17t & -17t \\
\hline
-\tfrac{1}{31} \cdot (-31t) = & -20 \cdot (-\tfrac{1}{31}) \\
t = & \tfrac{20}{31}
\end{array}
$$

2.5 $5(2t - 6) = 4(3t - 1)$

▌ First we eliminate the parentheses: $10t - 30 = 12t - 4$. Don't forget the distributive law! Now we proceed as we did in Probs. 2.1 to 2.4.

$$
\begin{array}{rl}
10t - 30 = & 12t - 4 \\
+30 & +30 \\
\hline
10t \;\; = & 12t + 26 \\
-12t & -12t \\
\hline
-2t \quad = & +26 \\
t \quad = & -13
\end{array}
$$
(Multiply both sides by $-\frac{1}{2}$.)

2.6 $2(3x - 6) + 4(3x - 5) = 14x$

▌ First, remove parentheses: $6x - 12 + 12x - 20 = 14x$. Next, collect like terms on the left-hand side (LHS): $(12x + 6x) - 12 - 20 = 14x$.

$$
\begin{array}{rl}
18x - 32 = & 14x \\
-18x & -18x \\
\hline
-32 = & -4x \\
\frac{32}{4} = & x \\
x = & 8
\end{array}
$$

2.7 $-2(4x + 2) = -3 - 4x - (x - 2)$

▌ Proceed as in Prob. 2.6. Do not attempt to solve this until you have simplified the LHS and RHS: $-8x - 4 = -3 - 4x - x + 2$.

$$
\begin{array}{rl}
-8x - 4 = & -1 \; - 5x \\
+1 & +1 \\
\hline
-8x - 3 = & -5x \\
8x & 8x \\
\hline
-3 = & 3x \\
x = & -1
\end{array}
$$

2.8 $(3x - 1)(4x + 3) = (2x + 3)(6x + 10)$

▌ Begin by eliminating the parentheses:

$$
\begin{array}{rl}
12x^2 + 5x - 3 \;\; = & 12x^2 + 38x + 30 \\
-5x - 30 & -5x - 30 \\
\hline
-33 = & 33x \\
-1 = & x
\end{array}
$$

2.9 $x/3 - 2 = \frac{1}{10} - x$

▌ Don't let the fractions bother you. Proceed in exactly the same way as in earlier examples.

$$
\begin{array}{rl}
x/3 - 2 = & \frac{1}{10} - x \\
+2 & +2 \\
\hline
x/3 \;\; = & \frac{21}{10} - x \qquad \left(\frac{21}{10} = 2 + \frac{1}{10}\right) \\
+x & +x \\
\hline
4x/3 \;\; = & \frac{21}{10} \\
& x = \frac{21}{10} \cdot \frac{3}{4} = \frac{63}{40} \qquad \text{(Multiply both sides by } \tfrac{3}{4}.\text{)}
\end{array}
$$

There is an alternative way to handle this kind of problem. See Prob. 2.11.

2.10 $(x-1)(2x+1) = (x+1)(2x-1)$

$$
\begin{array}{rcl}
2x^2 - x - 1 &=& 2x^2 + x - 1 \\
-2x^2 & & -2x^2 \\
\hline
-x - 1 &=& x - 1 \\
+1 & & +1 \\
\hline
-x &=& x \\
+x & & +x \\
\hline
0 &=& 2x \\
x &=& 0
\end{array}
$$

2.11 $x/3 - 2 = \frac{1}{10} - x$

Combine fractions first on the LHS and RHS: $\dfrac{x}{3} - \dfrac{2}{1} = \dfrac{1}{10} - \dfrac{x}{1}$, or $\dfrac{x-6}{3} = \dfrac{1-10x}{10}$. Cross-multiply: $10(x-6) = 3(1-10x)$; proceed as we did previously.

$$
\begin{array}{rcl}
10x - 60 &=& 3 - 30x \\
+30x & & +30x \\
\hline
40x - 60 &=& 3 \\
+60 & & +60 \\
\hline
40x &=& 63 \\
x &=& \frac{63}{40}
\end{array}
$$

2.12 $\dfrac{3}{2x-1} + 4 = \dfrac{6x}{2x-1}$

$\dfrac{3}{2x-1} + 4 = \dfrac{6x}{2x-1}$ is similar to the equation in Prob. 2.14. Combining fractions on the LHS, we get $\dfrac{3 + 4(2x-1)}{2x-1} = \dfrac{6x}{2x-1}$ or $\dfrac{-1+8x}{2x-1} = \dfrac{6x}{2x-1}$. Multiply both sides by $2x-1$, so $-1 + 8x = 6x$, $2x = 1$, or $x = \frac{1}{2}$.

2.13 $\dfrac{2}{x+3} = \dfrac{4}{x+4}$

Cross multiply: $2(x+4) = 4(x+3)$, $2x + 8 = 4x + 12$, $-4 = 2x$, or $-2 = x$. See Prob. 2.14.

2.14 $\dfrac{2x}{x+3} = \dfrac{4}{x+4} + 2$

$\dfrac{2x}{x+3} = \dfrac{4 + 2(x+4)}{x+4}$. (Do you see where this term came from?) $\dfrac{2x}{x+3} = \dfrac{2x+12}{x+4}$. Cross multiply: $2x(x+4) = (2x+12)(x+3)$, $2x^2 + 8x = 2x^2 + 18x + 36$, $-36 = 10x$, or $-3.6 = x$.

2.15 $4x + 9 = 4x + 11$

Subtract $4x$ from both sides (i.e., add $-4x$ to both sides). Then $9 = 11$. Since $9 \neq 11$, there is no solution.

2.16 $4t + 9 = -2(-2t - \frac{9}{2})$

$4t + 9 = 4t + 9$ (using the distributive law). Since $4t + 9 = 4t + 9$ for any value of t, any real number is a solution. There are infinitely many solutions.

2.17 $\dfrac{5}{x-3}=\dfrac{33-x}{x^2-6x+9}$

▮ $\dfrac{5}{x-3}=\dfrac{33-x}{x^2-6x+9}$. Cross multiply: $5(x-3)^2=(x-3)(33-x)$. Divide both sides by $x-3$. This is legal since $x\ne 3$ (why?). So $5(x-3)=33-x$, $5x-15=33-x$, $6x=48$, or $x=8$.

2.18 $\dfrac{n-5}{6n-6}=\dfrac{1}{9}-\dfrac{n-3}{4n-4}$

▮ $\dfrac{n-5}{6(n-1)}=\dfrac{4(n-1)-9(n-3)}{9(4)(n-1)}$, or $\dfrac{n-5}{6(n-1)}=\dfrac{-5n+23}{36(n-1)}$. Then $36(n-1)(n-5)=6(n-1)\times(-5n+23)$ and $6(n-5)=-5n+23$. Thus, $n=\frac{53}{11}$.

2.19 Check the result in Prob. 2.1.

▮ In Prob. 2.1, we obtained $x=-\frac{2}{13}$ as the solution for $6x+3=19x+5$. $6(-\frac{2}{13})+3\overset{?}{=}19(-\frac{2}{13})+5$, $-\frac{12}{13}+3\overset{?}{=}-\frac{38}{13}+5$, $\dfrac{-12+39}{13}=\dfrac{-38+65}{13}$, $\frac{27}{13}=\frac{27}{13}$. Thus, $x=-\frac{2}{13}$ is a correct solution. This technique works for all equations.

2.20 Check the result in Prob. 2.17.

▮ $\dfrac{5}{x-3}=\dfrac{33-x}{x^2-6x+9}$ where $x=8$. $\dfrac{5}{8-3}\overset{?}{=}\dfrac{33-8}{8^2-6(8)+9}$, $1=\dfrac{25}{25}$.

2.21 Solve $p+q=rs$ for q.

▮ $\begin{aligned}p+q&=rs\\ -p\quad\ &\ -p\\ \hline q&=rs-p\end{aligned}$

Notice that we treat r, s, and p as constants since we are solving for the unknown q.

2.22 Check the solution in Prob. 2.21.

▮ $p+q=rs$ where $q=rs-p$. $p+(rs-p)\overset{?}{=}rs$, $p+rs-p\overset{?}{=}rs$, $rs=rs$. The solution is correct.

2.23 Solve $p+q=rs$ for s, where $r\ne 0$.

▮ Multiply $p+q=rs$ on both sides by $1/r$: $(1/r)(p+q)=rs(1/r)$, or $(p+q)/r=s$ $(r\ne 0)$.

2.24 Solve $y=\dfrac{2x-3}{3x+5}$ for x.

▮ $y(3x+5)=2x-3$, $3yx+5y=2x-3$, $3yx-2x=-3-5y$, $x(3y-2)=-3-5y$, or $x=\dfrac{-3-5y}{3y-2}$. You should check this solution.

2.25 Let m and n be real numbers with m larger than n. Then there exists a positive real number p such that $m=n+p$. Find the fallacy in the following argument:

$$m=n+p$$
$$(m-n)m=(m-n)(n+p)$$

$$m^2 - mn = mn + mp - n^2 - np$$

$$m^2 - mn - mp = mn - n^2 - np$$

$$m(m - n - p) = n(m - n - p)$$

$$m = n$$

▪ To get to the last step from the one before it, we divided by $m - n - p$. But if $m = n + p$ (see step 1), then $m - n - p = 0$, and division by 0 is illegal.

For Probs. 2.26 to 2.30, find k if the given number is a solution of the given equation.

2.26 12, $2x + 5 = 3x + k$

▪ If $x = 12$ is a solution, then $2(12) + 5$ must equal $3(12) + k$. Then $24 + 5 = 36 + k$, $29 = 36 + k$, or $-7 = k$.

2.27 2, $x^2 + kx + 2 = 0$

▪ As we did in Prob. 2.26, we substitute 2 for x. Then $2^2 + k(2) + 2 = 0$, $4 + 2k + 2 = 0$, $2k = -6$, or $k = -3$.

2.28 $-\frac{1}{2}$, $(2x + 3)(4x + 5) = (4x + 1)(2x + k)$

▪ $[2(-\frac{1}{2}) + 3][4(-\frac{1}{2}) + 5] = [4(-\frac{1}{2}) + 1][2(-\frac{1}{2}) + k]$. Therefore, $2(3) = (-1)(-1 + k)$, $6 = 1 - k$, or $k = -5$. See Prob. 2.29.

2.29 Consider Prob. 2.28 again, but this time simplify before substituting $x = -\frac{1}{2}$.

▪ $8x^2 + 22x + 15 = 8x^2 + 2x + 4xk + k$. Then $8(-\frac{1}{2})^2 + 22(-\frac{1}{2}) + 15 = 8(-\frac{1}{2})^2 + 2(-\frac{1}{2}) + 4(-\frac{1}{2})k + k$. After doing the arithmetic, we find $k = -5$.

2.30 $-\frac{1}{2}$, $x^2(3kx - 6k - 1) + 3x + k = 0$

▪ $(-\frac{1}{2})^2[3k(-\frac{1}{2}) - 6k - 1] + 3(-\frac{1}{2}) + k = 0$, $\frac{1}{4}(-\frac{3}{2}k - 6k - 1) - \frac{3}{2} + k = 0$, $-\frac{3}{8}k - \frac{3}{2}k - \frac{1}{4} - \frac{3}{2} + k = 0$. $-\frac{7}{8}k = \frac{7}{4}$, $-28k = 56$, or $k = -2$.

2.31 The sum of two consecutive even integers is 10. Find the integers.

▪ Let $x =$ one of the integers; then $x + 2$ must be the other. (Consecutive even integers differ by 2.) Thus, $x + (x + 2) = 10$, $2x + 2 = 10$, $2x = 8$, or $x = 4$, then $x + 2 = 6$. The integers are 4 and 6. *Check*: 4 and 6 are consecutive even numbers. $4 + 6 = 10$. Thus, their sum is 10.

2.32 Find three consecutive even integers such that the sum of the first and third is twice the second.

▪ Let $x =$ the first, $x + 2 =$ the second, and $x + 4 =$ the third number. Then $x + (x + 4)$ (the sum of the first and third) $= 2(x + 2)$ (twice the second), $2x + 4 = 2x + 4$, or $0 = 0$. Thus, *any* three consecutive integers are a solution, such as 10, 12, 14.

2.33 Find three consecutive integers such that the sum of the first and twice the second is 30 more than the third.

▪ Let x, $x + 1$, and $x + 2$ be the integers. Then $x + 2(x + 1) = (x + 2) + 30$, $3x + 2 = x + 32$, $3x = 30$, or $x = 15$; thus, the integers are 15 (x), 16 $(x + 1)$, and 17 $(x + 2)$. *Check*: 15, 16, 17 are consecutive. $15 + 2(16) = 47$, and $30 + 17 = 47$.

2.34 The perimeter of a rectangle is 30 m, and its length is twice its width. Find the length and width of the rectangle.

▪ Let $x =$ width. Then $2x =$ length (i.e., twice the width). Then $2x + 2x + x + x = 30$, $6x = 30$, $x = 5$ m (width), and $2x = 10$ m (length). *Check*: $10 = 2(5)$. Also $10 + 10 + 5 + 5 = 30$.

2.35 A rectangle 24 m long has the same area as a square that is 12 m on a side. Find the dimension of the rectangle.

▌ If the side of the square is 12 m, then the area of the square $S^2 = 144$ m^2. The area of the rectangle $A = lw = 144$ m^2. It is known that $l = 24$ m. Thus, $24w = 144$, and $w = 6$. *Check*: 6 m $\cdot 24$ m $= 144$ m$^2 =$ area of square.

2.36 Train A leaves station Q at the same time as train B leaves station R. Train A travels at a rate of 30 miles per hour (mi/h) directly toward R, and B travels at 45 mi/h directly toward Q. How many miles must A travel before the trains meet, if the stations are 60 mi apart? (See Fig. 2.1.)

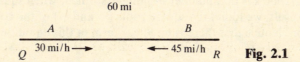

Fig. 2.1

▌ Let $x =$ distance A travels until the trains meet. Then $60 - x =$ distance B travels. Put that information in Fig. 2.2. Also insert the respective rates of the trains. Note that if $d = rt$, then $t = d/r$. Fill in these times. Notice that the two times are equal, since both trains must travel until they meet. So $x/30 = (60 - x)/45$, $45x = 1800 - 30x$, $75x = 1800$, $x = 24 =$ distance A travels. See Prob. 2.37.

	$d =$	r	$\times$	t
A	x	30		$\dfrac{x}{30}$
B	$60 - x$	45		$\dfrac{60 - x}{45}$

Fig. 2.2

2.37 Using Prob. 2.36, find out how long B travels before the trains meet.

▌ The time B travels is $(60 - x)/45$. We found $x = 24$ in Prob. 2.36. $(60 - 24)/45 = \frac{36}{45} = \frac{4}{5}$ h $= 48$ min.

2.38 The sale price on a camera after a 20 percent discount is $72. What was the price before the discount?

▌ Let $x =$ price before discount. Then $x - 20\%x =$ price after discount. $x - 0.2x = 72$, $0.8x = 72$, or $x = 72 \times \frac{10}{8} = \$90$.

2.39 Suppose Kate can paint a particular room in 6 h. At what minimum rate must her helper be able to paint the room alone if together they must complete the job in $3\frac{3}{7}$ h?

▌ Let $x =$ time in hours it would take Kate's helper to paint the room. If it takes x h to complete, then $1/x$ of the job could be completed in 1 h. Also if Kate can complete the job in 6 h, she will complete $\frac{1}{6}$ of it in 1 h. Then in 1 h $\frac{1}{6} + 1/x = 1/\frac{24}{7}$ $(\frac{24}{7} = 3\frac{3}{7})$, $\frac{1}{6} + 1/x = \frac{7}{24}$, $(x + 6)/(6x) = \frac{7}{24}$, $24x + 144 = 42x$, $144 = 18x$, or $x = 8$ h. Her helper must be able to paint the room working alone in 8 h. *Check*: $\frac{1}{6} + \frac{1}{8} = \frac{7}{24} = 1/\frac{24}{7}$.

2.40 Barbara is twice as old as Mary, and Dick is three times as old as Barbara. Their average age is 36. How old is Barbara?

▌ Let $x =$ Mary's age. Then $2x =$ Barbara's age, and $6x =$ Dick's age. $\dfrac{6x + 2x + x}{3} = 36$ (average of ages), $9x = 108$, so $x = 12$ years old, $2x = 24$ years old $=$ Barbara's age, and $6x = 72$ years old $=$ Dick's age.

2.41 Juanita has nickels and dimes in her pocket. Their total value is $1, and there are twice as many dimes as nickels. How many nickels does she have?

▌ Let x = number of nickels, $2x$ = number of dimes. Then, $5(x) + 10(2x) = 100$ ($100¢ = \$1$), $5x + 20x = 100$, $25x = 100$, or $x = 4$ nickels. *Check*: $2x = 8$ = number of dimes. $4(5¢) + 8(10¢) = 100¢ = \$1$.

2.2 NONLINEAR EQUATIONS

For Probs. 2.42 to 2.71, solve the given equation. Leave any answer containing a radical in simplest radical form.

2.42 $x^2 = 9$

▌ If $x^2 = 9$, then x is that number or numbers which, when squared, is 9. Thus, $x = \pm 3$. Notice that, to get this result, we found the square roots ($+$ and $-$) of the constant term isolated on one side of the equation. Notice that there is no linear term in this equation. *Check*: $3^2 = 9$; $(-3)^2 = 9$.

2.43 $4t^2 = 36$

▌ $4t^2 = 36$, $t^2 = 9$, or $t = \pm 3$. *Check*: $4(3)^2 = 36$; $4(-3)^2 = 36$.

2.44 $p^2 = 17$

▌ $p^2 = 17$, or $p = \pm\sqrt{17}$. *Check*: $(\sqrt{17})^2 = 17$; $(-\sqrt{17})^2 = 17$.

2.45 $2t^2 = 80$

▌ $2t^2 = 80$, $t^2 = 40$, or $t = \pm\sqrt{40} = \pm\sqrt{4 \cdot 10} = \pm\sqrt{4}\sqrt{10} = \pm 2\sqrt{10}$. *Check*: $2(2\sqrt{10})^2 = 2(4 \cdot 10) = 80$; $2(-2\sqrt{10})^2 = 2(4 \cdot 10) = 80$.

2.46 $x^2 - 10 = 0$

▌ $x^2 - 10 = 0$, $x^2 = 10$, or $x = \pm\sqrt{10}$.

2.47 $t^2 + 5 = 0$

▌ $t^2 + 5 = 0$, or $t^2 = -5$. But no real number squared is negative. Thus, the solutions must be complex numbers. If $t^2 = -5$, then $t = \pm\sqrt{-5} = \pm i\sqrt{5}$. (See Ch. 1 for a review of complex numbers.)

2.48 $4x^2 - 7 = 0$

▌ $4x^2 - 7 = 0$, $4x^2 = 7$, $x^2 = \frac{7}{4}$, $x = \pm\sqrt{\frac{7}{4}} = \pm\sqrt{7}/\sqrt{4} = \pm\sqrt{7}/2$.

2.49 $(n + 5)^2 = 9$

▌ $(n + 5)^2 = 9$, or $n + 5 = \pm 3$. Possibility 1: $n + 5 = 3$, or $n = -2$. Possibility 2: $n + 5 = -3$, or $n = 8$. So the answer is $n = -2, -8$.

2.50 $x^2 + 3x + 2 = 0$

▌ Factor the LHS: $(x + 2)(x + 1) = 0$ ($x + 2$ or $x + 1$ or both must be 0). If $x + 2 = 0$, then $x = -2$. If $x + 1 = 0$, then $x = -1$. *Check*: $(-2)^2 + 3(-2) + 2 = 4 - 6 + 2 = 0$, or $0 = 0$. Also $(-1)^2 + 3(-1) + 2 = 0$, $1 - 3 + 2 = 0$, or $0 = 0$.

2.51 $t^2 + 12t + 35 = 0$

▌ Factoring, we get $(t + 5)(t + 7) = 0$. If $t + 5 = 0$, then $t = -5$. If $t + 7 = 0$, then $t = -7$.

2.52 $4x^2 - 4x - 3 = 0$

▌ $(2x - 3)(2x + 1) = 0$. If $2x - 3 = 0$, then $x = \frac{3}{2}$. If $2x + 1 = 0$, then $x = -\frac{1}{2}$.

2.53 $2t^2 - 2t = 12$

▌ $2t^2 - 2t = 12$, $2t^2 - 2t - 12 = 0$, or $(t - 3)(2t + 4) = 0$. If $t - 3 = 0$, then $t = 3$. If $2t = -4$, $t = -2$. Note that we could have divided through by 2 initially. See Prob. 2.54.

2.54 $2t^2 - 2t = 12$

▌ $2t^2 - 2t = 12$, $t^2 - t = 6$, $t^2 - t - 6 = 0$, or $(t-3)(t+2) = 0$. If $t - 3 = 0$, then $t = 3$. If $t + 2 = 0$, then $t = -2$.

2.55 $10x^2 - 52x + 64 = 0$

▌ $(2x - 4)(5x - 16) = 0$. If $2x - 4 = 0$, then $x = 2$. If $5x - 16 = 0$, then $x = \frac{16}{5}$.

2.56 $p^4 - 2p^2 + 1 = 0$

▌ $(p^2 - 1)(p^2 - 1) = 0$, $p^2 = 1$, or $p = \pm 1$.

2.57 $x^3 + 3x^2 + 2x = 0$

▌ $x(x^2 + 3x + 2) = 0$, $x(x+2)(x+1) = 0$, or $x = 0, -2, -1$.

2.58 $x^2 + x - 1 = 0$

▌ This does not appear to be factorable. Using the quadratic formula $x = \dfrac{-b \pm \sqrt{b^2 - 4ac}}{2a}$ $(a \neq 0)$, we have $a = 1$, $b = 1$, $c = -1$, and $x = \dfrac{-1 \pm \sqrt{1^2 - 4(1)(-1)}}{2(1)} = \dfrac{-1 \pm \sqrt{5}}{2}$. Thus, $x = \dfrac{-1 + \sqrt{5}}{2}$ or $x = \dfrac{-1 - \sqrt{5}}{2}$.

2.59 $2t^2 + t - 4 = 0$

▌ Again, factoring does not appear to work. We have $a = 2$, $b = 1$, $c = -4$; $t = \dfrac{-1 \pm \sqrt{1^2 - 4(2)(-4)}}{2(2)} = \dfrac{-1 \pm \sqrt{33}}{4}$.

2.60 $2x^2 + 4x + 1 = 0$

▌ Here $a = 2$, $b = 4$, $c = 1$. Thus, $x = \dfrac{-4 \pm \sqrt{16 - 8}}{4} = \dfrac{-4 \pm \sqrt{8}}{4} = \dfrac{-4 \pm 2\sqrt{2}}{4} = \dfrac{-2 \pm \sqrt{2}}{2}$.

2.61 $2x^2 + 8x - 42 = 0$

▌ The LHS here is factorable into $(2x - 6)(x + 7)$. If, however, we had not noticed that, we could have solved the equation as follows:
$$x = \frac{-8 \pm \sqrt{64 - 4(2)(-42)}}{4} = \frac{-8 \pm \sqrt{400}}{4} = \frac{-8 \pm 20}{4}.$$
So $x = \dfrac{-8 + 20}{4} = 3$ or $x = \dfrac{-8 - 20}{4} = -7$.

2.62 $x^2 + 4x + 1 = 0$

▌ Here we illustrate the method of "completing the square." $x^2 + 4x + 4$ is a perfect square, since $x^2 + 4x + ④ = (x + 2)^2$. To obtain the circled 4, we find the coefficient of the x term (here it is 4), divide it by 2, and square it: $(4/2)^2 = 2^2 = 4$. $x^2 + 4x + 1 \neq x^2 + 4x + 4$, but $x^2 + 4x + 1 = (x^2 + 4x + 4) - 3$. So the equation $x^2 + 4x + 1 = 0$ is rewritten as $(x^2 + 4x + 4) - 3 = 0$, $(x + 2)^2 - 3 = 0$, $(x + 2)^2 = 3$, $x + 2 = \pm\sqrt{3}$, $x = \pm\sqrt{3} - 2$, or $x = \sqrt{3} - 2, -\sqrt{3} - 2$.

2.63 $t^2 - 6t + 2 = 0$

▌ $t^2 - 6t + 2 = (t^2 - 6t + 9) - 7$. Thus, $t^2 - 6t + 2 = 0$ is rewritten as $(t^2 - 6t + 9) - 7 = 0$ or $t^2 - 6t + 9 = 7$, $(t - 3)^2 = 7$, $t - 3 = \pm\sqrt{7}$, or $t = 3 \pm \sqrt{7}$.

2.64 $2x^2 + 6x - 9 = 0$

▌ $2(x^2 + 3x - \frac{9}{2}) = 0$ or $x^2 + 3x - \frac{9}{2} = 0$. Completing the square, we get $(x^2 + 3x + \frac{9}{4}) - \frac{9}{2} - \frac{9}{4} = 0$, $(x + \frac{3}{2})^2 = \frac{27}{4}$, $x + \frac{3}{2} = \pm\sqrt{\frac{27}{4}}$, $x + \frac{3}{2} = \pm\frac{\sqrt{27}}{2}$), $x = (\pm\frac{\sqrt{27}}{2})/2 - \frac{3}{2}$, $x = \frac{\sqrt{27}}{2} - \frac{3}{2} = (\sqrt{27} - 3)/2$, or $x = -\frac{\sqrt{27}}{2} - \frac{3}{2} = -(\sqrt{27} - 3)/2$.

2.65 $-x^2 + 3x - 2 = 0$

▌ By factoring we get $(-x + 1)(x - 2) = 0$, or $x = 1, 2$. Or we can rewrite the equation as $x^2 - 3x + 2 = 0$ and factor. See Prob. 2.66.

2.66 $-x^2 + 3x - 2 = 0$

▌ Using the quadratic formula we get $x = \dfrac{-3 \pm \sqrt{9 - 4(-1)(-2)}}{2} = \dfrac{-3 \pm \sqrt{1}}{-2} = 1$ or 2. See Prob. 2.67.

2.67 $-x^2 + 3x - 2 = 0$

▌ Rewrite the equation as $x^2 - 3x + 2 = 0$ and complete the square. $(x^2 - 3x + \frac{9}{4}) + 2 - \frac{9}{4} = 0$, $(x - \frac{3}{2})^2 = \frac{1}{4}$, $x - \frac{3}{2} = \pm\sqrt{\frac{1}{4}} = \pm\frac{1}{2}$, or $x = 1, 2$.

2.68 $(x^2 - 3)(x^2 - 4) = 0$

▌ Either $(x^2 - 3)$ or $(x^2 - 4)$ (or both) $= 0$. So $x^2 - 3 = 0$, $x = \pm\sqrt{3}$, or $x^2 - 4 = 0$, $x = \pm 2$. Thus, $x = 3, -3, 2,$ or -2.

2.69 $\left(\dfrac{2x + 1}{3x + 1}\right)^2 - 9 = 0$

▌ $\left(\dfrac{2x + 1}{3x + 1}\right)^2 = 9$, $\dfrac{2x + 1}{3x + 1} = \pm\sqrt{9} = \pm 3$. If $\dfrac{2x + 1}{3x + 1} = 3$, then $2x + 1 = 9x + 3$, $-7x = 2$, or $x = -\frac{2}{7}$. If $\dfrac{2x + 1}{3x + 1} = -3$, then $2x + 1 = -9x - 3$, $11x = -4$, or $x = -\frac{4}{11}$.

2.70 $(2x + 3)^2 + 4(2x + 3) + 3 = 0$

▌ Let $u = 2x + 3$. Then $u^2 + 4u + 3 = 0$. $(u + 1)(u + 3) = 0$, $u = -1, -3$. Thus, if $2x + 3 = -1$, then $x = -2$. If $2x + 3 = -3$, then $x = -3$.

2.71 $8y^{-2} + 6y^{-1} + 1 = 0$

▌ Let $u = 1/y = y^{-1}$. Then $u^2 = (1/y)^2 = y^{-2}$, $8u^2 + 6u + 1 = 0$. $(4u + 1)(2u + 1) = 0$, or $u = -\frac{1}{4}, -\frac{1}{2}$. Then $1/y = -\frac{1}{4}$ or $-\frac{1}{2}$, or $y = -4, -2$.

For Probs. 2.72 and 2.73, find the discriminant.

2.72 $Z^2 + 5Z - 6 = 0$

▌ The discriminant of a quadratic equation $ax^2 + bx + c = 0$ is $b^2 - 4ac$. In this case, $b^2 - 4ac = 5^2 - 4(1)(-6) = 25 + 24 = 49$.

2.73 $S^2 + 5S = 3$

▌ $a = 1, b = 5, c = -3, b^2 - 4ac = 25 - 4(1)(-3) = 25 + 12 = 37$.

For Probs. 2.74 to 2.76, determine whether the equation has real roots.

2.74 $x^2 + 11x + 11 = 0$

▌ Recall that when the discriminant is positive or zero, the equation has real roots. If the discriminant is zero, it has one real root; otherwise, it has two. If the discriminant is negative, the equation has imaginary roots. Here, $b^2 - 4ac = 11^2 - 4(1)(11) > 0$. Conclusion: Two real roots.

2.75 $x^2 - 3x + \frac{9}{4} = 0$

▮ $b^2 - 4ac = (-3)^2 - 4(1)(\frac{9}{4}) = 9 - 9 = 0$. Conclusion: One real root.

2.76 $2x^2 + x + 1 = 0$

▮ $b^2 - 4ac = 1^2 - 4(2)(1) < 0$. Conclusion: No real roots.

For Probs. 2.77 to 2.81, reduce the given equation to a quadratic and solve.

2.77 $\sqrt{x + 2} = x$

▮ Square both sides, $x + 2 = x^2$, $x^2 - x - 2 = 0$, $(x - 2)(x + 1) = 0$, or $x = 2, -1$.
Check: When $x = 2$, $\sqrt{x + 2} = \sqrt{2 + 2} = \sqrt{4} = 2$. This checks. When $x = -1$, $\sqrt{x + 2} = \sqrt{-1 + 2} = \sqrt{1} = 1 \neq -1$. This does not check. Conclusion: $x = 2$. (*Note*: When you square an equation, you may pick up extraneous roots. It is imperative that you check all solutions.)

2.78 $3 + \sqrt{2x - 1} = 0$

▮ $3 + \sqrt{2x - 1} = 0$, $\sqrt{2x - 1} = -3$, $2x - 1 = 9$, $2x = 10$, or $x = 5$. *Check*: $3 + \sqrt{10 - 1} = 3 + 3 = 6 \neq 0$. This equation has no solution. Notice that it did not reduce to a quadratic.

2.79 $\sqrt{3w - 2} - \sqrt{w} = 2$

▮ $\sqrt{3w - 2} = 2 + \sqrt{w}$, $3w - 2 = (2 + \sqrt{w})^2$ (squaring), $3w - 2 = 4 + 4\sqrt{w} + w$, $4\sqrt{w} = 6 - 2w$, $16w = (6 - 2w)^2$ (squaring), $16w = 36 - 24w + 4w^2$, $4w^2 - 40w + 36 = 0$, $w^2 - 10w + 9 = 0$, $(w - 9)(w - 1) = 0$, or $w = 9, 1$. *Check*: $\sqrt{3w - 2} - \sqrt{w} \stackrel{?}{=} 2$. If $w = 9$, $\sqrt{27 - 2} - \sqrt{9} = 5 - 3 = 2$. This checks. If $w = 1$, $\sqrt{3 - 2} - \sqrt{1} = \sqrt{1} - \sqrt{1} = 0$. This does not check. Conclusion: $w = 9$.

2.80 $y - 6 + \sqrt{y} = 0$

▮ $\sqrt{y} = 6 - y$, $y = (6 - y)^2 = 36 - 12y + y^2$, $y^2 - 13y + 36 = 0$, $(y - 9)(y - 4) = 0$, or $y = 9, 4$.
Check: $y = 9$ is extraneous. $y = 4$ is the solution. See Prob. 2.81.

2.81 $y - 6 + \sqrt{y} = 0$

▮ Let $u = \sqrt{y}$, then $u^2 = y$, so $u^2 - 6 + u = 0$, $u^2 + u - 6 = 0$, $(u + 3)(u - 2) = 0$, or $u = 3, 2$. $u = -3$ is extraneous, but $u = \sqrt{y}$, so $\sqrt{y} = 2$, and $y = 4$.

2.82 The height (in feet) of a ball thrown vertically upward above the ground in t seconds (s) is given by $h = 128t - 16t^2$. In how many seconds will the ball be 192 ft high?

▮ If $h = 192$ ft, then $192 = 128t - 16t^2$, $16t^2 - 128t + 192 = 0$, $t^2 - 8t + 12 = 0$, or $(t - 6)(t - 2) = 0$; after 2 s and after 6 s. *Note*: The ball goes upward, stops, and turns downward. That is why it reaches 192 ft after 2 s.

2.83 Find all values of R so that $x^2 + (R + 3)x + 4R = 0$ has one real root.

▮ The discriminant $(R + 3)^2 - 4(1)(4R) = 0$. Thus, $R^2 + 6R + 9 - 16R = 0$, $R^2 - 10R + 9 = 0$, $(R - 9)(R - 1) = 0$, or $R = 9, 1$.

2.84 Show that if $ax^2 + bx + a = 0$ has one real root, then $b = 2a$ or $b = -2a$.

▮ If $ax^2 + bx + a = 0$ has one real root, then $b^2 - 4(a)(a) = 0$. Then $b^2 - 4a^2 = 0$, $b^2 = 4a^2$, $b^2/a^2 = 4$ or $(b/a)^2 = 4$, so $b/a = 2$ or -2. Thus, $b = 2a$ or $-2a$.

2.85 Show that if $ax^2 + bx + c = 0$ has one real root, then $b = \pm\sqrt{4ac}$.

▌ If $ax^2 + bx + c = 0$ has one real root, then $b^2 - 4ac = 0$, $b^2 = 4ac$, or $b \pm \sqrt{4ac}$.

2.86 Solve for I: $P = EI - RI^2$

▌ $P = EI - RI^2$, $-RI^2 + EI - P = 0$, or $RI^2 - EI + P = 0$. Letting $a = R$, $b = -E$, $c = P$ gives
$I = \dfrac{E \pm \sqrt{E^2 - 4RP}}{2R}$.

2.87 If P dollars is invested at r percent compounded annually, at the end of 2 years the amount will be $A = P(1 + r)^2$. At what interest rate will \$1000 increase to \$1400 in 2 years?

▌ $A = P(1 + r)^2$, where $A = 1440$ and $P = 1000$. $1440 = 1000(1 + r)^2$, $(1 + r)^2 = 1.44$, $1 + r = \pm\sqrt{1.44}$ (reject the negative!), $1 + r = \sqrt{1.44} = 1.2$, or $r = 0.2$. The rate is 20 percent.

2.88 Solve for h: $h^2 + q^2 = 5x$

▌ $h^2 + q^2 = 5x$, $h^2 = 5x - q^2$, $h = \pm\sqrt{5x - q^2}$.

2.89 A number x has the property that the sum of x and twice its reciprocal is 3 times the number. Find x.

▌ Let $x =$ the number and $1/x =$ reciprocal. Then $x + 2(1/x) = 3x$, $x + 2/x = 3x$, $(x^2 + 2)/x = 3x$, $x^2 + 2 = 3x^2$, $2x^2 = 2$, $x^2 = 1$, or $x = \pm 1$. $x = 1$ and $x = -1$ are both solutions. *Check*: $1 + 2(1/1) = 3 = 3(1)$, $-1 + 2(1/-1) = -3 = 3(-1)$.

2.90 Find two consecutive positive integers whose product is 210.

▌ Let $x =$ first number. Then $x + 1 =$ second number. If their product is 210, then $x(x + 1) = 210$, $x^2 + x = 210$, $x^2 + x - 210 = 0$, $(x - 14)(x + 15) = 0$, $x = 14, -15$. $x = -15$ is extraneous (it is nonpositive). Thus, $x = 14$, $x + 1 = 15$ (check this!).

2.91 Find the length of a side of an equilateral triangle if the triangle's area is the same as the area of a square with a side of 5 m.

▌ If the square has side $x = 5$ m, then its area $= 25$ m^2. Let $S =$ length of side of triangle. Then the area of the triangle is $\frac{S^2}{4}\sqrt{3} = 25$, $S^2\sqrt{3} = 100$, $S^2 = 100/\sqrt{3} = 100\sqrt{3}/3$,
$S = \sqrt{\dfrac{100\sqrt{3}}{3}} = \sqrt{\dfrac{100}{3}}\sqrt{\sqrt{3}} = \dfrac{10}{\sqrt{3}} \cdot \sqrt[4]{3} = \dfrac{10}{3^{1/2}} \cdot 3^{1/4} = \dfrac{10}{3^{1/4}} = 10 \cdot 3^{-1/4}$ m.

2.3 LINEAR INEQUALITIES

For Probs. 2.92 to 2.96, write an algebraic expression for each statement.

2.92 $3y$ less than 4 times z is nonpositive.

▌ $4z - 3y \leq 0$

2.93 5 times t is less than or equal to 3 less than 3 times negative y.

▌ $5t \leq 3(-y) - 3$ or $5t \leq -3y - 3$

2.94 5 times t is more than 3 more than 3 times y.

▌ $5t > 3 + 3y$

2.95 5 times t is 3 more than 3 times y.

▌ $5t = 3 + 3y$

2.96 5 times t is more than 3 times y.

▌ $5t > 3y$

For Probs. 2.97 to 2.100, a and b are real numbers such that $a < b$. How are the given pairs of numbers related?

2.97 $a - 4$ and $b - 4$

   ▮ If $a < b$, then $a - 4 < b - 4$.

2.98 $-3a$ and $-3b$

   ▮ If $a < b$, then $3a < 3b$ and $-3a > -3b$ (the inequality sign reverses).

2.99 $2 - a$ and $2 - b$

   ▮ If $a < b$, then $-a > -b$, and $-a + 2 > -b + 2$, or $2 - a > 2 - b$.

2.100 $-a/4$ and $-b/4$

   ▮ If $a < b$, then $-a > -b$, and $-a \cdot \frac{1}{4} > -b \cdot \frac{1}{4}$, or $-a/4 > -b/4$.

For Probs. 2.101 to 2.103, is the given number a solution of the given inequality?

2.101 $-x + 5 \geq 0$; 3

   ▮ Substituting 3 for x, we get $-x + 5 = -3 + 5 = 2$ and $2 \geq 0$; 3 does satisfy the inequality.

2.102 $4/x + 3 \geq 1/x$; $\frac{1}{2}$

   ▮ When $x = \frac{1}{2}$, $4/x + 3 = 4/\frac{1}{2} + 3 = 11$, and $1/x = 2$, so $11 \geq 2$; $\frac{1}{2}$ does satisfy the inequality.

2.103 $x^{-1} + 1 < x^{-2} - 2$; 1

   ▮ When $x = 1$, $x^{-1} + 1 = 1^{-1} + 1 = 2$ and $x^{-2} - 2 = 1^{-2} - 2 = -1$, so $2 > -1$; 1 does not satisfy the inequality.

For Probs. 2.104 to 2.122, solve the given inequality (if possible).

2.104 $3t > 4 + 2t$

   ▮ $3t > 4 + 2t$, $t > 4$.

2.105 $14t \leq 3t - 4$

   ▮ $11t \leq -4$, or $t \leq -\frac{4}{11}$.

2.106 $-3s > 5s + 2$

   ▮ $-8s > 2$, or $s < -\frac{2}{8}$, or $s < -\frac{1}{4}$. (Notice the sign reversal!)

2.107 $4x + 9 \geq -7x - 16$

   ▮ $4x + 9 \geq -7x - 16$, $11x \geq -25$, or $x \geq -\frac{25}{11}$.

2.108 $-3(4t - 8) \leq 6(t + 5)$

   ▮ $-12t + 24 \leq 6t + 30$, $-6 \leq 18t$, $-\frac{6}{18} \leq t$, $-\frac{1}{3} \leq t$, or $t \geq -\frac{1}{3}$.

2.109 $-3(4t - 8) \leq 6(t + 5)$

   ▮ $-3(4t - 8) \leq 6(t + 5)$, $-12t + 24 \leq 6t + 30$, $-18t \leq 6$, $t \geq -\frac{6}{18}$, or $t \geq -\frac{1}{3}$. See Prob. 2.108.

Notice the difference between the methods in Probs. 2.108 and 2.109. It makes little difference whether the unknown is isolated on the RHS or LHS of the equation.

2.110 $-(x+3)-(2x+4)\le 2x+\frac{3}{2}$

▮ $-x+3-2x-4\le 2x+\frac{3}{2}$, $-3x-1\le 2x+\frac{3}{2}$, $-5x\le\frac{5}{2}$, or $x\le-\frac{1}{2}$.

2.111 $\dfrac{x-2}{6}+\dfrac{2x+3}{3}\ge 0$

▮ $\dfrac{x-2}{6}+\dfrac{2(2x+3)}{6}\ge 0$, $(x-2)+2(2x+3)\ge 0$ (here we multiply both sides by 6). $x-2+4x+6\ge 0$, $5x\ge -4$, or $x\ge-\frac{4}{5}$.

2.112 $\frac{1}{2}-\frac{1}{3}z\ge z/12+8$

▮ $\frac{1}{2}-\frac{1}{3}z\ge z/12+8$, $\frac{6}{12}-4z/12\ge z/12+\frac{96}{12}$, $-\frac{90}{12}\ge 5z/12$, $-90\ge 5z$, $-\frac{90}{5}\ge z$, $z\le-\frac{90}{5}$, or $z\le-18$.

2.113 $5<\dfrac{x-3}{4}\le 10$

▮ $20<x-3\le 40$, or $23<x\le 43$.

2.114 $-10<16-2x<10$

▮ $-26<-2x<-6$. Dividing by -2 gives $13>x>3$. Notice the reversal of both inequality signs.

2.115 $-6\le 2x-3<-1$

▮ See Prob. 2.116. Then $-\frac{3}{2}\le x<1$.

2.116 $-6\le 2x-3\le -1$

▮ $-3\le 2x\le 2$, or $-\frac{3}{2}\le x\le 1$. (See Prob. 2.115.)

2.117 $-6<x+2<10$

▮ $-6<x+2<10$, or $-8<x<8$ (adding -2 to all three parts of the inequality).

2.118 $-tx+q\le sx-r$. Solve for x.

▮ $-tx-sx\le -r-q$, or $x(-t-s)\le -r-q$. Case 1: If $-t-x>0$, then $x\le\dfrac{-r-q}{-t-s}$. Case 2: If $-t-s<0$, then $x\le\dfrac{-r-q}{-t-s}$. Case 3: If $-t-s=0$, then $0\le -r-q$, or $r\le -q$. Then if $r<-q$, any real x satisfies the equation. If $r=-q$, there is no solution.

2.119 $\dfrac{5x}{x+5}\le 2-\dfrac{25}{x+5}$, $x+5>0$

▮ Multiply both sides by $x+5$. Then $5x\le 2(x+5)-25$, $5x\le 2x+10-25$, $3x\le -15$, or $x\le -5$. But $x\ne -5$, since $x+5$ appears in the denominator. Thus, $x<-5$ is the only possible solution. Notice then that we can see that $\dfrac{5x}{x+5}\le 2-\dfrac{25}{x+5}$ has no solution.

2.120 $5x-3<\frac{1}{3}(15x-9)$

▮ $5x-3<5x-3$ or $0<0$. But $0=0$. Conclusion: No solution.

2.121 $2x/5-\frac{1}{2}(x-3)\le 2x/3-\frac{3}{10}(x+2)$

▮ $2x/5-x/2+\frac{3}{2}\le 2x/3-3x/10-\frac{6}{10}$, $\dfrac{12x-15x+45}{30}\le\dfrac{20x-9x-18}{30}$, $-3x+45\le 11x-18$, $-14x\le -63$, or $x\ge\frac{63}{14}$.

2.122 $2x - a < -3x + p$

▌ $-5x < p + a$, or $x > \dfrac{p + a}{-5}$.

For Probs. 2.123 to 2.127, an inequality is given. It is one of those in the above group. Graph the solution, and indicate the solution using interval notation.

2.123 $14t \le -3t - 4$ (2.105)

▌
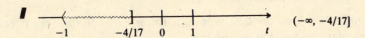
$(-\infty, -4/17]$

2.124 $4x + 9 \ge -7x - 16$ (2.107)

▌

$[-25/11, \infty)$

2.125 $\dfrac{5x}{x + 5} \le 2 - \dfrac{25}{x + 5}$, $x + 5 > 0$ (2.119)

▌ Recall that this inequality has no solution. One way to indicate $\varnothing$ using interval notation is $(1, 1)$. What would another way be?

2.126 $-6 < x + 2 < 10$ (2.117)

▌

$(-8, 8)$

2.127 $-6 \le 2x - 3 < -1$ (2.115)

▌
$[-3/2, 1)$

For Probs. 2.128 to 2.132, $a, b < 0$ and $a > b$. Tell whether each statement is true or false. If it is false, give a counterexample.

2.128 $a^2 > b^2$

▌ False; for example, $-5 > -6$, but $25 < 36$.

2.129 $ab < b^2$

▌ True. If $a < b$, then when we multiply by b on both sides, we reverse the inequality since $b < 0$.

2.130 $a(a - b) > b(a - b)$

▌ True; $a > b$ implies $a - b > 0$. Multiplying $a > b$ on both sides by $a - b$ maintains the inequality sign.

2.131 $ab(a - b) > 0$

▌ True; $a > b$ implies $a - b > 0$, which means $a(a - b) < 0$ and $ab(a - b) > 0$ (note that the signs are reversed).

2.132 $\dfrac{(a-b)^3}{b} > 0$

▮ False; $a - b > 0$, thus $(a - b)^3 > 0$ and $(a - b)^3/b < 0$. For example, let $a = -1$ and $b = -2$.

2.133 If both a and b are negative and b/a is greater than 1, is $a - b$ positive or negative?

▮ $b/a > 1$. Then $b < a$. (Careful! The sign reverses since $a < 0$.) So $0 < a - b$ and $a - b > 0$.

2.134 Assume that $m > n > 0$; then $mn > n^2$, $mn - m^2 > n^2 - m^2$, $m(n - m) > (n + m)(n - m)$, $m > n + m$, or $0 > n$. But we assumed that $n > 0$. Find the error.

▮ To get from $n(n - m) > (n + m)(n - m)$, we divided by $n - m$. But $n - m < 0$; thus, we needed to reverse the inequality sign. However, notice that we did not.

2.135 *Prove:* For any real numbers x, y, and z, if $x < y$, then $x + z < y + z$.

▮ If $x < y$, then $x - y < 0$. But $x - y = (x - y) = (z - z)$ which means $(x - y) + (z - z) < 0$, $(x + z) - (y + z) < 0$, or $x + z < y + z$.

2.136 *Prove:* If $a < b$ and $b < c$ (where a, b, and c are real numbers), then $a < c$.

▮ If $a < b$, then $a - b < 0$; if $b < c$, then $b - c < 0$. Then $(a - b) + (b - c) < 0$ and $(a - c) + (b - b) < 0$ or $a - c < 0$. Therefore, $a < c$.

2.137 The area of a square does not exceed 25 square centimeters (cm²). Find the possible integral values of a side of the square.

▮ $A = S^2 \leq 25$ cm². If $S = 1$ cm, then $S^2 = 1$ cm². If $S = 2$ cm, then $S^2 = 4$ cm². If $S = 5$ cm, then $S^2 = 25$ cm². Conclusion: 1 cm $\leq S \leq 5$ cm.

For Probs. 2.138 to 2.140, use the following: Linda bought some pizzas at \$5.40 apiece. The total bill did not exceed \$23.

2.138 Let $x =$ the number of pizzas bought. Use the above information to write an inequality in x.

▮ $x =$ number of pizzas bought. \$5.40 = price of 1 pizza. Then \$5.40$x =$ total amount spent. We know then that $5.40x \leq 23$.

2.139 Solve the inequality in Prob. 2.138.

▮ $5.4x < 23$ or $x < 23/5.4$.

2.140 Find how many pizzas Linda may have purchased.

▮ $x < 23/5.4$ where $x =$ number of pizzas bought. $23/5.4 = 4.25\ldots$ which means she bought 1, 2, 3, or 4. She could not have purchased 5 or more.

For Probs. 2.141 and 2.142, let a and b be positive integers whose sum is less than 12.

2.141 If we know that a is either 1 or 2, how many possible choices are there for b?

▮ $a + b < 12$ $(a, b > 0)$. If $a = 1$, then $1 + b < 12$ or $b < 11$, so b can be $1, 2, \ldots, 10$ (10 choices). If $a = 2$, then $2 + b < 12$ or $b < 10$, so b can be $1, 2, \ldots, 9$ (9 choices). Total number of choices = 19.

2.142 How many possibilities are there altogether for a and b?

▮
$$a = 1 \qquad 10 \text{ choices for } b$$
$$a = 2 \qquad 9 \text{ choices for } b$$
$$a = 3 \qquad 8 \text{ choices for } b$$

$$a = 4 \quad \text{7 choices for } b$$
$$a = 5 \quad \text{6 choices for } b$$
$$a = 6 \quad \text{5 choices for } b$$
$$a = 7 \quad \text{4 choices for } b$$
$$a = 8 \quad \text{3 choices for } b$$
$$a = 9 \quad \text{2 choices for } b$$
$$a = 10 \quad \text{1 choice for } b$$
Total: 55 possibilities

2.4   ABSOLUTE VALUE

For Probs. 2.143 to 2.192, solve each equation or inequality, if possible.

2.143 $|x| = 6$

▮ Recall that $|x|$ has two common equivalent definitions.

$$|x| = \begin{cases} x & x \geq 0 \\ -x & x < 0 \end{cases} \qquad \text{or} \qquad |x| = \sqrt{x^2}$$

If $|x| = 6$, then $x = 6$ or $x = -6$. *Check*: $|6| = 6$, $|-6| = 6$. Note that this check works given either of the definitions above: $\sqrt{6^2} = 6$ and $\sqrt{(-6)^2} = 6$.

2.144 $|x| < 5$

▮ Recall that $|x - a| < b$ can be interpreted as all x whose distance from a is less than b.

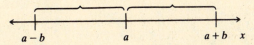

In Fig. 2.3, notice that the interval $(a - b, a + b)$ satisfies that criterion. In our problem, $a = 0$, $b = 5$. Thus $|x| < 5$ has the solution $-5 < x < 5$ ($-5 = 0 - 5$, $5 = 0 + 5$). More generally, let $a > 0$; then

$|x| = a$ has the solution $x = \pm a$
$|x| < a$ has the solution $-a < x < a$
$|x| > a$ has the solution $x > a$, $x < -a$
$|cx + d| < a$ can be written as $-a < cx + d < a$
$|cx + d| > a$ can be written as $cx + d > a$, $cx + d < -a$

Fig. 2.3

2.145 $|x| \geq 4$

▮ $|x| \geq 4$ has solutions $x \geq 4$ and $x \leq -4$. (See Fig. 2.3.) The graph of this solution set is

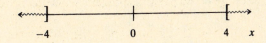

2.146 $|t| \geq 0$

▮ Then $t \geq 0$ which implies that t can be any real number. Note that this makes sense. $|t|$ is never negative!

2.147 $|x| < -10$

▮ This inequality has no solution: $|x| \geq 0$ for all x.

2.148 $|x - 4| = 3$

▌ Then $x - 4 = 3$ or $x - 4 = -3$. Then $x = 7$ or $x = 1$.

2.149 $|-S + 4| = 2$

▌ $-S + 4 = 2$ or $-S + 4 = -2$. Thus, $S = 2$ or $S = 6$.

2.150 $|2t| = 8$

▌ $2t = 8$ or $2t = -8$. Thus, $t = 4$ or $t = -4$.

2.151 $|-3x| = 21$

▌ (i) $-3x = 21$ or $-3x = -21$. Thus, $x = -7$ or $x = 7$. (ii) $|-3x| = |-3| \, |x| = 3 \, |x|$. $3 \, |x| = 21$. $|x| = 7$, so $x = 7$ or $x = -7$.

2.152 $|2x + 1| = 4$

▌ $2x + 1 = 4$ or $2x + 1 = -4$. Thus, $x = \frac{3}{2}$ or $x = -\frac{5}{2}$.

2.153 $|2s - 6| = 10$

▌ $2s - 6 = 10$ or $2s - 6 = -10$, $2s = 16$ or $2s = -4$, so $s = 8$ or $s = -2$.

2.154 $|-2t + 4| = -2$

▌ $-2t + 4 = -2$ or $-2t + 4 = 2$, $-2t = -6$ or $-2t = -2$, so $t = 3$ or $t = 1$.

2.155 $|2x| < 1$ (See Fig. 2.3.)

▌ Then $-1 < 2x < 1$ or $-\frac{1}{2} < x < \frac{1}{2}$. Using interval notation, $(-\frac{1}{2}, \frac{1}{2})$ is the solution set.

2.156 $|2s| \le 2$

▌ $-2 \le 2s \le 2$ or $-1 \le s \le 1$. The graph of the solution set is

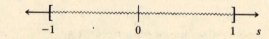

2.157 $|2t| > 1$ (See Fig. 2.3.)

▌ $2t > 1$ or $2t < -1$. Then $t > \frac{1}{2}$ or $t < -\frac{1}{2}$.

2.158 $|x - 1| < 2$

▌ $-2 < x - 1 < 2$ or $-1 < x < 3$. In interval form, $(-1, 3)$.

2.159 $|t - 2| \le 4$

▌ $-4 \le t - 2 \le 4$, or $-2 \le t \le 6$.

2.160 $|t - 2| \ge 5$

▌ $t - 2 \ge 5$ or $t - 2 \le -5$. Thus, $t \ge 7$ or $t \le -3$.

2.161 $|2x + 1| < 7$

▌ $-7 < 2x + 1 < 7$, $-8 < 2x < 6$, or $-4 < x < 3$.

2.162 $|2x - 5| < 10$

▌ $-10 < 2x - 5 < 10$, $-5 < 2x < 15$, or $-\frac{5}{2} < x < \frac{15}{2}$.

2.163 $|-2x + 2| < 2$

▮ $-2 < -2x + 2 < 2$, $-4 < -2x < 0$, or $2 > x > 0$. (Careful! The signs are reversed because of division by -2.)

2.164 $|-3t - 2| < 6$

▮ $-6 < -3t - 2 < 6$, $-4 < -3t < 8$, or $\frac{4}{3} > t > -\frac{8}{3}$.

2.165 $|2x + 1| > 6$

▮ $2x + 1 > 6$ or $2x + 1 < -6$, $2x > 5$ or $2x < -7$, so $x > \frac{5}{2}$ or $x < -\frac{7}{2}$.

2.166 $|2s - 2| > 8$

▮ $2s - 2 > 8$ or $2s - 2 < -8$, $2s > 10$ or $2s < -6$, so $s > 5$ or $s < -3$.

2.167 $|2x - 4| \geq 10$

▮ $2x - 4 \geq 10$ or $2x - 4 \leq -10$, $2x \geq 14$ or $2x \leq -6$, so $x \geq 7$ or $x \leq -3$.

2.168 $|-x + 1| \geq 5$

▮ $-x + 1 \geq 5$ or $-x + 1 \leq -5$, $-x \geq 4$ or $-x \leq -6$, so $x \leq -4$ or $x \geq 6$.

2.169 $|-s - 1| \geq 7$

▮ $-s - 1 \geq 7$ or $-s - 1 \leq -7$, $-s \geq 8$ or $-s \leq -6$, so $s \leq -8$ or $s \geq 6$.

2.170 $|-2x + 4| > 6$

▮ $-2x + 4 > 6$ or $-2x + 4 < -6$, $-2x > 2$ or $-2x < -10$, so $x < -1$ or $x > 5$.

2.171 $|-3x - 2| \geq 10$

▮ $-3x - 2 \geq 10$ or $-3x - 2 \leq -10$, $-3x \geq 12$ or $-3x \leq -8$, so $x \leq -4$ or $x \geq \frac{8}{3}$.

2.172 $|1 - x| < 5$

▮ $-5 < 1 - x < 5$, $-6 < -x < 4$, or $6 > x > -4$. (Careful!)

2.173 $|2 - t| \leq 6$

▮ $-6 \leq 2 - t \leq 6$, $-8 \leq -t \leq 4$, or $8 \geq t \geq -4$.

2.174 $|-3 - s| \leq 5$

▮ $-5 \leq -3 - s \leq 5$, $-2 \leq -s \leq 8$, or $2 \geq s \geq -8$.

2.175 $|1 - 2x| \leq 8$

▮ $-8 \leq 1 - 2x \leq 8$, $-9 \leq -2x \leq 7$, or $\frac{9}{2} \geq x \geq -\frac{7}{2}$.

2.176 $|2 - 3X| < 6$

▮ $-6 < 2 - 3X < 6$, $-8 < -3X < 4$, or $\frac{8}{3} > X > -\frac{4}{3}$.

2.177 $|-3 - 4x| \leq 12$

▮ $-12 \leq -3 - 4x \leq 12$, $-9 \leq -4x \leq 15$, or $\frac{9}{4} \geq x \geq -\frac{15}{4}$.

2.178 $|1 - t| > 11$

▮ $1 - t > 11$ or $1 - t < -11$, $-t > 10$ or $-t < -12$, so $t < -10$ or $t > 12$.

2.179 $|2 - s| > 10$

▮ $2 - s > 10$ or $2 - s < -10$, $-s > 8$ or $-s < -12$, so $s < -8$ or $s > 12$.

2.180 $|2 - 3x| \geq 13$

▮ $2 - 3x \geq 13$ or $2 - 3x \leq -13$, $-3x \geq 11$ or $-3x \leq -15$, or $x \leq -\frac{11}{3}$ or $x \geq 5$.

2.181 $|-1 - 2x| > 2$

▮ $-1 - 2x > 2$ or $-1 - 2x < -2$, $-2x > 3$, or $-2x < -1$, $x < -\frac{3}{2}$ or $x > \frac{1}{2}$.

2.182 $\left|\dfrac{x-1}{2}\right| < 3$

▮ (i) $-3 < \dfrac{x-1}{2} < 3$, $-3 < x/2 - \frac{1}{2} < 3$, $-\frac{5}{2} < x/2 < \frac{7}{2}$, $-5 < x < 7$. (ii) $|x - 1| < 6$ ($|2| = 2$, and cross multiply). Then $-6 < x - 1 < 6$, or $-5 < x < 7$.

2.183 $\left|\dfrac{2x-3}{4}\right| \leq 4$

▮ $|2x - 3| \leq 16$, $-16 \leq 2x - 3 \leq 16$, $-13 \leq 2x \leq 19$, or $-\frac{13}{2} \leq x \leq \frac{19}{2}$.

2.184 $|\frac{1}{2}(1 - 2x)| < 10$

▮ $|1 - 2x| < 20$, $-20 < 1 - 2x < 20$, $-21 < -2x < 19$, or $\frac{21}{2} > x > -\frac{19}{2}$.

2.185 $|x + 1| = x + 1$

▮ Recall that $|p| = p$ when $p \geq 0$. Thus, $|x + 1| = x + 1$ when $x + 1 \geq 0$, or $x \geq -1$.

2.186 $|2s - 3| = 3 - 2s$

▮ If $|2s - 3| = 3 - 2s$, then $|2s - 3| = -(2s - 3)$. But $|p| = -p$ only when $p \leq 0$. Thus, $2s - 3 \leq 0$, $2s \leq 3$, or $s \leq \frac{3}{2}$.

2.187 $|x + 2| < x + 2$

▮ No solution. $|x + 2| \geq x + 2$ or all x. Note that $|x + 2| < x + 2$ would be rewritten as $-(x + 2) < x + 2 < x + 2$ which is impossible.

2.188 $|t^2| = 9$

▮ Then $t^2 = 9$ or $t^2 = -9$, but $t^2 \neq -9$. So $t^2 = 9$, $t = \pm 3$.

2.189 $|x^2 - 2| = 7$

▮ Then $x^2 - 2 = \pm 7$. If $x^2 - 2 = 7$, then $x^2 = 9$, $x = \pm 3$. If $x^2 - 2 = -7$, then $x^2 = -5$ which is impossible.

2.190 $|2t - 1| < -3$

▮ No solution. $|2t - 1| \geq 0$ for all t.

2.191 $|-\frac{5}{2}(|x - 16|)| \geq 0$

▮ x can be any real number, since $|a| \geq 0$ for all a.

2.192 Solve $|2x - p| > s$ for x.

▮ $2x - p > s$ or $2x - p < -s$, $2x > p + s$ or $2x < -s + p$, so $x > \dfrac{p+s}{2}$ or $2x < \dfrac{-s+p}{2}$.

2.5 NONLINEAR INEQUALITIES

For Probs. 2.193 to 2.205, solve the given inequality. Graph the solution set for each even-numbered problem, and express each odd-numbered solution set in interval form.

2.193 $x^2 - 2x + 1 > 0$

▮ Factoring the LHS, we get $(x - 1)(x - 1) > 0$, which means that (1) both factors on the LHS are >0 or (2) both factors on the LHS are <0. Thus, (1) $x - 1 > 0$, $x > 1$, or (2) $x - 1 < 0$, $x < 1$. ($x - 1$ is the only factor!) Conclusion: $x > 1$ or $x < 1$. In interval form, $(-\infty, 1)$ or $(1, \infty)$. Using set notation: $\{(-\infty, 1), (1, \infty)\}$.

2.194 $x^2 - 3x + 2 > 0$

▮ $(x - 2)(x - 1) > 0$. (1) $x - 2 > 0$, $x - 1 > 0$; $x > 2$, $x > 1$, or (2) $x - 2 < 0$, $x - 1 < 0$; $x < 2$, $x < 1$. Simplify: (1) If $x > 2$ and $x > 1$, then $x > 2$, since $x > 2$ guarantees us that $x > 1$. (2) If $x < 2$ and $x < 1$, then $x < 1$, since $x < 1$ guarantees that $x < 2$. Conclusion: $x > 2$ or $x < 1$.

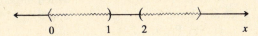

2.195 $x^2 - 3x + 2 \geq 0$

▮ See Prob. 2.194. All $>$ signs are to be replaced by $\geq$ signs. Conclusions: $x \geq 2$ or $x \leq 1$ or in interval form $[2, \infty) \cup (-\infty, 1]$.

2.196 $x^2 - 3x + 2 < 0$

▮ If $(x - 2)(x - 1) < 0$, then (1) $x - 2 > 0$, $x - 1 < 0$ or (2) $x - 2 < 0$, $x - 1 > 0$. In (1) $x > 2$ and $x < 1$, which is impossible. In (2) $x < 2$ and $x > 1$. Conclusion: $x < 2$ and $x > 1$, or $1 < x < 2$.

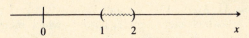

2.197 $x^2 - 3x + 2 \leq 0$

▮ See Prob. 2.196. Replacing $<$ by $\leq$, we get $x \leq 2$ and $x \geq 1$. In interval form, $[1, 2]$.

2.198 $x^2 + x - 6 \leq 0$

▮ $(x + 3)(x - 2) \leq 0$. Then (1) $x + 3 \leq 0$ and $x - 2 \geq 0$, or (2) $x + 3 \geq 0$ and $x - 2 \leq 0$. Thus, $x \leq -3$ and $x \geq 2$, which is impossible, or $x \geq -3$ and $x \leq 2$. Conclusion: $x \geq -3$ and $x \leq 2$, that is, $-3 \leq x \leq 2$.

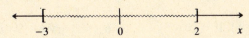

2.199 $x^2 + x - 6 > 0$

▮ $(x + 3)(x - 2) > 0$. Then (1) $x + 3 > 0$, $x - 2 > 0$ or (2) $x + 3 < 0$, $x - 2 < 0$. Thus, $x > -3$ and $x > 2$, which means $x > 2$, or $x < -3$ and $x < 2$, which means $x < -3$. Conclusion: $x > 2$ or $x < -3$ or, in interval form, $(\infty, -3) \cup (2, \infty)$.

2.200 $2x^2 - 5x - 3 \leq 0$

▮ $(2x + 1)(x - 3) \leq 0$. Then (1) $2x + 1 \leq 0$, $x - 3 \geq 0$ or (2) $2x + 1 \geq 0$, $x - 3 \leq 0$. Thus, $x \leq -\frac{1}{2}$ and $x \geq 3$ (which is impossible); or $x \geq -\frac{1}{2}$ and $x \leq 3$. Conclusion: $x \geq -\frac{1}{2}$ and $x \leq 3$.

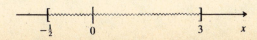

2.201 $3x^2 + x > 0$

▌ $x(3x + 1) > 0$. Then (1) $x > 0$ and $3x + 1 > 0$ or (2) $x < 0$ and $3x + 1 < 0$. Then $x > 0$ and $x > -\frac{1}{3}$, which means $x > 0$, or $x < 0$ and $x < -\frac{1}{3}$, which means $x < -\frac{1}{3}$. Conclusion: $x > 0$ or $x < -\frac{1}{3}$ or, in interval form, $(0, \infty) \cup (-\infty, -\frac{1}{3})$.

2.202 $3x^2 < -2x + 1$

▌ If $3x^2 < -2x + 1$, then $3x^2 + 2x - 1 < 0$. Thus, $(3x - 1)(x + 1) < 0$ which means (1) $3x - 1 < 0$, $x + 1 > 0$ or (2) $3x - 1 > 0$, $x + 1 < 0$. Then $x < \frac{1}{3}$ and $x > -1$, or $x > \frac{1}{3}$ and $x < -1$ which is impossible. Conclusion: $x < \frac{1}{3}$ and $x > -1$.

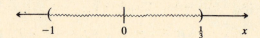

2.203 $S^2 \le 1$

▌ $S^2 \le 1$ means $S^2 - 1 \le 0$. Then $(S - 1)(S + 1) \le 0$, which means (1) $S - 1 \le 0$, $S + 1 \ge 0$ or (2) $S - 1 \ge 0$, $S + 1 \le 0$. Then $S \le 1$ and $S \ge -1$ or $S \ge 1$ and $S \le -1$ which is impossible. Conclusion: $S \le 1$ and $S \ge -1$ or $[-1, 1]$. See Prob. 2.204.

2.204 $S^2 \le 1$

▌ Here is another method. If $S^2 \le 1$, then $\sqrt{S^2} \le \sqrt{1}$. Recall that $\sqrt{S^2} = |S|$. Then $|S| \le 1$ or $-1 \le S \le 1$. Notice the result is the same as in Prob. 2.203.

2.205 $x^3 - x > 0$

▌ $x(x^2 - 1) > 0$. Thus, (1) $x > 0$, $x^2 - 1 > 0$ or (2) $x < 0$, $x^2 - 1 < 0$. If (1) $x > 0$, $(x - 1)(x + 1) > 0$, then $x - 1 > 0$, $x + 1 > 0$ which means $x > 1$, $x > -1$, which means $x > 1$; or $x - 1 < 0$, $x + 1 < 0$ which means $x < 1$, $x < -1$ which means $x < -1$. Thus, $x > 0$ and $(x > 1$ or $x < -1)$, or $x > 0$ and $x > 1$, or $x > 1$. If (2) $x < 0$, $x - 1 < 0$, then $x + 1 > 0$ which means $x < 1$ and $x > -1$; or $x - 1 > 0$, $x + 1 < 0$ which means $x > 1$ and $x < -1$ which is impossible. Thus, $x < 0$ and $(x < 1$ and $x > -1)$, which means $x > -1$ and $x < 0$. Conclusion: $x > 1$ or $(x > -1$ and $x < 0)$ or, in interval form, $(1, \infty) \cup (-1, 0)$.

2.6 MISCELLANEOUS PROBLEMS

For Probs. 2.206 through 2.212, solve by factoring.

2.206 $4x^2 - 5x = 0$

▌ $4x^2 - 5x = x(4x - 5)$. If $x(4x - 5) = 0$, then $x = 0$ or $4x - 5 = 0$; $x = 0, \frac{5}{4}$.

2.207 $4x^2 - 9 = 0$

▌ $4x^2 - 9$ is the difference of two perfect squares. Thus, if $4x^2 - 9 = 0$, $(2x - 3)(2x + 3) = 0$ and $2x - 3 = 0$ or $2x + 3 = 0$. $x = \pm\frac{3}{2}$.

2.208 $x^2 - 4x + 3 = 0$

▌ Factor the trinomial on the LHS: $x^2 - 4x + 3 = (x - 1)(x - 3)$. Then $(x - 1)(x - 3) = 0$

$$x - 1 = 0 \qquad x - 3 = 0$$
$$x = 1 \qquad x = 3$$

2.209 $x^2 - 6x + 9 = 0$

▮ Proceed as in Prob. 2.208: $x^2 - 6x + 9 = (x - 3)^2 = (x - 3)(x - 3)$. Then $(x - 3)(x - 3) = 0$

$$x - 3 = 0 \qquad x - 3 = 0$$
$$x = 3 \qquad x = 3$$

2.210 $4x^2 + 20x + 25 = 0$

▮ Proceed as in Prob. 2.209: $4x^2 + 20x + 25 = (2x + 5)^2 = (2x + 5)(2x + 5) = 0$. Then $2x + 5 = 0$, $2x = -5$, $x = -\frac{5}{2}$.

2.211 $3x^2 + 8ax - 3a^2 = 0$

▮ $3x^2 + 8ax - 3a^2 = (3x - a)(x + 3a)$ (Check that!) Thus, $(3x - a)(x + 3a) = 0$

$$3x - a = 0 \qquad x + 3a = 0$$
$$x = a/3 \qquad x = -3a$$

2.212 $10ax^2 + (15 - 8a^2)x - 12a = 0$

▮ $10ax^2 = (2ax)(5x)$ and $3 \cdot 5x - 4a(2ax) = (15 - 8a^2)x$. Thus, $10ax^2 + (15 - 8a^2)x - 12a = (2ax + 3)(5x - 4a)$. Then $(2ax + 3)(5x - 4a) = 0$

$$2ax + 3 = 0 \qquad 5x - 4a = 0$$
$$x = -3/2a \qquad x = 4a/5$$

For Probs. 2.213 through 2.215, solve by completing the square.

2.213 $x^2 - 2x - 1 = 0$

▮ If $x^2 - 2x - 1 = 0$, then $x^2 - 2x = 1$; $x^2 - 2x + 1 = 1 + 1$. [-2 is the coefficient in $-2x$; $-\frac{2}{2} = -1$; $(-1)^2 = 1$. Thus, we add 1 to both sides of the equation.] Then $(x - 1)^2 = 2$, $x - 1 = \pm\sqrt{2}$, $x = 1 \pm \sqrt{2}$.

2.214 $3x^2 + 8x + 7 = 0$

▮ See Prob. 2.213. If $3x^2 + 8x + 7 = 0$, then $3x^2 + 8x = -7$ and $x^2 + \frac{8}{3}x = -\frac{7}{3}$. (Divide by 3 to make it monic.) Then $x^2 + \frac{8}{3}x + \frac{16}{9} = -\frac{7}{3} + \frac{16}{9} = -\frac{5}{9}$

$$\left(x + \frac{4}{3}\right)^2 = -\frac{5}{9}$$
$$x + \frac{4}{3} = \pm\sqrt{-\frac{5}{9}} = \pm i\frac{\sqrt{5}}{3}$$
$$x = \frac{-4 \pm i\sqrt{5}}{3}.$$

2.215 $ax^2 - bx + c = 0$, $a \neq 0$

▮ We proceed as in Prob. 2.214: $x^2 + \frac{b}{a}x = -\frac{c}{a}$, $x^2 + \frac{b}{a}x + \frac{b^2}{4a^2} = \frac{b^2}{4a^2} - \frac{c}{a} = \frac{b^2 - 4ac}{4a^2}$. Then $x + \frac{b}{2a} = \pm\sqrt{\frac{b^2 - 4ac}{4a^2}} = \frac{\pm\sqrt{b^2 - 4ac}}{2a}$ and $x = \frac{-b \pm \sqrt{b^2 - 4ac}}{2a}$. Note that this is a formal derivation of the quadratic formula.

2.216 Solve for x: $9x^4 - 10x^2 + 1 = 0$.

▮ $9x^4 - 10x^2 + 1 = (x^2 - 1)(9x^2 - 1)$. (Check that by multiplication. What other factors were possibilities?) Then $(x^2 - 1)(9x^2 - 1) = 0$

$$x^2 - 1 = 0 \qquad 9x^2 - 1 = 0$$
$$x = \pm 1 \qquad x^2 = \frac{1}{9} \qquad x = \pm\frac{1}{3}$$
$$x + \pm 1, \pm\frac{1}{3}$$

2.217 Solve for x: $x^4 - 6x^3 + 12x^2 - 9x + 2 = 0$.

▮ Complete the square:

$$(x^4 - 6x^3 + 9x^2) + 3x^2 - 9x + 2 = 0$$
$$(x^2 - 3x)^2 + 3(x^2 - 3x) + 2 = 0$$

and $[(x^2 - 3x) + 2][(x^2 - 3x) + 1] = 0$, $x^2 - 3x + 2 = 0$, $x = 1, 2$ or $x^2 - 3x + 1 = 0$, $x = \dfrac{3 \pm \sqrt{5}}{2}$.

2.218 Solve for x: $\sqrt{5x - 1} - \sqrt{x} = 1$.

▮ Rewrite this equation by isolating one of the radicals, $\sqrt{5x - 1} = 1 + \sqrt{x}$, and square; then $5x - 1 = 1 + 2\sqrt{x} + x$, $4x - 2 = 2\sqrt{x}$, or $2x - 1 = \sqrt{x}$.

Square: $4x^2 - 4x + 1 = x$, $4x^2 - 5x + 1 = 0$, $(4x - 1)(x - 1) = 0$, and $x = \frac{1}{4}, 1$.

Check each prospective root; since we squared the equation, extraneous roots are likely. When $x = \frac{1}{4}$, $\sqrt{5(\frac{1}{4}) - 1} - \sqrt{\frac{1}{4}} = 0 \neq 1$. $x = 1$ checks and is the only solution.

2.219 Solve for x: $\sqrt{6x + 7} - \sqrt{3x + 3} = 1$.

▮ See Prob. 2.218.

$$\sqrt{6x + 7} = 1 + \sqrt{3x + 3} \qquad \text{(Transpose.)}$$
$$6x + 7 = 1 + 2\sqrt{3x + 3} + 3x + 3 \quad \text{(Square.)}$$
$$3x + 3 = 2\sqrt{3x + 3}$$

Squaring gives $9x^2 + 18x + 9 = 4(3x + 3) = 12x + 12$, $9x^2 + 6x - 3 = 3(3x - 1)(x + 1) = 0$, and $x = \frac{1}{3}, -1$. Check both of these. Neither one is extraneous.

For Probs. 2.220 through 2.223, determine the character of the roots.

2.220 $x^2 - 8x + 9 = 0$

▮ The discriminant $b^2 - 4ac = (-8)^2 - 4(1)^2(9) = 28$. Since $28 > 0$ and since 28 is not a perfect square, the roots are irrational and unequal.

2.221 $3x^2 - 8x + 9 = 0$

▮ $b^2 - 4ac = (-8)^2 - 4(3)(9) = -44$. Since $-44 < 0$, the roots are imaginary and unequal.

2.222 $6x^2 - 5x - 6 = 0$

▮ $b^2 - 4ac = 169$. $169 > 0$ and is a perfect square ($169 = 13^2$). Thus, the roots are rational and unequal.

2.223 $4x^2 - 4\sqrt{3}x + 3 = 0$

▮ $b^2 - 4ac = (-4\sqrt{3})^2 - 4(4)(3) = 0$. Thus, the roots are real and equal. (*Question*: What are they, and why are they not rational?)

2.224 Square each of the inequalities: (*a*) $-3 < 4$, (*b*) $-5 < 4$

▮ (*a*) $(-3)^2 = 9$, $4^2 = 16$; and $9 < 16$.
(*b*) $(-5)^2 = 25$, $4^2 = 16$, and $25 > 16$ (Be careful! The sign reverses.)

For Probs. 2.225 through 2.228, solve for x.

2.225 $3x + 4 > 5x + 2$

▌ $3x + 4 > 5x + 2$

$\qquad -2x > -2 \qquad$ (Subtract $5x + 4$.)

$\qquad x < 1 \qquad$ (Divide by -2.)

2.226 $2x - 8 < 7x + 12$

▌ $2x - 8 < 7x + 12$

$\qquad -5x < 20 \qquad$ (Subtract $7x - 8$.)

$\qquad x > -4 \qquad$ (Divide by -5.)

2.227 $x^2 > 4x + 5$

▌ $x^2 > 4x + 5$

$\qquad x^2 - 4x - 5 > 0$

If $x^2 - 4x - 5 = 0$, then $x = -1, 5$. On the interval $x < -1$, $f(-2) > 0$; on the interval $-1 < x < 5$, $f(0) < 0$; on the interval $x > 5$, $f(6) > 0$. Thus, the inequality is satisfied when $x < -1$ and $x > 5$.

2.228 $3x^2 + 2x + 2 < 2x^2 + x + 4$

▌ See Prob. 2.227. Then $x^2 + x - 2 < 0$ and if $x^2 + x - 2 = 0$, $x = -2, 1$.

$$f(-3) = 9 - 3 - 2 > 0$$
$$f(0) = -2 < 0$$
$$f(2) = 4 + 2 - 2 > 0$$

Thus, $-2 < x < 1$ is the solution interval.

For Probs. 2.229 through 2.235, solve for x or y and check the given equation.

2.229 $x - 2(1 - 3x) = 6 + 3(4 - x)$

▌ $x - 2 + 6x = 6 + 12 - 3x$

$\qquad 7x - 2 = 18 - 3x$

$\qquad 10x = 20$

$\qquad x = 2$

Check: $\qquad\qquad\qquad\qquad 2 - 2(1 - 6) \stackrel{?}{=} 6 + 3(4 - 2)$

$\qquad\qquad\qquad\qquad\qquad 12 = 12$

2.230 $ay + b = cy + d$

▌ $ay - cy = d - b$

$\qquad (a - c)y = d - b$

$\qquad y = \dfrac{d - b}{a - c}$

Check: $\qquad\qquad\qquad\qquad a\left(\dfrac{d - b}{a - c}\right) + b \stackrel{?}{=} c\left(\dfrac{d - b}{a - c}\right) + d$

$$\dfrac{ad - ab}{a - c} + b \stackrel{?}{=} \dfrac{cd - cb}{a - c} + d$$

$$\dfrac{ad - ab + b(a - c)}{a - c} \stackrel{?}{=} \dfrac{cd - cb + d(a - c)}{a - c}$$

$$\dfrac{ad - ab + ba - bc}{a - c} = \dfrac{cd - cb + da - dc}{a - c}$$

2.231 $\dfrac{3x-2}{5}=4-\tfrac{1}{2}x$

   ❚ $6x-4=40-5x$ (Multiplication by what?)

        $11x=44$

         $x=4$

   Check: $\dfrac{3(4)-2}{5}\overset{?}{=}4-\tfrac{1}{2}(4)$

                             $2=2$

2.232 $\dfrac{3x+1}{3x-1}=\dfrac{2x+1}{2x-3}$

   ❚ $(3x+1)(2x-3)=(2x+1)(3x-1)$

        $6x^2-7x-3=6x^2+x-1$

             $-8x=2$

              $x=-\tfrac{1}{4}$

   Check: $\dfrac{3(-\tfrac{1}{4})+1}{3(-\tfrac{1}{4})-1}\overset{?}{=}\dfrac{2(-\tfrac{1}{4})+1}{2(-\tfrac{1}{4})-3}$

                    $-\tfrac{1}{7}=-\tfrac{1}{7}$

2.333 $\dfrac{1}{x-3}-\dfrac{1}{x+1}=\dfrac{3x-2}{(x-3)(x+1)}$

   ❚ $x+1-(x-3)=3x-2$

            $-3x=-6$

              $x=2$

   Check: $-1-\tfrac{1}{3}\overset{?}{=}\dfrac{6-2}{-3}$

                    $-\tfrac{4}{3}=-\tfrac{4}{3}$

2.234 $\dfrac{x^2-2}{x-1}=x+1-\dfrac{1}{x-1}$

   ❚ $x^2-2=(x+1)(x-1)-1$

          $=x^2-2$

Thus, the solution set is the set of real numbers except $x=1$ (since division by 0 is not defined).

2.235 $\dfrac{1}{x-1}+\dfrac{1}{x-3}+\dfrac{2x-5}{(x-1)(x-3)}$

   ❚ $x-3+(x-1)=2x-5$

        $2x-4=2x-5$

But there is no x such that $2x-4=2x-5$. The solution set is $\varnothing$.

2.236 One number is 5 more than another, and the sum of the two is 71. Find the numbers.

   ❚ Let x be the smaller number and $x+5$ be the larger. Then $x+(x+5)=71$, $2x=66$, and $x=33$. The numbers are 33 and 38.

2.237 A father is now 3 times as old as his son. Twelve years ago he was 6 times as old as his son. Find the present age of each.

▮ Let x = age of son and $3x$ = age of father. Twelve years ago, the age of the son was $x - 12$, and the age of the father was $3x - 12$. Then $3x - 12 = 6(x - 12)$, $3x = 60$, and $x = 20$. The present age of the son is 20, and the father is 60.

2.238 Bleaching powder is obtained through the reaction of chlorine and slaked lime, with 74.10 pounds (lb) of lime and 70.91 lb of chlorine producing 127.00 lb of bleaching powder and 18.01 lb of water. How many pounds of lime will be required to produce 1000 lb of bleaching powder?

▮ Let x = pounds of lime required. Then $\dfrac{x \text{ lb of lime}}{1000 \text{ lb of powder}} = \dfrac{74.10 \text{ lb of lime}}{127 \text{ lb of powder}}$, $127x = 74{,}100$, and $x = 538.46$ lb

2.239 The pressure of a gas in a container at constant temperature varies inversely as the volume. If $p = 30$ when $v = 45$, find p when $v = 25$.

▮   First solution: Here $p = c/v$, $30 = c/45$, $c = 30 \cdot 45$; thus $p = 30 \cdot 45/v$.
   When $v = 25$, $p = 30 \cdot 45/25 = 54$.
Second solution: From $p_1 = c/v_1$ and $p_2 = c/v_2$, we obtain $p_1/p_2 = v_2/v_1$.
   Taking $v_1 = 25$, $p_2 = 30$, $v_2 = 45$, we obtain, as before, $p_1 = 54$.

2.240 Two pipes together can fill a reservoir in 6 h 40 min. Find the time each alone will take to fill the reservoir if one of the pipes can fill it in 3 h less time than the other.

▮ Let x = time (hours) required by smaller pipe, $x - 3$ = time required by larger pipe. Then $1/x$ = part filled in 1 h by smaller pipe, $\dfrac{1}{x-3}$ = part filled in 1 h by larger pipe. Since the two pipes together fill $1/\frac{20}{3} = \frac{3}{20}$ of the reservoir in 1 h, $\dfrac{1}{x} + \dfrac{1}{x-3} = \dfrac{3}{20}$, $20(x-3) + 20x = 3x(x-3)$, $3x^2 - 49x + 60 = (3x - 4)(x - 15) = 0$, and $x = \frac{4}{3}, 15$. The smaller pipe will fill the reservoir in 15 h and the larger in 12 h.

2.241 Express $x^2 + y^2 - 6x - 9y + 2 = 0$ in the form $a(x - h)^2 \pm b(y - k)^2 = c$.

▮ $(x^2 - 6x) + (y^2 - 9y) = -2$, $(x^2 - 6x + 9) + (y^2 - 9y + \frac{81}{4}) = -2 + 9 + \frac{81}{4} = \frac{109}{4}$, $(x - 3)^2 + (y - \frac{9}{2})^2 = \frac{109}{4}$.

2.242 Transform each of the following into the form $a\sqrt{(x - h)^2 + k}$ or $a\sqrt{k - (x - h)^2}$.

▮ **(a)** $\sqrt{4x^2 - 8x + 9} = 2\sqrt{x^2 - 2x + \frac{9}{4}} = 2\sqrt{(x^2 - 2x + 1) + \frac{5}{4}} = 2\sqrt{(x - 1)^2 + \frac{5}{4}}$
(b) $\sqrt{8x - x^2} = \sqrt{16 - (x - 4)^2}$
(c) $\sqrt{3 - 4x - 2x^2} = \sqrt{2}\,\sqrt{\frac{3}{2} - 2x - x^2} = \sqrt{2}\,\sqrt{\frac{5}{2} - (x^2 + 2x + 1)} = \sqrt{2}\,\sqrt{\frac{5}{2} - (x + 1)^2}$

2.243 If an object is thrown directly upward with initial speed v feet per second (ft/s), its distance s ft above the ground after t s is given by $s = vt - \frac{1}{2}gt^2$. Taking $g = 32.2$ ft/s^2 and the initial speed as 120 ft/s, find out **(a)** when the object is 60 ft above the ground, **(b)** when it is highest in its path and how high it is.

▮ **(a)** The equation of motion is $s = 120t - 16.1t^2$. When $s = 60$: $60 = 120t - 16.1t^2$ or $16.1t^2 - 120t + 60 = 0$. $t = \dfrac{120 \pm \sqrt{(120)^2 - 4(16.1)60}}{32.2} = \dfrac{120 \pm \sqrt{10{,}536}}{32.2} = \dfrac{120 \pm 102.64}{32.2} = 6.91, 0.54$. After $t = 0.54$ s the object is 60 ft above the ground and rising. After $t = 6.91$ s, the object is 60 ft above the ground and falling.

(b) The object is at its highest point when $t = \dfrac{-b}{2a} = \dfrac{-(-120)}{2(16.1)} = 3.72$ s. Its height is given by $120t - 16.1t^2 = 120(3.73) - 16.1(3.73)^2 = 223.6$ ft.

2.244 Without sketching, state whether the graph of each of the following functions crosses the x axis, is tangent to it, or lies wholly above or below it.

(a) $3x^2 + 5x - 2$ (b) $2x^2 + 5x + 4$
(c) $4x^2 - 20x + 25$ (d) $2x - 9 - 4x^2$

▌ (a) $b^2 - 4ac = 25 + 24 > 0$; the graph crosses the x axis.
(b) $b^2 - 4ac = 25 - 32 < 0$, and the graph is either wholly above or wholly below the x axis. Since $f(0) > 0$ (the value of the function for any other value of x would do equally well), the graph lies wholly above the x axis.
(c) $b^2 - 4ac = 400 - 400 = 0$; the graph is tangent to the x axis.
(d) $b^2 - 4ac = 4 - 144 < 0$ and $f(0) < 0$; the graph lies wholly below the x axis.

2.245 Form the quadratic equation whose roots x_1 and x_2 are (a) $3, \frac{2}{5}$; (b) $-2 + 3\sqrt{5}, -2 - 3\sqrt{5}$; (c) $(3 - i\sqrt{2})/2, (3 + i\sqrt{2})/2$.

▌ (a) $3, \frac{2}{5}$. Here $x_1 + x_2 = \frac{17}{5}$ and $x_1 x_2 = \frac{6}{5}$. The equation is $x^2 - \frac{17}{5}x + \frac{6}{5} = 0$ or $5x^2 - 17x + 6 = 0$.
(b) $-2 + 3\sqrt{5}, -2 - 3\sqrt{5}$. Here $x_1 + x_2 = -4$ and $x_1 x_2 = 4 - 45 = -41$. The equation is $x^2 + 4x - 41 = 0$.
(c) $\dfrac{3 - i\sqrt{2}}{2}, \dfrac{3 + i\sqrt{2}}{2}$. The sum of the roots is 3, and the product is $\frac{11}{4}$. The equation is
$x^2 - 3x + \frac{11}{4} = 0$ or $4x^2 - 12x + 11 = 0$.

2.246 Determine k so that the given equation will have the stated property, and write the resulting equation. (a) $x^2 + 4kx + k + 2 = 0$ has one root 0. (b) $4x^2 - 8kx - 9 = 0$ has one root the negative of the other. (c) $4x^2 - 8kx + 9 = 0$ has roots whose difference is 4.

▌ (a) $x^2 + 4kx + k + 2 = 0$ has one root 0. Since the product of the roots is to be 0, $k + 2 = 0$ and $k = -2$. The equation is $x^2 - 8x = 0$.
(b) $4x^2 - 8kx - 9 = 0$ has one root the negative of the other. Since the sum of the roots is to be 0, $2k = 0$ and $k = 0$. The equation is $4x^2 - 9 = 0$.
(c) $4x^2 - 8kx + 9 = 0$ has roots whose difference is 4. Denote the roots r and $r + 4$. Then $r + (r + 4) = 2r + 4 = 2k$ and $r(r + 4) = \frac{9}{4}$. Solving for $r = k - 2$ in the first and substituting in the second, we have $(k - 2)(k + 2) = \frac{9}{4}$; then $4k^2 - 16 = 9$ and $k = \pm\frac{5}{2}$. The equations are $4x^2 + 20x + 9 = 0$ and $4x^2 - 20x + 9 = 0$.

CHAPTER 3
Graphs, Relations, and Functions

3.1 CARTESIAN COORDINATE SYSTEM

3.1 Plot each of the following points in the cartesian coordinate system: $A(2, 0)$, $B(-2, 5)$, $C(-3, -2)$, $D(2, 1)$.

▮ See Fig. 3.1.

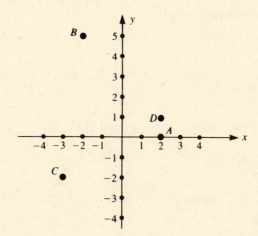

Fig. 3.1

For Probs. 3.2 to 3.5, find the distance between the given points.

3.2 $A(1, 0)$ and $B(0, 1)$

▮ The distance between two points A and B in the plane $= d(A, B) = AB = \sqrt{(x_2 - x_1)^2 + (y_2 - y_1)^2}$, where A has coordinates (x_1, y_1) and B has coordinates (x_2, y_2). Then, in this case, $d(A, B) = \sqrt{(0 - 1)^2 + (1 - 0)^2} = \sqrt{1 + 1} = \sqrt{2}$.

3.3 $A(0, 0)$ and $B(-3, 2)$

▮ $AB = \sqrt{(-3 - 0)^2 + (2 - 0)^2} = \sqrt{9 + 4} = \sqrt{13}$.

3.4 $A(1, 4)$ and $B(-1, 5)$

▮ $d(A, B) = \sqrt{[1 - (-1)]^2 + (4 - 5)^2} = \sqrt{2^2 + (-1)^2} = \sqrt{5}$. Notice that since $(x_1 - x_2)^2 = (x_2 - x_1)^2$, the order of subtraction is unimportant

3.5 $P(1, 3)$ and $Q(1, 5)$

▮ $PQ = \sqrt{(1 - 1)^2 + (3 - 5)^2} = \sqrt{0 + 4} = 2$. Notice that since P and Q have the same abscissa, we can also get the result by finding $|5 - 3| = |3 - 5| = 2$. This would also be the case if the ordinates were identical.

For Probs. 3.6 to 3.10, find the midpoint of the segment joining the two given points.

3.6 $(1, 2)$ and $(3, 2)$

▮ The midpoint M of the segment joining (x_1, y_1) and (x_2, y_2) has coordinates $\left(\dfrac{x_1 + x_2}{2}, \dfrac{y_1 + y_2}{2}\right)$. Thus, in this case M has coordinates $\left(\dfrac{1 + 3}{2}, \dfrac{2 + 2}{2}\right) = (2, 2)$.

Draw a picture. Does this result make sense?

3.7 $(-3, 1)$ and $(3, 1)$

▮ The midpoint has coordinates $\left(\dfrac{-3+3}{2}, \dfrac{1+1}{2}\right) = (0, 1)$.

3.8 $(-3, -5)$ and $(-4, -6)$

▮ M has coordinates $\left(\dfrac{-3-4}{2}, \dfrac{-5-6}{2}\right) = \left(\dfrac{-7}{2}, \dfrac{-11}{2}\right)$.

3.9 $(\sqrt{2}, 1)$ and $(2, \sqrt{3})$

▮ M is the point with coordinates $\left(\dfrac{\sqrt{2}+2}{2}, \dfrac{1+\sqrt{3}}{2}\right)$.

3.10 $(s, 4t)$ and $(-3s, -t)$

▮ M has coordinates $\left(\dfrac{s-3s}{2}, \dfrac{4t-t}{2}\right) = \left(-s, \dfrac{3t}{2}\right)$.

For Probs. 3.11 to 3.15, find the equation of the given circle.

3.11   The center is at $(0, 0)$, radius $= 4$.

▮ The equation of a circle with center (h, k) and radius r is $(x - h)^2 + (y - k)^2 = r^2$. In this case, $h = 0$, $k = 0$, $r = 4$. The equation of the circle is $(x - 0)^2 + (y - 0)^2 = 4^2$, or $x^2 + y^2 = 16$.

3.12   The center is at $(1, 2)$, radius $= 6$.

▮ $h = 1$, $k = 2$, $r = 6$. The equation of the circle is $(x - 1)^2 + (y - 2)^2 = 6^2$, or $(x - 1)^2 + (y - 2)^2 = 36$.

3.13   The center is at $(-2, -3)$, radius $= 1$.

▮ $h = -2$, $k = -3$, $r = 1$. The equation of the circle is $[x - (-2)]^2 + [y - (-3)]^2 = 1^2$, or $(x + 2)^2 + (y + 3)^2 = 1$.

3.14   The center is at $(-2, 3)$, radius $= \sqrt{2}$.

▮ $h = -2$, $k = 3$, $r = \sqrt{2}$. $(x + 2)^2 + (y - 3)^2 = 2$.

3.15   The center is at $A(0, 0)$, and the point $B(1, 1)$ lies on the circle.

▮ Since $(1, 1)$ lies on the circle, $d(A, B) =$ radius of the circle. $d(A, B) = \sqrt{1^2 + 1^2} = \sqrt{2}$. Then $h = 0$, $k = 0$, $r = \sqrt{2}$. The equation is $x^2 + y^2 = 2$.

For Probs. 3.16 and 3.17, tell whether the points all lie on a circle with given center P.

3.16 $P(0, 0)$; $(0, 1)$, $(-1, 0)$, $(\sqrt{2}/2, \sqrt{2}/2)$

▮ We need to find the distance from each given point to $(0, 0)$. If these distances are all the same, the points all lie on a circle with P as the center. $d((0, 1), P) = \sqrt{1^2 + 0^2} = 1$. $d((-1, 0), P) = \sqrt{(-1)^2 + 0^2} = 1$. $d((\sqrt{2}/2, \sqrt{2}/2), P) = \sqrt{(\sqrt{2}/2)^2 + (\sqrt{2}/2)^2} = 1$. Thus, the three points do lie on the circle.

3.17 $P(1, 2)$; $(4, 7)$, $(-2, -3)$, $(6, -1)$

▮ $d((4, 7), P) = \sqrt{3^2 + 5^2}$. $d((-2, -3), P) = \sqrt{3^2 + 5^2}$. $d((6, -1), P) = \sqrt{5^2 + 3^2}$. Since the three distances are the same, the points do lie on the same circle with center P.

For Probs. 3.18 to 3.21, find the center and radius of the given circle.

3.18 $(x-1)^2 + y^2 = 4$

▮ Since $(x-h)^2 + (y-k)^2 = r^2$ is the equation of the circle, in this case $h = 1$, $k = 0$, $r = 2$. The center is at $(1, 0)$, and the radius is 2.

3.19 $(x+2)^2 + (y+3)^2 = 3$

▮ $h = -2$, $k = -3$, $r = \sqrt{3}$. The center is at $(-2, -3)$, and the radius is $\sqrt{3}$.

3.20 $x^2 + y^2 - 4y = 0$

▮ We must put this equation in the form $(x-h)^2 + (y-k)^2 = r^2$. Complete the square on $y^2 - 4y$, making sure you add the required 4 to both sides: $x^2 + (y^2 - 4y + 4) = 0 + 4$. $x^2 + (y-2)^2 = 4$. The center is at $(0, 2)$, and the radius is 2.

3.21 $x^2 + y^2 + 4x - 4y = 1$

▮ Regrouping terms, we obtain $(x^2 + 4x) + (y^2 - 4y) = 1$. Completing the square, $(x^2 + 4x + 4) + (y^2 - 4y + 4) = 1 + (4 + 4)$, or $(x+2)^2 + (y-2)^2 = 9$. The center is at $(-2, 2)$, and the radius is 3.

3.22   Graph the equation in Problem 3.21.

▮ See Fig. 3.2.

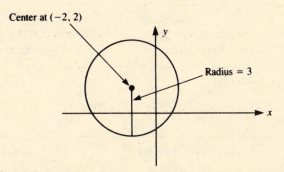

Center at $(-2, 2)$

Radius = 3

Fig. 3.2

For Probs. 3.23 to 3.30, decide whether the graph of the equation exhibits symmetry with respect to the x axis or origin. Do not graph the equation to determine whether the symmetry exists.

3.23 $y = x + 2$

The tests for symmetry are the following:
(**a**)   Replace x with $-x$. If the equation *does not* change, the graph is symmetric with respect to the y axis.
(**b**)   Replace y with $-y$. If the equation *does not* change, the graph is symmetric with respect to the x axis.
(**c**)   Replace x with $-x$ and y with $-y$. If the equation *does not* change, the graph is symmetric with respect to the origin.

▮ (**a**)   In $y = x + 2$, replace x with $-x$; then $y = -x + 2$. This is a change. The graph is not symmetric with respect to the y axis. (**b**)   Replace y with $-y$. Then $-y = x + 2$ or $y = -x - 2$. Not symmetric with respect to the x axis. (**c**)   Replace x with $-x$ and y with $-y$. Then $-y = -x + 2$, or $y = x - 2$. Not symmetric with respect to the origin.

3.24 $y = x$

▮

x axis	y axis	Origin
$-y = x$	$y = -x$	$-y = -x$
Not symmetric	Not symmetric	Symmetric

3.25 $x^2 + y^2 = 4$

▌

x axis	y axis	Origin
$x^2 + (-y)^2 = 4$	$(-x)^2 + y^2 = 4$	$(-x)^2 + (-y)^2 = 4$
$x^2 + y^2 = 4$	$x^2 + y^2 = 4$	$x^2 + y^2 = 4$
Symmetric	Symmetric	Symmetric

3.26 $x^2 + (y - 1)^2 = 9$

▌

x axis	y axis	Origin
$x^2 + (-y - 1)^2 = 9$	$(-x)^2 + (y - 1)^2 = 9$	$(-x)^2 + (-y - 1)^2 = 9$
$x^2 + (y + 1)^2 = 9$	$x^2 + (y - 1)^2 = 9$	$x^2 + (y + 1)^2 = 9$
Not symmetric	Symmetric	Not symmetric

3.27 $y = |2x|$

▌

x axis	y axis	Origin
$-y = |2x|$	$y = |2(-x)|$	$-y = |2(-x)|$
or $y = -|2x|$	or $y = |-2x|$	$-y = |2x|$
Not symmetric	Symmetric	Not symmetric

3.28 $|x| + 1 = y$

▌

x axis	y axis	Origin
$-y = |x| + 1$	$y = |-x| + 1$	$-y = |-x| + 1$
	or $y = |x| + 1$	$-y = |x| + 1$
Not symmetric	Symmetric	Not symmetric

3.29 $y = x^3$

▌

x axis	y axis	Origin
$-y = x^3$	$y = (-x)^3$	$-y = (-x)^3$
	or $y = -x^3$	$y = x^3$
Not symmetric	Not symmetric	Symmetric

3.30 $y = x^2 + 1$

▌ Since $(-x)^2 = x^2$, we see y axis symmetry. It does not exhibit x axis or origin symmetry.

3.31 Graph $y = x + 2$.

▌ See Fig. 3.3. See Prob. 3.23 above, Notice the lack of symmetry first uncovered in Prob. 3.23.

x	0	1	2
y	2	3	4

3.32 Graph $x^2 + (y - 1)^2 = 4$.

▌ See Fig. 3.4. See Prob. 3.26 above. Here we have a circle with center $(0, 1)$ and radius 2. Notice y axis symmetry.

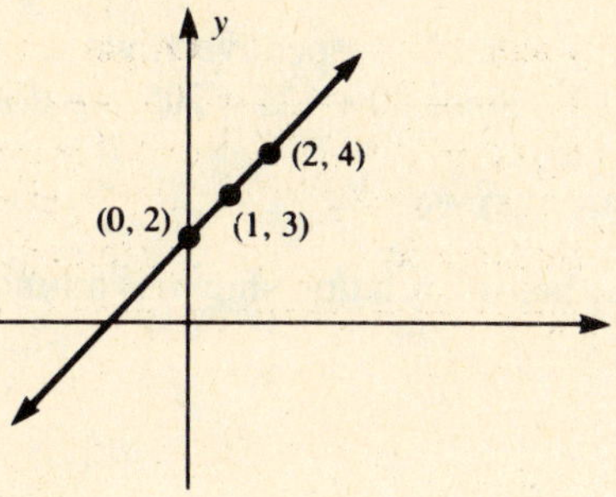

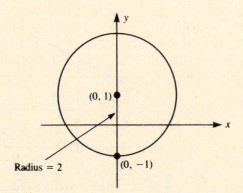

Fig. 3.3

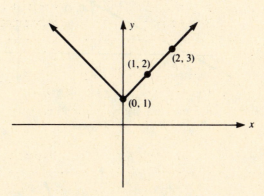

Fig. 3.4

3.33 Graph $y = |x| + 1$.

▮ See Fig. 3.5 and Prob. 3.28. Notice that we calculated y only for positive x values since we already knew the graph had y axis symmetry.

x	0	1	2
y	1	2	3

Fig. 3.5

3.34 Show that $(5, 2\sqrt{3})$, $(2, -\sqrt{3})$, and $(8, -\sqrt{3})$ form an equilateral triangle.

▮ Let A be the point $(5, 2\sqrt{3})$, let B be the point $(2, -\sqrt{3})$, and let C be the point $(8, -\sqrt{3})$. Then $d(A, B) = \sqrt{(5-2)^2 + (3\sqrt{3})^2} = \sqrt{9 + 27} = 6$, $d(B, C) = \sqrt{(8-2)^2 + (0)^2} = \sqrt{36} = 6$, and $d(A, C) = \sqrt{3^2 + (3\sqrt{3})^2} = 6$. Since the three lengths are equal, it is an equilateral triangle.

For Probs. 3.35 and 3.36, are the given points the vertices of a right triangle?

3.35 $(-3, 2)$, $(1, -2)$, $(8, 5)$

▮ Call the three points A, B, and C, respectively. Then $AB = \sqrt{(-4)^2 + (4)^2} = \sqrt{32}$, $BC = \sqrt{(-7)^2 + (7)^2} = \sqrt{98}$, $AC = \sqrt{(-11)^2 + (-3)^2} = \sqrt{130}$. Then $AB^2 + BC^2 = 32 + 98 = 130 = AC^2$. Since these lengths satisfy the pythagorean theorem, the three points are the vertices of a right triangle.

3.36 $(-4, -1), (0, 7), (6, -6)$

▮ Calling the points A, B, and C, respectively, we see that $AB^2 = 16 + 64 = 80$, $BC^2 = 36 + 169 = 205$, $AC^2 = 100 + 25 = 125$. Since $80 + 125 = 205$, we do have a right triangle.

3.2 RELATIONS AND FUNCTIONS

For Probs. 3.37 to 3.46, tell whether the relation shown is a function. In all cases, x is the independent variable.

3.37   See Fig. 3.6.

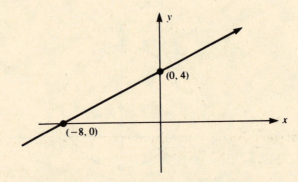

Fig. 3.6

▮ Figure 3.6 is a function. No two ordered pairs have the same abscissa. Notice (see Fig. 3.7) that any line drawn at x_0 perpendicular to the x axis on the x axis intersects the function only once.

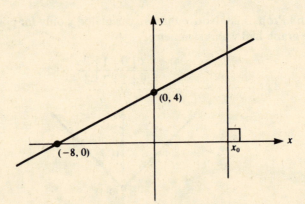

Fig. 3.7

3.38   See Fig. 3.8.

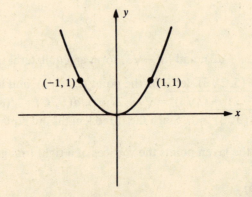

Fig. 3.8

▮ Be careful here! Figure 3.8 exhibits a function. For each x value, there is only one y value; thus, no two ordered pairs have the same abscissa. However, unlike Prob. 3.37, two ordered pairs share an ordinate, for example, $(1, 1)$ and $(-1, 1)$.

3.39 See Fig. 3.9.

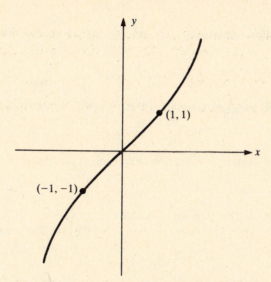

Fig. 3.9

❚ Figure 3.9 is a function. All ordered pairs have different abscissas. Notice that this graph reminds us of $y = x^3$.

3.40 See Fig. 3.10.

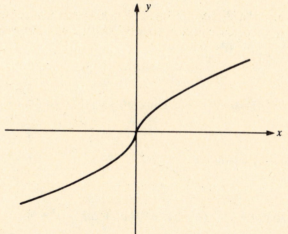

Fig. 3.10

❚ Figure 3.10 shows a function. Draw a perpendicular to the x axis at any point on the x axis, and note that it intersects the function only once.

3.41 See Fig. 3.11.

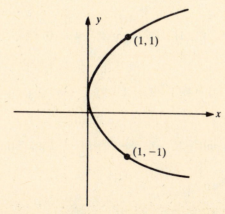

Fig. 3.11

❚ Figure 3.11 is not a function. Since $(1, 1)$ and $(1, -1)$ have the same first element, it violates the definition of a function.

3.42 $f(x) = 2x - 10$.

▮ f is a function. For every x, $2x - 40$ has only one value. It is impossible to get two values of f from one x.

3.43 $f(x) = -2x^2 + 8$

▮ f is a function. If we plug in a value for x, we get only one value for f.

3.44 $y = |x|$

▮ This equation does define a function. For each x, $|x|$ has only one value.

3.45 $x = |y|$

▮ Remember that x is the independent variable. But for $x = 1$ (for example), if $1 = |y|$, then $y = \pm 1$. Thus, $(1, 1)$ and $(1, -1)$ are both in this relation. The relation is not a function.

3.46 $x = 2y^2$

▮ This is not a function. Two ordered pairs share an x value, for example, $(8, 2)$ and $(8, -2)$.

For Probs. 3.47 to 3.56, find the indicated value if $f(x) = x - 9$, $g(x) = x^2 - 9$, $h(x) = |x|$, and $F(x) = x^3 - x + 4$.

3.47 $g(0)$

▮ $g(x) = x^2 - 9$, so $g(0) = 0^2 - 9 = -9$.

3.48 $f(0)$

▮ $f(x) = x - 9$, so $f(0) = 0 - 9 = -9$. Thus, $g(0) = f(0)$, but $g(x) \neq f(x)$.

3.49 $h(-6)$

▮ $h(x) = |x|$, so $h(-6) = |-6| = 6$.

3.50 $g(a + b)$

▮ $g(x) = x^2 - 9$, so $g(a + b) = (a + b)^2 - 9 = a^2 + 2ab + b^2 - 9$.

3.51 $F(2) - g(3)$

▮ $F(2) - g(3) = (2^3 - 2 + 4) - (3^2 - 9) = 10 - 0 = 10$.

3.52 $F(1) + F(2)$

▮ $F(1) + F(2) = (1^3 - 1 + 4) + (2^3 - 2 + 4) = 4 + 10 = 14$.

3.53 $F(0) \cdot f(0)$

▮ $F(0) \cdot f(0) = (0^3 - 0 + 4)(0 - 9) = 4(-9) = -36$.

3.54 $3g(a)[-2h(-b)]$

▮ $3g(a) = 3(a^2 - 9)$ and $-2h(-b) = -2|-b| = -2|b|$. Thus, $3g(a)[-2h(-b)] = -6|b|(a^2 - 9)$.

3.55 $\dfrac{f(0)}{F(0)}$

▮ $\dfrac{f(0)}{F(0)} = \dfrac{0 - 9}{0^3 - 0 + 4} = -\dfrac{9}{4}$.

3.56 $\dfrac{g(1+h)-g(1)}{h}$

▮ $\dfrac{g(1+h)-g(1)}{h}=\dfrac{(1+h)^2-9-(1^2-9)}{h}=\dfrac{(1+h)^2-1^2}{h}=\dfrac{1+2h+h^2-1}{h}=\dfrac{h^2+2h}{h}=h+2.$

For Probs. 3.57 to 3.66, find the domain of the given relation. In all cases, x is the independent variable.

3.57 $2x+5=y.$

▮ For any real value x, $2x+5=y$ is real. Since any real number can replace x, the domain is the set of all real numbers $\mathcal{R}$.

3.58 $y=x^2$

▮ Since any real number squared yields a number, y is defined for any real x. The domain equals the set of all real numbers.

3.59 $y=\sqrt{3-x^2}$

▮ We must ensure that $3-x^2\geq0$. Then $3\geq x^2$ or $x^2\leq3$. Then $-\sqrt{3}\leq x\leq\sqrt{3}$ is the domain.

3.60 $y=|x|-1$

▮ For any real x, $|x|-1$ is real. The domain equals the set of all real numbers.

3.61 $y=\dfrac{3}{2-x}$

▮ We need to be certain that $2-x\neq0$. The domain is the set of all real numbers except for $x=2$, that is, $(-\infty,2)\cup(2,\infty)$ or $\mathcal{R}-\{2\}$.

3.62 $x^2+y^2=8$

▮ Since x is the independent variable, we need to solve for y to examine the domain. If $x^2+y^2=8$, then $y^2=8-x^2$ or $y=\pm\sqrt{8-x^2}$. Then $8-x^2\geq0$ if and only if $8\geq x^2$ or $x^2\leq8$; that is, $-2\sqrt{2}<x<2\sqrt{2}$ is the domain of this relation. Note that this relation is not a function.

3.63 $y=\sqrt{\dfrac{2}{x-2}}$

▮ We need to be cautious here. We must be certain that $x-2\neq0$ and $\dfrac{2}{x-2}\geq0$. But $\dfrac{2}{x-2}\geq0$ means that $x-2\geq0$, since the numerator is ≥0 always. Thus, $x-2\neq0$ and $x-2\geq0$. Thus, $x-2>0$, or $x>2$ is the domain.

3.64 $2x^2+y^2=0$

▮ Since $2x^2\geq0$ and $y^2\geq0$ for all real x and y, the only way $2x^2+y^2=0$ is for $x=0$ and $y=0$. Thus, the domain is $\{0\}$.

3.65 $\{(x,y)\,|\,x\in\mathcal{R},y\in\mathcal{R},y=3\}$

▮ The domain is $\mathcal{R}$. For any real x, y is that value and is therefore always defined.

3.66 $\{(1,4),(2,2),(3,8)\}$

▮ The domain is the set of x values; in this case, the domain $=\{1,2,3\}$.

For Probs. 3.67 to 3.76, find the range of the given function.

3.67 $\{(1, 0), (0, 1), (2, a)\}$

▊ The range is the set of ordinates, which in this case is $\{0, 1, a\}$.

3.68 $f(x) = 2x - 3$

▊ Any real y can be expressed as $2x - 3$ for some real x. For example, if $-4 = 2x - 3$, $2x = -1$, or $x = -\frac{1}{2}$. The range $= \mathcal{R}$.

3.69 $f(x) = x^2$

▊ For every $x \in \mathcal{R}$, $x^2 \geq 0$. The range is the nonnegative real numbers, or $\mathcal{R}^+ \cup \{0\}$ or $[0, \infty)$.

3.70 $f(x) = x^4 - 2$

▊ For any real x, $x^4 - 2 \geq -2$, since $x^4 \geq 0$. The range is $[-2, \infty)$.

3.71 $g(x) = 2 - |x|$

▊ $|x| \geq 0$; thus, $-|x| \leq 0$ and $2 - |x| \leq 2$. Range $= (-\infty, 2]$.

3.72 $f(x) = |x| + 5$

▊ $|x| \geq 0$; thus, $|x| + 5 \geq 5$. Range $= [5, \infty)$.

3.73 $f(x) = \sqrt{2 - x}$

▊ If $x = 2$, $f(x) = 0$. If $x < 2$, $f(x) > 0$. (Note that x cannot be > 2.) The range is $[0, \infty)$.

3.74 $h(x) = 4$

▊ Since 4 is the function's value for any x, the range is $\{4\}$.

3.75 $f(x) = 1/x$

▊ $x \neq 0$. For $x > 0$, $0 < 1/x < 1$. For $x < 0$, $-1 < 1/x < 0$. The range is $(-1, 0) \cup (0, 1)$.

3.76 $\{(x, y) \mid x \in \mathcal{R}, y \in \mathcal{R}, y = x^8\}$

▊ For every real x, y is defined. Also for every real y, y is the eighth power of some real x. Range $= \mathcal{R}$.

For Probs. 3.77 to 3.81 tell whether the function is one-to-one.

3.77 $f(x) = 3x + 4$

▊ A function is one-to-one if whenever $x_1 \neq x_2$, $f(x_1) \neq f(x_2)$. In this case, if $x_1 \neq x_2$, $3x_1 + 4 \neq 3x_2 + 4$, so the function is one-to-one.

3.78 $f(x) = |x| + 1$

▊ $|x| + 1$ has the same value for x and $-x$. Thus, f is not one-to-one.

3.79 $g(x) = 3\sqrt{1 - x}$

▊ $3\sqrt{1 - x}$ is not the same value for two different x choices. This function is one-to-one.

3.80 $f(x) = \{(1, 2), (2, 1)\}$

▊ This is one-to-one. No two ordered pairs have the same second element.

3.81 $h(x) = \{(1, 3), (2, 4), (3, 1), (3, 4)\}$

▮ This function is not one-to-one. Notice that (2, 4) and (3, 4) have the same second element.

3.82 Give an example of two functions that are unequal, but whose domains and range are the same.

▮ Let $f(x) = x$ and $g(x) = x^2$. Then $f \neq g$, but the domain of f = domain of g and the range of f = range of g.

3.3 GRAPH OF A FUNCTION

For Probs. 3.83 to 3.86, refer to Fig. 3.12 which is the graph of $y = f(x)$.

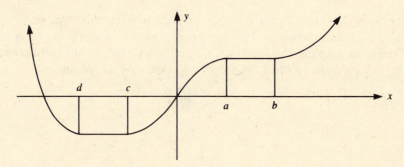

Fig. 3.12

3.83 Is the relation $y = f(x)$ a function?

▮ Yes; for every x value there is exactly one y value.

3.84 Find all intervals over which $y = f(x)$ is constant.

▮ Notice the horizontal segments. They occur on $[d, c]$ and $[a, b]$. These are the only intervals over which $y = f(x)$ is constant.

3.85 Find all intervals over which $y = f(x)$ is increasing and all intervals over which it is decreasing.

▮ Approaching d from the left, we see that the y values are decreasing. From c to a, they are increasing, and from b on (infinitely far!), f is increasing. Decreasing: $(-\infty, d)$. Increasing: $(c, a), (b, \infty)$.

3.86 Find all intervals over which $y = f(x)$ is nonincreasing and those over which it is nondecreasing.

▮ Nonincreasing means decreasing or constant; nondecreasing means increasing or constant. Refer to Probs. 3.84 and 3.85. Nonincreasing: $(-\infty, c]$. Nondecreasing: $[d, \infty)$.

For Probs. 3.87 and 3.88, answer true or false, and explain your answer.

3.87 If $y = f(x)$ is an increasing function, then it is a nondecreasing function.

▮ True. Since nondecreasing means increasing or constant, increasing implies nondecreasing.

3.88 If $y = g(x)$ is a nonincreasing function, then $y = g(x)$ is a decreasing function.

▮ False. $y = 3$ is nonincreasing (it's constant), but it is not decreasing.

For Probs. 3.89 to 3.109, graph the given function.

3.89 $f(x) = 3$

▮ See Fig. 3.13. Notice that this is of the form $f(x) = a$, which is a constant function. The vertical axis is the y or $f(x)$ axis.

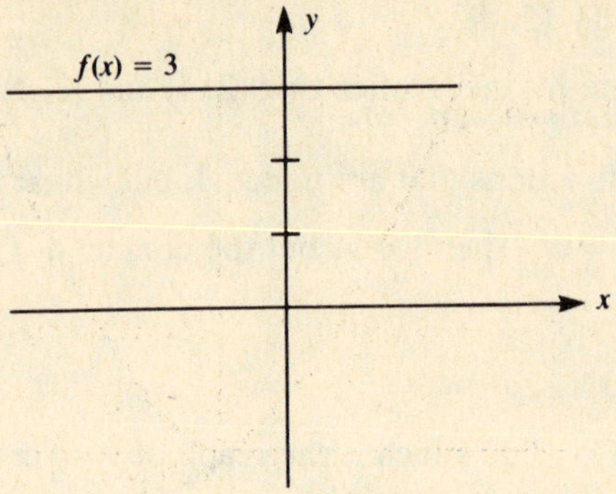

$f(x) = 3$

Fig. 3.13

3.90 $f(x) = 2x - 3$

▮ See Fig. 3.14. This function is of the form $f(x) = ax + b$, which is the form for a linear function. Since we know the graph is a straight line, we find two points. Find a third point as well to make certain you did not make an error. The third point must lie on the line.

x	0	1
y	-3	-1

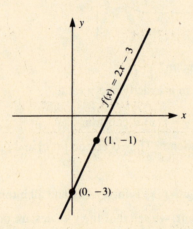

Fig. 3.14

3.91 $x = 4$

▮ See Fig. 3.15. Notice that this is not a function of x, the independent variable. It is in the form $x = a$ for a vertical line.

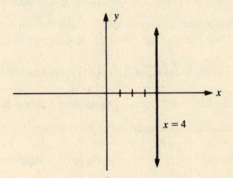

Fig. 3.15

3.92 $f(x) = x^2$

▮ See Fig. 3.16. Notice that since $f(-x) = x^2 = f(x)$, this function's graph is symmetric about the y axis. Also $f(x) \geq 0$ for all x. This graph is a parabola. See Chap. 14 for more about the conic sections in general.

x	-2	-1	0	1	2
y	4	1	0	1	4

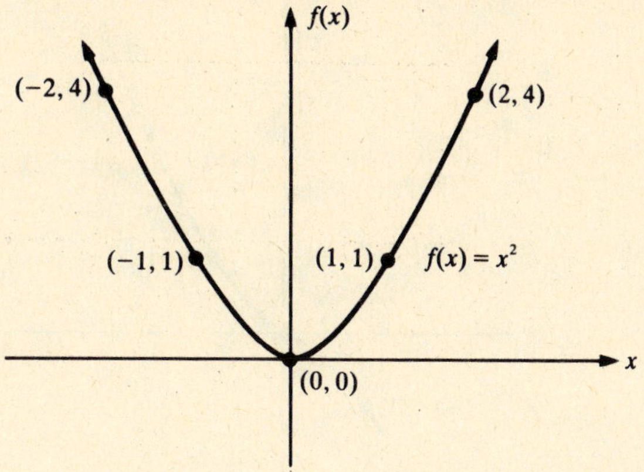

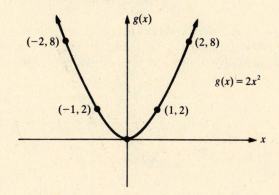

Fig. 3.16

3.93 $g(x) = 2x^2$

▮ See Fig. 3.17 and Prob. 3.92. This function has the property that each y value is double what the y value is for $f(x)$ in Prob. 3.92 (for corresponding x values).

Fig. 3.17

3.94 $h(x) = -\frac{1}{3}x^2$

▮ See Fig. 3.18. Again we look at $f(x) = x^2$ in Prob. 3.92. Notice that for corresponding x values we multiply the y values by $-\frac{1}{3}$. Thus, $(0, 0)$ remains $(0, 0)$, $(1, 1)$ becomes $(1, -\frac{1}{3})$, etc.

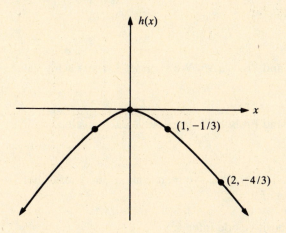

Fig. 3.18

3.95 $f(x) = x^3$

▮ See Fig. 3.19. This function is symmetric about the origin, since $y = x^3$ and $-y = (-x)^3$.

x	0	1	-1	2	-2
y	0	1	-1	8	-8

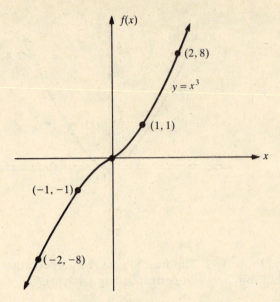

Fig. 3.19

3.96 $k(x) = x^2 - 1$

∎ See Fig. 3.20. Take the ordered pairs in $f(x) = x^2$ and subtract 1 from each y value [since $k(x) = x^2 - 1$]. $(0, 0)$ becomes $(0, -1)$, $(1, 1)$ becomes $(1, 0)$, etc. Note that we could have used y axis symmetry and found several ordered pairs to plot the graph directly, instead of using Prob. 3.92.

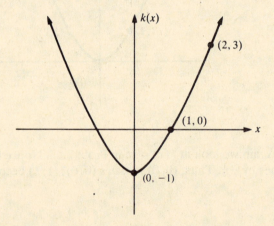

Fig. 3.20

3.97 $g(x) = x^3 - 1$

∎ See Fig. 3.21 and Prob. 3.95. We subtract 1 from each y value. You can also graph g directly.

3.98 $h(x) = 2(x^3 - 1)$

∎ See Fig. 3.22 and Prob. 3.97. We double each y value.

3.99 $f(x) = x^4$

∎ See Fig. 3.23. This graph is symmetric about the y axis since $f(-x) = f(x)$. Notice that it is just a very steep version of $y = x^2$.

3.100 $g(x) = x^n$, n is a positive odd integer

∎ See Fig. 3.24. Compare this to Prob. 3.101 below. Since n is odd, $g(-x) = -g(x) = -x^n$; thus, the graph exhibits origin symmetry.

x	-2	-1	0	1	2
y	-2^n	-1	0	1	2^n

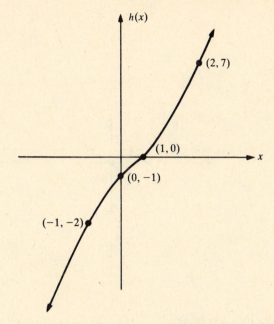

Fig. 3.21

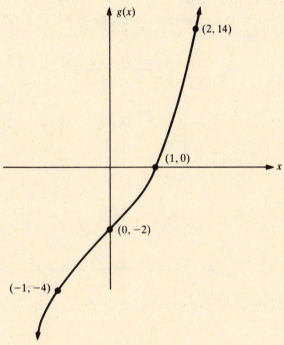

Fig. 3.22

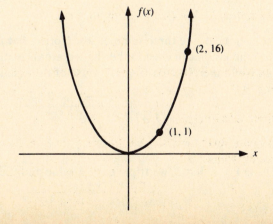

Fig. 3.23

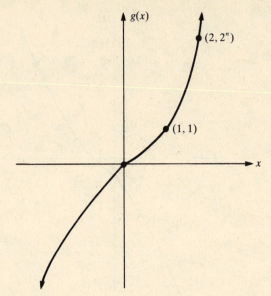

Fig. 3.24

3.101 $f(x) = x^m$, m is a positive even integer

▮   See Fig. 3.25. If m is a positive even integer, then $f(-x) = (-x)^m = x^m$; thus, the graph is symmetric about the y axis.

x	-2	-1	0	1	2
y	2^m	1	0	1	2^m

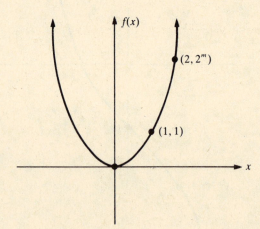

Fig. 3.25

3.102 $f(x) = \sqrt{x+2}$

▮   Figure 3.26 exhibits none of the three symmetries. Notice that $f(x) \geq 0$ for all x and that $x + 2 \geq 0$ or $x \geq -2$. No part of the graph can exist in the shaded areas. Notice that the graph is one-half of a parabola. The "negative branch" is "missing" because of the square root.

x	-2	-1	0	2
$f(x)$	0	1	$\sqrt{2}$	2

3.103 $g(x) = 3\sqrt{x+2} - 1$

▮   See Fig. 3.27. Using Prob. 3.102, we triple each y value and then subtract 1.

3.104 $f(x) = |x|$

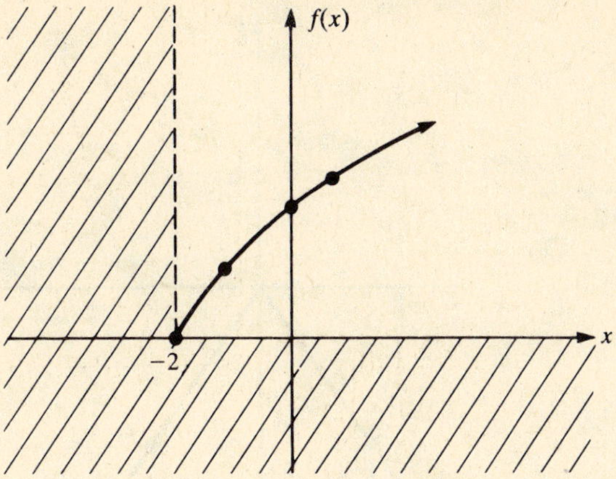

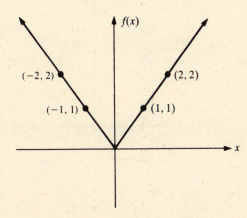

Fig. 3.26

Fig. 3.27

▮ Figure 3.28 is symmetric about the y axis. Also $f(x) \geq 0$ for all x.

x	-1	0	1	2	-2
$f(x)$	1	0	1	2	2

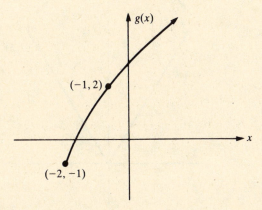

Fig. 3.28

3.105 $g(x) = -2 \, |x|$

▮ See Fig. 3.29. Using Prob. 3.104, we multiply each y value by -2. Notice the y axis symmetry.

3.106 $h(x) = 2 \, |x| + 1$

▮ See Fig. 3.30. Again we used Prob. 3.104 to find the graph. Note that you could have found ordered pairs directly.

3.107 $f(x) = [x]$

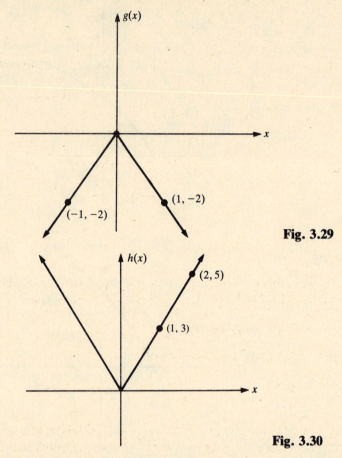

Fig. 3.29

Fig. 3.30

▌ See Fig. 3.31. Here f is the greatest-integer function: For every x, $f(x)$ is the "greatest integer that is not greater than x." Then $[0] = 0$, $[1] = 1$, $[-1] = -1$, $[2] = 2$, $[-2] = -2$, etc. However, $[\frac{1}{2}] = 0$, $[\frac{7}{8}] = 0$, $[4.2] = 4$, $[-2.5] = -3$. The graph then looks as follows:

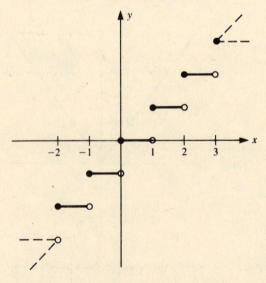

Fig. 3.31

3.108 $g(x) = 2[x]$

▌ See Fig. 3.32. We will do this in two ways. See Prob. 3.109 for the second way where we will use Prob. 3.107. Here we plot points directly.

x	-3	-2.5	-1	0	1	1.5	1.7	3.5
$g[x] = 2[x]$	-6	-6	-2	0	2	2	2	2

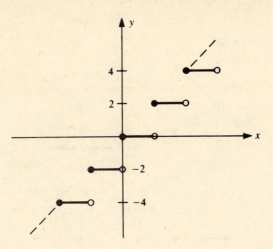

Fig. 3.32

3.109 $g(x) = 2[x]$

❚ Here, we simply double each y value in Prob. 3.107 and obtain the same graph as Fig. 3.32.
$2 \cdot [0] = 0, 2 \cdot [1] = 2, 2 \cdot [-1] = -2, 2 \cdot [2] = 4, 2 \cdot [-2] = -4, 2 \cdot [\frac{1}{2}] = 0, 2 \cdot [\frac{7}{8}] = 0, 2 \cdot [4] = 8$, etc.

3.4 STEP FUNCTIONS AND CONTINUITY

For Probs. 3.110 to 3.121 graph the given function.

3.110 $f(x) = \begin{cases} 1 & x \geq 1 \\ 0 & x < 1 \end{cases}$

❚ This is an example of a step function. The graph of the function (Fig. 3.33) consists of horizontal segments (or steps). In this case, in the xy coordinate system, $y = 1$ at 1 and $x > 1$; $y = 0$ for all other x.

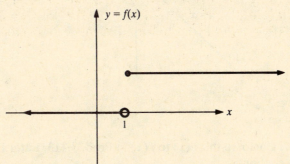

Fig. 3.33

3.111 $g(x) = \begin{cases} 1 & x > 1 \\ -3 & x \leq 1 \end{cases}$

❚ See Fig. 3.34.

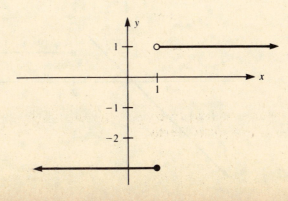

Fig. 3.34

3.112 $h(x) = \begin{cases} 1 & x > 1 \\ 2 & -2 < x \le 1 \\ -1 & x \le -2 \end{cases}$

▮ See Fig. 3.35.

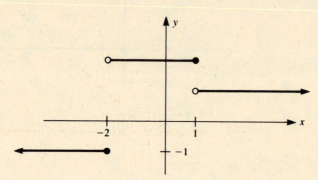

Fig. 3.35

3.113 $f(x) = \begin{cases} 1 & x \ge 1 \\ x & x < 1 \end{cases}$

▮ See Fig. 3.36.

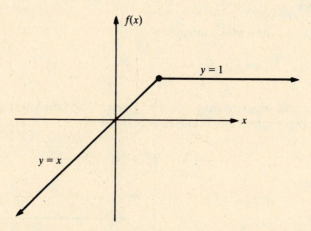

Fig. 3.36

3.114 $g(x) = \begin{cases} 1 & x \ge 2 \\ x & x < 2 \end{cases}$

▮ See Fig. 3.37. Compare this $g(x)$ to $f(x)$ in Prob. 3.113. Later in this section we will look at these again to discuss continuity.

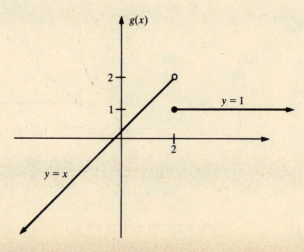

Fig. 3.37

3.115 $h(x) = \begin{cases} x & x > 0 \\ x^2 & x \le 0 \end{cases}$

▮ See Fig. 3.38.

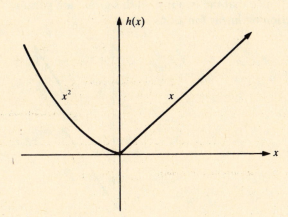

Fig. 3.38

3.116 $f(x) = \begin{cases} x & x > 2 \\ x^2 & x \le 2 \end{cases}$

▮ See Fig. 3.39.

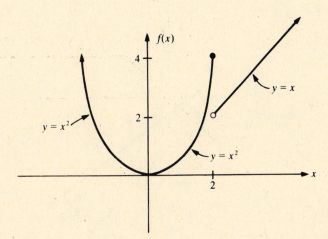

Fig. 3.39

3.117 $g(x) = \begin{cases} x & x > 2 \\ x^3 & x \le 2 \end{cases}$

▮ See Fig. 3.40.

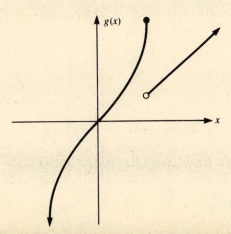

Fig. 3.40

3.118 $g(x) = \begin{cases} x+1 & x>2 \\ x-1 & x\le 2 \end{cases}$

▮ See Fig. 3.41. Notice that the two rays in this group are parallel. We will look at other examples like this in the section on linear functions.

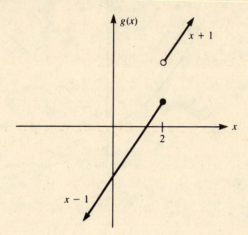

Fig. 3.41

3.119 $h(x) = \begin{cases} x-2 & x>2 \\ -x+1 & x\le 2 \end{cases}$

▮ See Fig. 3.42.

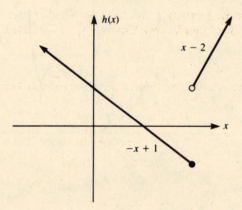

Fig. 3.42

3.120 $f(x) = \begin{cases} x & x>0 \\ x+1 & -1<x\le 0 \\ x^2 & x\le -1 \end{cases}$

▮ See Fig. 3.43.

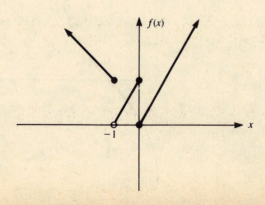

Fig. 3.43

3.121 $g(x) = \begin{cases} 1 & x \text{ an integer} \\ -1 & \text{otherwise} \end{cases}$

▮ See Fig. 3.44.

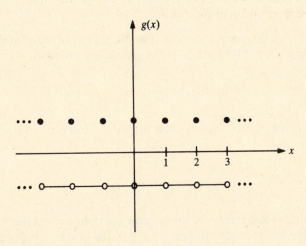

Fig. 3.44

For Probs. 3.122 to 3.131, tell us whether the given function is continuous for every x in its domain. If it is not, tell where it is discontinuous. Some of these are from Probs. 3.110 to 3.121 above.

3.122 $f(x) = x^2$

▮ This function is continuous everywhere. There are no breaks in the graph of this function. Until we study continuity more formally (see Chap. 15), we will use this simple criterion.

3.123 $f(x) = x^3 - 1$

▮ This function is continuous everywhere. Clearly there are no breaks in the graph of this function.

3.124 $g(x) = ax^2 + bx + c$; $a, b, c \in \mathcal{R}$; $b \neq 0$.

▮ In general, all quadratic and linear functions are continuous everywhere. This is a quadratic function (or linear if $a = 0$). It is continuous everywhere.

3.125 $f(x) = \begin{cases} 1 & x \geq 1 \\ 0 & x < 1 \end{cases}$

▮ See Prob. 3.110 above. Figure 3.33 clearly indicates a break at $x = 1$. This function is continuous for all x except $x = 1$; it is discontinuous at $x = 1$. See Prob. 3.126.

3.126 $f(x) = \begin{cases} 1 & x \geq 1 \\ 0 & x < 1 \end{cases}$

▮ Without looking at the graph, we note that the place at which we might have a discontinuity is $x = 1$. Since $1 \neq 0$, there is clearly going to be a jump at $x = 1$.

3.127 $g(x) = \begin{cases} 1 & x > 1 \\ -1 & x \leq 1 \end{cases}$

▮ Discontinuous at $x = 1$. You can look at Fig. 3.34 (Prob. 3.111 above) or use the exact same argument as in Prob. 3.126.

3.128 $h(x) = \begin{cases} 1 & x \geq 1 \\ x & x < 1 \end{cases}$

∎ Looking at Fig. 3.37 (Prob. 3.114 above), we see that $h(x)$ is continuous everywhere. See Prob. 3.130.

3.129 $h(x) = \begin{cases} 1 & x \geq 1 \\ x & x < 1 \end{cases}$

∎ Without looking at the graph, we note that the only candidate for discontinuity is $x = 1$. Since $1 = x$, there clearly will not be a jump on the graph. $h(x)$ is continuous everywhere.

3.130 $g(x) = \begin{cases} 1 & x \geq 2 \\ x & x < 2 \end{cases}$

∎ See Fig. 3.37 (Prob. 3.114 above). Clearly the function is discontinuous when $x = 2$. See Prob. 3.131.

3.131 $g(x) = \begin{cases} 1 & x \geq 2 \\ x & x < 2 \end{cases}$

∎ Without looking at the graph, we investigate $g(x)$ when $x = 2$ (the only possible discontinuity for g). Since $1 \neq x$ when $x = 2$, there will be a jump in the graph. g is discontinuous when $x = 2$.

3.5 LINEAR FUNCTIONS

For Probs. 3.132 to 3.137, find the slope and y intercept (if they exist).

3.132 $y = 3x + 1$

∎ Since when $y = mx + b$, $m = $ slope and $b = y$ intercept, in this case m (slope) $= 3$ and b (y intercept) $= 1$.

3.133 $2y = 3x + 1$

∎ Solve for y; $y = \frac{3}{2}x + \frac{1}{2}$, $m = \frac{3}{2}$ (slope); $b = \frac{1}{2}$ (y intercept).

3.134 $-3y = x + 6$

∎ If $-3y = x + 6$, then $y = -x/3 - 2$. Slope $= -\frac{1}{3}$, y intercept $= -2$.

3.135 $x + y + 4 = 0$

∎ $y = -x - 4$. Slope $= -1$, y intercept $= -4$.

3.136 $y = 3$

∎ Then $y = mx + b$ where $m = 0$, $b = 3$. Note that this is a horizontal line.

3.137 $x = -2$

∎ This is a vertical line; it has no slope and no y intercept.

For Probs. 3.138 to 3.143, graph the linear functions given. Use the slope and y intercept graphing technique.

3.138 $y = 3x + 2$

∎ See Fig. 3.45. After plotting the y intercept, make use of the fact that the slope is 3. That means that $(y_2 - y_1)/(x_2 - x_1) = 3$ for any (x_1, y_1), (x_2, y_2) on the line. Thus, if x changes by 1, then y changes by 3. In the graph we went from $(0, 2)$ to $(1, 5)$. Therefore, $y_2 - y_1 = 5 - 2$ is a change of 3, and $x_2 - x_1 = 1 - 0$ is a change of 1.

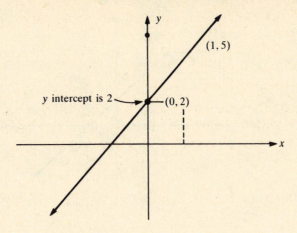

Fig. 3.45

3.139 $y = 2x + 3$

▮   See Fig. 3.46. To get $(1, 5)$, we note that $m = 2$. Thus, for a change of 1 in x, we have a change of 2 in y.

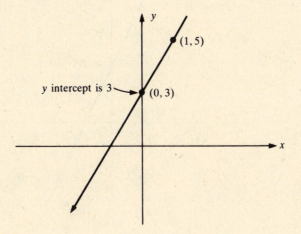

Fig. 3.46

3.140 $3y = 3x + 6$

▮   See Fig. 3.47. Then $y = x + 2$. We get the point $(1, 3)$ from $(0, 2)$, using the fact that $m = 1$.

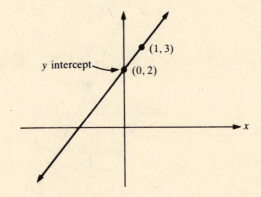

Fig. 3.47

3.141 $y = -x + 1$

▮   See Fig. 3.48. Notice that $b = 1$ and $m = -1$; then for each change of 1 in y, the x change is -1. Notice that in the graph we went from $(0, 1)$ to $(1, 0)$. Therefore, $y_2 - y_1 = 0 - 1$ is a change of -1, and $x_2 - x_1 = 1 - 0$ is a change of 1.

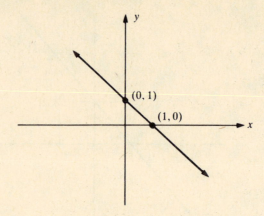

Fig. 3.48

3.142 $y = -2x + 1$

▮ See Fig. 3.49. If x changes by 1, then y changes by -2. See Prob. 3.143.

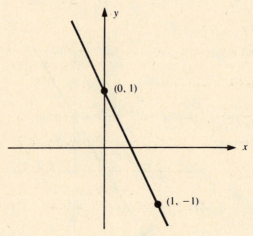

Fig. 3.49

3.143 $y = -2x + 1$

▮ See Fig. 3.50. Alternatively, we can use $(0, 1)$ together with an x change of -1 and a y change of 2. That gives us $(0, 1)$ and $(-1, 3)$. This is the same line as in Prob. 3.142.

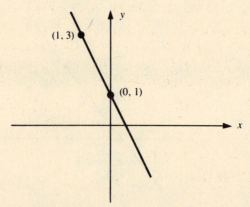

Fig. 3.50

For Probs. 3.144 to 3.161, find an equation of the line that satisfies the given conditions.

3.144 Slope $m = 1$, y intercept $(b) = 2$

▮ $y = mx + b$ is slope-intercept form, $y = 1x + 2 = x + 2$, $y = x + 2$.

3.145 Slope 2, y intercept $= 3$

▮ $y = mx + b = 2x + 3$, $y = 2x + 3$.

3.146 Slope -1, y intercept $= \frac{1}{2}$

▮ $y = mx + b = -x + \frac{1}{2}$, $y = -x + \frac{1}{2}$.

3.147 Slope $= 0$, y intercept $= 3$

▮ $y = 0x + 3 = 3$, $y = 3$.

3.148 No slope, x intercept is 7

▮ "No slope" signals a vertical line. This one is $x = 7$.

3.149 No slope, passes through the origin

▮ Vertical line, x intercept 0. This is $x = 0$ (the y axis!).

3.150 The line is the x axis.

▮ This is a horizontal line with $y =$ intercept $= 0$. Then $y = 0$.

3.151 Contains the points $(0, 0)$ and $(1, 1)$

▮ Since $m = \dfrac{y_2 - y_1}{x_2 - x_1} = \dfrac{1 - 0}{1 - 0} = 1$, $y - y_1 = m(x - x_1)$ or $y - 0 = 1(x - 0)$ or $y = x$. [*Note*: we could have used $y - y_2 = m(x - x_2)$ and obtained the same result.]

3.152 Contains the points $(1, -1)$ and $(2, 3)$

▮ $m = \dfrac{3 - (-1)}{2 - 1} = \dfrac{4}{1} = 4$. Thus, $y - 3 = 4(x - 2)$, $y = 4x - 5$.

3.153 Contains the point $(2, 1)$ and is parallel to $y = x - 3$

▮ If the line is parallel to $y = x - 3$, then its slope $m = 1$. Thus, $y - y_1 = m(x - x_1)$, or $y - 1 = 1(x - 2)$, $y = x - 1$.

3.154 Contains the point $(3, 1)$ and is perpendicular to $2x + y = 3$

▮ $2x + y = 3$ is the equivalent of a line with slope -2. The slope of our line here must be the negative reciprocal of -2, or $\frac{1}{2}$. Thus, $y - 1 = \frac{1}{2}(x - 3)$, or $y = x/2 - \frac{1}{2}$.

3.155 Parallel to $x + 2y = 12$ and has y intercept $= -5$.

▮ $x + 2y = 12$, so $2y = -x + 12$, or $y = -x/2 + 6$. Our line has $m = -\frac{1}{2}$. Then $b = -5$, and $y = -\frac{1}{2}x - 5$.

3.156 Perpendicular to $x + 2y = 12$, y intercept $= -5$

▮ See Prob. 3.155. Then $m = 2$, $b = -5$. $y = 2x - 5$.

3.157 Vertical line containing $(7, -6)$

▮ The line must be of the form $x = a$. Since $x = 7$ is on the line, the line is $x = 7$.

3.158 Horizontal line containing $(7, -6)$

▮ The line must be of the form $y = a$. Since $y = -6$ is on the line, the line is $y = -6$.

3.159 Horizontal line containing the y intercept of $x - y = 3$

▌ The y intercept of $x - y = 3$ is -3. Thus, our line has equation $y = -3$.

3.160 x intercept is 3; y intercept is 2

▌ The line contains $(3, 0)$ and $(0, 2)$. Then $m = (2 - 0)/(0 - 3) = -\frac{2}{3}$ and $y - 2 = -\frac{2}{3}(x - 0)$, or $y = -\frac{2}{3}x + 2$.

3.161 x intercept is 4, y intercept is 6

▌ $m = (6 - 0)/(0 - 4) = -\frac{3}{2}$ [points [4, 0] and (0, 6) are on the line]. Then $y = -\frac{3}{2}x + 6$.

For Probs. 3.162 to 3.164, evaluate the given line or function at the given value.

3.162 $f(x) = 3x - 5$. Find $f(0)$.

▌ $f(0) = 3(0) - 5 = -5$.

3.163 $f(t) = 2t + 1$. Find $f(-3)$.

▌ $f(-3) = 2(-3) + 1 = -5$.

3.164 $f(t) = t - 5$. Find $f(a + b)$.

▌ $f(a + b) = (a + b) - 5 = a + b - 5$.

3.165 Find the point of intersection of $2x + 1 = y$ and $x - y = 3$.

▌ Since $y = 2x + 1$ and $y = x - 3$, they will meet when $2x + 1 = x - 3$. Then $x = -4$; thus, since $y = x - 3$, $y = -4 - 3 = -7$. The intersection point is $(-4, 7)$.

3.166 Find all points of intersection of $y = x^2$ and $y = x$.

▌ Since $x^2 = x$ when the graphs intersect, $x^2 - x = 0$, $x(x - 1) = 0$, $x = 0$, $x = 1$. When $x = 0$, $y = 0$. When $x = 1$, $y = 1$. Thus, $(0, 0)$ and $(1, 1)$ are the intersection points.

For Probs. 3.167 to 3.171, use the quadrilateral with vertices $A(0, 2)$, $B(4, -1)$, $C(1, 5)$, and $D(-3, -2)$.

3.167 Show $\overline{AB} \parallel \overline{DC}$.

▌ m for $\overline{AB}$ is $\dfrac{2 - (-1)}{0 - 4} = -\dfrac{3}{4}$, and m for $\overline{CD}$ is $-\dfrac{5 - (-2)}{1 - (-3)} = -\dfrac{3}{4}$. They are parallel since their slopes are equal and they are not the same line.

3.168 Show that $\overline{DA} \parallel \overline{CB}$.

▌ m for $\overline{AD}$ is $-\dfrac{2 - 2}{-3 - 0} = -\dfrac{4}{-3} = \dfrac{4}{3}$ and m for $\overline{CB} = -\dfrac{1 - (-5)}{4 - 1} = \dfrac{4}{3}$. The slopes are the same, and they are parallel.

3.169 Show that $\overline{AB} \perp \overline{BC}$.

▌ m for $\overline{AB} = -\frac{3}{4}$ and m for $\overline{BC} = \frac{4}{3}$. Since $-\frac{3}{4} = -1/\frac{4}{3}$, the lines are perpendicular.

3.170 Show that $\overline{AD} \perp \overline{DC}$.

▌ m for $\overline{AD} = \frac{4}{3}$ and m for $\overline{DC} = -\frac{3}{4}$. Since $\frac{4}{3} = -1/-\frac{3}{4}$, they are perpendicular.

3.171 Find the perpendicular bisector of $\overline{AD}$.

▌ $\overline{AD}$ has slope $\frac{4}{3}$ and passes through $(0, 2)$. Its equation is then $y - 2 = \frac{4}{3}(x - 0)$ or $y = \frac{4}{3}x + 2$. Then the perpendicular bisector has slope $-\frac{3}{4}$. The midpoint of $\overline{AD}$ is $\left(\dfrac{0-3}{2}, \dfrac{2-2}{2}\right) = (-\frac{3}{2}, 0)$. The perpendicular bisector has $m = -\frac{3}{4}$ and passes through $(-\frac{3}{2}, 0)$. Its equation is $y = -\frac{3}{4}(x + \frac{3}{2})$.

3.172 If $f(-1) = 3$, $f(3) = 5$, and f is linear, find $f(x)$.

▌ The line contains $(-1, 3)$ and $(3, 5)$. Thus, $m = \dfrac{5-3}{3-(-1)} = \dfrac{1}{2}$ and $y - 5 = \frac{1}{2}(x - 3)$ or $y = x/2 + \frac{7}{2}$.

3.173 Determine whether $y = 3/x + 1$ is a linear function.

▌ It is not. Notice that equations of the form $xy = a$ are nonlinear unless $a = 0$. If $a = 0$, we find the graph is two lines (see Prob. 3.172). Any function otherwise containing an xy term is nonlinear.

3.174 Discuss the graph of $xy = 0$.

▌ If $xy = 0$, then $x = 0$ or $y = 0$. Thus, the graph is the x axis and the y axis.

3.6 ALGEBRA OF FUNCTIONS

For Probs. 3.175 to 3.184, find $(f + g)(x)$, $(f - g)(x)$, and their respective domains.

3.175 $f(x) = x$, $g(x) = 2x$

▌ $(f + g)(x) = x + 2x = 3x$; $(f - g)(x) = x - 2x = -x$. Domain of both $= \mathcal{R}$ (all real numbers).

3.176 $f(x) = x^2$, $g(x) = x - 1$

▌ $(f + g)(x) = x^2 + (x - 1) = x^2 + x - 1$; $(f - g)(x) = x^2 - (x - 1) = x^2 - x + 1$. Domain of both $= \mathcal{R}$.

3.177 $f(x) = x - 2$, $g(x) = 1/x$

▌ $(f + g)(x) = x - 2 + 1/x$; $(f - g)(x) = x - 2 - 1/x$. Domain of both $= (-\infty, 0) \cup (0, \infty)$ (x can't be zero).

3.178 $f(x) = \dfrac{1}{x - 1}$, $g(x) = \sqrt{x}$

▌ $(f + g)(x) = \dfrac{1}{x - 1} + \sqrt{x}$; $(f - g)(x) = \dfrac{1}{x - 1} - \sqrt{x}$. Domain of both $= \{x \in \mathcal{R} \mid x \geq 0, x \neq 1\}$.

3.179 $f(x) = \sqrt{x + 1}$, $g(x) = \sqrt{x - 1}$

▌ $(f + g)(x) = \sqrt{x + 1} + \sqrt{x - 1}$; $(f - g)(x) = \sqrt{x + 1} - \sqrt{x - 1}$. Domain of both $= \{x \in \mathcal{R} \mid x \geq 1\}$. Note that if $x < 1$, $\sqrt{x - 1}$ is not defined.

3.180 $f(x) = |x|$, $g(x) = [x]$

▌ $(f + g)(x) = |x| + [x]$; $(f - g)(x) = |x| - [x]$. Domain for both is $\mathcal{R}$.

3.181 $f(x) = x^2$, $g(x) = \dfrac{1}{x - x^2}$

▌ $(f + g)(x) = x^2 + \dfrac{1}{1 - x^2}$; $(f - g)(x) = x^2 - \dfrac{1}{1 - x^2}$. Note that we need $1 - x^2 \neq 0$. Then $x^2 \neq 1$. Then $x^2 \neq 1$ or $x \neq 1, -1$. Domain (for both) $= \{x \in \mathcal{R} \mid x \neq \pm 1\}$.

3.182 $f(x) = x^3 + 1$, $g(x) = 0$

▌ $(f + g)(x) = x^3 + 1 + 0 = x^3 + 1$; $(f - g)(x) = x^3 + 1 - 0 = x^3 + 1$. Notice that if $g(x) = 0$, then $f(x) + g(x) = f(x)$ for all $f(x)$. $g(x) = 0$ is the zero function, behaving for functions the way 0 does for real numbers.

3.183 $f(x) = \sqrt[4]{x - 9}$, $g(x) = x$

▌ $(f + g)(x) = \sqrt[4]{x - 9} + x$; $(f - g)(x) = \sqrt[4]{x - 9} - x$. Domain $= \{x \in \mathcal{R} \mid x \geq 9\}$.

3.184 $f(x) = x + \sqrt{x}$, $g(x) = f(x) + x^2$

▌ $(f + g)(x) = \underbrace{(x + \sqrt{x})}_{f(x)} + \underbrace{(x + \sqrt{x} + x^2)}_{f(x) \ + x^2} = 2x + 2\sqrt{x} + x^2$

$(f - g)(x) = (x + \sqrt{x}) - (x + \sqrt{x} + x^2) = -x^2$
Domains: For $f + g$, $[0, \infty)$, for $f - g$, $\mathcal{R}$.

For Probs. 3.185 to 3.189, find $fg(x)$ and $(f/g)(x)$ and their respective domains.

3.185 $f(x) = 3x$, $g(x) = 2x$

▌ $fg(x) = f(x) \cdot g(x) = 3x \cdot 2x = 6x^2$; $(f/g)(x) = f(x)/g(x) = 3x/(2x) = \frac{3}{2} \ (2x \neq 0)$. Domain $fg = \mathcal{R}$. Domain $f/g = (-\infty, 0) \cup (0, \infty)$. Notice that $g(x) = 2x$ and $g(x)$ is the denominator of f/g.

3.186 $f(x) = 1 + x$, $g(x) = x$

▌ $fg(x) = (1 + x)x = x^2 + x$; $\dfrac{f}{g}(x) = \dfrac{1 + x}{x} \ (x \neq 0)$. Domain $fg = \mathcal{R}$. Domain $f/g = \mathcal{R} - \{0\}$.

3.187 $f(x) = \sqrt{x + 1}$, $g(x) = x - 3$

▌ $fg(x) = (x - 3)\sqrt{x + 1}$; $\dfrac{f}{g}(x) = \dfrac{\sqrt{x + 1}}{x - 3} \ (x \neq 3)$. Domain $fg = \{x \in \mathcal{R} \mid x \geq 1\}$. Domain $f/g = \{x \in \mathcal{R} \mid x \geq -1 \text{ and } x \neq 3\}$.

3.188 $f(x) = \sqrt{x - 2}$, $g(x) = \sqrt{x - 2}$

▌ $fg(x) = \sqrt{x - 2} \sqrt{x - 2} = (\sqrt{x - 2})^2 = x - 2$; $\dfrac{f}{g}(x) = \dfrac{\sqrt{x - 2}}{\sqrt{x - 2}} = 1 \ (x \neq 2)$. Domain $fg = \{x \in \mathcal{R} \mid x \geq 2\}$. Domain $f/g = \{x \in \mathcal{R} \mid x > 2\}$.

3.189 $f(x) = x^3$, $g(x) = 1$

▌ $fg(x) = x^3 \cdot 1 = x^3$; $(f/g)(x) = x^3/1 = x^3$. Domain $fg = $ domain $f/g = \mathcal{R}$. Notice that $g(x) = 1$ functions as a multiplicative identity for function multiplication.

For Probs. 3.190 to 3.199, find $f \circ g(x)$.

3.190 $f(x) = x$, $g(x) = 2x + 1$

▌ $f \circ g(x) = f(g(x)) = f(2x + 1) = 2x + 1$.

3.191 $f(x) = x + 3$, $g(x) = x^2$

▌ $f \circ g(x) = f(x^2) = x^2 + 3$.

3.192 $f(x) = x^2$, $g(x) = 2x + 5$

▌ $f \circ g(x) = f(2x + 5) = (2x + 5)^2 = 4x^2 + 10x + 25$.

3.193 $f(x) = 1$, $g(x) = 3x^3$

▮ $f \circ g(x) = f(3x^3) = 1$. [Do you see this? $f(x) = 1$ for all x!]

3.194 $f(x) = x^3 + 4$, $g(x) = 3$

▮ $f \circ g(x) = f(3) = 3^3 + 4 = 31$.

3.195 $f(x) = a + x^2$, $g(x) = x$

▮ $f \circ g(x) = f(x) = a + (x)^2 = a + x^2$.

3.196 $f(x) = x^2 + x$, $g(x) = x^2 + x$

▮ $f \circ g(x) = f(x^2 + x) = (x^2 + x)^2 + (x^2 + x) = (x^2 + x)(x^2 + x + 1)$.

3.197 $f(x) = \dfrac{1}{1 + x}$, $g(x) = \dfrac{1}{2 + x}$

▮ $f \circ g(x) = f\left(\dfrac{1}{2 + x}\right) = \dfrac{1}{1 + 1/(2 + x)} = \dfrac{2 + x}{3 + x}$.

3.198 $f(x) = \frac{1}{2}$, $g(x) = 2$

▮ $f \circ g(x) = f(2) = \frac{1}{2}$.

3.199 $f(x) = x/2$, $g(x) = 2x$

▮ $f \circ g(x) = f(2x) = 2x/2 = x$. Look at the results in Probs. 3.198 and 3.199. Don't be tricked in the future.

For Probs. 3.200 to 3.206, find $g \circ f(x)$. These are functions from Probs. 3.190 through 3.199 above. Check the results to notice that $f \circ g(x)$ and $g \circ f(x)$ are not always the same function.

3.200 $f(x) = x$, $g(x) = 2x + 1$

▮ $g \circ f(x) = g(x) = 2x + 1$. [See Prob. 3.190; $f \circ g(x) = g \circ f(x)$ in this case.]

3.201 $f(x) = x + 3$, $g(x) = x^2$

▮ $g \circ f(x) = g(x + 3) = (x + 3)^3$. [See Prob. 3.191; $f \circ g(x) \neq g \circ f(x)$ in this case.]

3.202 $f(x) = x^2$, $g(x) = 2x + 5$

▮ $g \circ f(x) = g(x^2) = 2(x^2) + 5 = 2x^2 + 5$. See Prob. 3.192.

3.203 $f(x) = 1$, $g(x) = 3x^3$

▮ $g \circ f(x) = g(1) = 3(1^3) = 3$. See Prob. 3.193.

3.204 $f(x) = a + x^2$, $g(x) = x$

▮ $g \circ f(x) = g(a + x^2) = a + x^2$. See Prob. 3.195.

3.205 $f(x) = \dfrac{1}{1 + x}$, $g(x) = \dfrac{1}{2 + x}$

▮ $g \circ f(x) = g\left(\dfrac{1}{1 + x}\right) = \dfrac{1}{2 + 1/(1 + x)} = \dfrac{1 + x}{3 + 2x}$. See Prob. 3.197.

3.206 $f(x) = x/2$, $g(x) = 2x$

▌ $g \circ f(x) = g(x/2) = 2(x/2) = x$. See Prob. 3.199, and notice that $f \circ g(x) = g \circ f(x) = x$.

For Probs. 3.207 to 3.211, let $f(x) = 2x - 4$.

3.207 Find domain and range of f and f^{-1}.

▌ First note that f is one-to-one, so we are assured of the existence of f^{-1}. Also the domain of f = range of f^{-1}, and the range of f = domain of f^{-1}. Since the domain f = range $f = \mathscr{R}$, the domain f^{-1} = range $f^{-1} = \mathscr{R}$.

3.208 Find $f^{-1}(x)$.

▌ $y = 2x - 4$. Reverse the x and y: $x = 2y - 4$, $2y = x + 4$, or $y = x/2 + 2$, $f^{-1}(x) = x/2 + 2$.

3.209 Find $f^{-1} \circ f(x)$.

▌ $f^{-1} \circ f(x) = f^{-1}(2x - 4) = \dfrac{2x - 4}{2} + 2 = x - 2 + 2 = x$. $f^{-1} \circ f(x) = x$.

3.210 Find $f \circ f^{-1}(x)$.

▌ $f \circ f^{-1}(x) = f(x/2 + 2) = 2(x/2 + 2) - 4 = x + 4 - 4 = x$. Thus, $f \circ f^{-1}(x) = x$.

3.211 Graph $f(x)$ and $f^{-1}(x)$ in the same coordinate system. Show that these two graphs are symmetric about $y = x$.

▌ See Fig. 3.51.

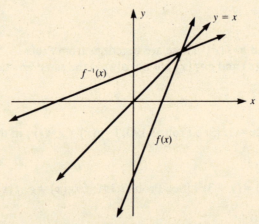

Fig. 3.51

For Probs. 3.212 to 3.216, let $f(x) = x^2 + 1$, $x \ge 0$.

3.212 Find the domain and range of f^{-1}.

▌ Notice first that $f(x)$ here is one-to-one. Note that if we did not add the restriction $x \ge 0$, f would not be one-to-one and we would not have an inverse function. The domain of f^{-1} = range of $f = [1, \infty)$ since $x^2 + 1 \ge 1$ for all x. The range of f^{-1} = domain of $f = [0, \infty)$ since we are given that $x \ge 0$.

3.213 Find $f^{-1}(x)$.

▌ $y = x^2 + 1$; $x \ge 0$. Thus if $x = y^2 + 1$, $y \ge 0$, then $y^2 = x - 1$, or $y = \sqrt{x - 1} = f^{-1}(x)$.

3.214 Find $f^{-1} \circ f(x)$.

▌ $f^{-1} \circ f(x) = f^{-1}(x^2 + 1) = \sqrt{x^2 + 1 - 1} = \sqrt{x^2} = |x|$. But $x \ge 0$, so $|x| = x$.

3.215 Find $f \circ f^{-1}(x)$.

▮ $f \circ f^{-1}(x) = f(\sqrt{x-1}) = (\sqrt{x-1})^2 + 1 = x - 1 + 1 = x$.

3.216 Graph $f(x)$, $f^{-1}(x)$, and $y = x$ on the same axis system.

▮ See Fig. 3.52.

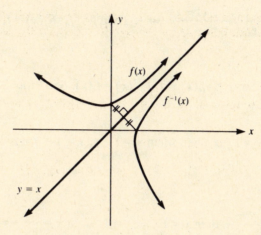

Fig. 3.52

For Probs. 3.217 to 3.227, let D be the set of real numbers other than 0 and 1. Consider $f_1, f_2, \ldots, f_6$ defined on D as follows:

$$f_1(x) = x \qquad f_2(x) = \frac{1}{x} \qquad f_3(x) = 1 - x$$

$$f_4(x) = \frac{x}{x-1} \qquad f_5(x) = \frac{1}{1-x} \qquad f_6(x) = \frac{x-1}{x}$$

Find the following functions.

3.217 $f_2 \circ f_3(x)$

▮ $f_2 \circ f_3(x) = f_2(1 - x) = \dfrac{1}{1 - x} = f_5(x)$.

3.218 $f_3 \circ f_5(x)$

▮ $f_3 \circ f_5(x) = f_3\left(\dfrac{1}{1-x}\right) = 1 - \dfrac{1}{1-x} = \dfrac{-x}{1-x} = \dfrac{x}{x-1} = f_4(x)$.

3.219 $f_1 \circ f_2(x)$

▮ $f_1 \circ f_2(x) = f_1(1/x) = 1/x = f_2(x)$.

3.220 $f_2 \circ f_1(x)$

▮ $f_2 \circ f_1(x) = f_2(x) = 1/x = f_2(x)$.

3.221 $f_4 \circ f_4(x)$

▮ $f_4 \circ f_4(x) = f_4\left(\dfrac{x}{x-1}\right) = \dfrac{x/(x-1)}{x/(x-1) - 1} \cdot \dfrac{x-1}{x-1} = \dfrac{x}{x - (x-1)} = x = f_1(x)$. Notice that f_4 is its own inverse.

3.222 $f_3 \circ f_3(x)$

▮ $f_3 \circ f_3(x) = f_3(1 - x) = 1 - (1 - x) = 1 - 1 + x = x = f_1(x)$.

3.223 $f_5 \circ f_6(x)$

▮ $f_5 \circ f_6(x) = f_5\left(\dfrac{x-1}{x}\right) = \dfrac{1}{1-(x-1)/x} = \dfrac{x}{x-(x-1)} = \dfrac{x}{x-x+1} = x = f_1(x).$

3.224 $f_3 \circ (f_4 \circ f_5)(x)$

▮ $f_4 \circ f_5(x) = f_4\left(\dfrac{1}{1-x}\right) = \dfrac{1/(1-x)}{1/(1-x)-1} = \dfrac{1}{x}.$ Then $f_3 \circ (f_4 \circ f_5)(x) = f_3 \circ f_2(x) = 1 - \dfrac{1}{x} = \dfrac{x-1}{x} = f_6(x).$

3.225 $f_4^{-1}(x)$

▮ See Prob. 3.221. Since $f_4 \circ f_4(x) = x$, $f_4^{-1}(x) = f_4(x)$. See Prob. 3.226 for an alternative method.

3.226 $f_3^{-1}(x)$

▮ $f_3(x) = 1 - x$. If $y = 1 - x$, then when $x = 1 - y$, $y = 1 - x = f^{-1}(x)$. Notice that $f_3(x) = f_3^{-1}(x)$. See Prob. 3.222.

3.227 $f_6^{-1}(x)$

▮ $y = \dfrac{x-1}{x}$. If $x = \dfrac{y-1}{y}$, then $x = 1 - \dfrac{1}{y}$. $\dfrac{1}{y} = 1 - x$ or $y = \dfrac{1}{1-x} = f_6^{-1}(x) = f_5(x).$

3.7 PROBLEM SOLVING AND FORMULAS

3.228 A piece of wire 30 inches (in) long is bent to form a rectangle. If one of its dimensions is x in, express the area as a function of x.

▮ Since the semiperimeter of the rectangle is $\frac{1}{2} \cdot 30 = 15$ in and one dimension is x in, the other is $(15 - x)$ in. Thus, $A = x(15 - x)$.

3.229 An open box is to be formed from a rectangular sheet of tin 20×32 in by cutting equal squares, x in on a side, from the four corners and turning up the sides. Express the volume of the box as a function of x.

▮ From Fig. 3.53 we see that the base of the box has dimensions $20 - 2x$ by $32 - 2x$ in and the height is x in. Then $V = x(20 - 2x)(32 - 2x) = 4x(10 - x)(16 - x)$.

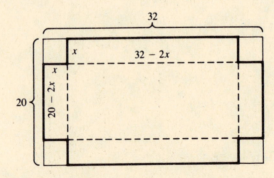

Fig. 3.53

3.230 A closed box is to be formed from the sheet of tin of Prob. 3.229 by cutting equal squares, x in on a side, from two corners of the short side and two equal rectangles of width x in from the other two corners, and folding along the dotted lines, as shown in Fig. 3.54. Express the volume of the box as a function of x.

▮ One dimension of the base of the box is $(20 - 2x)$ in; let y in be the other, Then $2x + 2y = 32$ and $y = 16 - x$. Thus, $V = x(20 - 2x)(16 - x) = 2x(10 - x)(16 - x)$.

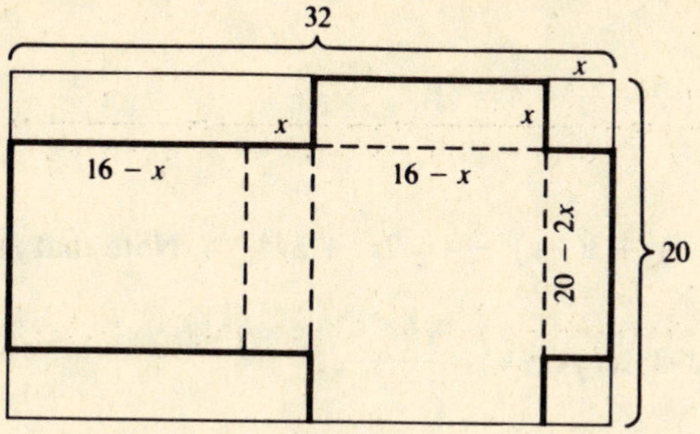

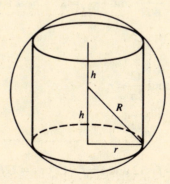

Fig. 3.54

3.231 A farmer has 600 ft of woven wire fencing available to enclose a rectangular field and to divide it into three parts by two fences parallel to one end. If x ft of stone wall is used as one side of the field, express the area enclosed as a function of x when the dividing fences are parallel to the stone wall. Refer to Fig. 3.55.

▮ The dimensions of the field are x and y ft, where $3x + 2y = 600$. Then $y = \frac{1}{2}(600 - 3x)$, and the required area is

$$A = xy = x \cdot \tfrac{1}{2}(600 - 3x) = \tfrac{3}{2}x(200 - x)$$

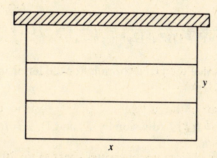

Fig. 3.55

3.232 A right circular cylinder is said to be inscribed in a sphere if the circumference of the base of the cylinder is in the surface of the sphere. If the sphere has radius R, express the volume of the inscribed right circular cylinder as a function of the radius r of its base.

▮ Let the altitude of the cylinder be denoted by $2h$. From the adjoining Fig. 3.56, $h = \sqrt{R^2 - r^2}$ and the required volume is $V = \pi r^2 \cdot 2h = 2\pi r^2 \sqrt{R^2 - r^2}$.

Fig. 3.56

For Probs. 3.233 through 3.235, let $z = f(x, y) = 2x^2 + 3y^2 - 4$. Find:

3.233 $f(0, 0)$

▮ $f(0, 0) = 2(0)^2 + 3(0)^2 - 4 = -4$.

3.234 $f(2, -3)$

▐ $f(2, -3) = 2(2)^2 + 3(-3)^3 - 4 = 31$.

3.235 $f(-x, -y)$

▐ $f(-x, -y) = 2(-x)^2 + 3(-y)^2 - 4 = 2x^2 + 3y^2 - 4$. Note that $f(x, y) = f(-x, -y)$.

For Probs. 3.236 through 3.238, let $f(x, y) = \dfrac{x^2 + y^2}{x^2 - y^2}$.

3.236 Find $f(x, -y)$.

▐ $f(x, -y) = \dfrac{x^2 + (-y)^2}{x^2 - (-y)^2} = \dfrac{x^2 + y^2}{x^2 - y^2} = f(x, y)$

3.237 Find $f(-x, y)$.

▐ $f(-x, y) = \dfrac{(-x)^2 + y^2}{(-x)^2 - y^2} = \dfrac{x^2 + y^2}{x^2 - y^2} = f(x, -y) = f(x, y)$.

3.238 Find $f(1/x, 1/y)$.

▐ $f\left(\dfrac{1}{x}, \dfrac{1}{y}\right) = \dfrac{(1/x)^2 + (1/y)^2}{(1/x)^2 - (1/y)^2} = \dfrac{1/x^2 + 1/y^2}{1/x^2 - 1/y^2} = \dfrac{y^2 + x^2}{y^2 - x^2} = -f(x, y)$.

For Probs. 3.239 through 3.248, solve the given formula for the letter indicated.

3.239 $D = RT$ for R

▐ If $D = RT$, then $D/T = RT/T$ (division by T) and $D/T = R$. Note that we divided by T to isolate R, where R is the unknown.

3.240 $D = RT$ for T

▐ See Prob. 3.239 above. If $D = RT$, then $D/R = RT/R$ (division by R) and $D/R = T$. Note again that the goal of the division was to isolate the unknown.

3.241 $V = LWH$ for L

▐ If $V = LWH$, then $\dfrac{V}{WH} = \dfrac{LWH}{WH}$ and $\dfrac{V}{WH} = L$. *Check*: $V = \dfrac{V}{WH}(WH)$

$$V = V$$

3.242 $C = 2\pi r$ for r

▐ If $C = 2\pi r$, then $\dfrac{C}{2\pi} = \dfrac{2\pi r}{2\pi}$ and $\dfrac{C}{2\pi} = r$.

3.243 $\dfrac{i}{12} = f$ for i

▐ If $i/12 = f$, then $i/12 \cdot 12 = f \cdot 12$ and $i = f \cdot 12$ or $i = 12f$.

3.244 $\dfrac{b}{2} = \dfrac{A}{h}$ for A

▐ Cross multiply: $bh = 2A$ and $\dfrac{bh}{2} = \dfrac{2A}{2}$ or $\dfrac{bh}{2} = A$.

3.245 $C = 10d + 25q$ for q

▮ $C - 10d = 25q$

$$\frac{C - 10d}{25} = \frac{25q}{25}$$

$$\frac{C - 10d}{25} = q$$

Check: $C = 10d + 25\left(\dfrac{C - 10d}{25}\right)$

$$C = 10d + C - 10d$$

$$C = C$$

3.246 $A = \dfrac{h}{2}(b + b^1)$ for h

▮ $\dfrac{A}{b + b^1} = \dfrac{h}{2}$ (division by $b + b^1$) and $h = 2\left(\dfrac{A}{b + b^1}\right) = \dfrac{2A}{b + b^1}$.

3.247 $V = \frac{1}{3}Bh$ for B

▮ $3V = 3(\frac{1}{3}Bh)$

$$3V = Bh$$

$$B = 3V/h$$

3.248 $S = \dfrac{n}{2}(a + 1)$ for a

▮ $2S = n(a + 1)$

$$\frac{2S}{n} = a + 1$$

$$\frac{2S}{n} - 1 = a$$

3.8 MISCELLANEOUS PROBLEMS

3.249 Show that the points $A(1, 2)$, $B(0, -3)$, and $C(2, 7)$ are on the graph of $y = 5x - 3$.

▮ The point $A(1, 2)$ is on the graph since $2 = 5(1) - 3$, $B(0, -3)$ is on the graph since $-3 = 5(0) - 3$, and $C(2, 7)$ is on the graph since $7 = 5(2) - 3$.

3.250 Show that the points $D(0, 0)$ and $E(-1, -2)$ are not on the graph of $y = 5x - 3$.

▮ The point $D(0, 0)$ is not on the graph since $0 \neq 5(0) - 3$, and $E(-1, -2)$ is not on the graph since $-2 \neq 5(-1) - 3$.

3.251 Sketch the graph of the function $2x$.

▮ This is a linear function, and its graph is a straight line. For this locus only two points are necessary. Three points are used to provide a check. See Fig. 3.57.

x	0	1	2
$y = f(x)$	0	2	4

The equation of the line is $y = 2x$.

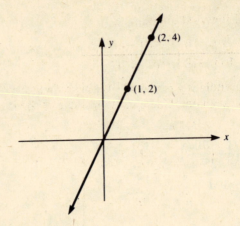

Fig. 3.57

3.252 Sketch the graph of the function $6 - 3x$.

▮ $\dfrac{x}{y = f(x)}$ $\begin{array}{|c|c|c|} 0 & 2 & 3 \\ \hline 6 & 0 & -3 \end{array}$. See Fig. 3.58. The equation of the line is $y = 6 - 3x$.

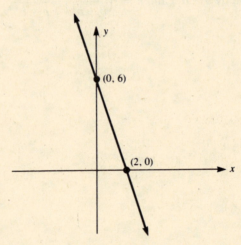

Fig. 3.58

3.253 Sketch the graph of the function x^2.

▮ $\dfrac{x}{y = f(x)}$ $\begin{array}{|c|c|c|c|c|} 3 & 1 & 0 & -2 & -3 \\ \hline 9 & 1 & 0 & 4 & 9 \end{array}$. See Fig. 3.59. The equation of this locus, called a *parabola*, is $y = x^2$. Note for $x \neq 0$, $x^2 \neq 0$. Thus, the curve is never below the x axis. Moreover, as $|x|$ increases, x^2 increases; i.e., as we move from the origin along the x axis in either direction, the curve moves farther and farther from the axis. Hence, in sketching parabolas sufficient points must be plotted so that its U shape can be seen.

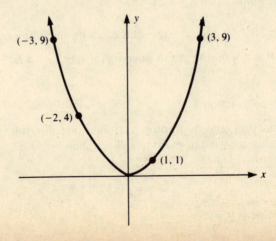

Fig. 3.59

3.254 Sketch the graph of the function $x^2 + x - 12$.

x	4	3	1	0	-1	-4	-5
$y = f(x)$	8	0	-10	-12	-12	0	8

The equation of the parabola is $y = x^2 + x - 12$. Note that the points $(0, -12)$ and $(-1, -12)$ are not joined by a straight line segment. Check that the value of the function is $-12\frac{1}{4}$ when $x = -\frac{1}{2}$. See Fig. 3.60.

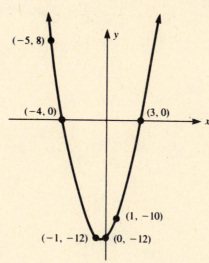

Fig. 3.60

3.255 Sketch the graph of the function $-2x^2 + 4x + 1$.

x	3	2	1	0	-1
$y = f(x)$	-5	1	3	1	-5

. See Fig. 3.61.

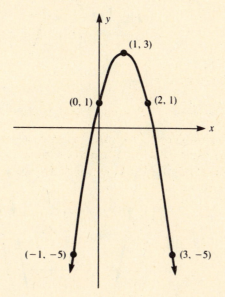

Fig. 3.61

3.256 Sketch the graph of the function $(x + 1)(x - 1)(x - 2)$.

x	3	2	$\frac{3}{2}$	1	0	-1	-1
$y = f(x)$	8	0	$-\frac{5}{8}$	0	2	0	-12

This is a cubic curve of equation $y = (x + 1)(x - 1)(x - 2)$. It crosses the x axis where $x = -1$, 1, and 2. See Fig. 3.62.

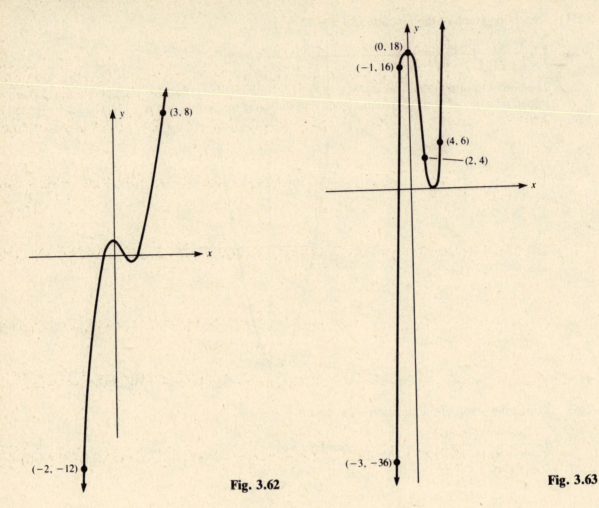

Fig. 3.62

Fig. 3.63

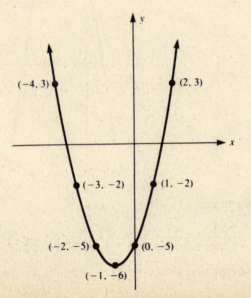

Fig. 3.64

3.257 Sketch the graph of the function $(x + 2)(x - 3)^2$.

x	5	4	$\frac{7}{2}$	3	2	1	0	−1	−2	−3
$y = f(x)$	28	6	$\frac{11}{8}$	0	4	12	18	16	0	−36

This cubic crosses the x axis where $x = -2$ and is tangent to the x axis where $x = 3$. Note that for $x > -2$, the value of the function is positive except for $x = 3$, where it is 0. Thus, to the right of $x = -2$, the curve is never below the x axis. See Fig. 3.63.

3.258 Sketch the graph of the function $x^2 + 2x - 5$, and use it to determine the real roots of $x^2 + 2x - 5 = 0$.

x	2	1	0	−1	−2	−3	−4
$y = f(x)$	3	−2	−5	−6	−5	−2	3

The parabola cuts the x axis at a point whose abscissa is between 1 and 2 (the value of the function changes sign) and at a point whose abscissa is between −3 and −4.
Reading from the graph in Fig. 3.64, we see the roots are $x = 1.5$ and $x = -3.5$ approximately.

CHAPTER 4
Polynomial and Rational Functions /

4.1 POLYNOMIAL FUNCTIONS

For Probs. 4.1 to 4.10, write the given equation in standard form.

4.1 $4x^2 + 2x^3 - 6 + 5x = 0$.

▮ In standard form, polynomial equations are written from the highest power of x to the lowest. In this case, the standard form is $2x^3 + 4x^2 + 5x - 6 = 0$.

4.2 $-3x^3 + 6x - 4x^2 + 2 = 0$.

▮ $3x^3 + 4x^2 - 6x - 2 = 0$.

4.3 $2x^5 + x^3 + 4 = 0$

▮ $2x^5 + 0x^4 + x^3 + 0x^2 + 4 = 0$.

4.4 $x^3 + \frac{1}{2}x^2 - x + 2 = 0$

▮ $x^3 + \frac{1}{2}x^2 - x + 2 = 0$ is rewritten as $2x^3 + x^2 - 2x + 4 = 0$ since in standard form all coefficients are integers.

4.5 $4x^4 + 6x^3 - 8x^2 + 12x - 10 = 0$

▮ $2x^4 + 3x^3 - 4x^2 + 6x - 5 = 0$, since in standard form $(a_0, \ldots, a_n) = 1$ for all coefficients $a_0, a_1, \ldots, a_n$.

4.6 $2x - 4x^4 = 0$

▮ $-4x^4 + 0x^3 + 0x^2 + 2x + 0 = 0$, and then notice that gcd $(-4, 2) \neq 1$. So the *answer* is $-2x^4 + 0x^3 + 0x^2 + x + 0 = 0$.

4.7 $\frac{x^2}{3} + \frac{x^4}{12} + 3 = 0$

▮ Multiplying by 12 gives $x^4 + 4x^2 + 36 = 0$. Inserting zeros, $x^4 + 0x^3 + 4x^2 + 0x + 36 = 0$.

4.8 $\frac{x^5}{2} - 1 = 0$

▮ $x^5 + 0x^4 + 0x^3 + 0x - 2 = 0$.

4.9 $2x^4 - 6x^3 + 2 = 0$

▮ $x^4 - 3x^2 + 1 = 0$. Insert zeros: $x^4 - 3x^2 + 0x + 1 = 0$.

4.10 $(x + 2)^2 + 5 = 0$

▮ $(x^2 + 4x + 4) + 5 = 0$, or $x^2 + 4x + 9 = 0$.

For Probs. 4.11 to 4.15, rewrite the equation in the standard form $y - k = a(x - h)^2$.

4.11 $y = x^2 + 1$

▮ $y = x^2 + 1$ is written as $y - 1 = x^2$, or $y - 1 = 1(x - 0)^2$.

4.12 $y = -2x^2 + 3$

▌ $y - 3 = -2x^2$, or $y - 3 = -2(x - 0)^2$.

4.13 $y = 4 - (x - 1)^2$

▌ $y - 4 = -(x - 1)^2$, or $y - 4 = -1(x - 1)^2$.

4.14 $y = x^2 - 4x + 6$

▌ $y = (x^2 - 4x + 4) + 2$ (by completing the square); $y = (x - 2)^2 + 2$, or $y - 2 = 1(x - 2)^2$.

4.15 $y = (1 - x)(2 - x)$

▌ $y = 2 - 3x + x^2 = (x^2 - 3x + \frac{9}{4}) + (-\frac{1}{4})$ (since $\frac{9}{4} - \frac{1}{4} = 2$); $y = (x - \frac{3}{2})^2 - \frac{1}{4}$, or $y + \frac{1}{4} = 1(x - \frac{3}{2})^2$.

For Probs. 4.16 to 4.20, find the vertex of the given parabola.

4.16 $y + 1 = 2(x - 3)^2$

▌ If $y - k = a(x - h)^2$, then (h, k) is the vertex of the given parabola. In this case, $k = -1$ and $h = 3$. $(3, -1)$ is the vertex.

4.17 $y = -2x^2 + 3$

▌ Then $y - 3 = -2(x - 0)^2$. Thus, $k = 3$ and $h = 0$. The vertex is at $(0, 3)$.

4.18 $y = x^2 - 4x + 6$

▌ $y = (x^2 - 4x + 2) + 4 = (x - 2)^2 + 4$; $y - 4 = (x - 2)^2$. The vertex is at $(2, 4)$.

4.19 $y = x(x + 2)$

▌ $y = x^2 + 2x = (x^2 + 2x + 1) - 1$ (since $1 - 1 = 0$); $y + 1 = (x + 1)^2$. The vertex is at $(-1, -1)$.

4.20 $y = 2x^2 - 8x + 5$

▌ $y - 5 = 2(x^2 - 4x) = 2(x^2 - 4x + 4) - 8$; $y + 3 = 2(x - 2)^2$. The vertex is at $(2, -3)$.

For Probs. 4.21 to 4.24, identify the given parabola as opening upward or downward.

4.21 $y = x^2 - 5x + 6$

▌ If $y = ax^2 + bx + c$, then the parabola opens upward if $a > 0$ and downward if $a < 0$. In this case, $a = 1 > 0$, and the parabola opens upward.

4.22 $y = 2 - x^2$

▌ If $y = 2 - x^2$, then $y = -x^2 + 2$ and $a = -1 < 0$; the parabola open downward.

4.23 $y = x(1 - x)$

▌ If $y = x - x^2$, then $y = -x^2 + x$ and $a = -1 < 0$, and the parabola opens downward.

4.24 $y = (2 - x)(3 - x)$

▌ $y = 6 - 5x + x^2$; $a = 1$ and the parabola opens upward.

For Probs. 4.25 to 4.29, find the maximum (or minimum) value of the quadratic function.

4.25 $f(x) = 2(x - 3)^2 - 1$

▌ If $y = 2(x-3)^2 - 1$, then $y + 1 = 2(x-3)^2$. The vertex is at $(3, -1)$, and the parabola is opening upward (since $a > 0$, where a is the coefficient of x^2). Thus, the function has a minimum value of -1.

4.26 $f(x) = 2x^2 + 3$

▌ If $y = -2x^2 + 3$, then $y - 3 = -2(x - 0)^2$. The vertex is at $(0, 3)$, and since $a < 0$, the maximum value is 3.

4.27 $f(x) = 4 - (x-1)^2$

▌ If $y = 4 - (x-1)^2$, then $y - 4 = -(x-1)^2$. Thus, f has a maximum value of 4.

4.28 $f(x) = x^2 - 4x + 6$

▌ $y = (x-2)^2 + 2$ (since $4 + 2 = 6$). Thus, $a > 0$ (f has a minimum), and the minimum value is 2.

4.29 $f(x) = (x-1)(x+3)$

▌ $y = x^2 + 2x - 3$; $a > 0$, so f has a minimum. $y = (x+1)^2 - 4$ (since $1 - 4 = -3$), so $y + 4 = (x+1)^2$, and the minimum value is -4.

For Probs. 4.30 to 4.35, is the given function a polynomial function?

4.30 $y = x^2$

▌ This is a polynomial function, since $y = a_0 x^n + a_1 x^{n-1} + \cdots + a_{n-1} x + a_n$, where n is a positive integer or zero and all a_i are complex numbers.

4.31 $y = 1/x^2$

▌ This is not a polynomial function since $y = x^{-2}$, and -2 is not a positive integer or zero.

4.32 $y = (x+2)^x$

▌ This is not a polynomial function. Note that x is the independent variable and thus may not be a positive integer or zero.

4.33 $y = 2x + 4 - 2i$

▌ This is a polynomial function. If $y = 2x + (4 - 2i)$, then all conditions of the definition in Prob. 4.30 above are satisfied. Note that the constant term $a_n = 4 - 2i$.

4.34 $y = 2x^2 + \sqrt{3}x - i$

▌ This is a polynomial function. the coefficients 2 and $\sqrt{3}$ are complex as in the constant term $-i$.

4.35 $y = ix^3 - \sqrt{4x} - 2$

▌ This is not a polynomial function since $x^{1/2}$ appears and $\frac{1}{2}$ is nonintegral.

For Probs. 4.36 to 4.39, determine whether the graph of the polynomial will cross or touch the x axis at each zero.

4.36 $f(x) = x^2(x-1)$

▌ The zeros here are 0 with a multiplicity of 2 and 1 with a multiplicity of 1. Thus, the graph touches at 0, and the graph crosses at 1. (Recall that the graph of a polynomial crosses at a zero of odd multiplicity and touches at a zero of even multiplicity.)

4.37 $f(x) = x(x-1)^2(x+2)$

▮ The zeros here are 0 with a multiplicity of 1 (crosses), 1 with a multiplicity of 2 (touches), and −2 with a multiplicity of 1 (crosses).

4.38 $f(x) = (x^2 - 1)(x + 2)^3$

▮ The zeros here are 1 with a multiplicity of 1 (crosses), −1 with a multiplicity of 1 (crosses), and −2 with a multiplicity of 3 (crosses). (Do you see where 1 and −1 come from? If $x^2 - 1 = 0$, then $x = \pm 1$.)

4.39 $f(x) = (x + 1)^3(x + 2)^4(x + 3)^5$

▮ Touches at −2; crosses at −1 and −3.

For Probs. 4.40 to 4.42, find the remainder by actually performing the division.

4.40 $(2x^2 - 4x + 8) \div (x - 2)$

▮
$$
\begin{array}{r}
2x \\
x - 2 \overline{)2x^2 - 4x + 8} \\
\underline{2x^2 - 4x} \\
+ 8
\end{array}
$$
Remainder = 8.

4.41 $(3x^2 + x - 7) \div (x - 1)$

▮
$$
\begin{array}{r}
3x + 4 \\
x - 1 \overline{)3x^2 + x - 7} \\
\underline{3x^2 - 3x} \\
4x - 7 \\
\underline{4x - 4} \\
- 3
\end{array}
$$
Remainder = −3.

4.42 $(3x^3 + 4x^2 - 7x + 5) \div (x + 1)$

▮
$$
\begin{array}{r}
3x^2 + x - 8 \\
x + 1 \overline{)3x^3 + 4x^2 - 7x + 5} \\
\underline{3x^3 + 3x^2} \\
x^2 - 7x \\
\underline{x^2 + x} \\
- 8x + 5 \\
\underline{- 8x - 8} \\
13
\end{array}
$$
Remainder = 13.

For Probs. 4.43 and 4.44, find the remainder by using the remainder theorem.

4.43 $(x^3 - 5x^2 - 3x + 15) \div (x + 2)$

▮ The remainder theorem tells us that the remainder upon division by $x - r$ will be $f(r)$. In this case, $r = -2$, so $f(-2) = (-2)^3 - 5(-2)^2 - 3(-2) + 15 = -7$.

4.44 $(2x^4 + 6x^3 + 3x - 4) \div (x + 3)$

▮ $f(r) = f(-3) = 2(-3)^4 + 6(-3)^3 + 3(-3) - 4 = -13$.

For Probs. 4.45 to 4.52, show that the second expression is a factor of the first by means of the factor theorem.

4.45 $2x^3 - 3x^2 + 2x - 8, \ x - 2$

▮ By the factor theorem, we must show that r is a root of $f(x) = 0$ to show that $x - r$ is a factor of $f(x)$. In this case, $f(2) = 2 \cdot 2^3 - 3 \cdot 2^2 + 2 \cdot 2 - 8 = 0$. Thus, $x - 2$ is a factor.

4.46 $3x^3 + 5x^2 - 6x + 18, \ x + 3$

▮ $f(x) = 3x^3 + 5x^2 - 6x + 18; \ r = -3$. Then $f(-3) = 3 \cdot (-3)^3 + 5 \cdot (-3)^2 - 6 \cdot (-3) + 18 = 0$. Thus $x + 3$ is a factor.

4.47 $2x^3 - 3x^2 + x - 6, \ x - 2$

▮ $f(x) = 2x^3 - 3x^2 + x - 6; \ r = 2$. Then $f(2) = 2 \cdot 2^3 - 3 \cdot 2^2 + 2 - 6 = 0$. Thus, $x - 2$ is a factor.

4.48 $2x^3 - 7x^2 - 3x - 4, \ x - 4$

▮ $f(x) = 2x^3 - 7x^2 - 3x - 4; \ r = 4$. Then $f(4) = 2 \cdot 4^3 - 7 \cdot 4^2 - 3 \cdot 4 - 4 = 0$. Thus, $x - 4$ is a factor.

4.49 $3x^4 - x^3 - 3x^2 + \frac{1}{3}, \ x - \frac{1}{3}$

▮ $f(x) = 3x^4 - x^3 - 3x^2 + \frac{1}{3}; \ r = \frac{1}{3}$. Then $f(\frac{1}{3}) = 3 \cdot (\frac{1}{3})^4 - (\frac{1}{3})^3 - 3(\frac{1}{3})^2 + \frac{1}{3} = 0$. Thus, $x - \frac{1}{3}$ is a factor.

4.50 $x^4 - 81, \ x + 3$

▮ $f(-3) = -3^4 - 81 = 0$. Thus, $x + 3$ is a factor. See Prob. 4.51.

4.51 $x^9 + a^9, \ x + a$

▮ $f(x) = x^9 + a^9; \ r = -a$. Then $f(-a) = (-a)^9 + a^9 = -a^9 + a^9 = 0$. Thus, $x + a$ is a factor. See Prob. 4.52.

4.52 $x^n + a^n, \ x + a$, where n is an odd positive integer

▮ $f(x) = x^n + a^n; \ r = -a$. Then $f(-a) = (-a)^n + a^n = -a^n + a^n = 0$ (since n is an odd positive integer). Thus, $x + a$ is a factor.

For Probs. 4.53 to 4.56, write equations with integral coefficients having the given numbers and no others as roots.

4.53 $1, 2, -3$

▮ By the factor theorem, $x - 1, \ x - 2, \ x + 3$ must be factors. Then $f(x) = (x - 1)(x - 2)(x + 3) = (x^2 - 3x + 2)(x + 3) = x^3 - 7x + 6 = 0$.

4.54 $-1, -2, -3$

▮ By the factor theorem, $x + 1, \ x + 2, \ x + 3$ must be factors. Then $f(x) = (x + 1)(x + 2)(x + 3) = (x^2 + 3x + 2)(x + 3) = x^3 + 6x^2 + 11x + 6 = 0$.

4.55 $-4, 3, 0$

▮ $(x + 4)(x - 3)(x) = 0; \ x^3 + x^2 - 12x = 0$.

4.56 $\sqrt{3}, -\sqrt{3}, 2$

▮ $(x - \sqrt{3})(x + \sqrt{3})(x - 2) = 0; \ (x^2 - 3)(x - 2) = 0; \ x^3 - 2x^2 - 3x + 6 = 0$.

4.2 GRAPHING POLYNOMIAL FUNCTIONS

For Probs. 4.57 to 4.83, sketch the graph of the given polynomial function.

4.57 $f(x) = x^3$

▮ See Fig. 4.1. If $f(x) = x^3 = 0$, then $x = 0$. The graph crosses the x axis at $x = 0$ since multiplicity is odd. When $x = 0$, $y = 0$, so the graph crosses the y axis when $x = 0$. Since $f(-x) = -f(x)$, the graph is symmetric about the origin. As x grows larger, so does y; as x grows smaller, so does y. Notice the steps we have gone through here. We will use them throughout this section.

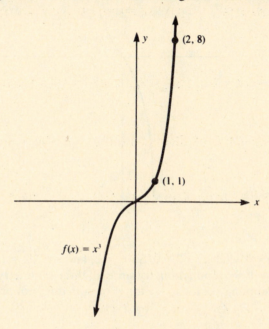

Fig. 4.1

4.58 $f(x) = -x^3$

▮ See Fig. 4.2. If $f(x) = -x^3 = 0$, then $x = 0$. The graph crosses the x axis at $(0, 0)$. If $x = 0$, then $f(0) = 0$. If $f(-x) = -f(x)$, then f is symmetric about the origin; $f(1) = -1$, $f(-1) = 1$, $f(2) = -8$, $f(-2) = 8$.

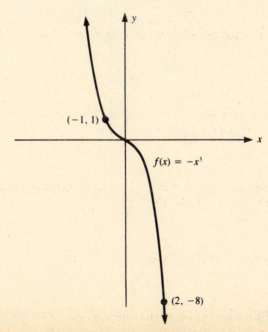

Fig. 4.2

4.59 $f(x) = 3x^3$

▮ See Fig. 4.3. $f(0) = 0$ and $f(-x) = -f(x)$, so the graph is symmetric about the origin. $f(1) = 3$, $f(-1) = -3$, $f(2) = 24$.

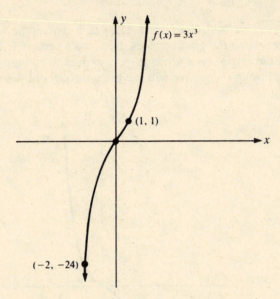

Fig. 4.3

4.60 $f(x) = x^3 - 2$

▮ See Fig. 4.4. $f(x) = 0$ when $x^3 - 2 = 0$. So $x^3 = 2$, and $x = \sqrt[3]{2} \approx 1.26$. Note that the multiplicity is odd. Also the graph will show no symmetry since all three symmetry tests fail: $f(-x) \neq f(x)$ and $f(-x) \neq -f(x)$. Substituting $-x$ for x, $-y$ for y, or both does not produce equivalent equations.

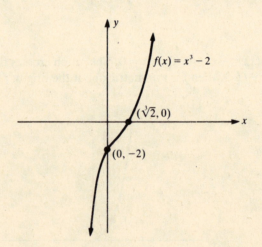

Fig. 4.4

4.61 $f(x) = 2(x^3 - 2)$.

▮ See Fig. 4.5 and Prob. 4.60. We can sketch this graph by observing that all y values in Prob. 4.60 are doubled.

4.62 $f(x) = -2(x^3 - 2)$

▮ See Fig. 4.6 and Probs. 4.60 and 4.61.

4.63 $f(x) = (x - 2)^3$

▮ See Fig. 4.7. $f(0) = -8$, and when $(x - 2)^3 = 0$, $x = 2$.

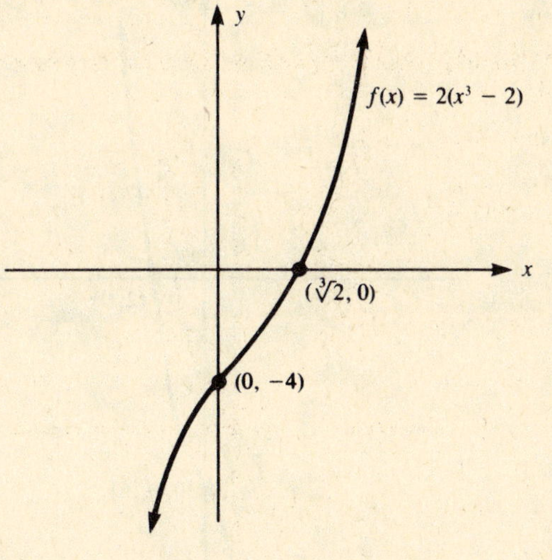

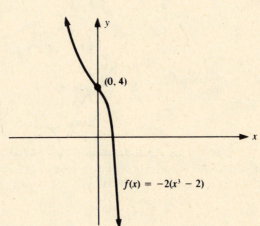

$f(x) = 2(x^3 - 2)$

$(\sqrt[3]{2}, 0)$

$(0, -4)$

Fig. 4.5

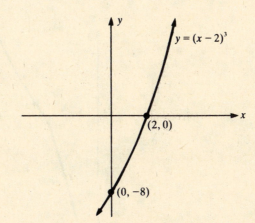

$(0, 4)$

$f(x) = -2(x^3 - 2)$

Fig. 4.6

$y = (x - 2)^3$

$(2, 0)$

$(0, -8)$

Fig. 4.7

4.64 $f(x) = 2(x - 2)^3$

▌ See Fig. 4.8 and Prob. 4.63.

4.65 $f(x) = -2(x - 2)^3$

▌ See Fig. 4.9 and Prob. 4.64.

4.66 $f(x) = 2(x - 2)^3 + 1$

▌ See Fig. 4.10 and Prob. 4.64.

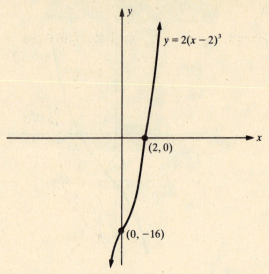

Fig. 4.8

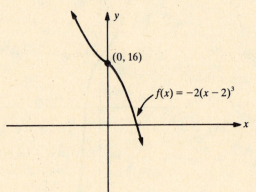

Fig. 4.9

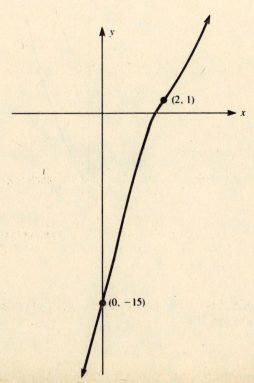

Fig. 4.10

4.67 $f(x) = x^3 - 4$

▮ See Fig. 4.11. When $x^3 - 4 = 0$, $x = \sqrt[3]{4} \approx 1.587$. $f(0) = -4$.

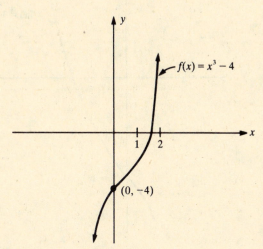

Fig. 4.11

4.68 $f(x) = 3(x^3 - 4)$

▮ See Fig. 4.12 and Prob. 4.67. Triple each y value.

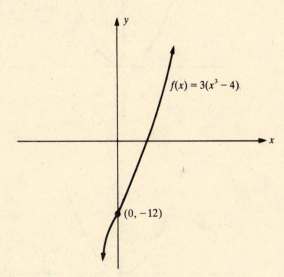

Fig. 4.12

4.69 $f(x) = 3(x^3 - 4) + 2$

▮ See Fig. 4.13 and Probs. 4.67 and 4.68. Add 2 to each y value in Prob. 4.68.

4.70 $f(x) = x^4$

▮ See Fig. 4.14. If $y = x^4$, then $y = (-x)^4 = x^4$; the graph is symmetric about the y axis. $f(0) = 0$, $f(1) = 1$, $f(-1) = 1$, $f(2) = 16$, $f(-2) = 16$.

4.71 $f(x) = 3x^4$

▮ See Fig. 4.15 and Prob. 4.70. This graph will have all ordinate values in Prob. 4.70 multiplied by 3.

4.72 $f(x) = 3x^4 - 2$

▮ See Fig. 4.16 and Prob. 4.71. Subtract 2 from each ordinate in the graph of $y = 3x^4$.

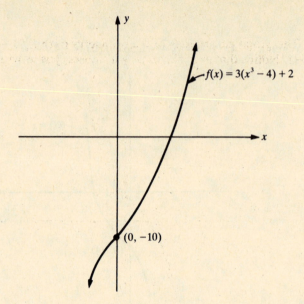

$f(x) = 3(x^3 - 4) + 2$

$(0, -10)$

Fig. 4.13

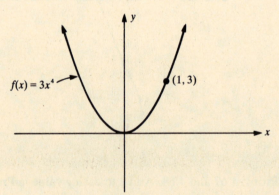

$(-2, 16)$

$y = x^4$

$(1, 1)$

Fig. 4.14

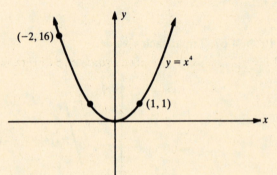

$f(x) = 3x^4$

$(1, 3)$

Fig. 4.15

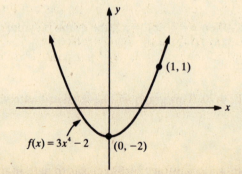

$(1, 1)$

$f(x) = 3x^4 - 2$

$(0, -2)$

Fig. 4.16

4.73 $f(x) = x^5 + 1$

▌ See Fig. 4.17. This graph passes none of our symmetry tests. $f(0) = 1$ and if $x^5 + 1 = 0$, then $x^5 = -1$ and $x = -1$. Notice that as x increases, y increases, just as with $y = x^3$.

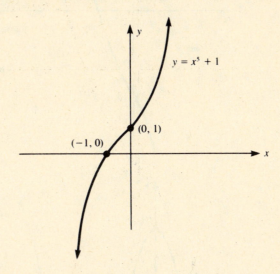

Fig. 4.17

4.74 $f(x) = 3(x^5 + 1)$

▌ See Fig. 4.18 and Prob. 4.73. We triple each ordinate in the graph of $y = x^5 + 1$.

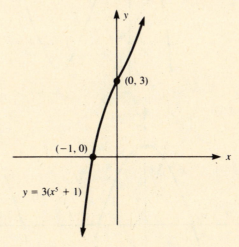

Fig. 4.18

4.75 $f(x) = 3(x^5 + 1) - 6$

▌ See Fig. 4.19 and Prob. 4.74. We subtract 6 for each ordinate in the graph of $y = 3(x^5 + 1)$.

4.76 $y = (x - 3)(x^2 - 1)$

▌ See Fig. 4.20. If $(x - 3)(x^2 - 1) = 0$, then $x = 3, \pm 1$. Each of these zeros has a multiplicity of 1, so the graph will cross the x axis at these points. Also the three symmetry tests fail. Investigating the behavior of this function, we find that **(a)** it is negative for $x < -1$, **(b)** it is positive for $-1 < x < 1$, **(c)** it is negative for $1 < x < 3$, **(d)** it is positive for $x > 3$.

4.77 $y = 2(x - 3)(x^2 - 1)$

▌ See Fig. 4.21 and Prob. 4.76.

4.78 $y = 2(x - 3)(x^2 - 1) + 1$

▌ See Fig. 4.22 and Prob. 4.77. Add 1 to each y value in the graph of $y = 2(x - 3)(x^2 - 1)$.

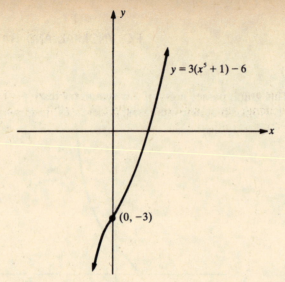

$y = 3(x^5 + 1) - 6$

$(0, -3)$

Fig. 4.19

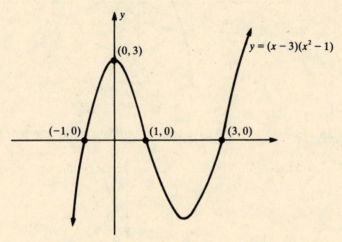

$(0, 3)$

$y = (x - 3)(x^2 - 1)$

$(-1, 0)$ $(1, 0)$ $(3, 0)$

Fig. 4.20

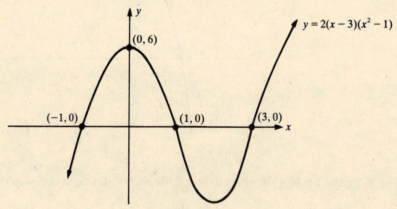

$(0, 6)$

$y = 2(x - 3)(x^2 - 1)$

$(-1, 0)$ $(1, 0)$ $(3, 0)$

Fig. 4.21

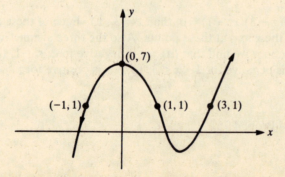

$(0, 7)$

$(-1, 1)$ $(1, 1)$ $(3, 1)$

Fig. 4.22

4.79 $y = (x^2 - 1)(x^2 - 9)$

▌ See Fig. 4.23. This graph has four zeros: $x = \pm 1$, $x = \pm 3$, all with odd multiplicity. Also replacing x with $-x$ does not change the equation, so the graph exhibits y-axis symmetry. $f(0) = 9$. $f > 0$ for $x < -3$, $x > 3$, $-1 < x < 1$; and $f < 0$ for $-3 < x < -1$, $1 < x < 3$.

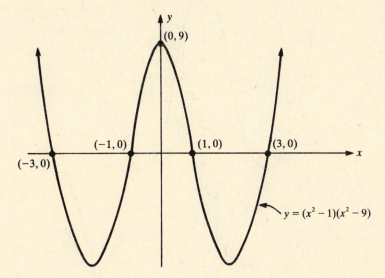

Fig. 4.23

4.80 $y = 2(x^2 - 1)(x^2 - 9)$

▌ See Fig. 4.24 and Prob. 4.79. Multiply all y values in the graph of $y = (x^2 - 1)(x^2 - 9)$ by 2.

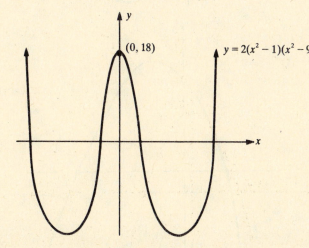

Fig. 4.24

4.81 $y = 2(x^2 - 1)(x^2 - 9) + 1$

▌ See Fig. 4.25 and Prob. 4.80. Add 1 to each y value in the graph of $y = 2(x^2 - 1)(x^2 - 9)$.

4.82 $f(x) = x^2 - 6x + 5$

▌ See Fig. 4.26. Since this is in nonfactored form, we first factor the equation's RHS: $x^2 - 6x + 5 = (x - 1)(x - 5)$. Clearly, 1 and 5 are zeros. Also $f(0) = 5$, and the function is negative between 1 and 5.

4.83 $y = -x^2 + 2x + 8$

▌ See Fig. 4.27. Factoring, we find $-x^2 + 2x + 8 = -(x^2 - 2x - 8) = -(x^2 - 2x + 1 - 9) =$ $-(x^2 - 2x + 1) + 9 = -(x - 1)^2 + 9$. $f(0) = 8$, and if $f(x) = 0$, $-(x - 1)^2 + 9 = 0$, $-(x - 1)^2 = -9$, $(x - 1)^2 = 9$. Then $x - 1 = 3$ when $x = 4$, and $x - 1 = -3$ when $x = -2$.

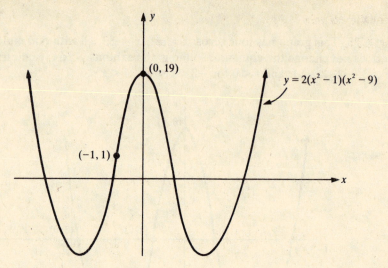

$$y = 2(x^2 - 1)(x^2 - 9)$$

(0, 19)

(−1, 1)

Fig. 4.25

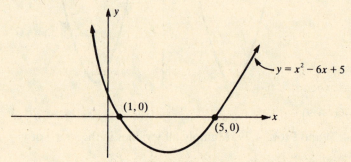

$$y = x^2 - 6x + 5$$

(1, 0)

(5, 0)

Fig. 4.26

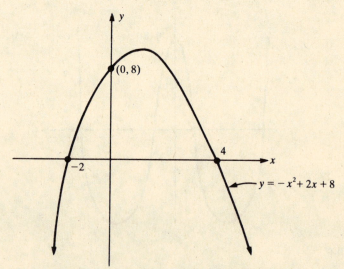

(0, 8)

−2

4

$$y = -x^2 + 2x + 8$$

Fig. 4.27

For Probs. 4.84 to 4.87, refer to the following situation: A parcel delivery service will deliver only packages with length plus girth not exceeding 108 in. A packaging company wishes to design a box with a square base that will have a maximum volume and will meet the delivery service's restrictions.

4.84 Write the volume of the box $V(x)$ in terms of x.

▮ Volume $= l \cdot w \cdot h = x \cdot x \cdot (108 - x - x - x - x) = x^2(108 - 4x)$.

4.85 What is the domain of V in Prob. 4.84?

▮ $108 - 4x$ must be nonnegative in order that $V \geq 0$. Thus, $0 \leq x \leq 27$. If $x = 0$ or $x = 27$, the volume of the box is 0.

4.86 Graph V for the domain in Prob. 4.85.

▌ See Fig. 4.28.

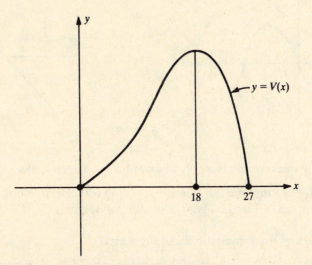

Fig. 4.28

4.87 From Fig. 4.28 estimate (to the nearest inch) the dimensions of the box with maximum volume. What is this maximum volume?

▌ To the nearest inch, the maximum occurs when $x \approx 18$. $V(18) = 18 \cdot 18 \cdot 36 = 11,664$ in$^2 \approx$ maximum volume.

For Probs. 4.88 to 4.96, refer to the graphs in Figs. 4.29 to 4.34.

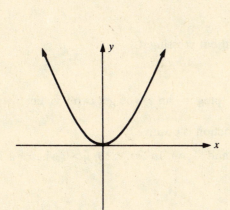

Fig. 4.29

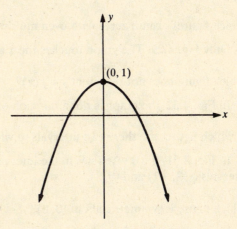

Fig. 4.30

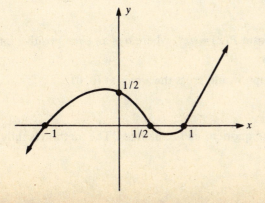

Fig. 4.31

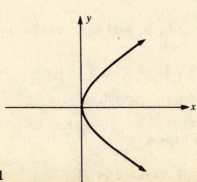

Fig. 4.32

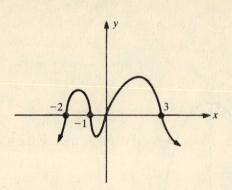

Fig. 4.33

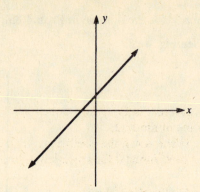

Fig. 4.34

4.88   Which of the figures represents a polynomial function which is always increasing?

▮ Figure 4.34. Note that if $a > b$, then $f(a) > f(b)$ for all a, b. None of the other graphs have this property. The graph of Fig. 4.34 is a polynomial of degree 1.

4.89   Which of the figures are symmetric about the y axis?

▮ Figures 4.29 and 4.30. For both of these we see that if (x, y) is on the graph, so is $(-x, y)$.

4.90   Which of the figures exhibits x axis symmetry?

▮ Only Fig. 4.32. Notice that if (x, y) is on the graph, so is $(x, -y)$.

4.91   Which figure has four zeros?

▮ Figure 4.33. Notice that the graph crosses the x axis four times. Each zero must be odd for this to happen.

4.92   Which figures exhibit zeros with even multiplicity?

▮ Only Fig. 4.29. The graph touches the x axis at 0 without crossing it.

4.93   Which figure does not represent $y = f(x)$?

▮ In Fig. 4.32, the graph is not a function of x. If you "plug in" an x, you get two y values.

4.94   In which figures are there two intervals in which the function is positive?

▮ In Fig. 4.31, $f > 0$ when x is in the intervals $(-1, \frac{1}{2})$ and $(1, \infty)$. In Fig. 4.33, $f > 0$ when x is in the intervals $(-2, -1)$ and $(0, 3)$.

4.95   Which figure is symmetric about $(0, 0)$?

▮ None; there is no figure in the collection where, for all (x, y) on the graph, $(-x, -y)$ is on the graph.

4.96   Which figures must represent a function of the form $P(x) = ax^n$, where n is an even positive integer and $a > 0$?

▮ The graph in Fig. 4.29. If n were odd, the graph would cross the x axis at $(0, 0)$.

4.3 SYNTHETIC DIVISION

For Probs. 4.97 to 4.122, use synthetic division to find the quotient and remainder. The method is fully described in Prob. 4.97.

4.97 $(x^3 + x^2 + x + 3) \div (x - 1)$

$$
\begin{array}{rrrr|r}
1 & 1 & 1 & 3 & \underline{1} \\
  & 1 & 2 & 3 & \\
\hline
1 & +2 & +3 & +6 &
\end{array}
$$

$\underbrace{\qquad\qquad}_{\text{Quotient}}\quad\underbrace{\quad}_{\text{remainder}}$

The quotient is $1x^2 + 2x + 3 = x^2 + 2x + 3$, and the remainder is 6.

Outline of method:

(a) Write down the coefficients of the dividend from left to right in decreasing order of the powers of x. Insert 0 for any missing terms.

$$1 \quad 1 \quad 1 \quad 3$$

(b) To the right of this, write the a of the divisor $x - a$.

$$1 \quad 1 \quad 1 \quad 3 \ \underline{1}$$

(c) On the *third* line (skip one line) rewrite the leading coefficient.

$$1 \quad 1 \quad 1 \quad 3 \ \underline{1}$$

$$\overline{}$$
$$1$$

(d) Multiply this leading coefficient by a, and write it on the second line in the second column as shown. Then add the terms in column 2, and write the sum on the third line as shown.

$$
\begin{array}{rrrr|r}
1 & 1 & 1 & 3 & \underline{1} \\
  & 1 &   &   & \\
\hline
1 & 2 &   &   &
\end{array}
$$

(e) Multiply the sum obtained in (d) by a, and repeat the procedure in (d) for each column.

$$
\begin{array}{rrrrr|r}
1 & 1 & 1 & & 3 & \underline{1} \\
  & 1 & 2 & & 3 & \\
\hline
1 & 2 & 3 & & 6 &
\end{array}
$$

$\underbrace{\qquad\qquad}_{\text{Coefficients of quotient}}\quad\underbrace{\quad}_{\text{remainder}}$

The quotient is $(1)x^2 + (2)x + 3 = x^2 + 2x + 3$, and the remainder is 6.

4.98 $(x^3 - x - 1) \div (x - 1)$

$$
\begin{array}{rrrr|r}
1 & 0 & -1 & -1 & \underline{1} \\
  & 1 & 1 & 0 & \\
\hline
1 & 1 & 0 & -1 &
\end{array}
$$

The quotient is $x^2 + x$, and the remainder is -1.

4.99 $(x^3 + x^2 + x - 1) \div (x - 1)$

$$
\begin{array}{rrrr|r}
1 & 1 & 1 & -1 & \underline{-1} \\
  & -1 & 0 & -1 & \\
\hline
1 & 0 & 1 & -2 &
\end{array}
$$

The quotient is $x^2 + 1$, and the remainder is -2.

4.100 $(x^3 + x^2 + x + 1) \div (x - 1)$

$$
\begin{array}{rrrr|r}
1 & 1 & 1 & 1 & \underline{1} \\
  & 1 & 2 & 3 & \\
\hline
1 & 2 & 3 & 4 &
\end{array}
$$

The quotient is $x^2 + 2x + 3$, and the remainder is 4.

4.101 $(2x^3 + x^2 + x + 1) \div (x + 1)$

$$\begin{array}{rrrr|r} 2 & 1 & 1 & 1 & \underline{-1} \\ & -2 & 1 & -2 & \\ \hline 2 & -1 & 2 & -1 & \end{array}$$

The quotient is $2x^2 - x + 2$, and the remainder is -1.

4.102 $(2x^3 + 2x^2 + x + 1) \div (x + 1)$

$$\begin{array}{rrrr|r} 2 & 2 & 1 & 1 & \underline{-1} \\ & -2 & 0 & -1 & \\ \hline 2 & 0 & 1 & 0 & \end{array}$$

The quotient is $2x^2 + 1$, and the remainder is 0. (What does that mean?)

4.103 $(2x^3 + 2x^2 + 2x + 1) \div (x + 1)$

$$\begin{array}{rrrr|r} 2 & 2 & 2 & 1 & \underline{-1} \\ & -2 & 0 & -2 & \\ \hline 2 & 0 & 2 & -1 & \end{array}$$

The quotient is $2x^2 + 2$, and the remainder is -1.

4.104 $(2x^3 + 2x^2 + 2x + 2) \div (x + 1)$

$$\begin{array}{rrrr|r} 2 & 2 & 2 & 2 & \underline{-1} \\ & -2 & 0 & -2 & \\ \hline 2 & 0 & 2 & 0 & \end{array}$$

The quotient is $2x^2 + 2$, and the remainder is 0.

4.105 $(2x^3 + 2x^2 + 2x + 2) \div (x + 2)$

$$\begin{array}{rrrr|r} 2 & 2 & 2 & 2 & \underline{-2} \\ & -4 & 4 & -12 & \\ \hline 2 & -2 & 6 & -10 & \end{array}$$

The quotient is $2x^2 - 2x + 6$, and the remainder is -10.

4.106 $(2x^3 + 2x^2 + 2x + 2) \div (x + 3)$

$$\begin{array}{rrrr|r} 2 & 2 & 2 & 2 & \underline{-3} \\ & -6 & 12 & -42 & \\ \hline 2 & -4 & 14 & -40 & \end{array}$$

The quotient is $2x^2 - 4x + 14$, and the remainder is -40.

4.107 $(2x^3 + 2x^2 + 2x + 2) \div (x - 3)$

$$\begin{array}{rrrr|r} 2 & 2 & 2 & 2 & \underline{3} \\ & 6 & 24 & 78 & \\ \hline 2 & 8 & 26 & 80 & \end{array}$$

The quotient is $2x^2 + 8x + 26$, and the remainder is 80.

4.108 $(5x^3 - 2x^2 - 2x - 2) \div (x + 1)$

|

$$
\begin{array}{rrrr|r}
5 & -2 & -2 & -2 & \underline{-1} \\
 & -5 & 7 & -5 & \\
\hline
5 & -7 & 5 & -7 &
\end{array}
$$

The quotient is $5x^2 - 7x + 5$, and the remainder is -7.

4.109 $(5x^3 - 2x^2 - 2x - 2) \div (x - 1)$

|

$$
\begin{array}{rrrr|r}
5 & -2 & -2 & -2 & \underline{1} \\
 & 5 & 3 & 1 & \\
\hline
5 & 3 & 1 & -1 &
\end{array}
$$

The quotient is $5x^2 = 3x + 1$, and the remainder is -1. ·

4.110 $(5x^3 - 2x^2 - 2x - 2) \div (x + 3)$

|

$$
\begin{array}{rrrr|r}
5 & -2 & -2 & -2 & \underline{-3} \\
 & -15 & 51 & -147 & \\
\hline
5 & -17 & 49 & -149 &
\end{array}
$$

The quotient is $5x^2 - 17x + 49$, and the remainder is -149.

4.111 $(5x^3 - 2x^2 - 2x - 2) \div (x + 2)$

|

$$
\begin{array}{rrrr|r}
5 & -2 & -2 & -2 & \underline{-2} \\
 & -10 & 24 & -44 & \\
\hline
5 & -12 & 22 & -46 &
\end{array}
$$

The quotient is $5x^2 - 12x + 22$, and the remainder is -46.

4.112 $(5x^3 - 2x^2 - 2x - 2) \div (x - 2)$

|

$$
\begin{array}{rrrr|r}
5 & -2 & -2 & -2 & \underline{2} \\
 & 10 & 16 & 28 & \\
\hline
5 & 8 & 14 & 26 &
\end{array}
$$

The quotient is $5x^2 + 8x + 14$, and the remainder is 26.

4.113 $(5x^3 - 3x^2 - 2x - 2) \div (x + 2)$

|

$$
\begin{array}{rrrr|r}
5 & -3 & -2 & -2 & \underline{-2} \\
 & -10 & 26 & -48 & \\
\hline
5 & -13 & 24 & -50 &
\end{array}
$$

The quotient is $5x^2 - 13x + 24$, and the remainder is -50.

4.114 $(x^4 + x^3 + x^2 + x + 1) \div (x - 1)$

|

$$
\begin{array}{rrrrr|r}
1 & 1 & 1 & 1 & 1 & \underline{1} \\
 & 1 & 2 & 3 & 4 & \\
\hline
1 & 2 & 3 & 4 & 5 &
\end{array}
$$

The quotient is $x^3 + 2x^2 + 3x + 4$, and the remainder is 5.

4.115 $(x^4 + x^3 + x^2 + x + 1) \div (x + 1)$

$$
\begin{array}{r}
1 \quad 1 \ 1 \quad 1 \ 1 \ \underline{|-1} \\
-1 \ 0 \ -1 \ 0 \\
\hline
1 \quad 0 \ 1 \quad 0 \ 1
\end{array}
$$

The quotient is $x^3 + x$, and the remainder is 1.

4.116 $(x^4 + 3x^2 + 2x + 2) \div (x - 1)$

$$
\begin{array}{r}
1 \ 0 \ 3 \ 2 \ 2 \ \underline{|1} \\
1 \ 1 \ 4 \ 6 \\
\hline
1 \ 1 \ 4 \ 6 \ 8
\end{array}
$$

The quotient is $x^3 + x^2 + 4x + 6$, and the remainder is 8.

4.117 $(x^4 + 3x^2 + 2x + 2) \div (x + 1)$

$$
\begin{array}{r}
1 \quad 0 \ 3 \quad 2 \ 2 \ \underline{|-1} \\
-1 \ 1 \ -4 \ 2 \\
\hline
1 \ -1 \ 4 \ -2 \ 4
\end{array}
$$

The quotient is $x^3 - x^2 + 4x - 2$, and the remainder is 4.

4.118 $(x^4 + 2x + 2) \div (x - 1)$

$$
\begin{array}{r}
1 \ 0 \ 0 \ 2 \ 2 \ \underline{|1} \\
1 \ 1 \ 1 \ 3 \\
\hline
1 \ 1 \ 1 \ 3 \ 5
\end{array}
$$

The quotient is $x^3 + x^2 + x + 3$, and the remainder is 5.

4.119 $(x^4 + x + 2) \div (x - 3)$

$$
\begin{array}{r}
1 \ 0 \ 0 \ 1 \quad 2 \ \underline{|3} \\
3 \ 9 \ 27 \ 84 \\
\hline
1 \ 3 \ 9 \ 28 \ 86
\end{array}
$$

The quotient is $x^3 + 3x^2 + 9x + 28$, and the remainder is 86.

4.120 $(x^4 + 1) \div (x + 1)$

$$
\begin{array}{r}
1 \quad 0 \ 0 \quad 0 \ 1 \ \underline{|-1} \\
-1 \ 1 \ -1 \ 1 \\
\hline
1 \ -1 \ 1 \ -1 \ 2
\end{array}
$$

The quotient is $x^3 - x^2 + x - 1$, and the remainder is 2.

4.121 $(x^4 + 1) \div (x - 1)$

$$
\begin{array}{r}
1 \ 0 \ 0 \ 0 \ 1 \ \underline{|1} \\
1 \ 1 \ 1 \ 1 \\
\hline
1 \ 1 \ 1 \ 1 \ 2
\end{array}
$$

The quotient is $x^3 + x^2 + x + 1$, and the remainder is 2.

4.122 $(x^4 - 1) \div (x + 1)$

I Using the method in Probs. 4.155 to 4.16
Using synthetic division

$$2$$

$$\underline{}$$

$$2$$

we find that 1 is not a zero (remainder $\neq 0$).
rational zero, we find $\frac{1}{2}$ and 2 are the rationa

4.162 $x^3 + 3x^2 + x - 2$

I The possible zeros are $\pm 2, \pm 1$.

$$1 \quad$$

$$\underline{}$$

$$1 \quad$$

Thus, -2 is a zero. Checking with synthetic
remainder. Thus, -2 is the only rational zerc

4.163 $t^3 + 1$

I The only possibilities are ± 1.

$$1 \quad$$

$$\underline{}$$

$$1 \quad$$

-1 is the only rational root. Check to see tha

4.164 $x^4 - x^3 - 5x^2 + 3x + 2$

I The possible rational zeros are $\pm 2, \pm 1$.

$$1 \quad -1$$

$$1$$

$$\underline{}$$

$$1 \quad 0$$

1 is a rational root. Checking, you will find th

4.165 $2x^3 - x^2 - 5x - 2$

I Possible rational zeros are $+\frac{1}{2}, \pm 2, \pm 1$. U
zeros.

$$2 \quad -1$$

$$-1$$

$$\underline{}$$

$$2 \quad -2$$

Check the other roots, using synthetic division

4.166 $32u^5 - 1$

I The possibilities are $\pm 1, \pm\frac{1}{2}, \pm\frac{1}{4}, \pm\frac{1}{8}, \pm\frac{1}{16}$.
$32(\frac{1}{2})^5 - 1 = 0$, we find $\frac{1}{2}$ to be the only rationa

For Probs. 4.167 to 4.169, find all roots for the given μ

4.167 $x^3 + 3x^2 + x - 2$

$$
\begin{array}{rrrrr|r}
1 & 0 & 0 & 0 & -1 & \underline{-1} \\
 & -1 & 1 & -1 & 1 & \\
\hline
1 & -1 & 1 & -1 & 0 &
\end{array}
$$

The quotient is $x^3 - x^2 + x - 1$, and the remainder is 0.

For Probs. 4.123 to 4.126, evaluate by synthetic division.

4.123 $f(-2)$ if $f(x) = x^3 - 7x^2 + 12x - 3$

I We will divide by $x + 2$; the remainder will be $f(-2)$.

$$
\begin{array}{rrrr|r}
1 & -7 & 12 & -3 & \underline{-2} \\
 & -2 & 18 & -60 & \\
\hline
1 & -9 & 30 & -63 &
\end{array}
$$

The remainder is $-63 = f(-2)$.

4.124 $f(3)$ if $f(x) = x^3 - 7x^2 + 12x - 3$

I Divide by $x - 3$

$$
\begin{array}{rrrr|r}
1 & -7 & 12 & -3 & \underline{3} \\
 & 3 & -12 & 0 & \\
\hline
1 & -4 & 0 & -3 &
\end{array}
$$

The remainder is $-3 = f(3)$.

4.125 $f(5)$ if $f(x) = x^3 - 4x^2 - 20x + 55$

I

$$
\begin{array}{rrrr|r}
1 & -4 & -20 & 55 & \underline{5} \\
 & 5 & 5 & -75 & \\
\hline
1 & 1 & -15 & -20 &
\end{array}
$$

The remainder is $-20 = f(5)$.

4.126 $f(\frac{1}{2})$ if $f(x) = x^6 + 1$

I

$$
\begin{array}{rrrrrrr|r}
1 & 0 & 0 & 0 & 0 & 0 & 1 & \underline{\frac{1}{2}} \\
 & \frac{1}{2} & \frac{1}{4} & \frac{1}{8} & \frac{1}{16} & \frac{1}{32} & \frac{1}{64} & \\
\hline
1 & \frac{1}{2} & \frac{1}{4} & \frac{1}{8} & \frac{1}{16} & \frac{1}{32} & \frac{65}{64} &
\end{array}
$$

The remainder is $f(\frac{1}{2}) = \frac{65}{64}$.

For Probs. 4.127 to 4.129, use synthetic division to show that the first polynomial is a factor of the second.

4.127 $x - 3, \ x^3 - 18x + 27$

I We will divide and show that we obtain a zero remainder.

$$
\begin{array}{rrrr|r}
1 & 0 & -18 & 27 & \underline{3} \\
 & 3 & 9 & -27 & \\
\hline
1 & 3 & -9 & 0 &
\end{array}
$$

The remainder is 0, so $x - 3$ is a factor.

4.128 $x + 1, \ x^4 + 2x^3 + 2x^2 + 2x + 1$

▮ If 1 is a zero, then $x - 1$ is a factor. Using

$$1 \quad -3$$

$$1 \quad -2$$

$x^2 - 2x - 1$ is a factor; but if $x^2 - 2x - 1 = 0$,
the remaining zeros are $1 + \sqrt{2}$ and $1 - \sqrt{2}$.

4.152 $x^3 - 3x^2 + 2, \; 1 + \sqrt{3}$

▮ Then $1 - \sqrt{3}$ must be a zero, so $x - (1 - \sqrt{3})$
$[x - (1 + \sqrt{3})][x - (1 - \sqrt{3})] = x^2 - 2x - 2$. Si
are the other zeros.

4.153 $x^4 + 5x^2 + 4, \; i$

▮ If i is a zero, then $-i$ is a zero as well. $(x -$
$x^2 + 1$ and obtain $x^2 + 2$. When $x^2 + 2 = 0$, x^2

4.154 $x^4 - x^3 + 6x^2 - 26x + 20, \; -1 - 3i$

▮ If $-1 - 3i$ is a zero, then $-1 + 3i$ is as well
$x^2 + 2x + 10$. We divide $x^4 - x^3 + 6x^2 - 26x +$
$(x - 2)(x - 1)$. Thus, the other zeros are 2, 1,

For Probs. 4.155 to 4.160, find all possible candidates
4.155 for a description of the method.

4.155 $2x^2 - 5x + 2$

▮ If p/q (a rational number in lowest terms)
constant term. In this case, any zero p/q mus
$q \mid 2$ (q divides the coefficient of x^2). Thus, p

4.156 $3x^2 - 7$

▮ If p/q is a zero, then $p \mid 7$ and $q \mid 3$. Thus,

4.157 $x^2 + 2x + 4$

▮ If p/q is a zero, then $p \mid 4$ and $q \mid 1$. Thus,

4.158 $3x^3 + x^2 + 2x + 6$

▮ If p/q is a zero, then $p \mid 6$ and $q \mid 3$. Thus,
$\pm 2, \; \pm\frac{2}{3}, \; \pm 3, \; \pm 6$.

4.159 $x^5 - 4$

▮ If p/q is a zero, then $p \mid 4$ and $q \mid 1$. Thus,

4.160 $3u^4 + u^2 - 15$

▮ If p/q is a zero, then $p \mid 15$ and $q \mid 3$. Thu
$\pm\frac{1}{3}, \; \pm 3, \; \pm 5, \; \pm\frac{5}{3}, \; \pm 15$.

For Probs. 4.161 to 4.166, find the rational zeros of th

4.161 $2x^2 - 5x + 2$

$$
\begin{array}{rrrr|r}
1 & -2 & 0 & 3 & \underline{-1} \\
  & -1 & 3 & -3 & \\
\hline
1 & -3 & 3 & 0 &
\end{array}
\qquad
\begin{array}{rrrr|r}
1 & -2 & 0 & 3 & \underline{-2} \\
  & -2 & 8 & -16 & \\
\hline
1 & -4 & 8 & -13 & \text{(alternates)}
\end{array}
$$

-2 is a lower bound.

4.179 $M(x) = x^4 - x^2 + 3x + 2$

▮

$$
\begin{array}{rrrrr|r}
1 & 0 & -1 & 3 & 2 & \underline{1} \\
  & 1 & 1 & 0 & 3 & \\
\hline
1 & 1 & 0 & 3 & 5 & \text{(nonnegative)}
\end{array}
$$

1 is an upper bound.

$$
\begin{array}{rrrrr|r}
1 & 0 & -1 & 3 & 2 & \underline{-1} \\
  & -1 & 1 & 0 & -3 & \\
\hline
1 & -1 & 0 & 3 & -1 &
\end{array}
\qquad
\begin{array}{rrrrr|r}
1 & 0 & -1 & 3 & 2 & \underline{-2} \\
  & -2 & 4 & -6 & 6 & \\
\hline
1 & -2 & 3 & -3 & 8 & \text{(alternates)}
\end{array}
$$

-2 is a lower bound.

4.180 $P(x) = x^3 + 7x^2 - x + 2$

▮

$$
\begin{array}{rrrr|r}
1 & 7 & -1 & 2 & \underline{1} \\
  & 1 & 8 & 7 & \\
\hline
1 & 8 & 7 & 9 & \text{(nonnegative)}
\end{array}
$$

1 is an upper bound.

$$
\begin{array}{rrrr|r}
1 & 7 & -1 & 2 & \underline{-1} \\
  & -1 & -6 & 7 & \\
\hline
1 & 6 & -7 & 9 &
\end{array}
\quad
\begin{array}{rrrr|r}
1 & 7 & -1 & 2 & \underline{-2} \\
  & -2 & -10 & 22 & \\
\hline
1 & 5 & -11 & 24 &
\end{array}
\quad
\begin{array}{rrrr|r}
1 & 7 & -1 & 2 & \underline{-8} \\
  & -8 & 8 & -56 & \\
\hline
1 & -1 & 7 & -54 &
\end{array}
$$

-8 is a lower bound. (Do you see why we jumped from dividing by -2 to dividing by -8?)

For Probs. 4.181 to 4.183, show that there is at least one real zero between the given values of a and b.

4.181 $P(x) = x^2 - 3x - 2; \; a = 3, \; b = 4$

▮ $P(3) = 3^2 - 3(3) - 2 = -2$, and $P(4) = 4^2 - 3(4) - 2 = 2$; since $P(3)$ and $P(4)$ are opposite in sign, there must be at least 1 real zero between them.

4.182 $P(x) = x^3 - 3x + 5; \; a = -3, \; b = -2$

▮ $P(-3) = -13$, and $P(-2) = 3$. Since $P(-3)$ and $P(-2)$ are opposite in sign, there must be a real root between them.

4.183 $Q(x) = x^3 - 3x^2 - 3x + 9; \; a = 1, \; b = 2$

▮ $Q(a) = Q(1) = 4$, and $Q(b) = Q(2) = -1$. 4 and -1 are opposite in sign, so there is a real root between them.

For Probs. 4.184 to 4.187, write an equation whose roots are the negatives of those of $f(x) = 0$.

4.184 $x^3 - 8x^2 + x - 1 = 0$

▮ We change the signs of alternating terms, beginning with the second: $x^3 + 8x^2 + x + 1 = 0$.

4.185 $x^4 + 3x^2 + 2x + 1 = 0$

▮ $x^4 + 0x^3 + 3x^2 + 2x + 1 = 0$ is changed to $x^4 - 0x^3 + 3x^2 - 2x + 1 = 0$, or $x^4 + 3x^2 - 2x + 1 = 0$. Be careful with equations that are not in standard form.

4.186 $2x^4 - 5x^2 + 8x - 3 = 0$

┃ This is $2x^4 + 0x^3 - 5x^2 + 8x - 3 = 0$; change it to $2x^4 - 0x^3 - 5x^2 - 8x - 3 = 0$, or $2x^4 - 5x^2 - 8x - 3 = 0$.

4.187 $x^5 + x + 2 = 0$

┃ This is $x^5 + 0x^4 + 0x^3 + 0x^2 + x + 2 = 0$; we change it to $x^5 - 0x^4 + 0x^3 - 0x^2 + x - 2 = 0$, or $x^5 + x - 2 = 0$.

4.5 RATIONAL AND ALGEBRAIC FUNCTIONS

For Probs. 4.188 to 4.198, tell whether the given function is rational and why.

4.188 $f(x) = cx + d$

┃ This is rational, since $f(x) = cx + d = \dfrac{cx + d}{1}$, where $cx + d$ and 1 are polynomials.

4.189 $f(x) = \dfrac{(x-2)^{3/2}}{x-2}$

┃ This is nonrational. The numerator cannot be written as a polynomial. If we divide, we find $f(x) = \sqrt{x-2}$.

4.190 $g(x) = |x|$

┃ This is nonrational. $|x|$ is not a polynomial. Look at Fig. 4.35 which is the graph of $g(x)$. Do you notice the sharp peak? You will learn in calculus that this ensures that the function is nonrational.

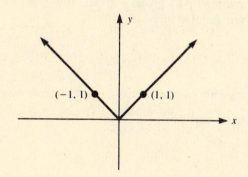

Fig. 4.35

4.191 $f(x) = \dfrac{1}{\pi x}$

┃ 1 and πx are both polynomials. This is a rational function. Do not be misled by the irrational coefficient of x.

4.192 $h(x) = 3^{x^2} + 2$

┃ This is nonrational. The power to which 3 is raised is variable.

4.193 $y = \dfrac{x^2 + 3x - 7}{x^2 + 5x - 6}$

┃ The numerator and denominator are each a polynomial of degree 2. This is a rational function.

4.194 $f(x) = (x + 2)^x$

▮ This is nonrational. The exponent x is variable.

4.195 $\quad y = \dfrac{x-1}{x}$

▮ This function is rational. The numerator and denominator are each a polynomial of degree 1.

4.196 $\quad f(x) = \sqrt{x-2}$

▮ $\sqrt{x-2} = (x-2)^{1/2}$ which is not a polynomial (the exponent is nonintegral).

4.197 $\quad h(x) = \dfrac{(x+1)^2}{2x+i}$

▮ This is rational. The numerator and denominator are each polynomials.

4.198 $\quad f(x) = \dfrac{x^n}{3^x}$

▮ This is nonrational. The denominator is not a polynomial.

For Probs. 4.199 to 4.207, find all vertical and horizontal asymptotes of the graph of the given function. Do *not* graph the function.

4.199 $\quad f(x) = \dfrac{x}{x+1}$

▮ Vertical: Since $f(x) = \dfrac{x}{x+1}$, as $x+1 \to 0$, f grows without bound. Thus, $x = -1$ is a vertical asymptote. Horizontal: As x grows without bound, $f(x) = \dfrac{x}{x+1}$ gets nearer and nearer to 1. $y = 1$ is a horizontal asymptote.

4.200 $\quad f(x) = \dfrac{x+1}{x}$

▮ As $x \to 0$, $\dfrac{x+1}{x}$ grows without bound. $x = 0$ is a vertical asymptote. As $x \to \infty$, $\dfrac{x+1}{x}$ grows closer to 1. $y = 1$ is a horizontal asymptote.

4.201 $\quad f(x) = \dfrac{x^2}{x^2-4}$

▮ As $x^2 - 4 \to 0$, $\dfrac{x^2}{x^2-4} \to \infty$, which implies that $x^2 - 4 = 0$ will give us vertical asymptotes. $x = 2$, $x = -2$ are vertical asymptotes. As $x \to \infty$, $\dfrac{x^2}{x^2-4} \to 1$. $y = 1$ is a horizontal asymptote.

4.202 $\quad f(x) = \dfrac{2x^2}{3x^2+4}$

▮ Since $3x^2 + 4$ does not approach 0 for any choices of x, there is no vertical asymptote. As $x \to \infty$, $f(x) \to 1$. $y = 1$ is a horizontal asymptote. ˙

4.203 $\quad f(x) = \dfrac{x}{(x-1)(x+2)}$

▮ As $x - 1 \to 0$ or $x + 2 \to 0$, $f(x) \to \infty$. Thus, $x = 1$ and $x = -2$ are vertical asymptotes. As $x \to \infty$, $\dfrac{x}{(x-1)(x+2)} \to 0$. $y = 0$ is a horizontal asymptote.

4.204 $f(x) = \dfrac{x^2 + 2}{x^4 + x}$

▮ $x^4 + x = x(x^3 + 1)$, and $x(x^3 + 1) \to 0$ if $x \to 0$ or $x^3 + 1 \to 0$. Thus, $x = 0$ and $x = -1$ are vertical asymptotes. As $x \to \infty$, $f(x) \to 0$. $y = 0$ is a horizontal asymptote.

4.205 $f(x) = \dfrac{x + 3}{(x - 2)(x + 1)(x + 2)}$

▮ If $x \to 2$, $x \to -1$, or $x \to -2$, the denominator approaches 0. Thus, $x = 2$, $x = -1$, $x = -2$ are vertical asymptotes. As $x \to \infty$, $f(x) \to 0$, and $y = 0$ is a horizontal asymptote.

4.206 $f(x) = \dfrac{1}{(1 - x)^3}$

▮ As $x \to 1$, $(1 - x)^3 \to 0$ and $f \to \infty$. Thus, $x = 1$ is a vertical asymptote. As $x \to \infty$, $f(x) \to 0$, and $y = 0$ is a horizontal asymptote.

4.207 $f(x) = 1 + 3/4x$

▮ As $x \to 0$, $3/4x \to \infty$. $x = 0$ is a vertical asymptote. As $x \to \infty$, $1 + 3/4x \to 1$. $y = 1$ is a horizontal asymptote.

For Probs. 4.208 to 4.213, sketch the graph of the given rational function.

4.208 $f(x) = 1/x$

▮ If $x = 0$, then we have a vertical asymptote. As $x \to \infty$, $1/x \to 0$, so $y = 0$ is a vertical asymptote. There are no x or y intercepts. $y = 1/x$ shows origin symmetry. We plot a few points: $(1, 1)$, $(-1, -1)$, $(2, \frac{1}{2})$. See Fig. 4.36.

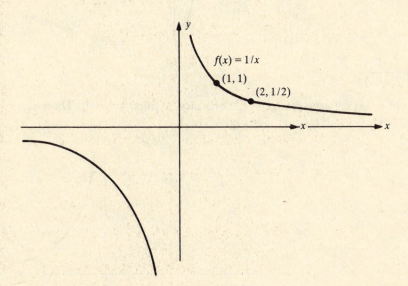

$f(x) = 1/x$
$(1, 1)$
$(2, 1/2)$

Fig. 4.36

4.209 $f(x) = -2/x$

▮ This is similar to Prob. 4.208, but we multiply each ordinate by -2. See Fig. 4.37.

4.210 $f(x) = 2 + 1/x$

▮ If $x \to 0$, $1/x$ grows without bound. $x = 0$ is a vertical asymptote. As $x \to \infty$, $f \to 2$, so $y = 2$ is a horizontal asymptote. There is no symmetry. We plot the points $(1, 3)$ and $(\frac{1}{2}, 4)$. See Fig. 4.38.

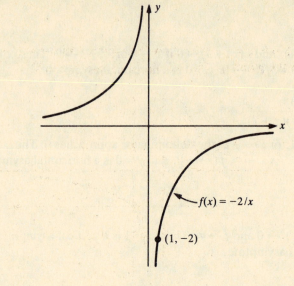

$f(x) = -2/x$

$(1, -2)$

Fig. 4.37

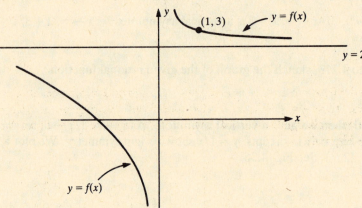

$(1, 3)$ $y = f(x)$

$y = 2$

$y = f(x)$

Fig. 4.38

4.211 $f(x) = \dfrac{1}{2(x - 1)}$

▮ We find vertical asymptote $x = 1$, horizontal asymptote $y = 0$. There is no symmetry. We plot the points $(2, \frac{1}{2})$, $(-2, -\frac{1}{6})$, $(3, \frac{1}{4})$. See Fig. 4.39.

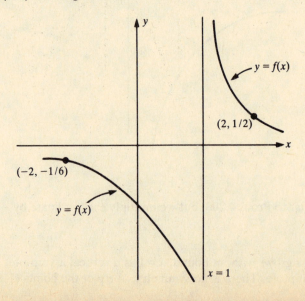

$y = f(x)$

$(2, 1/2)$

$(-2, -1/6)$

$y = f(x)$

$x = 1$

Fig. 4.39

4.212 $f(x) = \dfrac{x}{x+1}$

▮ As $x \to -1$, $x + 1 \to 0$. $x = -1$ is a vertical asymptote. As $x \to \infty$, $f \to 1$, and $y = 1$ is a horizontal asymptote. We plot the points $(1, \frac{1}{2})$, $(-2, 2)$, $(2, \frac{2}{3})$. See Fig. 4.40.

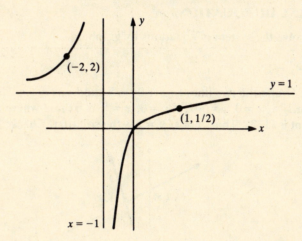

Fig. 4.40

4.213 $f(x) = \dfrac{x^2}{x^2 - 4}$

▮ If $x^2 - 4 = 0$, $x = \pm 2$. As $x \to 2$ or $x \to -2$, f grows without bound; $x = \pm 2$ are vertical asymptotes. As $x \to \infty$, $f \to 1$, so $y = 1$ is a horizontal asymptote. Replacing x by $-x$ leaves the equation line unchanged: the graph exhibits y-axis symmetry. We plot the points $(0, 0)$, $(1, -\frac{1}{3})$, $(3, \frac{9}{5})$. See Fig. 4.41.

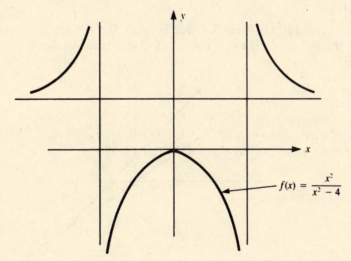

Fig. 4.41

CHAPTER 5
Systems of Equations and Inequalities

5.1 SYSTEMS OF LINEAR EQUATIONS

For Probs. 5.1 to 5.10, solve the system of equations by graphing.

5.1 $x + 2y = 5$, $3x - y = 1$

▮ See Fig. 5.1. Graph the two lines l_1: $x + 2y = 5$ and l_2: $3x - y = 1$. For l_1, if $x = 0$, $y = 2.5$; if $y = 0$, $x = 5$. For l_2, if $x = 0$, $y = -1$; if $y = 0$, $x = \frac{1}{3}$. Next find where these lines intersect. The intersection point is $(1, 2)$. Thus, $x = 1$, $y = 2$ is the solution. *Check*: $1 + 2(2) = 5$, $3(1) - 2 = 1$.

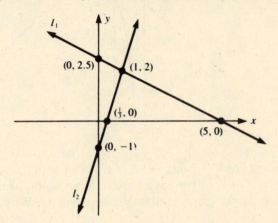

Fig. 5.1

5.2 $x + y = 1$, $2x + 3y = 0$

▮ Follow the procedures in Prob. 5.1. See Fig. 5.2. For $x + y = 1$, if $x = 0$, $y = 1$; if $y = 0$, $x = 1$. For $2x + 3y = 0$, if $x = 0$, $y = 0$; if $x = 3$, $y = -2$. The intersection point is $(3, -2)$, thus, $x = 3$, $y = -2$ is the solution.

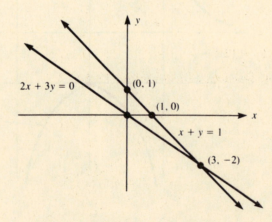

Fig. 5.2

5.3 $2x + 4y = -7$, $x - 3y = 9$

▮ See Fig. 5.3. Let $2x + 4y = -7$ be l_1 and $x - 3y = 9$ be l_2. For l_1, if $x = 0$, $y = -\frac{7}{4}$; if $y = 0$, $x = -\frac{7}{2}$. For l_2, if $x = 0$, $y = -3$; if $y = 0$, $x = 9$. The lines intersect at $(1.5, -2.5)$: $x = 1.5$, $y = -2.5$.

5.4 $x - 2y = 3$, $3x - 6y = 9$

▮ See Fig. 5.4. For l_1 ($x - 2y = 3$), if $x = 0$, $y = -\frac{3}{2}$; if $y = 0$, $x = 3$. For l_2 ($3x - 6y = 9$), if $x = 0$, $y = -\frac{9}{6} = -\frac{3}{2}$; if $y = 0$, $x = 3$. Then l_1 and l_2 coincide, and the system is dependent: Every (x, y) will satisfy the system.

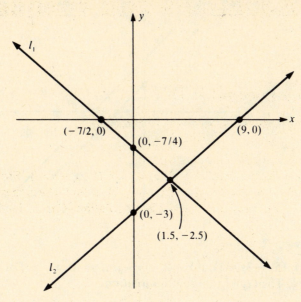

Fig. 5.3

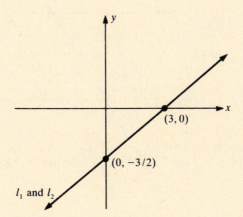

Fig. 5.4

5.5 $4x + 5y = 6$, $8x + 10y = 12$

　▮ See Fig. 5.5. The system is dependent. See Prob. 5.4.

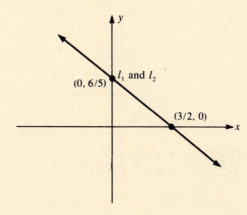

Fig. 5.5

5.6 $2x + y = 2$, $6x + 3y = 5$

　▮ See Fig. 5.6. For l_1 $(2x + y = 2)$, if $x = 0$, $y = 2$; if $y = 0$, $x = 1$. For l_2 $(6x + 3y = 5)$, if $x = 0$, $y = \frac{5}{3}$; if $y = 0$, $x = \frac{5}{6}$. Then $l_1 \parallel l_2$; there is no solution (the system is inconsistent).

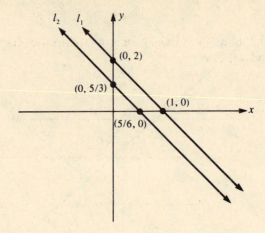

Fig. 5.6

5.7 $5x - 3y = 1$, $10x - 6y = -2$

▌ See Fig. 5.7. For l_1 ($5x - 3y = 1$), if $x = 0$, $y = -\frac{1}{3}$; if $y = 0$, $x = \frac{1}{5}$. For l_2 ($10x - 6y = -2$), if $x = 0$, $y = \frac{1}{3}$; if $y = 0$, $x = -\frac{1}{5}$. Then $l_1 \parallel l_2$; there is no solution.

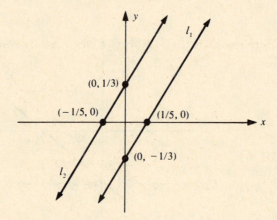

Fig. 5.7

5.8 $x = y$, $x = -y$

▌ See Fig. 5.8. for l_1 ($x = y$) we plot the points $(1, 1)$ and $(2, 2)$. For l_2 ($x = -y$) we plot the points $(1, -1)$ and $(2, -2)$. Then $l_1 \cap l_2 = \{(0, 0)\}$, and $x = 0$, $y = 0$.

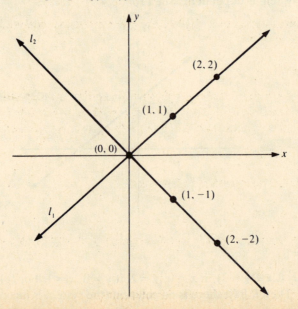

Fig. 5.8

5.9 $x + y = \frac{3}{2}, y = \frac{3}{2}$

  ▌ See Fig. 5.9. Let l_1 be $x + y = \frac{3}{2}$ and l_2 be $y = \frac{3}{2}$. Then l_1 contains the points $(0, \frac{3}{2})$ and $(\frac{3}{2}, 0)$. Then $l_1 \cap l_2 = \{(0, \frac{3}{2})\}$, and $x = 0, y = \frac{3}{2}$.

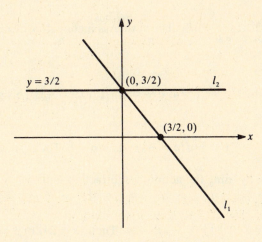

Fig. 5.9

5.10 $x + y = 1, y = |x|$

  ▌ See Fig. 5.10. Here $x = \frac{1}{2}$ and $y = \frac{1}{2}$. Note that, for $x < 0$, $(x + y = 1) \| (y = |x|)$.

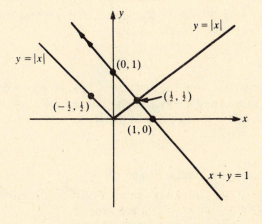

Fig. 5.10

For Probs. 5.11 to 5.18, without solving, tell whether the given system is independent, dependent, or inconsistent.

5.11 $3x + y = 6, 2x + y = 5$

  ▌ Recall that if $ax + by = c$ and $dx + ey = f$, then the system is independent if $d/a \neq e/b$, dependent if $d/a = e/b = f/c$, and inconsistent if $d/a = e/b \neq f/c$. Here $a = 3, b = 1, c = 6, d = 2, e = 1, f = 5$. Then $d/a = \frac{2}{3}$; $e/b = 1/1 = 1$; thus, $d/a \neq e/b$, and the system is independent.

5.12 $3x - y = 4, x + y = 7$

  ▌ $d/a = \frac{3}{1}$; $e/b = -1/1$. The system is independent.

5.13 $2x + y = 10, -4x - 2y = -20$

  ▌ $d/a = 2/-4$; $e/b = 1/-2$; $f/c = 10/-20$; $d/a = e/b = f/c$, and the system is dependent.

5.14 $2x - y = -10, -4x - y = 10$

  ▌ $d/a = 2/-4$, $e/b = -1/-1$. The system is independent.

5.15 $y + x = 5,\ y + x = 8$

▮ $d/a = 1/1;\ e/b = 1/1;\ f/c = \frac{5}{8}$. Then $d/a = e/b \neq f/c$, and the system is inconsistent.

5.16 $y + x = 3,\ y + x = 9$

▮ $d/a = 1/1;\ e/b = 1/1;\ f/c = 3/9$. The system is inconsistent. Compare this to Prob. 5.15. Can you generalize this result?

5.17 $x = 3y + 8,\ 2y = x + 4$

▮ Then $x - 3y = 8$ and $-x + 2y = 4$. $d/a = 1/-1;\ e/b = -\frac{3}{2}$. This system is independent.

5.18 $y = x,\ 2y = x + 6$

▮ $-x + y = 0$ and $-x + 2y = 6$. $d/a = -1/-1;\ e/b = \frac{1}{2}$. The system is independent.

For Probs. 5.19 to 5.30, solve by using the addition/subtraction method.

5.19 $x + y = 10,\ x - y = 14$

▮ Adding, we get $(x + x) + [y + (-y)] = 24$. Then $2x = 24$, or $x = 12$. Replace x by 12 in either equation. Then $12 - y = 14$, or $y = -2$. Thus, $x = 12,\ y = -2$ is the solution. *Check*: $12 + (-2) = 10$, and $12 - (-2) = 14$.

5.20 $2x + y = 20,\ -x - y = 10$

▮ Adding gives $(2x - x) + 0 = 30$, or $x = 30$. Substituting, we find $2(30) + y = 20$, or $y = -40$. *Check*: $2(30) - 40 = 20$, and $-30 - (-40) = 10$.

5.21 $x - \frac{1}{2}y = 6,\ 2x - \frac{1}{2}y = 10$

▮ Subtract the second equation from the first: $(x - 2x) - \frac{1}{2}y + \frac{1}{2}y = 6 - 10$. (Note the $+\frac{1}{2}y$ term. It is positive because we are subtracting a $-\frac{1}{2}y$.) Then $-x = -4$, or $x = 4$. Substituting gives $4 - \frac{1}{2}y = 6$, $\frac{1}{2}y = -2$, or $y = -4$. *Check*: $4 - \frac{1}{2}(-4) = 6$, and $8 - \frac{1}{2}(-4) = 10$.

5.22 $3x + y = 5,\ x + 2y = 5$

▮ Multiply the first equation by 2 and subtract:

$$\begin{array}{r} 6x + 2y = 10 \\ x + 2y = \ 5 \\ \hline 5x \quad\ \ = \ 5 \end{array}$$

Then $x = 1$. Substituting 1 for x in $x + 2y = 5$, we get $1 + 2y = 5$, $2y = 5 - 1 = 4$, or $y = 2$. Check this.

5.23 $4x - y = 10,\ x + 3y = 9$

▮ Multiply the first equation by 3 and add:

$$\begin{array}{r} 12x - 3y = 30 \\ x + 3y = \ 9 \\ \hline 13x \quad\ \ = 39 \end{array}$$

Then $x = 3$. Substituting into $x + 3y = 9$ yields $3 + 3y = 9$, so $3y = 9 - 3 = 6$. Thus, $y = 2$. See Prob. 5.24.

5.24 $4x - y = 10,\ x + 3y = 9$

▮ This time, we multiply the second equation by 4 and subtract:

$$\begin{array}{r} 4x - \quad y = \quad 10 \\ 4x + 12y = \quad 36 \\ \hline -13y = -26 \end{array}$$

Then $y = 2$. Substituting, we get $x + 3(2) = 9$, or $x = 9 - 6 = 3$. Note that we get the same result as in Prob. 5.23. We should, since these are equivalent techniques.

5.25 $2x - 3y = 10$, $3x - 2y = 5$

❚ Multiply the first equation by 3, multiply the second equation by 2, and subtract:

$$\begin{array}{r} 6x - 9y = 30 \\ 6x - 4y = 10 \\ \hline -5y = 20 \end{array}$$

Then $y = -4$. Substituting gives $2x + 12 = 10$, $2x = -2$, or $x = -1$.

5.26 $2x + y = 5$, $4x + 2y = 10$

❚ Multiply the first equation by 2 and subtract:

$$\begin{array}{r} 4x + 2y = 10 \\ 4x + 2y = 10 \\ \hline 0 = 0 \end{array}$$

The system is dependent.

5.27 $3x + y = 20$, $2x + y = 20$

❚ Subtracting, we get $(3x - 2x) + 0 = 0$, or $x = 0$. Then, substituting gives $0 + y = 20$, or $y = 20$.

5.28 $3x + y = 20$, $3x + y = 30$

❚ Subtracting, we get $(3x - 3x) + (y - y) = 20 - 30$, or $0 = -10$. The system is inconsistent.

5.29 $0.4x + 0.8y = 0$, $0.1x + y = 16$

❚ Multiply the second equation by 4 and subtract:

$$\begin{array}{r} 0.4x + 0.8y = 0 \\ 0.4x + 4y = 64 \\ \hline -3.2y = -64 \end{array}$$

Then $y = 20$. Substituting into $0.1x + y = 16$, we get $0.1x + 20 = 16$, $0.1x = 16 - 20 = -4$; or $x = -40$.

5.30 $x/2 + y/3 = 6$, $-x/3 + y/2 = 8$

❚ Multiply both equations by 6: $3x + 2y = 36$ and $-2x + 3y = 48$. Multiply the first equation by 2, multiply the second equation by 3, and add:

$$\begin{array}{r} 6x + 4y = 72 \\ -6x + 9y = 144 \\ \hline 13y = 216 \end{array}$$

Then $y = 16.62$ (rounded). Substituting, we find $3x + 33.24 = 36$, $3x = 2.76$, or $x = 0.92$.

For Probs. 5.31 to 5.41, solve by substitution.

5.31 $x + y = 16$, $x - y = 10$

❚ If $x + y = 16$, then $x = 16 - y$. Substitute $16 - y$ for x in the second equation. Then $(16 - y) - y = 10$. $16 - 2y = 10$, $2y = 6$, or $y = 3$. Substituting for y, we get $x + 3 = 16$, or $x = 16 - 3 = 13$. Check this.

5.32 $x + 2y = 10$, $2x + 3y = 8$

$x = 10 - 2y$. Substitute $10 - 2y$ for x in the second equation. Then $2(10 - 2y) + 3y = 8$, $20 - 4y + 3y = 8$, $-y = -12$, or $y = 12$. We substitute for y: $x = 10 - 2y = 10 - 24 = -14$.

5.33 $-x - y = 15$, $-2x - 4y = 4$

▮ $-x = 15 + y$, or $x = -y - 15$. Substituting $-y - 15$ for x in the second equation gives $-2(-y - 15) - 4y = 4$, $2y + 30 - 4y = 4$, $-2y = -26$, or $y = 13$. Substituting for y, $x = -13 - 15 = -28$. See Prob. 5.34.

5.34 $-x + y = 15$, $-2x + 4y = 4$

▮ Solve the second equation for x: $-2x = 4 - 4y$, or $x = -2 + 2y$. Substituting for x in the first equation, we get $-(-2 + 2y) + y = 15$, $2 - 2y + y = 15$, or $y = -13$. Then $x = -2 + 2(-13)$, $x = -28$.

5.35 $x = 5 - 2y$, $y = 10x - 6$

▮ $y = 10(5 - 2y) - 6$, or $y = 50 - 20y - 6$, $21y = 44$, or $y = \frac{44}{21}$. We substitute $\frac{44}{21}$ for y in the first equation, $x = 5 - 2(\frac{44}{21}) = 5 - \frac{88}{21} = \frac{105}{21} - \frac{88}{21} = \frac{17}{21}$.

5.36 $\frac{x}{2} + y = 6$, $\frac{y}{2} + x = 10$

▮ $y = 6 - \frac{x}{2}$. Substituting $6 - \frac{x}{2}$ for y in the second equation, $\frac{6 - x/2}{3} + x = 10$, $6 - \frac{x}{2} + 3x = 30$, $\frac{5x}{2} = 24$, $5x = 48$, or $x = \frac{48}{5}$. Substituting for x, $y = 6 - \frac{48}{5}/2 = 6 - \frac{48}{10} = \frac{60}{10} - \frac{48}{10} = \frac{12}{10} = \frac{6}{5}$.
See Prob. 5.37.

5.37 $\frac{x}{2} + y = 6$, $\frac{y}{3} + x = 10$

▮ $x/2 = 6 - y$, or $x = 12 - 2y$. Substituting $12 - 2y$ for x in the second equation, we get $y/3 + (12 - 2y) = 10$, $y + 36 - 6y = 30$, $-5y = -6$, or $y = \frac{6}{5}$. Substituting for y gives $x = 12 - 2(\frac{6}{5}) = 12 - \frac{12}{5} = \frac{48}{5}$. See Prob. 5.36.

5.38 $\frac{x}{2} + \frac{y}{3} = 8$, $\frac{x}{3} + \frac{y}{2} = 10$

▮ $x/2 = 8 - y/3$ or $x = 16 - 2y/3$. Substituting $16 - 2y/3$ for x in the second equation, we find $x/3 = \frac{16}{3} - 2y/9$, $\frac{16}{3} - 2y/9 + y/2 = 10$, $\frac{16}{3} + 5y/18 = 10$, $5y/18 = \frac{30}{3} - \frac{16}{3} = \frac{14}{3}$, $15y = 252$, or $y = 16.8$. Substituting for y gives $x = 16 - \frac{2}{3}(16.8) = 4.8$.

5.39 $0.2x + 0.5y = 10$, $x - 0.3y = 15$

▮ $x = 15 + 0.3y$. Substitute for x: $0.2(15 + 0.3y) + 0.5y = 10$, $3.0 + 0.06y + 0.5y = 10$, $0.56y = 7$, or $y = 12.5$. Substituting for y gives $x = 15 + 0.3(12.5) = 18.75$. See Prob. 5.40.

5.40 $0.2x + 0.5y = 10$, $x - 0.3y = 15$

▮ $0.2x = 10 - 0.5y$, $2x = 100 - 5y$, or $x = 50 - 2.5y$. Substituting for x, we get $50 - 2.5y - 0.3y = 15$, $50 - 2.8y = 15$, $2.8y = 35$, or $y = 12.5$. We substitute for y: $x = 50 - 2.5(12.5) = 18.75$. See Prob. 5.39.

5.41 $0.2x - \frac{1}{2}y = 10$, $0.6y - x/3 = 12$

▮ $0.2x = 10 + \frac{1}{2}y$, $2x = 100 + 5y$, or $x = 50 + \frac{5}{2}y$. Substituting for x, we get $0.6y - (50 + \frac{5}{2}y)/3 = 12$, $1.8y - 50 - \frac{5}{2}y = 36$, $-0.7y = 86$, or $y = -86/0.7 = -860/7$. Substituting for y yields $x = 50 + \frac{5}{2}(-\frac{860}{7}) = 50 + (-\frac{4300}{14}) = \frac{700}{14} - \frac{4300}{14} = \frac{3600}{14} = \frac{1800}{7}$.

For Probs. 5.42 to 5.44, solve the given problem by using a system of linear equations.

5.42 The sum of two numbers is 20, and their difference is 10. Find the two numbers.

▮ Let x = one number and y = the other. Then $x + y = 20$ and $x - y = 10$. Thus, $2x = 30$, or $x = 15$, and $y = x - 10 = 5$. *Check*: Sum = 20; difference = 10.

5.43 The length of a rectangle is 10 in more than the width. The perimeter of the rectangle is 60 in. Find the dimensions of the rectangle.

▮ Let l = length and w = width. Then $l = 10 + w$ and $2l + 2w = 60$. Substituting for l gives $2(10 + w) + 2w = 60$, $20 + 2w + 2w = 60$, $4w = 40$, or $w = 10$ in. Then $l = 10$ in + 10 in = 20 in. *Check*: $20 = 10 + 10$ (length is 10 more), and $P = 2(20) + 2(10) = 60$.

5.44 The sum of two numbers is 15. The larger minus thrice the smaller is the negative of the larger. Find the numbers.

▮ Let x = smaller and y = larger. Then $x + y = 15$ and $y - 3x = -y$ or $2y - 3x = 0$. Multiply the first equation by 2 and subtract:

$$\begin{aligned} 2y + 2x &= 30 \\ 2y - 3x &= 0 \\ \hline 5x &= 30 \end{aligned}$$

Then $x = 6$ and $y = 9$.

For Probs. 5.45 to 5.47, solve the given system.

5.45 $x + y + z = 5$, $x - y + z = 10$, $x + y = 3$

▮ Add the first two equations: $2x + 2z = 15$. Add the third and the second: $2x + z = 13$. Then subtract:

$$\begin{aligned} 2x + 2z &= 15 \\ 2x + z &= 13 \\ \hline z &= 2 \end{aligned}$$

Substituting gives $2x + 4 = 15$, or $x = \frac{11}{2}$. Then $y = 3 - x = 3 - \frac{11}{2} = -\frac{5}{2}$.

5.46 $x + y + 2z = 10$, $x - y - 2z = 5$, $x + y + z = 20$

▮ Add the first two equations: $2x = 15$, or $x = \frac{7}{2}$. Subtract the third from the first: $2z - z = -10$, or $z = -10$. Then from $x + y + z = 20$, $y = 20 - x - z = 20 - \frac{7}{2} - (-10) = 30 - \frac{7}{2} = \frac{53}{2}$.

5.47 $x + 2y + z = -5$, $2x - 2y + 2z = 10$, $-x - y - z = 5$

▮ Add the first two equations: $3x + 3z = 5$. Multiply the third by -2, and add to the second:

$$\begin{aligned} 2x - 2y + 2z &= 10 \\ 2x + 2y + 2z &= -10 \\ \hline 4x + 4z &= 0 \end{aligned}$$

Then $4x + 4z = 0$ and $3x + 3z = 5$, so $12x + 12z = 0$ and $12x + 12z = 20$. No solution; inconsistent.

5.48 If the numerator of a fraction is increased by 2, the fraction is $\frac{1}{4}$; if the denominator is decreased by 6, the fraction is $\frac{1}{6}$. Find the fraction.

▮ Let x/y be the original fraction. Then

(1) $$\frac{x + 2}{y} = \frac{1}{4} \qquad \text{or} \qquad 4x - y = -8$$

(2)
$$\frac{x}{y-6} = \frac{1}{6} \quad \text{or} \quad 6x - y = -6$$

Subtract (1) from (2): $2x = 2$ and $x = 1$
Substitute $x = 1$ in (1): $4 - y = -8$ and $y = 12$

The fraction is $\frac{1}{12}$.

5.49   A woman can row downstream 6 mi in 1 h and return in 2 h. Find her rate in still water and the rate of the river.

❚   Let $x =$ rate in still water (mi/h), $y =$ rate of the river (mi/h). Then $x + y =$ rate downstream and $x - y =$ rate upstream.
Now $x + y = 6$ (1)   Add (1) and (2): $2x = 9$ and $x = 4\frac{1}{2}$
      $x - y = \frac{6}{2} = 3$ (2)   Subtract (2) from (1): $2y = 3$ and $y = 1\frac{1}{2}$

The rate in still water is $4\frac{1}{2}$ mi/h, and the rate of the river is $1\frac{1}{2}$ mi/h.

5.50   Solve the system $\begin{cases} x - 5y + 3z = 9 & (1) \\ 2x - y + 4z = 6 & (2) \\ 3x - 2y + z = 2 & (3) \end{cases}$

❚   Eliminate z:

Rewrite (1): $x - 5y + 3z = 9$ Rewrite (2): $2x - y + 4z = 6$
Multiply (3) by -3: $-9x + 6y - 3z = -6$ Multiply (3) by -4: $-12x + 8y - 4z = -8$

Add: $-8x + y = 3$ (4)   Add: $-10x + 7y = -2$ (5)

Multiply (4) by -7: $56x - 7y = -21$ Substitute $x = -\frac{1}{2}$ in (4):
Rewrite (5): $-10x + 7y = -2$ $-8(-\frac{1}{2}) + y = 3$ and $y = -1$

Add: $46x = -23$ Substitute $x = -\frac{1}{2}$, $y = -1$ in (1):
                  $x = -\frac{1}{2}$ $-\frac{1}{2} - 5(-1) + 3z = 9$ and $z = \frac{3}{2}$.

Check: Using (2), $2(-\frac{1}{2}) - (-1) + 4(\frac{3}{2}) = -1 + 1 + 6 = 6$.

5.51   A parabola $y = ax^2 + bx + c$ passes through the points $(1, 0)$, $(2, 2)$, and $(3, 10)$. Determine its equation.

❚   Since $(1, 0)$ is on the parabola: $a + b + c = 0$ (1)
Since $(2, 2)$ is on the parabola: $4a + 2b + c = 2$ (2)
Since $(3, 10)$ is on the parabola: $9a + 3b + c = 10$ (3)
Subtract (1) from (2): $3a + b = 2$ (4)
Subtract (1) from (3): $8a + 2b = 10$ (5)

Multiply (4) by -2 and add to (5): $2a = 6$ and $a = 3$
Substitute $a = 3$ in (4): $3(3) + b = 2$ and $b = -7$
Substitute $a = 3$, $b = -7$ in (1): $3 - 7 + c = 0$ and $c = 4$

The equation of the parabola is $y = 3x^2 - 7x + 4$.

5.2 MATRICES AND DETERMINANTS

For Probs. 5.52 to 5.55, find a, b, c, d.

5.52 $[a \quad b \quad c \quad d] = [2 \quad 3 \quad -1 \quad 0]$

❚   Two matrices are equal if and only if their corresponding elements are equal. Thus, $a = 2$, $b = 3$, $c = -1$, $d = 0$.

5.53 $\begin{bmatrix} a & b \\ c & d \end{bmatrix} = \begin{bmatrix} 1 & 5 \\ -2 & 3 \end{bmatrix}$

▮ Corresponding elements must be equal. Thus, $a = 1$, $b = 5$, $c = -2$, $d = 3$.

5.54 $\begin{bmatrix} a+2 & 3c-1 \\ 2b+3 & d-2 \end{bmatrix} = \begin{bmatrix} 5 & 8 \\ 7 & 0 \end{bmatrix}$

▮ Then $a + 2 = 5$, $a = 3$; $3c - 1 = 8$; $c = 3$; $2b + 3 = 7$, $b = 2$; $d - 2 = 0$, $d = 2$.

5.55 $\begin{bmatrix} a+2b & 2a-b & c+2 \\ a & 3d-4 & b+1 \end{bmatrix} = \begin{bmatrix} 5 & 0 & 6 \\ 1 & 2 & 3 \end{bmatrix}$

▮ $a + 2b = 5$, so $2b = 4$, $b = 2$, since $a = 1$. Since $c + 2 = 6$, $c = 4$; since $3d - 4 = 2$, $d = 2$.

For Probs. 5.56 to 5.70, refer to the following matrices.

$$A = \begin{bmatrix} 1 & -1 \\ 3 & 1 \end{bmatrix} \qquad B = \begin{bmatrix} -3 & 2 \\ -2 & -3 \end{bmatrix} \qquad C = \begin{bmatrix} -2 \\ -3 \\ 1 \end{bmatrix}$$

$$D = \begin{bmatrix} 2 \\ 3 \\ 5 \end{bmatrix} \qquad E = [-4 \quad -1 \quad 0 \quad -2] \qquad F = \begin{bmatrix} -2 & -3 \\ -2 & 0 \\ 1 & -2 \\ 3 & -5 \end{bmatrix}$$

5.56 What are the dimensions of B and of E?

▮ A matrix is $m \times n$ if it has m rows (horizontal) and n columns (vertical). In this case, B has two rows and two columns, so it is 2×2; E has one row and four columns, so it is 1×4.

5.57 What are the dimensions of F and of D?

▮ See Prob. 5.56. F is 4×2; D is 3×1.

5.58 What element is in the second row, first column of F?

▮ $[-2, 0]$ is the second row. Then -2 is in the second-row, first-column position.

5.59 Write a zero matrix of the same dimension as B.

▮ A zero matrix has all entries zero. Thus $\begin{bmatrix} 0 & 0 \\ 0 & 0 \end{bmatrix}$ is the matrix we are looking for.

5.60 How many additional columns would F need to be square?

▮ F is 4×2. If it were 4×4, it would be square. It needs two additional columns.

5.61 Find $A + B$.

▮ $A + B = \begin{bmatrix} 1 & -1 \\ 3 & 1 \end{bmatrix} + \begin{bmatrix} -3 & 2 \\ -2 & -3 \end{bmatrix} = \begin{bmatrix} -2 & 1 \\ 1 & -2 \end{bmatrix}$

5.62 Find $C + D$.

▮ $C + D = \begin{bmatrix} -2 \\ -3 \\ 1 \end{bmatrix} + \begin{bmatrix} 2 \\ 3 \\ 5 \end{bmatrix} = \begin{bmatrix} -2+2 \\ -3+3 \\ 1+5 \end{bmatrix} = \begin{bmatrix} 0 \\ 0 \\ 6 \end{bmatrix}$

5.63 Find the negative of matrix B.

▍ For any matrix B, $-B$ is the matrix such that $B + (-B) =$ zero matrix. All the entries of $-B$ must be the negative of the corresponding entries in B.

$$\begin{bmatrix} -(-3) & -2 \\ -(-2) & -(-3) \end{bmatrix} = \begin{bmatrix} 3 & -2 \\ 2 & 3 \end{bmatrix}$$

Note that $B + (-B) = 0$ ($0 =$ the zero matrix).

5.64 Find $D - C$.

▍ $D - C = D + (-C) = \begin{bmatrix} 2 \\ 3 \\ 5 \end{bmatrix} + \begin{bmatrix} 2 \\ 3 \\ -1 \end{bmatrix} + \begin{bmatrix} 4 \\ 6 \\ 4 \end{bmatrix}$

5.65 Find $A - A$.

▍ $A - A = A + (-A) = \begin{bmatrix} 1 & -1 \\ 3 & 1 \end{bmatrix} + \begin{bmatrix} -1 & 1 \\ -3 & -1 \end{bmatrix} = \begin{bmatrix} 0 & 0 \\ 0 & 0 \end{bmatrix} = 0$

5.66 Find $5B$.

▍ $5B = 5\begin{bmatrix} -3 & 2 \\ -2 & -3 \end{bmatrix} = \begin{bmatrix} 5(-3) & 5(2) \\ 5(-2) & 5(-3) \end{bmatrix} = \begin{bmatrix} -15 & 10 \\ -10 & -15 \end{bmatrix}$

5.67 Find $-2E$.

▍ $-2E = -2[-4 \quad -1 \quad 0 \quad -2] = [8 \quad 2 \quad 0 \quad 4]$.

5.68 Find $2A + B$.

▍ $2A + B = \begin{bmatrix} 2 & -2 \\ 6 & 2 \end{bmatrix} + \begin{bmatrix} -3 & 2 \\ -2 & -3 \end{bmatrix} = \begin{bmatrix} -1 & 0 \\ 4 & -1 \end{bmatrix}$

5.69 Find $B + 2A$.

▍ $B + 2A = \begin{bmatrix} -3 & 2 \\ -2 & -3 \end{bmatrix} + \begin{bmatrix} 2 & -2 \\ 6 & 2 \end{bmatrix} = \begin{bmatrix} -1 & 0 \\ 4 & -1 \end{bmatrix}$. See Prob. 5.68.

5.70 Find $3D - 4C$.

▍ $3D - 4C = 3D + (-4C) = \begin{bmatrix} 6 \\ 9 \\ 15 \end{bmatrix} + \begin{bmatrix} 8 \\ 12 \\ -4 \end{bmatrix} = \begin{bmatrix} 14 \\ 21 \\ 11 \end{bmatrix}$.

For Probs. 5.71 to 5.78, let $A = \begin{bmatrix} 2 & 3 & 1 \\ 0 & -4 & 5 \end{bmatrix}$ and $B = \begin{bmatrix} -5 & 2 & 4 \\ 3 & 0 & -1 \end{bmatrix}$.

5.71 Find $2B$.

▍ $2B = 2\begin{bmatrix} -5 & 2 & 4 \\ 3 & 0 & -1 \end{bmatrix} = \begin{bmatrix} -10 & 4 & 8 \\ 6 & 0 & -2 \end{bmatrix}$

5.72 Find $0b$.

▍ $0B = 0\begin{bmatrix} -5 & 2 & 4 \\ 3 & 0 & -1 \end{bmatrix} = \begin{bmatrix} 0 & 0 & 0 \\ 0 & 0 & 0 \end{bmatrix}$

5.73 Find A^T.

▌ A^T = the transpose of A = that matrix which has rows and columns of A interchanged.

$$A^T = \begin{bmatrix} 2 & 3 & 1 \\ 0 & -4 & 5 \end{bmatrix}^T = \begin{bmatrix} 2 & 0 \\ 3 & -4 \\ 1 & 5 \end{bmatrix}$$

Row 1 of A is column 1 of A^T.

5.74 Find $2B^T$.

▌ $2B^T = 2 \begin{bmatrix} -5 & 3 \\ 2 & 0 \\ 4 & -1 \end{bmatrix}$ (see Prob. 5.73) $= \begin{bmatrix} -10 & 6 \\ 4 & 0 \\ 8 & -2 \end{bmatrix}$.

5.75 Find $3A^T$.

▌ See Prob. 5.73. $A^T = \begin{bmatrix} 2 & 0 \\ 3 & -4 \\ 1 & 5 \end{bmatrix}$. Thus $3A^T = 3\begin{bmatrix} 2 & 0 \\ 3 & -4 \\ 1 & 5 \end{bmatrix} = \begin{bmatrix} 6 & 0 \\ 9 & -12 \\ 3 & 15 \end{bmatrix}$.

5.76 Find $A^T - 2B^T$.

▌ $A^T - 2B^T = A^T + (-2)B^T = \begin{bmatrix} 2 & 0 \\ 3 & -4 \\ 1 & 5 \end{bmatrix} + \begin{bmatrix} 10 & -6 \\ -4 & 0 \\ -8 & 2 \end{bmatrix} = \begin{bmatrix} 12 & -6 \\ -1 & -4 \\ -7 & 7 \end{bmatrix}$.

5.77 Find $A + B^T$.

▌ A is 2×1, and B^T is 1×2. Thus, $A + B^T$ is not defined.

5.78 Find $A^T + B^T$.

▌ $A^T + B^T = \begin{bmatrix} 2 & 0 \\ 3 & -4 \\ 1 & 5 \end{bmatrix} + \begin{bmatrix} -5 & 3 \\ 2 & 0 \\ 4 & -1 \end{bmatrix} = \begin{bmatrix} -3 & 3 \\ 5 & -4 \\ 5 & 4 \end{bmatrix}$

For Probs. 5.79 to 5.82, perform the operation indicated.

5.79 $\begin{bmatrix} 2 & 1 & 5 \end{bmatrix} + 2\begin{bmatrix} 1 & 0 & 5 \end{bmatrix}$

▌ $\begin{bmatrix} 2 & 1 & 5 \end{bmatrix} + 2\begin{bmatrix} 1 & 0 & 5 \end{bmatrix} = \begin{bmatrix} 2 & 1 & 5 \end{bmatrix} + \begin{bmatrix} 2 & 0 & 10 \end{bmatrix} = \begin{bmatrix} 4 & 1 & 15 \end{bmatrix}$.

5.80 $2\begin{bmatrix} 4 & -1 \end{bmatrix} - 5\begin{bmatrix} 9 & -1 \end{bmatrix}$

▌ $2\begin{bmatrix} 4 & -1 \end{bmatrix} - 5\begin{bmatrix} 9 & -1 \end{bmatrix} = \begin{bmatrix} 8 & -2 \end{bmatrix} + (-5)\begin{bmatrix} 9 & -1 \end{bmatrix} = \begin{bmatrix} 8 & -2 \end{bmatrix} + \begin{bmatrix} -45 & 5 \end{bmatrix} = \begin{bmatrix} -37 & 3 \end{bmatrix}$.

5.81 $-\begin{bmatrix} 1 \\ 4 \end{bmatrix} + 2\begin{bmatrix} 0 & 1 \end{bmatrix}^T$

▌ $\begin{bmatrix} 0 & 1 \end{bmatrix}^T = \begin{bmatrix} i \end{bmatrix}$. Thus, $-\begin{bmatrix} 1 \\ 4 \end{bmatrix} + 2\begin{bmatrix} 0 & 1 \end{bmatrix}^T = \begin{bmatrix} -1 \\ -4 \end{bmatrix} + \begin{bmatrix} 0 \\ 2 \end{bmatrix} = \begin{bmatrix} -1 \\ -2 \end{bmatrix}$.

5.82 $\left(-\begin{bmatrix} 1 \\ 4 \end{bmatrix} + 2\begin{bmatrix} 0 & 1 \end{bmatrix}^T \right)^T$

▌ See Prob. 5.81: $-\begin{bmatrix} 1 \\ 4 \end{bmatrix} + 2\begin{bmatrix} 0 & 1 \end{bmatrix}^T = \begin{bmatrix} -1 \\ -2 \end{bmatrix}$; thus, the result is $\begin{bmatrix} -1 \\ -2 \end{bmatrix}^T = \begin{bmatrix} -1 & -2 \end{bmatrix}$.

For Probs. 5.83 to 5.85, solve for x and y, or s and t.

5.83 $[x + 2 \quad 6 + y] = 2\begin{bmatrix} 1 \\ 0 \end{bmatrix}^T$

▮ $2\begin{bmatrix} 1 \\ 0 \end{bmatrix}^T = 2[1 \quad 0] = [2 \quad 0]$. Thus, $x + 2 = 2$, $x = 0$; and $6 + y = 0$, $y = -6$.

5.84 $\begin{bmatrix} 1 & 0 \\ 0 & 1 \end{bmatrix} = -2\begin{bmatrix} s - 3 & 0 \\ 0 & t + 2 \end{bmatrix}$

▮ Then $1 = -2s + 6$, $2s = 5$, or $s = \frac{5}{2}$; and $1 = -2t - 4$, $2t = -5$, or $t = -\frac{5}{2}$.

5.85 $\begin{bmatrix} a & b \\ -x & -y \end{bmatrix} = -2\begin{bmatrix} e & f \\ 11 & 12 \end{bmatrix}$

▮ Then $-x = -2(11)$, or $x = 22$; and $-y = -2(12)$, or $y = 24$. *Question*: What is the relationship between a and e? b and f?

For Probs. 5.86 to 5.90, let $A_i = \begin{bmatrix} a_i & b_i & c_i \\ d_i & e_i & f_i \end{bmatrix}$ where $i = 1, 2, 3$ are any three matrices and $h, k \in \mathcal{R}$.

5.86 Prove $A_1 + A_2 = A_2 + A_1$.

▮ $A_1 + A_2 = \begin{bmatrix} a_1 & b_1 & c_1 \\ d_1 & e_1 & f_1 \end{bmatrix} + \begin{bmatrix} a_2 & b_2 & c_2 \\ d_2 & e_2 & f_2 \end{bmatrix} = \begin{bmatrix} a_1 + a_2 & b_1 + b_2 & c_1 + c_2 \\ d_1 + d_2 & e_1 + e_2 & f_1 + f_2 \end{bmatrix}$

$= \begin{bmatrix} a_2 + a_1 & b_2 + b_1 & c_2 + c_1 \\ d_2 + d_1 & e_2 + e_1 & f_2 + f_1 \end{bmatrix} = A_2 + A_1$

(Note that $a_1 + a_2 = a_2 + a_1$ since $a_i \in \mathcal{R}$.)

5.87 $h(A_1 + A_3) = hA_1 + hA_3$

▮ $h(A_1 + A_2) = h\begin{bmatrix} a_1 + a_3 & b_1 + b_3 & c_1 + c_3 \\ d_1 + d_3 & e_1 + e_3 & f_1 + f_3 \end{bmatrix}$

$= \begin{bmatrix} h(a_1 + a_3) & h(b_1 + b_3) & h(c_1 + c_3) \\ h(d_1 + d_3) & h(e_1 + e_3) & h(f_1 + f_3) \end{bmatrix} = \begin{bmatrix} ha_1 + ha_3 & hb_1 + hb_3 & hc_1 + hc_3 \\ hd_1 + hd_3 & he_1 + he_3 & hf_1 + hf_3 \end{bmatrix}$

$= \begin{bmatrix} ha_1 & hb_1 & hc_1 \\ hd_1 & he_1 & hf_1 \end{bmatrix} + \begin{bmatrix} ha_3 & hb_3 & hc_3 \\ hd_3 & he_3 & hf_3 \end{bmatrix} = hA_1 + hA_3$

5.88 $(h + k)A_1 = hA_1 + kA_1$

▮ $(h + k)A_1 = (h + k)\begin{bmatrix} a_1 & b_1 & c_1 \\ d_1 & e_1 & f_1 \end{bmatrix} = \begin{bmatrix} (h + k)a_1 & (h + k)b_1 & (h + k)c_1 \\ (h + k)d_1 & (h + k)e_1 & (h + k)f_1 \end{bmatrix}$

$= \begin{bmatrix} ha_1 + ka_1 & hb_1 + kb_1 & hc_1 + kc_1 \\ hd_1 + kd_1 & he_1 + ke_1 & hf_1 + kf_1 \end{bmatrix} = hA_1 + kA_1$

5.89 $A_2 + (-A_2) = 0$

▮ $A_2 + (-A_2) = \begin{bmatrix} a_2 & b_2 & c_2 \\ d_2 & e_2 & f_2 \end{bmatrix} + \begin{bmatrix} -a_2 & -b_2 & -c_2 \\ -d_2 & -e_2 & -f_2 \end{bmatrix} = \begin{bmatrix} 0 & 0 & 0 \\ 0 & 0 & 0 \end{bmatrix} = 0.$

5.90 $A_1 + 0 = A_1$

▮ $A_1 + 0 = \begin{bmatrix} a_1 & b_1 & c_1 \\ d_1 & e_1 & f_1 \end{bmatrix} + \begin{bmatrix} 0 & 0 & 0 \\ 0 & 0 & 0 \end{bmatrix} = \begin{bmatrix} a_1 + 0 & b_1 + 0 & c_1 + 0 \\ d_1 + 0 & e_1 + 0 & f_1 + 0 \end{bmatrix} = \begin{bmatrix} a_1 & b_1 & c_1 \\ d_1 & e_1 & f_1 \end{bmatrix} = A_1.$

For Probs. 5.91 to 5.93, find the indicated matrix, where $A = \begin{bmatrix} 1 & 3 \\ 2 & 4 \end{bmatrix}$ and $B = \begin{bmatrix} 1 & 0 \\ 0 & 1 \end{bmatrix}$.

5.91 Find B^T.

▌ $B = \begin{bmatrix} 1 & 0 \\ 0 & 1 \end{bmatrix}$ and $B^T = \begin{bmatrix} 1 & 0 \\ 0 & 1 \end{bmatrix} = B$. The first row of b is the first column of B^T.

5.92 Find $(A^T)^T$.

▌ $A^T = \begin{bmatrix} 1 & 2 \\ 3 & 4 \end{bmatrix}$ and $(A^T)^T = \begin{bmatrix} 1 & 3 \\ 2 & 4 \end{bmatrix} = A$. The first row of A^T is the first column of $(A^T)^T$.

5.93 Find $[(A^T)^T]^T$.

▌ $(A^T)^T = \begin{bmatrix} 1 & 3 \\ 2 & 4 \end{bmatrix}$ (see Prob. 5.92). Then $[(A^T)^T]^T = \begin{bmatrix} 1 & 2 \\ 3 & 4 \end{bmatrix} = A^T$. The first row of $(A^T)^T$ is the first column of $[(A^T)^T]^T$. What would $\{[(A^T)^T]^T\}^T$ be? Can you generalize?

For Probs. 5.94 to 5.96, find the matrix X.

5.94 $X - \begin{bmatrix} 2 & 3 \\ -1 & 4 \end{bmatrix} = \begin{bmatrix} 5 & 8 \\ 2 & 1 \end{bmatrix}$

▌ Let $X = \begin{bmatrix} a & b \\ c & d \end{bmatrix}$. Then $a - 2 = 5$, $a = 7$; $b - 3 = 8$, $b = 11$; $c + 1 = 2$, $c = 1$; $d - 4 = 1$, $d = 5$. Thus, $X = \begin{bmatrix} 7 & 11 \\ 1 & 5 \end{bmatrix}$.

5.95 $X + \begin{bmatrix} 3 & 2 \\ 0 & 1 \end{bmatrix} = 2 \begin{bmatrix} -1 & 3 \\ 2 & 4 \end{bmatrix}$

▌ $\begin{bmatrix} a & b \\ c & d \end{bmatrix} + \begin{bmatrix} 3 & 2 \\ 0 & 1 \end{bmatrix} = \begin{bmatrix} -2 & 6 \\ 4 & 8 \end{bmatrix}$. $a + 3 = -2$, $a = -5$; $b + 2 = 6$, $b = 4$; $c + 0 = 4$, $c = 4$; $d + 1 = 8$, $d = 7$. Thus, $X = \begin{bmatrix} -5 & 4 \\ 4 & 7 \end{bmatrix}$.

5.96 $2X + \begin{bmatrix} 2 & 4 \\ 0 & 6 \end{bmatrix} = 3 \begin{bmatrix} -2 & 0 \\ 2 & 4 \end{bmatrix}$

▌ $\begin{bmatrix} 2a & 2b \\ 2c & 2d \end{bmatrix} + \begin{bmatrix} 2 & 4 \\ 0 & 6 \end{bmatrix} = \begin{bmatrix} -6 & 0 \\ 6 & 12 \end{bmatrix}$. $2a + 2 = -6$, $2a = -8$, $a = -4$; $2b + 4 = 0$, $2b = -4$, $b = -2$; $2c = 6$, $c = 3$; $2d + 6 = 12$, $2d = 6$, $d = 3$. Thus, $X = \begin{bmatrix} -4 & -2 \\ 3 & 3 \end{bmatrix}$.

For Probs. 5.97 to 5.106, find the product indicated.

5.97 $\begin{bmatrix} 1 & 2 \\ 3 & 1 \end{bmatrix} \begin{bmatrix} -1 & 2 \\ -1 & 2 \end{bmatrix}$

▌ This product is defined since the number of columns of $\begin{bmatrix} 1 & 2 \\ 3 & 1 \end{bmatrix}$ equals the number of rows of $\begin{bmatrix} -1 & 2 \\ -1 & 2 \end{bmatrix}$. We then find the product as follows:

$$\begin{bmatrix} 1 & 2 \\ 3 & 1 \end{bmatrix} \begin{bmatrix} -1 & 2 \\ -1 & 2 \end{bmatrix} = \begin{bmatrix} 1(-1) + 2(-1) & 1 \cdot 2 + 2 \cdot 2 \\ 3(-1) + 1(-1) & 3 \cdot 2 + 1 \cdot 2 \end{bmatrix}$$

In general, $(AB)_{m \times n}$ of the matrices $A_{m \times s}$ and $B_{s \times n}$ is the matrix whose ijth element is $a_{i1}b_{1j} + a_{i2}b_{2j} + \cdots + a_{is}b_{sj}$, where i ranges from 1 to m and j ranges from 1 to n. In this case, then, the product is $\begin{bmatrix} -1-2 & 2+4 \\ -3-1 & 6+2 \end{bmatrix} = \begin{bmatrix} -3 & 6 \\ -4 & 8 \end{bmatrix}$.

5.98 $\begin{bmatrix} -2 & 3 \\ 1 & 4 \end{bmatrix} \begin{bmatrix} 0 & 1 \\ 6 & -2 \end{bmatrix}$

▮ The product is $\begin{bmatrix} -2 \cdot 0 + 3 \cdot 6 & -2 \cdot 1 + 3(-2) \\ 1 \cdot 0 + 4 \cdot 6 & 1 \cdot 1 + 4(-2) \end{bmatrix} = \begin{bmatrix} 18 & -8 \\ 24 & -7 \end{bmatrix}$.

5.99 $\begin{bmatrix} 0 & 1 \\ 6 & 2 \end{bmatrix} \begin{bmatrix} 2 & 3 \\ 1 & 4 \end{bmatrix}$

▮ $\begin{bmatrix} 0 \cdot 2 + 1 \cdot 1 & 0 \cdot 3 + 1 \cdot 4 \\ 6 \cdot 2 + 2 \cdot 1 & 6 \cdot 3 + 2 \cdot 4 \end{bmatrix} = \begin{bmatrix} 1 & 4 \\ 14 & 24 \end{bmatrix}$.

5.100 $\begin{bmatrix} 1 \\ 1 \\ 0 \end{bmatrix} [2 \quad 3 \quad 1]$

▮ Let $A = \begin{bmatrix} 1 \\ 1 \\ 0 \end{bmatrix}$ and $B = [2 \quad 3 \quad 1]$. Is AB defined? Yes; the number of columns of A = the number of rows of B = 1. Then $\begin{bmatrix} 1 \\ 1 \\ 0 \end{bmatrix} [2 \quad 3 \quad 1] = \begin{bmatrix} 1 \cdot 2 & 1 \cdot 3 & 1 \cdot 1 \\ 1 \cdot 2 & 1 \cdot 3 & 1 \cdot 1 \\ 0 \cdot 2 & 0 \cdot 3 & 0 \cdot 1 \end{bmatrix} = \begin{bmatrix} 2 & 3 & 1 \\ 2 & 3 & 1 \\ 0 & 0 & 0 \end{bmatrix}$.

5.101 $[1 \quad 5] \begin{bmatrix} 5 & 6 \\ 0 & 1 \end{bmatrix}$

▮ $[1 \quad 5]$ has two columns; $\begin{bmatrix} 5 & 6 \\ 0 & 1 \end{bmatrix}$ has two rows. The product is $[1 \cdot 5 + 5 \cdot 0 \quad 1 \cdot 6 + 5 \cdot 1] =$ $[5 \quad 11]$. Be very careful. $1 \cdot 6 + 5 \cdot 1$ is the first-row, second-column position. Since $[1 \quad 5]$ has only one row, the product has only one row.

5.102 $[2 \quad 1 \quad 3] \begin{bmatrix} 2 & 1 & 0 \\ 0 & 0 & 0 \\ 1 & 0 & 2 \end{bmatrix}$

▮ The product is $[2 \cdot 2 + 1 \cdot 0 + 3 \cdot 1 \quad 2 \cdot 1 + 1 \cdot 0 + 3 \cdot 0 \quad 2 \cdot 0 + 1 \cdot 0 + 3 \cdot 2] = [7 \quad 2 \quad 6]$.

5.103 $\begin{bmatrix} 4 \\ 1 \\ 6 \end{bmatrix} \begin{bmatrix} 1 & 0 & 1 \\ 0 & 2 & 0 \\ 0 & 0 & 1 \end{bmatrix}$

▮ The product is not defined. Do you see why? The number of columns of $\begin{bmatrix} 4 \\ 1 \\ 6 \end{bmatrix} \neq$ the number of rows of $\begin{bmatrix} 1 & 0 & 1 \\ 0 & 2 & 0 \\ 0 & 0 & 1 \end{bmatrix}$. Before going on, compose another example in which the product of two matrices is not defined.

5.104 $\begin{bmatrix} 1 & 2 & 3 \\ 1 & 0 & 1 \end{bmatrix} \begin{bmatrix} 3 & 1 \\ 4 & 2 \\ 6 & 0 \end{bmatrix}$

▮ The product is $\begin{bmatrix} 1 \cdot 3 + 2 \cdot 4 + 3 \cdot 6 & 1 \cdot 1 + 2 \cdot 2 + 3 \cdot 0 \\ 1 \cdot 3 + 0 \cdot 4 + 1 \cdot 6 & 1 \cdot 1 + 0 \cdot 2 + 1 \cdot 0 \end{bmatrix} = \begin{bmatrix} 29 & 5 \\ 9 & 1 \end{bmatrix}$.

5.105 $\begin{bmatrix} -6 & 4 \\ 193 & 8 \end{bmatrix} \begin{bmatrix} 1 & 0 \\ 0 & 1 \end{bmatrix}$

▮ The product is $\begin{bmatrix} -6 \cdot 1 + 4 \cdot 0 & -6 \cdot 0 + 4 \cdot 1 \\ 193 \cdot 1 + 8 \cdot 0 & 193 \cdot 0 + 8 \cdot 1 \end{bmatrix} = \begin{bmatrix} -6 & 4 \\ 193 & 8 \end{bmatrix}$. Notice that $\begin{bmatrix} -6 & 4 \\ 193 & 8 \end{bmatrix}$ was unchanged by multiplication by $\begin{bmatrix} 1 & 0 \\ 0 & 1 \end{bmatrix}$. See Prob. 5.106.

5.106 $\begin{bmatrix} 5 & -2 \\ 1 & 6 \end{bmatrix} \begin{bmatrix} 1 & 0 \\ 0 & 1 \end{bmatrix}$

▮ The product is $\begin{bmatrix} 5 \cdot 1 + (-2) \cdot 0 & 5 \cdot 0 + (-2) \cdot 1 \\ 1 \cdot 1 + 6 \cdot 0 & 1 \cdot 0 + 6 \cdot 1 \end{bmatrix} = \begin{bmatrix} 5 & -2 \\ 1 & 6 \end{bmatrix}$. Do you see that $\begin{bmatrix} 1 & 0 \\ 0 & 1 \end{bmatrix}$ is a good candidate for multiplicative identity?

For Probs. 5.107 to 5.110, perform the indicated operations. The · represents the *dot product*.

5.107 $[1 \quad 1] \cdot \begin{bmatrix} 1 \\ 5 \end{bmatrix}$

▮ Here, $A \cdot B = 1 \cdot 1 + 1 \cdot 5 = 6$.

5.108 $[1 \quad -3] \cdot \begin{bmatrix} 1 \\ -6 \end{bmatrix}$

▮ $A \cdot B = 1 \cdot 1 + (-3)(-6) = 1 + 18 = 19$.

5.109 $[1 \quad 6] \cdot \begin{bmatrix} 0 \\ 1 \end{bmatrix} + [2 \quad 3 \quad 4] \cdot \begin{bmatrix} 1 \\ 2 \\ 3 \end{bmatrix}$

▮ $[1 \quad 6] \cdot \begin{bmatrix} 0 \\ 1 \end{bmatrix} = 1 \cdot 0 + 6 \cdot 1 = 6$; $[2 \quad 3 \quad 4] \cdot \begin{bmatrix} 1 \\ 2 \\ 3 \end{bmatrix} = 2 \cdot 1 + 3 \cdot 2 + 4 \cdot 3 = 2 + 6 + 12 = 20$. So

$[1 \quad 6] \cdot \begin{bmatrix} 0 \\ 1 \end{bmatrix} + [2 \quad 3 \quad 4] \cdot \begin{bmatrix} 1 \\ 2 \\ 3 \end{bmatrix} = 6 + 20 = 26$.

5.110 $[1 \quad 2 \quad 3 \quad \cdots \quad n] \cdot \begin{bmatrix} 1 \\ 2 \\ \vdots \\ n \end{bmatrix}$

▮ $A \cdot B = 1 \cdot 1 + 2 \cdot 2 + \cdots + n \cdot n = n^2 + (n-1)^2 + \cdots + 2^2 + 1^2$. $1^2 + 2^2 + \cdots + n^2 = \dfrac{n(n+1)(2n+1)}{6}$. If you don't recognize that formula, you may leave the answer as $1^2 + 2^2 + \cdots + n^2$.

For Probs. 5.111 to 5.114, verify the given statements, where $A = \begin{bmatrix} 1 & 2 \\ 0 & 1 \end{bmatrix}$, $B = \begin{bmatrix} 1 & 1 \\ 2 & 3 \end{bmatrix}$, $C = \begin{bmatrix} -3 & 1 \\ -1 & 2 \end{bmatrix}$.

5.111 $AB \neq BA$.

▮ $AB = \begin{bmatrix} 1 \cdot 1 + 2 \cdot 2 & 1 \cdot 1 + 2 \cdot 3 \\ 0 \cdot 1 + 1 \cdot 2 & 0 \cdot 2 + 1 \cdot 3 \end{bmatrix} = \begin{bmatrix} 5 & 7 \\ 2 & 3 \end{bmatrix}$

$BA = \begin{bmatrix} 1 \cdot 1 + 1 \cdot 0 & 1 \cdot 2 + 1 \cdot 1 \\ 2 \cdot 1 + 3 \cdot 0 & 2 \cdot 2 + 3 \cdot 1 \end{bmatrix} \neq \begin{bmatrix} 5 & 7 \\ 2 & 3 \end{bmatrix}$

$AB \neq BA$.

5.112 $(AB)C = A(BC)$.

▮ $AB = \begin{bmatrix} 5 & 7 \\ 2 & 3 \end{bmatrix}$ (see Prob. 5.111)

$(AB)C = \begin{bmatrix} 5 & 7 \\ 2 & 3 \end{bmatrix} \begin{bmatrix} -3 & 1 \\ -1 & 2 \end{bmatrix} = \begin{bmatrix} 5(-3) + 7(-1) & 5 \cdot 1 + 7 \cdot 2 \\ 2(-3) + 3(-1) & 2 \cdot 1 + 3 \cdot 2 \end{bmatrix} = \begin{bmatrix} -22 & 19 \\ -9 & 8 \end{bmatrix}$

$BC = \begin{bmatrix} 1 & 1 \\ 2 & 3 \end{bmatrix} \begin{bmatrix} -3 & 1 \\ -1 & 2 \end{bmatrix} = \begin{bmatrix} -4 & 3 \\ -9 & 8 \end{bmatrix}$

$A(BC) = \begin{bmatrix} 1 & 2 \\ 0 & 1 \end{bmatrix} \begin{bmatrix} -4 & 3 \\ -9 & 8 \end{bmatrix} = \begin{bmatrix} -22 & 19 \\ -9 & 8 \end{bmatrix} = (AB)C$

5.113 $A(B + C) = AB + AC$.

▮ $A(B + C) = \begin{bmatrix} 1 & 2 \\ 0 & 1 \end{bmatrix} \begin{bmatrix} -2 & 2 \\ 1 & 5 \end{bmatrix} = \begin{bmatrix} 0 & 12 \\ 1 & 5 \end{bmatrix}$. $AB = \begin{bmatrix} 5 & 7 \\ 2 & 3 \end{bmatrix}$. $AC = \begin{bmatrix} -5 & 5 \\ -1 & 2 \end{bmatrix}$.

$AB + AC = \begin{bmatrix} 0 & 12 \\ 1 & 5 \end{bmatrix} = A(B + C)$.

5.114 $(B + C)A = BA + CA$.

▮ $B + C = \begin{bmatrix} -2 & 2 \\ 1 & 5 \end{bmatrix}$. $(B + C)A = \begin{bmatrix} -2 & 2 \\ 1 & 5 \end{bmatrix} \begin{bmatrix} 1 & 2 \\ 0 & 1 \end{bmatrix} = \begin{bmatrix} -2 & -2 \\ 1 & 7 \end{bmatrix}$. $BA = \begin{bmatrix} 1 & 3 \\ 2 & 7 \end{bmatrix}$ (see Prob. 5.111).

$CA = \begin{bmatrix} -3 & 1 \\ -1 & 2 \end{bmatrix} \begin{bmatrix} 1 & 2 \\ 0 & 1 \end{bmatrix} = \begin{bmatrix} -3 & -5 \\ -1 & 0 \end{bmatrix}$. $BA + CA = \begin{bmatrix} -2 & -2 \\ 1 & 7 \end{bmatrix} = (B + C)A$.

For Probs. 5.115 to 5.118, evaluate the given determinants.

5.115 $\begin{vmatrix} 1 & 6 \\ 5 & 4 \end{vmatrix}$

▮ Remember that $\begin{vmatrix} a_{11} & a_{12} \\ a_{21} & a_{22} \end{vmatrix} = a_{11}a_{22} - a_{21}a_{12}$. In this case $\begin{vmatrix} 1 & 6 \\ 5 & 4 \end{vmatrix} = 1 \cdot 4 - 5 \cdot 6 = 4 - 30 = -26$.

5.116 $\begin{vmatrix} 1 & 4 \\ -3 & 8 \end{vmatrix}$

▮ $\begin{vmatrix} 1 & 4 \\ -3 & 8 \end{vmatrix} = 1 \cdot 8 - (4)(-3) = 8 + 12 = 20$.

5.117 $\begin{vmatrix} 1 & \frac{1}{2} \\ \frac{1}{8} & \frac{1}{4} \end{vmatrix}$

▮ $\begin{vmatrix} 1 & \frac{1}{2} \\ \frac{1}{8} & \frac{1}{4} \end{vmatrix} = 1 \cdot \frac{1}{4} - \frac{1}{2} \cdot \frac{1}{8} = \frac{1}{4} - \frac{1}{16} = \frac{3}{16}$.

5.118 $\begin{vmatrix} 1 & 4 \\ 0.1 & 0.7 \end{vmatrix}$

▮ $\begin{vmatrix} 1 & 4 \\ 0.1 & 0.7 \end{vmatrix} = 1(0.7) - (4)(0.1) = 0.7 - 0.4 = 0.3.$

For Probs. 5.119 to 5.123, evaluate the given determinant by using cofactors.

5.119 $\begin{vmatrix} 3 & -2 & -8 \\ -2 & 0 & 2 \\ 1 & 0 & -4 \end{vmatrix}$

▮ Use column 2 since it has two zero entries.

$\begin{vmatrix} 3 & -2 & -8 \\ -2 & 0 & 2 \\ 1 & 0 & -4 \end{vmatrix} = (-2)(-1)^{1+2}\begin{vmatrix} -2 & 2 \\ 1 & -4 \end{vmatrix} + 0 + 0 = 2\begin{vmatrix} -2 & 2 \\ 1 & -4 \end{vmatrix} = 2(8-2) = 12.$

5.120 $\begin{vmatrix} 1 & 4 & 1 \\ 1 & 1 & -2 \\ 2 & 1 & -1 \end{vmatrix}$

▮ Using row 1, we get $1(-1)^{1+1}\begin{vmatrix} 1 & -2 \\ 1 & -1 \end{vmatrix} + 4(-1)^{1+2}\begin{vmatrix} 1 & -2 \\ 2 & -1 \end{vmatrix} + 1(-1)^{1+3}\begin{vmatrix} 1 & 1 \\ 2 & 1 \end{vmatrix}$

$= \begin{vmatrix} 1 & -2 \\ 1 & -1 \end{vmatrix} - 4\begin{vmatrix} 1 & -2 \\ 2 & -1 \end{vmatrix} + \begin{vmatrix} 1 & 1 \\ 2 & 1 \end{vmatrix} = -12.$

5.121 $\begin{vmatrix} 1 & 0 & 0 \\ -2 & 5 & 3 \\ 5 & -2 & 1 \end{vmatrix}$

▮ Use row 1 (it has two zeros). $\begin{vmatrix} 1 & 0 & 0 \\ -2 & 5 & 3 \\ 5 & -2 & 1 \end{vmatrix} = 1\begin{vmatrix} 5 & 3 \\ -2 & 1 \end{vmatrix} + 0 + 0 = 5 - (-6) = 11.$

5.122 $\begin{vmatrix} 0 & 1 & 5 \\ 3 & -7 & 6 \\ 0 & -2 & 3 \end{vmatrix}$

▮ Use column 1. Then $\begin{vmatrix} 0 & 1 & 5 \\ 3 & -7 & 6 \\ 0 & -2 & 3 \end{vmatrix} = 0 + 3(-1)^{2+1}\begin{vmatrix} 1 & 5 \\ -2 & 3 \end{vmatrix} + 0 = -39.$

5.123 $\begin{vmatrix} 2 & 6 & 1 & 7 \\ 0 & 3 & 0 & 0 \\ 3 & 4 & 2 & 5 \\ 0 & 9 & 0 & 2 \end{vmatrix}$

▮ Use row 2; then the determinant $= 3(-1)^{2+2}\begin{vmatrix} 2 & 1 & 7 \\ 3 & 2 & 5 \\ 0 & 0 & 2 \end{vmatrix}$. (The others are all zero.) Now use row 3

(which has two zeros). $\begin{vmatrix} 2 & 1 & 7 \\ 3 & 2 & 5 \\ 0 & 0 & 2 \end{vmatrix} = 2(-1)^{3+3}\begin{vmatrix} 2 & 1 \\ 3 & 2 \end{vmatrix} = 2 \cdot 1 \cdot 1 = 2; \; 3(-1)^{2+2} \cdot 2 = 6.$

For Probs. 5.124 to 5.127, show that each statement is true. All letters represent real numbers.

5.124 $\begin{vmatrix} a & b \\ ka & kb \end{vmatrix} = 0$

∎ $\begin{vmatrix} a & b \\ ka & kb \end{vmatrix} = a(kb) - b(ka) = abk - abk = 0.$

5.125 $\begin{vmatrix} a & b \\ c & d \end{vmatrix} = - \begin{vmatrix} b & a \\ d & c \end{vmatrix}$

∎ $\begin{vmatrix} a & b \\ c & d \end{vmatrix} = ad - bc$ and $- \begin{vmatrix} b & a \\ d & c \end{vmatrix} = -(bc - ad) = bc + ad = ad - bc = \begin{vmatrix} a & b \\ c & d \end{vmatrix}.$

5.126 $\begin{vmatrix} a & b \\ c & d \end{vmatrix} = \begin{vmatrix} a & c \\ b & d \end{vmatrix}$

∎ $\begin{vmatrix} a & b \\ c & d \end{vmatrix} = ad - bc = ad - cb = \begin{vmatrix} a & c \\ b & d \end{vmatrix}.$

5.127 $\begin{vmatrix} ka & kb \\ c & d \end{vmatrix} = k \begin{vmatrix} a & b \\ c & d \end{vmatrix}$

∎ $\begin{vmatrix} ka & kb \\ c & d \end{vmatrix} = (ka)d - (kb)c = kad - kbc.$ Then $k \begin{vmatrix} a & b \\ c & d \end{vmatrix} = k(ad - bc) = kad - kbc = \begin{vmatrix} ka & kb \\ c & d \end{vmatrix}.$

For Probs. 5.128 to 5.134, state the theorem that can be used to justify the given statement.

5.128 $\begin{vmatrix} 6 & 8 \\ 0 & -1 \end{vmatrix} = 2 \begin{vmatrix} 3 & 4 \\ 0 & -1 \end{vmatrix}$

∎ If every element in a column or row of a determinant is multiplied by a constant (here, row 1 is multiplied by 2), the new determinant is the constant times the original.

5.129 $\begin{vmatrix} 1 & 3 & 6 \\ 0 & 0 & 0 \\ 4 & 9 & 1 \end{vmatrix} = 0$

∎ In a given determinant, if any column or row is all zeros, the determinant is 0. Here, row 2 is all zeros.

5.130 $\begin{vmatrix} 0 & 4 & 6 & 3 & 8 \\ 0 & 1 & 2 & 1 & 0 \\ 0 & 1 & 6 & 18 & 9 \\ 0 & 1 & 4 & 0 & 1 \\ 0 & 0 & 1 & 1 & 0 \end{vmatrix} = 0$

∎ See Prob. 5.129. All entries here in column 1 are zero. Notice how much work is saved by using this theorem.

5.131 $\begin{vmatrix} 4 & 3 \\ 1 & 2 \end{vmatrix} = \begin{vmatrix} 4-3 & 3-6 \\ 1 & 2 \end{vmatrix}$

∎ If a constant multiple of a row or column is added to another row or column, the determinant's value is unchanged. Here we multiplied row 2 by -3 and added it to row 1.

5.132 $\begin{vmatrix} 3 & 2 \\ 5 & 1 \end{vmatrix} = \begin{vmatrix} 3+4 & 2 \\ 5+2 & 1 \end{vmatrix}$

∎ See Prob. 5.131. We multiply column 2 by 2 and add it to column 1. The determinant's value does not change.

5.133
$$\begin{vmatrix} 1 & 2 & 4 & 1 \\ 1 & 3 & 4 & 2 \\ 1 & 2 & 4 & 1 \\ 0 & 1 & 4 & 1 \end{vmatrix} = 0$$

▮ If two rows or columns in a given determinant are identical, then the determinant's value $= 0$. Here row 1 = row 3.

5.134
$$\begin{vmatrix} 2 & 1 & 3 \\ 4 & 6 & 9 \\ 2 & 8 & 1 \end{vmatrix} = - \begin{vmatrix} 4 & 6 & 9 \\ 2 & 1 & 3 \\ 2 & 8 & 1 \end{vmatrix}$$

▮ If two rows or columns are interchanged in a given determinant, the resulting determinant is the negative of the original. Here, rows 1 and 2 were interchanged.

For Probs. 5.135 and 5.136, prove the given statement.

5.135
$$\begin{vmatrix} a & b & c \\ e & f & g \\ q & r & t \end{vmatrix} = - \begin{vmatrix} e & f & g \\ a & b & c \\ q & r & t \end{vmatrix} = \begin{vmatrix} e & g & f \\ a & c & b \\ q & t & r \end{vmatrix}$$

▮ Here A is the first determinant, B is the second, C is the third. $A = -B = C$ since A and B are the same except rows 1 and 2 are interchanged; thus, $A = -B$. B and C have columns 2 and 3 interchanged; thus, $B = -C$ or $-B = C$; $A = -B = C$

5.136
$$\begin{vmatrix} a & b & c \\ c & a & e \\ a & b & c \end{vmatrix} = 0$$

▮ In the given determinant, row 1 = row 3. Thus, the value of the determinant = 0.

5.137 Show that $(2, 5)$ and $(-3, 4)$ satisfy the equation $\begin{vmatrix} x & y & 1 \\ 2 & 5 & 1 \\ -3 & 4 & 1 \end{vmatrix} = 0$.

▮ If $x = 2$, $y = 5$, then we have $\begin{vmatrix} x & y & 1 \\ 2 & 5 & 1 \\ -3 & 4 & 1 \end{vmatrix} = \begin{vmatrix} 2 & 5 & 1 \\ 2 & 5 & 1 \\ -3 & 4 & 1 \end{vmatrix}$ which is 0 since row 1 = row 2.

Similarly, if $x = -3$ and $y = 4$, then row 1 = row 3 and the determinant is zero.

5.138 Show that $\begin{vmatrix} x & y & 1 \\ 2 & 3 & 1 \\ -1 & 2 & 1 \end{vmatrix} = 0$ is a line passing through $(2, 3)$, $(-1, 2)$.

▮ Note that $(2, 3)$ and $(-1, 2)$ satisfy the equation, since we would have equal rows. Next, expanding, we get $-2 \begin{vmatrix} y & 1 \\ 2 & 1 \end{vmatrix} + (-1) \begin{vmatrix} y & 1 \\ 3 & 1 \end{vmatrix} + x \begin{vmatrix} 3 & 1 \\ 2 & 1 \end{vmatrix} = 0$. Then
$-2(y - 2) - 1(y - 3) + x(3 - 2) = 0$, $-2y + 4 - y + 3 + x = 0$, and $-3y + x = -7$. This is a line.

For Probs. 5.139 to 5.141, solve for x.

5.139 $\begin{vmatrix} 2 & 1 \\ -6 & x \end{vmatrix} = 0$

▮ $2x - (-6) = 0$, $2x = -6$, or $x = -3$.

5.140 $\begin{vmatrix} 1 & x \\ 2-x & -3 \end{vmatrix} = 0$

▮ $-3 - x(2-x) = 0$, $3 - 2x + x^2 = 0$, $x^2 - 2x - 3 = 0$, $(x-3)(x+1) = 0$, or $x = 3$, $x = -1$.

5.141 $\begin{vmatrix} x & 1 & 3 \\ 1 & x & 2 \\ 1 & 1 & 2 \end{vmatrix} = 0$

▮ Using cofactors, we get $x \begin{vmatrix} x & 2 \\ 1 & 2 \end{vmatrix} - 1 \begin{vmatrix} 1 & 3 \\ 1 & 2 \end{vmatrix} + 1 \begin{vmatrix} 1 & 3 \\ x & 2 \end{vmatrix} = 0$. Then

$x(2x - 2) - (2 - 3) + (2 - 3x) = 0$, $2x^2 - 2x + 1 + 2 - 3x = 0$, $2x^2 - 5x + 3 = 0$,
$(2x - 3)(x - 1) = 0$, or $x = \frac{3}{2}$, $x = 1$.

For Probs. 5.142 and 5.143, show that the two matrices are inverses of each other.

5.142 $\begin{bmatrix} 3 & -4 \\ -2 & 3 \end{bmatrix} \begin{bmatrix} 3 & 4 \\ 2 & 3 \end{bmatrix}$

▮ The product is $\begin{bmatrix} 3 \cdot 3 + (-4)(2) & 3 \cdot 4 + (-4)(3) \\ (-2)(3) + 3 \cdot 2 & (-2)(4) + 3 \cdot 3 \end{bmatrix} = \begin{bmatrix} 1 & 0 \\ 0 & 1 \end{bmatrix} = I$, the identity matrix. Thus, they
are inverses of each other. If $AB = I$, then $A = B^{-1}$.

5.143 $\begin{bmatrix} 1 & -1 & 1 \\ 0 & 2 & -1 \\ 2 & 3 & 0 \end{bmatrix} \begin{bmatrix} 3 & 3 & -1 \\ -2 & -2 & 1 \\ -4 & -5 & 2 \end{bmatrix}$

▮ The product here is

$\begin{bmatrix} 1 \cdot 3 + (-1)(-2) + (1)(-4) & \cdots \\ 0 \cdot 3 + 2(-2) + (-1)(-4) & \cdots \\ \cdots\cdots\cdots\cdots\cdots\cdots\cdots\cdots\cdots \end{bmatrix} = \begin{bmatrix} 1 & 0 & 0 \\ 0 & 1 & 0 \\ 0 & 0 & 1 \end{bmatrix} = I$. So $A = B^{-1}$.

For Probs. 5.144 to 5.147, a matrix M is given. Find M^{-1}.

5.144 $\begin{bmatrix} 1 & 2 \\ 1 & 3 \end{bmatrix}$

▮ We write the augmented matrix $\begin{bmatrix} 1 & 2 & | & 1 & 0 \\ 1 & 3 & | & 0 & 1 \end{bmatrix}$ and transform it to a matrix of the form:

$\begin{bmatrix} 1 & 0 \\ 0 & 1 \end{bmatrix} S \end{bmatrix}$. When we do this, $S = A^{-1}$, where $A = \begin{bmatrix} 1 & 2 \\ 1 & 3 \end{bmatrix}$. We convert one matrix to the other,
using elementary row operations: (1) interchange two rows, (2) multiply a row by $k \neq 0$,
(3) add kR_i to R_j, where R is a row in the matrix. In this case we have the following:

Add $-R_1$ to R_2 and replace R_2 by $R_2 - R_1$. Then $\begin{bmatrix} 1 & 2 & | & 1 & 0 \\ 1 & 3 & | & 0 & 1 \end{bmatrix} \sim \begin{bmatrix} 1 & 2 & | & 1 & 0 \\ 0 & 1 & | & -1 & 1 \end{bmatrix}$.

Replace R_1 by $R_1 + (-2)R_2$. Then $\begin{bmatrix} 1 & 2 & | & 1 & 0 \\ 0 & 1 & | & -1 & 1 \end{bmatrix} \sim \begin{bmatrix} 1 & 0 & | & 3 & -2 \\ 0 & 1 & | & -1 & 1 \end{bmatrix}$.

Thus, $M^{-1} = \begin{bmatrix} 3 & -2 \\ -1 & 1 \end{bmatrix}$. Check it! $\begin{bmatrix} 3 & -2 \\ -1 & 1 \end{bmatrix} \begin{bmatrix} 1 & 2 \\ 1 & 3 \end{bmatrix} = \begin{bmatrix} 1 & 0 \\ 0 & 1 \end{bmatrix}$.

5.145 $\begin{bmatrix} 1 & 3 \\ 2 & 7 \end{bmatrix}$

▮ Replace R_2 by $R_2 + (-2)R_1$. Then $\begin{bmatrix} 1 & 3 & | & 1 & 0 \\ 2 & 7 & | & 0 & 1 \end{bmatrix} \sim \begin{bmatrix} 1 & 3 & | & 1 & 0 \\ 0 & 1 & | & -2 & 1 \end{bmatrix}$. Then if we replace R_1 by $R_1 + (-3)R_2$, we get $\begin{bmatrix} 1 & 0 & | & 7 & -3 \\ 0 & 1 & | & -2 & 1 \end{bmatrix}$. Thus, $M^{-1} = \begin{bmatrix} 7 & -3 \\ -2 & 1 \end{bmatrix}$.

5.146 $\begin{bmatrix} 1 & -3 & 0 \\ 0 & 3 & 1 \\ 2 & -1 & 2 \end{bmatrix}$

▮ $\begin{bmatrix} 1 & -3 & 0 & | & 1 & 0 & 0 \\ 0 & 3 & 1 & | & 0 & 1 & 0 \\ 2 & -1 & 2 & | & 0 & 0 & 1 \end{bmatrix} \sim \begin{bmatrix} 1 & -3 & 0 & | & 1 & 0 & 0 \\ 0 & 3 & 1 & | & 0 & 1 & 0 \\ 0 & 5 & 2 & | & -2 & 0 & 1 \end{bmatrix}$ (We replaced R_3 by $R_3 - 2R_2$.)

$\sim \begin{bmatrix} 1 & -3 & 0 & | & 1 & 0 & 0 \\ 0 & 1 & \frac{1}{3} & | & 0 & \frac{1}{3} & 0 \\ 0 & 5 & 2 & | & -2 & 0 & 1 \end{bmatrix}$ (We replaced R_2 by $\frac{1}{3}R_2$.)

$\sim \begin{bmatrix} 1 & 0 & 1 & | & 1 & 1 & 0 \\ 0 & 1 & \frac{1}{3} & | & 0 & \frac{1}{3} & 0 \\ 0 & 0 & \frac{1}{3} & | & -2 & -\frac{5}{3} & 1 \end{bmatrix}$ (We replaced R_1 by $R_1 + 3R_2$, and R_3 by $R_3 - 5R_2$.)

$\sim \begin{bmatrix} 1 & 0 & 1 & | & 1 & 1 & 0 \\ 0 & 1 & \frac{1}{3} & | & 0 & \frac{1}{3} & 0 \\ 0 & 0 & 1 & | & -6 & -5 & 3 \end{bmatrix}$ (We replaced R_3 by $3R_3$.)

$\sim \begin{bmatrix} 1 & 0 & 0 & | & 7 & 6 & -3 \\ 0 & 1 & 0 & | & 2 & 2 & -1 \\ 0 & 0 & 1 & | & -6 & -5 & 3 \end{bmatrix}$ (We replaced R_1 by $R_1 - R_2$, R_2 by $R_2 - \frac{1}{3}$.)

Then $M^{-1} = \begin{bmatrix} 7 & 6 & -3 \\ 2 & 2 & -1 \\ -6 & -5 & 3 \end{bmatrix}$.

5.147 Show that if $M = \begin{bmatrix} 3 & 9 \\ 2 & 6 \end{bmatrix}$, then M^{-1} does not exist.

▮ Replace R_2 by $R_2 - \frac{2}{3}R_1$. Then $\begin{bmatrix} 3 & 9 & | & 1 & 0 \\ 2 & 6 & | & 0 & 1 \end{bmatrix} \sim \begin{bmatrix} 3 & 9 & | & 1 & 0 \\ 0 & 0 & | & -\frac{2}{3} & 1 \end{bmatrix}$. But then M^{-1} does not exist since we get a row of all zeros on the left-hand side.

For Probs. 5.148 to 5.158, use Cramer's rule to solve the given system.

5.148 $x + 2y = 6$, $3x - 5y = 10$

▮ By Cramer's rule

$$x = \frac{\begin{vmatrix} k_1 & a_{12} \\ k_2 & a_{22} \end{vmatrix}}{D} \qquad y = \frac{\begin{vmatrix} a_{11} & k_1 \\ a_{21} & k_2 \end{vmatrix}}{D}$$

where $D = \begin{vmatrix} a_{11} & a_{12} \\ a_{21} & a_{22} \end{vmatrix}$ $(\neq 0)$. Here $a_{11} = 1$, $a_{12} = 2$, $a_{21} = 3$, $a_{22} = -5$, $k_1 = 6$, and $k_2 = 10$.

$$x = \frac{\begin{vmatrix} 6 & 2 \\ 10 & -5 \end{vmatrix}}{D} \qquad y = \frac{\begin{vmatrix} 1 & 6 \\ 3 & 10 \end{vmatrix}}{D} \qquad D = \begin{vmatrix} 1 & 2 \\ 3 & -5 \end{vmatrix} = -5 - 6 = -11$$

Since $\begin{vmatrix} 6 & 2 \\ 10 & -5 \end{vmatrix} = -30 - 20 = -50$ and $\begin{vmatrix} 1 & 6 \\ 3 & 10 \end{vmatrix} = 10 - 18 = -8$, $x = \frac{-50}{-11} = \frac{50}{11}$, and $y = \frac{-8}{-11} = \frac{8}{11}$.

5.149 $x - y = 8, 2x + 3y = 10$

▎ $a_{11} = 1, a_{12} = -1, a_{21} = 2, a_{22} = 3, k_1 = 8,$ and $k_2 = 10.$

$$x = \frac{\begin{vmatrix} 8 & -1 \\ 10 & 3 \end{vmatrix}}{D} \qquad y = \frac{\begin{vmatrix} 1 & 8 \\ 2 & 10 \end{vmatrix}}{D} \qquad D = \begin{vmatrix} 1 & -1 \\ 2 & 3 \end{vmatrix} = 3 - (-2) = 5$$

Then $x = \dfrac{24 - (-10)}{5} = \dfrac{34}{5}; y = \dfrac{10 - 16}{5} = \dfrac{-6}{5}.$

5.150 $3x - 5y = 8, 2x + y = 2$

▎ $x = \dfrac{\begin{vmatrix} 8 & -5 \\ 2 & 1 \end{vmatrix}}{D} \qquad y = \dfrac{\begin{vmatrix} 3 & 8 \\ 2 & 2 \end{vmatrix}}{D} \qquad D = \begin{vmatrix} 3 & -5 \\ 2 & 1 \end{vmatrix} = 3 - (-10) = 13$

Then $x = \dfrac{8 + 10}{13} = \dfrac{18}{13}; y = \dfrac{6 - 16}{13} = \dfrac{-10}{13}.$

5.151 $2x - 6y = 1, 3x + y = 2$

▎ $x = \dfrac{\begin{vmatrix} 1 & -6 \\ 2 & 1 \end{vmatrix}}{D} \qquad y = \dfrac{\begin{vmatrix} 2 & 1 \\ 3 & 2 \end{vmatrix}}{D} \qquad D = \begin{vmatrix} 2 & -6 \\ 3 & 1 \end{vmatrix} = 2 - (-18) = 20$

Then $x = \dfrac{1 - (-12)}{20} = \dfrac{13}{20}; y = \dfrac{4 - 3}{20} = \dfrac{1}{20}.$

5.152 $-x - y = 40, 2x - y = 35$

▎ $x = \dfrac{\begin{vmatrix} 40 & -1 \\ 35 & -1 \end{vmatrix}}{D} \qquad y = \dfrac{\begin{vmatrix} -1 & 40 \\ 2 & 35 \end{vmatrix}}{D} \qquad D = \begin{vmatrix} -1 & -1 \\ 2 & -1 \end{vmatrix} = 1 - (-2) = 3$

Then $x = \dfrac{-40 - (-35)}{3} = \dfrac{-5}{3}; y = \dfrac{-35 - 80}{3} = \dfrac{-115}{3}.$

5.153 $3x - y = 10, 40x - 11y = -100$

▎ $x = \dfrac{\begin{vmatrix} 10 & -1 \\ -100 & -11 \end{vmatrix}}{D} \qquad y = \dfrac{\begin{vmatrix} 3 & 10 \\ 40 & -100 \end{vmatrix}}{D} \qquad D = \begin{vmatrix} 3 & -1 \\ 40 & -11 \end{vmatrix} = -33 - (-40) = 7$

Then $x = \dfrac{10(-11) - 100}{7} = \dfrac{-210}{7} = -30; y = \dfrac{-300 - 400}{7} = \dfrac{-700}{7} = -100.$

5.154 $2x + y = 5, 4x + 2y = 10$

▎ $D = \begin{vmatrix} 2 & 1 \\ 4 & 2 \end{vmatrix} = 4 - 4 = 0.$ Thus, Cramer's rule will not work. This means that the system is either dependent or inconsistent. Here we notice that the problem is dependency.

5.155 $x + y = 0, 2y + z = -5, -x + z = -3$

▎ Given a system of equations

$$a_{11}x + a_{12}y + a_{13}z = k_1$$
$$a_{21}x + a_{22}y + a_{23}z = k_2$$
$$a_{31}x + a_{32}y + a_{33}z = k_3$$

then

$$x = \frac{\begin{vmatrix} k_1 & a_{12} & a_{13} \\ k_2 & a_{22} & a_{23} \\ k_3 & a_{32} & a_{33} \end{vmatrix}}{D} \qquad y = \frac{\begin{vmatrix} a_{11} & k_1 & a_{13} \\ a_{21} & k_2 & a_{23} \\ a_{31} & k_3 & a_{33} \end{vmatrix}}{D} \qquad z = \frac{\begin{vmatrix} a_{11} & a_{12} & k_1 \\ a_{21} & a_{22} & k_2 \\ a_{31} & a_{32} & k_3 \end{vmatrix}}{D}$$

$$D = \begin{vmatrix} a_{11} & \cdots & a_{13} \\ \vdots & & \vdots \\ a_{31} & \cdots & a_{33} \end{vmatrix} \quad (\neq 0)$$

Here,
$$D = \begin{vmatrix} 1 & 1 & 0 \\ 0 & 2 & 1 \\ -1 & 0 & 1 \end{vmatrix} = 1$$

$$x = \frac{\begin{vmatrix} 0 & 1 & 0 \\ -5 & 2 & 1 \\ -3 & 0 & 1 \end{vmatrix}}{D} = \frac{2}{1} = 2 \qquad y = \frac{\begin{vmatrix} 1 & 0 & 1 \\ 0 & -5 & 1 \\ -1 & -3 & 1 \end{vmatrix}}{D} = \frac{-2}{1} = -2 \qquad z = \frac{\begin{vmatrix} 1 & 1 & 0 \\ 0 & 2 & -5 \\ -1 & 0 & -3 \end{vmatrix}}{D} = \frac{-1}{1} = -1$$

5.156 $x + y = 1,\ 2y + z = 0,\ -x + z = 0$

▐ $D = \begin{vmatrix} 1 & 1 & 0 \\ 0 & 2 & 1 \\ -1 & 0 & 1 \end{vmatrix} = 1.$ Then
$$x = \frac{\begin{vmatrix} 1 & 1 & 0 \\ 0 & 2 & 0 \\ 0 & 0 & 1 \end{vmatrix}}{1} = 2;\ y = \frac{\begin{vmatrix} 1 & 1 & 0 \\ 0 & 0 & 1 \\ -1 & 0 & 1 \end{vmatrix}}{1} = -1;\ z = \frac{\begin{vmatrix} 1 & 1 & 1 \\ 0 & 2 & 0 \\ -1 & 0 & 0 \end{vmatrix}}{1} = 2$$

5.157 $y + z = -4,\ x + 2z = 0,\ x - y = 5$

▐ $D = \begin{vmatrix} 0 & 1 & 1 \\ 1 & 0 & 2 \\ 1 & -1 & 0 \end{vmatrix} = 1.$ Then

$$x = \begin{vmatrix} -4 & 1 & 1 \\ 0 & 0 & 2 \\ 5 & -1 & 0 \end{vmatrix} = 2; \qquad y = \begin{vmatrix} 0 & -4 & 1 \\ 1 & 0 & 2 \\ 1 & 5 & 0 \end{vmatrix} = -3; \qquad z = \begin{vmatrix} 0 & 1 & -4 \\ 1 & 0 & 0 \\ 1 & -1 & 5 \end{vmatrix} = -1$$

5.158 $2y - z = -4,\ x - y - z = 0,\ x - y + 2z = 6$

▐ $D = \begin{vmatrix} 0 & 2 & -1 \\ 1 & -1 & -1 \\ 1 & -1 & 2 \end{vmatrix} = -6.$ Then

$$x = \frac{\begin{vmatrix} -4 & 2 & -1 \\ 0 & -1 & -1 \\ 6 & -1 & 2 \end{vmatrix}}{-6} = \frac{-6}{-6} = 1; \qquad y = \frac{\begin{vmatrix} 0 & -4 & -1 \\ 1 & 0 & -1 \\ 1 & 6 & 2 \end{vmatrix}}{-6} = -1; \qquad z = \frac{\begin{vmatrix} 0 & 2 & -4 \\ 1 & -1 & 0 \\ 1 & -1 & 6 \end{vmatrix}}{-6} = -1$$

5.3 SYSTEMS OF NONLINEAR EQUATIONS

For Probs. 5.159 to 5.184, solve the given system algebraically.

5.159 $3x - y = 8,\ 3x^2 - y^2 = 26$

▮ $y = 3x - 8$. Substituting, we get $26 = 3x^2 - (3x - 8)^2 = 3x^2 - (9x^2 - 48x + 64) =$
$-6x^2 + 48x - 64 = 26$. Then $-6x^2 + 48x - 90 = 0$, $x^2 - 8x + 15 = 0$, $(x - 5)(x - 3) = 0$, and $x = 5$,
$x = 3$. If $x = 5$, $y = 3(5) - 8 = 7$. If $x = 3$, $y = 3(3) - 8 = 1$. Check to see that these solutions are
correct. They are!

5.160 $3x - 2y = 5$, $3x^2 - 2y^2 = 19$

▮ $2y = 3x - 5$, or $y = \frac{3}{2}x - \frac{5}{2}$. Substituting gives $19 = 3x^2 - 2(\frac{3}{2}x - \frac{5}{2})^2 =$
$3x^2 - 2(\frac{9}{4}x^2 - \frac{30}{4}x + \frac{25}{4}) = 3x^2 - \frac{18}{4}x^2 + \frac{60}{4}x - \frac{50}{4} = -\frac{3}{2}x^2 + \frac{60}{4}x - \frac{50}{4} = -\frac{3}{2}x^2 + 15x - \frac{25}{2}$.
Multiplying both sides by 2, we get $-3x^2 + 30x - 25 = 38$. Subtracting 38 from both sides, we get
$3x^2 - 30x + 63 = 0$, $x^2 - 10x + 21 = 0$, $(x - 7)(x - 3) = 0$, and $x = 7$, $x = 3$. If $x = 7$, $y = \frac{3}{2} \cdot 7 - \frac{5}{2} = 8$.
If $x = 3$, $y = \frac{3}{2} \cdot 3 - \frac{5}{2} = 2$.

5.161 $2x - y = -1$, $16x^2 - 3y^2 = -11$

▮ $y = 2x + 1$. Substituting gives $-11 = 16x^2 - 3(2x + 1)^2 = 16x^2 - 3(4x^2 + 4x + 1) = 16x^2 - 12x^2 -$
$12x - 3$. Adding 11 to both sides, we get $4x^2 - 12x + 8 = 0$, $x^2 - 3x + 2 = 0$, $(x - 2)(x - 1) = 0$, and
$x = 2$, $x = 1$. If $x = 2$, $y = 2(2) + 1 = 5$. If $x = 1$, $y = 2(1) + 1 = 3$.

5.162 $4x - y = 11$, $8x^2 + 5y^2 = 77$

▮ $y = 4x - 11$. Substituting yields $77 = 8x^2 + 5(4x - 11)^2 = 8x^2 + 5(16x^2 - 88x + 121) = 8x^2 + 80x^2 -$
$440x + 605$. We subtract 77 from both sides: $88x^2 - 440x + 528 = 0$, $x^2 - 5x + 6 = 0$, $(x - 6)(x + 1) =$
0, and $x = 6$, $x = -1$. If $x = 6$, $y = 24 - 11 = 13$. If $x = -1$, $y = -4 - 11 = -15$.

5.163 $x^2 + 2y^2 = 9$, $xy = 2$

▮ $x = 2/y$. Substituting yields $(2/y)^2 + 2y^2 = 9$, $4/y^2 + 2y^2 = 9$, $4 + 2y^4 = 9y^2$, $2y^4 - 9y^2 + 4 = 0$. Let
$s = y^2$; then $2S^2 - 9S + 4 = 0$. $S = \dfrac{9 \pm \sqrt{81 - 4(2)(4)}}{4} = \dfrac{9 \pm \sqrt{49}}{4} = \dfrac{9 \pm 7}{4}$. Thus $S = (9 + 7)/4 = 4$ or
$S = (9 - 7)/4 = \frac{1}{2}$. Then $y^2 = S = 4$, and $y = \pm 2$; or $y^2 = S = \frac{1}{2}$, and $y = \pm\sqrt{2}/2$. If $y = 2$, $x = 1$; if
$y = -2$, $x = -1$; if $y = \sqrt{2}/2$, $x = 2\sqrt{2}$; if $y = -\sqrt{2}/2$, $x = -2\sqrt{2}$.

5.164 $2x^2 + 4y^2 = 19$, $3x^2 - 8y^2 = 25$

▮   Multiply the first equation by 2, and add:

$$\begin{array}{r} 4x^2 + 8y^2 = 38 \\ 3x^2 - 8y^2 = 25 \\ \hline 7x^2 = 63 \end{array}$$

Then $x^2 = \frac{63}{7} = 9$, $x = \pm 3$. If $x = 3$, $2(9) + 4y^2 = 19$, $4y^2 = 1$, and $y = \pm\frac{1}{2}$. If $x = -3$, $2(9) + 4y^2 = 19$,
and $y = \pm\frac{1}{2}$. Thus, the solutions are $(3, \pm\frac{1}{2})$, $(-3, \pm\frac{1}{2})$.

5.165 $x^2 + 2y^2 = 3$, $3x^2 - y^2 = 2$

▮   Multiplying the first equation by 3 and subtracting gives

$$\begin{array}{r} 3x^2 + 6y^2 = 9 \\ 3x^2 - y^2 = 2 \\ \hline 7y^2 = 7 \end{array}$$

Then $y = \pm 1$. If $y = 1$, $x^2 + 2 = 3$, and $x = \pm 1$. If $y = -1$, $x = \pm 1$. Thus, the solutions are $(\pm 1, 1)$,
$(\pm 1, -1)$.

5.166 $x^2 + 2y^2 = 3$, $3x^2 - y^2 = 2$

▮   Multiply the second equation by 2, and add:

$$\begin{array}{r} x^2 + 2y^2 = 3 \\ 6x^2 - 2y^2 = 4 \\ \hline 7x^2 = 7 \end{array}$$

Then $x^2 = 1$, or $x = \pm 1$. If $x = 1$, $2y^2 = 2$, and $y = \pm 1$. If $x = -1$, $2y^2 = 2$, and $y = \pm 1$. Thus, the solutions are $(1, \pm 1)$, $(-1, \pm 1)$.

5.167 $4x^2 + 3y^2 = 4$, $8x^2 + 5y^2 = 7$

❙ Multiply the first equation by 2, and subtract:

$$\begin{array}{r} 8x^2 + 6y^2 = 8 \\ 8x^2 + 5y^2 = 7 \\ \hline y^2 = 1 \end{array}$$

Then $y = \pm 1$. If $y = 1$, $4x^2 = 1$, and $x = \pm \frac{1}{2}$; If $y = -1$, $x = \pm \frac{1}{2}$. Thus, the solutions are $(\pm \frac{1}{2}, 1)$, $(\pm \frac{1}{2}, -1)$.

5.168 $2x^2 + 3y^2 = 7$, $3x^2 + 2y^2 = 8$

❙ Multiply the first equation by 3, the second equation by 2, and subtract:

$$\begin{array}{r} 6x^2 + 9y^2 = 21 \\ 6x^2 + 4y^2 = 16 \\ \hline 5y^2 = 5 \end{array}$$

Then $y = \pm 1$. If $y = 1$, $6x^2 = 12$, $x^2 = 2$, and $x = \pm \sqrt{2}$. If $y = -1$, $x = \pm \sqrt{2}$. Thus, the solutions are $(\pm \sqrt{2}, 1)$, $(\pm \sqrt{2}, -1)$.

5.169 $\dfrac{1}{x} + \dfrac{2}{y} = 1$, $\dfrac{3}{x} + \dfrac{10}{y} = 4$

❙ If $1/x = 1 - 2/y$, then $3/x = 3 - 6/y$. Thus, $3 - 6y + 10/y = 4$, $4/y = 1$, or $y = 4$. If $y = 4$, $1/x + \frac{1}{2} = 1$, $1/x = \frac{1}{2}$, or $x = 2$. The solution is $(2, 4)$.

5.170 $x^2 + 4y^2 - 2x = 1$, $3x^2 + 8y^2 - 6x = 2$

❙ Multiply the first equation by 3 and subtract:

$$\begin{array}{r} 3x^2 + 12y^2 - 6x = 3 \\ 3x^2 + 8y^2 - 6x = 2 \\ \hline 4y^2 = 1 \end{array}$$

Then $y = \pm \frac{1}{2}$. If $y = \frac{1}{2}$, $x^2 - 2x = 1 - 1 = 0$, $x(x - 2) = 0$, and $x = 0, 2$. If $y = -\frac{1}{2}$, $x = 0, 2$. Thus, the solutions are $(0, \frac{1}{2})$, $(2, \frac{1}{2})$, $(0, -\frac{1}{2})$, $(2, -\frac{1}{2})$.

5.171 $9x^2 - 41y^2 + 60y = 100$, $x^2 - 5y^2 + 8y = 12$

❙

$$\begin{array}{r} 9x^2 - 41y^2 + 60y = 100 \\ 9x^2 - 45y^2 + 72y = 108 \\ \hline 4y^2 - 12y = -8 \end{array}$$

Then $y^2 - 3y + 2 = 0$, $y = \dfrac{3 \pm \sqrt{9 - 4(1)(2)}}{2} = \dfrac{3 \pm \sqrt{9 - 8}}{2} = \dfrac{3 \pm 1}{2}$, and $y = 2$, $y = 1$. If $y = 2$, $x^2 - 20 + 16 = 12$, $x^2 = 16$, and $x^2 = 16$, and $x = \pm 4$. If $y = 1$, $x^2 - 5 + 8 = 12$, $x^2 = 9$, and $x = \pm 3$. Thus, the solutions are $(3, 1)$, $(-3, 1)$, $(\pm 4, 2)$.

5.172 $2x^2 - y^2 = 14$, $x - y = 1$

❙ Substituting for y, we get $2x^2 - (x - 1)^2 = 14$, $2x^2 - (x^2 - 2x + 1) = 14$, $x^2 + 2x - 15 = 0$, $(x - 3)(x + 5) = 0$, and $x = 3$, $x = -5$. If $x = 3$, $y = 2$. If $x = -5$, $y = -6$.

5.173 $xy + x^2 = 24$, $y - 3x + 4 = 0$

▮ $y = 3x - 4$. Substituting for y gives $x(3x - 4) + x^2 = 24$, $3x^2 - 4x + x^2 = 24$, $4x^2 - 4x - 24 = 0$, $x^2 - x - 6 = 0$, $(x - 3)(x + 2) = 0$, and $x = 3$, $x = -2$. If $x = 3$, $y = 5$; if $x = -2$, $y = -10$.

5.174 $3xy - 10x = y$, $2 - y + x = 0$

▮ $y = 2 + x$. Substituting for y, we get $3x(2 + x) - 10x = 2 + x$, $6x + 3x^2 - 10x = 2 + x$, $3x^2 - 5x - 2 = 0$, $(3x + 1)(x - 2) = 0$, and $x = -\frac{1}{3}$, $x = 2$. If $x = -\frac{1}{3}$, $y = \frac{5}{3}$; if $x = 2$, $y = 4$.

5.175 $xy = -3$, $xy = -6$

▮ If $xy = -3$, then, since $xy = -6$, we get $-3 = -6$. No solution. (Graphically, what does this say about these two hyperbolas?)

5.176 $4x + 5y = 6$, $xy = -2$

▮ $y = -2/x$. Substituting for y, we get $4x + 5(-2/x) = 6$, $4x - 10/x = 6$, $4x^2 - 10 = 6x$, $2x^2 - 3x - 5 = 0$, $(2x - 5)(x + 1) = 0$, and $x = \frac{5}{2}$, $x = -1$. If $x = \frac{5}{2}$, $y = -\frac{4}{5}$; if $x = -1$, $y = 2$.

5.177 $\dfrac{9}{x^2} + \dfrac{16}{y^2} = 5$, $\dfrac{18}{x^2} + \dfrac{12}{y^2} = -1$

▮ $9/x^2 = 5 - 16/y^2$. Thus, $18/x^2 = 10 - 32/y^2$. But $18/x^2 = -1 + 12/y^2$. Then $-1 + 12/y^2 = 10 - 32/y^2$, $44/y^2 = 11$, $11y^2 = 44$, and $y = \pm 2$. If $y = 2$, $9/x^2 + \frac{16}{4} = 5$, $9/x^2 = 1$, $x^2 = 9$, and $x = \pm 3$. If $y = -2$, $x = \pm 3$. Thus, the solutions are $(3, 2)$, $(-3, 2)$, $(3, -2)$, $(-3, -2)$.

5.178 $x^2 - xy = 12$, $xy - y^2 = 3$

▮ $x(x - y) = 12$ and $y(x - y) = 3$. Then $x/y = 4$ ($x \neq y$) and $(4y)^2 - (4y)(y) = 12$. So $16y^2 - 4y^2 = 12$, $12y^2 = 12$, and $y = \pm 1$. If $y = 1$, $x = 4$; if $y = -1$, $x = -4$. Since we divided by $x - y$, check both solutions.

5.179 $x^3 - y^3 = 9$, $x - y = 3$

▮ $x = y + 3$. Substituting for x, we get $(y + 3)^3 - y^3 = 9$, $y^3 + 9y^2 + 27y + 27 - y^3 = 9$, $9y^2 + 27y + 18 = 0$, $y^2 + 3y + 2 = 0$, $(y + 2)(y + 1) = 0$, and $y = -1$, $y = -2$. If $y = -1$, $x = 2$; if $y = -2$, $x = 1$.

5.180 $9x^2 + y^2 = 90$, $x^2 + 9y^2 = 90$

▮ Subtract:

$$\begin{array}{r} 9x^2 + y^2 = 90 \\ 9x^2 + 81y^2 = 810 \\ \hline -80y^2 = -720 \end{array}$$

Then $y^2 = 9$, and $y = \pm 3$. If $y = 3$, $x = \pm 3$. If $y = -3$, $x = \pm 3$.

5.181 $\dfrac{2}{x^2} - \dfrac{3}{y^2} = 5$, $\dfrac{1}{x^2} + \dfrac{2}{y^2} = 6$

▮ $1/x^2 = 6 - 2/y^2$. Thus, $2/x^2 = 12 - 4/y^2$, $12 - 4/y^2 = 5 + 3/y^2$, $12y^2 - 4 = 5y^2 + 3$, $7y^2 = 7$, and $y = \pm 1$. If $y = 1$, $x = \pm \frac{1}{2}$; if $y = -1$, $x = \pm \frac{1}{2}$.

5.182 $y^2 = 4x - 8$, $y^2 = -6x + 32$

▮ Then $-6x + 32 = 4x - 8$, $10x = 40$, and $x = 4$. Then $y^2 = 16 - 8 = 8$, and $y = \pm 2\sqrt{2}$. Thus, the solutions are $(4, \pm 2\sqrt{2})$.

5.183 $x^2 - y^2 = 16$, $y^2 = 2x - 1$

▮ From the first equation, $y^2 = x^2 - 16$. Then $y^2 = x^2 - 16 = 2x - 1$, $x^2 - 2x - 15 = 0$, $x = [2 \pm \sqrt{4 + 4(1)(15)}]/2 = (2 \pm 8)/2$, and $x = 5$ or $x = -3$. [We also could have factored:

$(x-5)(x+3) = 0.$] If $x = 5$, $y^2 = 10 - 1 = 9$, and $y = \pm 3$. If $x = -3$, $y^2 = -6 - 1 = -7$, and $y = \pm i\sqrt{7}$. Notice the imaginary solution here.

5.184 $2x^2 + y^2 = 6$, $x^2 + y^2 + 2x = 3$

▌ Subtract:

$$
\begin{array}{r}
2x^2 + y^2 \quad\quad = 6 \\
x^2 + y^2 + 2x = 3 \\
\hline
x^2 \quad\quad - 2x = 3
\end{array}
$$

Then $x^2 - 2x - 3 = 0$, $(x-3)(x+1) = 0$ and $x = 3$ or $x = -1$. If $x = -1$, $y = \pm\sqrt{3}$. If $x = 3$, $y = \pm\sqrt{6 - 2x^2} = \pm 2i\sqrt{3}$.

5.185 Solve the system

$$
\begin{cases}
2y^2 - 3x = 0 & (1) \\
4y - x = 6 & (2)
\end{cases}
$$

▌ Solve (2) for x: $x = 4y - 6$. Substitute in (1): $2y^2 - 3(4y - 6) = 2(y - 3)^2 = 0$; $y = 3, 3$. When $y = 3$, $x = 4y - 6 = 12 - 6 = 6$. The solutions are $x = 6$, $y = 3$; $x = 6$, $y = 3$. The straight line is tangent to the parabola at $(6, 3)$. See Fig. 5.11.

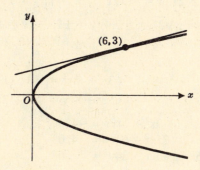

(6, 3)

Fig. 5.11

5.186 Solve the system

$$
\begin{cases}
y^2 - 4y - 3x + 1 = 0 & (1) \\
3y - 4x = 7 & (2)
\end{cases}
$$

▌ Solve (2) for x: $x = \frac{1}{4}(3y - 7)$. Substitute in (1): $y^2 - 4y - \frac{3}{4}(3y - 7) + 1 = 0$, $4y^2 - 16y - 9y + 21 + 4 = 4y^2 - 25y + 25 = (y - 5)(4y - 5) = 0$ or $y = 5$ and $y = \frac{5}{4}$. When $y = 5$, $x = \frac{1}{4}(3y - 7) = 2$; when $y = \frac{5}{4}$, $x = \frac{1}{4}(3y - 7) = -\frac{13}{16}$. The solutions are $x = 2$, $y = 5$, $x = -\frac{13}{16}$, $y = \frac{5}{4}$. The straight line intersects the parabola in the points $(2, 5)$ and $(-\frac{13}{16}, \frac{5}{4})$.

5.187 Solve the system

$$
\begin{cases}
3x^2 - y^2 = 27 & (1) \\
x^2 - y^2 = -45 & (2)
\end{cases}
$$

▌ Subtracting, we get $2x^2 = 72$, $x^2 = 36$, and $x = \pm 6$. When $x = 6$, $y^2 = x^2 + 45 = 36 + 45 = 81$ and $y = \pm 9$. When $x = -6$, $y^2 = x^2 + 45 = 36 + 45 = 81$ and $y = \pm 9$. The solutions are $x = +6$, $y = \pm 9$; $x = -6$, $y = \pm 9$. The two hyperbolas intersect in the points $(6, 9)$, $(-6, 9)$, $(-6, -9)$, and $(6, -9)$.

5.188 Solve the system

$$
\begin{cases}
5x^2 + 3y^2 = 92 & (1) \\
2x^2 + 5y^2 = 52 & (2)
\end{cases}
$$

▌ Multiply (1) by 5: $\quad 25x^2 + 15y^2 = 460$
Multiply (2) by -3: $\quad -6x^2 - 15y^2 = -156$

Add: $19x^2 = 304$; $x^2 = 16$ and $x = \pm 4$. When $x = \pm 4$, $3y^2 = 92 - 5x^2 = 92 - 80 = 12$, $y^2 = 4$, and $y = \pm 2$. The solutions are $x = +4$, $y = \pm 2$; $x = -4$, $y = \pm 2$. See Fig. 5.12.

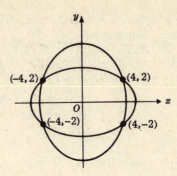

Fig. 5.12

5.189 Solve the system

$$\begin{cases} x^2 + 4xy = 0 & (1) \\ x^2 - xy + y^2 = 21 & (2) \end{cases}$$

▮ Solve (1) for x: $x(x + 4y) = 0$ and $x = 0$, $x = -4y$. Solve the systems:

$$\begin{array}{cc} x^2 - xy + y^2 = 21 & x^2 - xy + y^2 = 21 \\ x = 0 & x = -4y \\ y^2 = 21,\ y = \pm\sqrt{21} & y^2 = 1,\ y = \pm 1;\ x = -4y = \pm 4 \end{array}$$

The solutions are $x = 0$, $y = \pm\sqrt{21}$; $x = \pm 4$, $y = \pm 1$.

5.190 Solve the system

$$\begin{cases} 3x^2 + 8y^2 = 140 & (1) \\ 5x^2 + 8xy = 84 & (2) \end{cases}$$

▮ Multiply (1) by -3: $\qquad -9x^2 - 24y^2 = -420$
Multiply (2) by 5: $\qquad 25x^2 + 40xy = 420$
Add: $\qquad 16x^2 + 40xy - 24y^2 = 0$

Then $(2x - 3y^2) = 8(2x - y)(x + 3y) = 0$ and $x = \frac{1}{2}y$, $x = -3y$. Solve the systems:

$$\begin{array}{cc} 3x^2 + 8y^2 = 140 & 3x^2 + 8y^2 = 140 \\ x = \frac{1}{2}y & x = -3y \\ \frac{3}{4}y^2 + 8y^2 = \frac{35}{4}y^2 = 140 & 27y^2 + 8y^2 = 35y^2 = 140 \\ y^2 = 16,\ y = \pm 4;\ x = \frac{1}{2}y = \pm 2 & y^2 = 4,\ y = \pm 2;\ x = -3y = \pm 6 \end{array}$$

The solutions are $x = \pm 2$, $y = \pm 4$; $x = \pm 6$, $y = \pm 2$.

5.191 Solve the system

$$\begin{cases} x^2 - 3xy + 2y^2 = 15 & (1) \\ 2x^2 + y^2 = 6 & (2) \end{cases}$$

▮ Multiply (1) by -2: $\quad -2x^2 + 6xy - 4y^2 = -30$
Multiply (2) by 5: $\quad 10x^2 + 5y^2 = 30$

Add: $8x^2 + 6xy + y^2 = (4x + y)(2x + y) = 0$. Then $y = -4x$ and $y = -2x$. Solve the systems:

$$\begin{array}{cc} 2x^2 + y^2 = 6 & 2x^2 + y^2 = 6 \\ y = -4x & y = -2x \\ 2x^2 + 16x^2 = 18x^2 = 6,\ x^2 = \frac{1}{3} & 2x^2 + 4x^2 = 6x^2 = 6,\ x^2 = 1 \end{array}$$

$x = \pm\sqrt{3}/3$ and $y = -4x = \pm 4\sqrt{3}/3$; $x = \pm 1$, $y = \pm 2$.

5.192 Solve the system

$$\begin{cases} x^2 + y^2 + 3x + 3y = 8 \\ xy + 4x + 4y = 2 \end{cases}$$

▌ Substitute $x = u + v$, $y = u - v$ in the given system:

(1) $\qquad (u + v)^2 + (u - v)^2 + 3(u + v) - 3(u - v) = 2u^2 + 2v^2 + 6u = 8$

(2) $\qquad (u + v)(u - v) + 4(u + v) + 4(u - v) = u^2 - v^2 + 8u = 2$

Add (1) and 2 times (2):

$$4u^2 + 22u - 12 = 2(2u - 1)(u + 6) = 0 \qquad u = \tfrac{1}{2}, \ -6.$$

For $u = \tfrac{1}{2}$, (2) yields $v^2 = u^2 + 8u - 2 = \tfrac{1}{4} + 4 - 2 = \tfrac{9}{4}$; $v = \pm\tfrac{3}{2}$.

When $u = \tfrac{1}{2}$, $v = \ \tfrac{3}{2}$; $\quad x = u + v = \ \ 2$, $\quad y = u - v = -1$.

When $u = \tfrac{1}{2}$, $v = -\tfrac{3}{2}$; $\quad x = u + v = -1$, $\quad y = u - v = \ \ 2$.

For $u = -6$, (2) yields $v^2 = u^2 + 8u - 2 = 36 - 48 - 2 = -14$; $v = \pm i\sqrt{14}$.

When $u = -6$, $v = \ \ i\sqrt{14}$; $\quad x = u + v = -6 + i\sqrt{14}$, $\quad y = u - v = -6 - i\sqrt{14}$.

When $u = -6$, $v = -i\sqrt{14}$; $\quad x = u + v = -6 - i\sqrt{14}$, $\quad y = u - v = -6 + i\sqrt{14}$.

The solutions are $x = 2$, $y = -1$; $x = -1$, $y = 2$; $x = -6 \pm i\sqrt{14}$, $y = -6 \pm i\sqrt{14}$.

5.193 Solve the system

$$\begin{cases} x^2 + y^2 = 25 & (1) \\ \qquad xy = 12 & (2) \end{cases}$$

▌ Multiply (2) by 2 and add to (1):

$$x^2 + 2xy + y^2 = 49 \qquad \text{or} \qquad x + y = \pm 7$$

Multiply (2) by -2 and add to (1):

$$x^2 - 2xy + y^2 = 1 \qquad \text{or} \qquad x = y = \pm 1$$

Solve the systems:

$$
\begin{array}{ll}
\begin{array}{l}
x + y = 7 \\
x - y = 1 \\
\overline{\quad 2x = 8} \\
\quad\ x = 4 \\
\quad\ y = 7 - x = 3
\end{array}
&
\begin{array}{l}
x + y = 7 \\
x - y = -1 \\
\overline{\quad 2x = 6} \\
\quad\ x = 3 \\
\quad\ y = 7 - x = 4
\end{array}
\\[3em]
\begin{array}{l}
x + y = -7 \\
x - y = 1 \\
\overline{\quad 2x = -6} \\
\quad\ x = -3 \\
\quad\ y = -7 - x = -4
\end{array}
&
\begin{array}{l}
x + y = -7 \\
x - y = -1 \\
\overline{\quad 2x = -8} \\
\quad\ x = -4 \\
\quad\ y = -7 - x = -3
\end{array}
\end{array}
$$

The solutions are $x = \pm 4$, $y = \pm 3$; $x = \pm 3$, $y = \pm 4$. See Fig. 5.13.

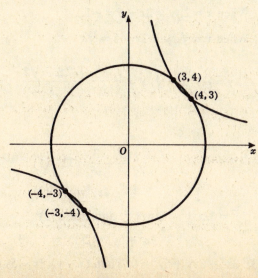

Fig. 5.13

5.194 Solve the system

$$\begin{cases} x^2 - xy - 12y^2 = 8 & (1) \\ x^2 + xy - 10y^2 = 20 & (2) \end{cases}$$

▮ This system may be solved by the procedure used in Probs. 5.190 and 5.191. Here we give an alternative solution. When $m = -\frac{3}{8}$, $x^2 = \dfrac{8}{1 - m - 12m^2} = -\dfrac{128}{5}$; $x = \pm\dfrac{8i\sqrt{10}}{5}$, $y = mx = \pm\dfrac{3i\sqrt{10}}{5}$. The solutions are $x = \pm5$, $y = \pm1$; $x = \pm\dfrac{8i\sqrt{10}}{5}$, $y = \pm\dfrac{3i\sqrt{10}}{5}$.

5.195 Solve the system

$$\begin{cases} x^3 - y^3 = 19 & (1) \\ x^2 + xy + y^2 = 19 & (2) \end{cases}$$

▮ Divide (1) by (2): $x - y = 1$. Solve the system

$$\begin{cases} x^2 + xy + y^2 = 19 & (2) \\ x - y = 1 & (3) \end{cases}$$

Solve (3) for x: $x = y + 1$. substitute in (2): $(y + 1)^2 + (y + 1)y + y^2 = 3y^2 + 3y + 1 = 19$. then $3y^2 + 3y - 18 = 3(y + 3)(y - 2) = 0$ and $y = -3, 2$. When $y = -3$, $x = y + 1 = -2$; when $y = 2$, $x = y + 1 = 3$. The solutions are $x = -2, y = -3$; $x = 3, y = 2$.

5.196 Solve the system

$$\begin{cases} (2x - y)^2 - 4(2x - y) = 5 & (1) \\ x^2 - y^2 = 3 & (2) \end{cases}$$

▮ Factor (1): $(2x - y)^2 - 4(2x - y) - 5 = (2x - y - 5)(2x - y + 1) = 0$. Then $2x - y = 5$ and $2x - y = -1$. Solve the systems:

$$\begin{cases} x^2 - y^2 = 3 \\ 2x - y = 5 \end{cases} \qquad \begin{cases} x^2 - y^2 = 3 \\ 2x - y = -1 \end{cases}$$

$$y = 2x - 5 \qquad\qquad y = 2x + 1$$
$$x^2 - (2x - 5)^2 = 3 \qquad x^2 - (2x + 1)^2 = 3$$
$$3x^2 - 20x + 28 = (x - 2)(3x - 14) = 0 \qquad 3x^2 + 4x + 4 = 0$$

$$x = 2, \tfrac{14}{3} \qquad\qquad x = \dfrac{-4 \pm \sqrt{16 - 48}}{6} = \dfrac{-2 \pm 2i\sqrt{2}}{3}$$

When $x = 2$, $y = 2x - 5 = -1$.

When $x = \frac{14}{3}$, $y = 2x - 5 = \frac{13}{3}$. $\qquad\qquad y = 2x + 1 = \dfrac{-1 \pm 4i\sqrt{2}}{3}$

The solutions are $x = 2, y = -1$; $x = \frac{14}{3}, y = \frac{13}{3}$; $x = \dfrac{-2 \pm 2i\sqrt{2}}{3}, y = \dfrac{-1 \pm 4i\sqrt{2}}{3}$.

5.197 Solve the system

$$\begin{cases} \dfrac{5}{x^2} + \dfrac{3}{y^2} = 32 & (1) \\ 4xy = 1 & (2) \end{cases}$$

▮ Write (1) as $3x^2 + 5y^2 = 32x^2y^2 = 2(4xy)^2$. Substitute (2):

$$3x^2 + 5y^2 = 2(1)^2 = 2 \quad (3)$$

Subtract 2(2) from (3): $3x^2 - 8xy + 5y^2 = 0$. Then $(x - y)(3x - 5y) = 0$ and $x = y$, $x = 5y/3$. Solve the systems

$$\begin{cases} 4xy = 1 \\ \quad x = y \end{cases} \qquad\qquad \begin{cases} 4xy = 1 \\ \quad x = \dfrac{5y}{3} \end{cases}$$

$$4y^2 = 1, \ y = \pm\tfrac{1}{2}, \ x = y = \pm\tfrac{1}{2} \qquad \tfrac{20}{3}y^2 = 1 \qquad y^2 = \tfrac{3}{20} = \tfrac{15}{100}$$

$$y = \pm\dfrac{\sqrt{15}}{10} \qquad x = \dfrac{5}{3}y = \pm\dfrac{\sqrt{15}}{6}$$

For Probs. 5.198 to 5.202, solve the given problem by using a system of equations.

5.198 Two numbers differ by 2, and their squares differ by 48. Find the numbers

▮ Let x and y be the numbers. Then $x - y = 2$ (differ by 2) and $x^2 - y^2 = 48$ (squares differ by 48). Thus, $x = y + 2$. Substituting for x, we get $(y + 2)^2 - y^2 = 48$, $y^2 + 4y + 4 - y^2 = 48$, $4y = 44$, and $y = 11$. If $y = 11$, $x = 13$.

5.199 The sum of the circumferences of two circles is 88 in, and the sum of their areas is $\tfrac{2200}{7}$ in^2 when $\pi \approx \tfrac{22}{7}$. Find the radius of each circle.

▮ Let r_1 and r_2 be the radii. Thus, $2(\tfrac{22}{7})r_1 + 2(\tfrac{22}{7})r_2 = 88$ and $\tfrac{22}{7}r_1^2 + \tfrac{22}{7}r_2^2 = \tfrac{2200}{7}$. Then $44r_1 + 44r_2 = 616$, $r_1 + r_2 = 14$, $r_1 = 14 - r_2$. Also $r_1^2 + r_2^2 = 100$. Substituting for r_1 in this equation gives $(14 - r_2)^2 + r_2^2 = 100$, $196 + r_2^2 - 28r_2 + r_2^2 = 100$, $2r_2^2 - 28r + 96 = 0$, $r_2^2 - 14r + 48 = 0$, and $r = 6$ in, 8 in.

5.200 A party costing \$30 is planned. It is found that by adding 3 more to the group, the cost per person would be reduced by 50¢. For how many people was the party originally planned?

▮ Let x = cost per person and y = number of people. Then $xy = 30$ and $x = 30/y$. Thus, $x - 0.5 = 30/(y + 3)$ (since the cost per person was reduced by 50¢ and the number of people was increased by 3). Then $30/y - 0.5 = 30/(y + 3)$, and $y = 12$.

5.201 The square of a certain number exceeds twice the square of another number by 16. Find the numbers if the sum of their squares is 208.

▮ Let x and y be the numbers. Then $x^2 = 2y^2 + 16$ and $x^2 + y^2 = 208$. Thus, $x^2 = 208 - y^2 = 2y^2 + 16$, $2y^2 + 16 = 208 - y^2$, $3y^2 = 192$, $y^2 = 64$, and $y = \pm 8$. If $y = 8$, $x = \pm 12$. If $y = -8$, $x = \pm 12$. Thus the solutions are $x = 12$, $y = 8$; $x = 12$, $y = -8$; $x = -12$, $y = 8$; $x = -12$, $y = -8$.

5.202 The diagonal of a rectangle is 85 ft. If the short side is increased by 11 ft and the long side decreased by 7 ft, the length of the diagonal remains the same. Find the original dimensions.

▮ See Fig. 5.14a. Let x = width and y = length. Then $d = \sqrt{x^2 + y^2}$, and $d' = \sqrt{(x + 11)^2 + (y - 7)^2}$ (see Fig. 5.14b). If $d = d'$, then $\sqrt{x^2 + y^2} = \sqrt{(x + 11)^2 + (y - 7)^2}$. Since $\sqrt{x^2 + y^2} = 85$, $\sqrt{(x + 11)^2 + (y - 7)^2} = 85$. Then we have the equations $(x + 11)^2 + (y - 7)^2 = 7225$ and $x^2 + y^2 = 7225$. Solving these two equations, we get $x = 40$ ft and $y = 75$ ft.

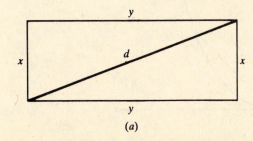
(a)

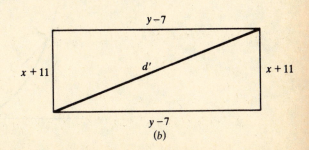
(b)

Fig. 5.14

For Probs. 5.203 to 5.208, solve the given system graphically.

5.203 $y = x^2$, $y = x^3$

    ❚ See Fig. 5.15. For both equations if $x = 1$, $y = 1$.

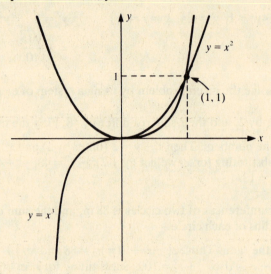

Fig. 5.15

5.204 $y = x^2$, $y = x^4$

    ❚ See Fig. 5.16, For $y = x^2$ if $x = 1$, $y = 1$. For $y = x^4$, if $x = -1$, $y = 1$.

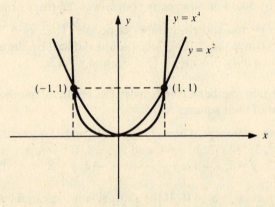

Fig. 5.16

5.205 $y = x^2$, $y = |x|$

    ❚ See Fig. 5.17. For $y = x^2$, if $x = 1$, $y = 1$. For $y = |x|$, if $x = -1$, $y = 1$.

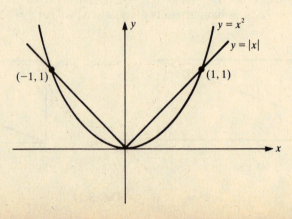

Fig. 5.17

5.206 $y = x^2$, $2y = 3x + 2$

▎ See Fig. 5.18. For $y = x^2$, if $x = 2$, $y = 4$. For $2y = 3x + 2$, if $x = -\frac{1}{2}$, $y = \frac{1}{4}$. Notice that a carefully drawn graph is crucial here.

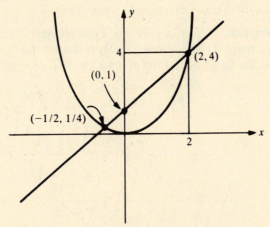

Fig. 5.18

5.207 $xy = 1$, $x = y$

▎ See Fig. 5.19. For $xy = 1$, if $x = 1$, $y = 1$. For $x = y$, if $x = -1$, $y = -1$.

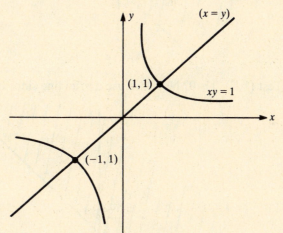

Fig. 5.19

5.208 $x^2 + y^2 = 25$, $y = x - 5$

▎ See Fig. 5.20. For $x^2 + y^2 = 25$, if $x = 5$, $y = 0$. For $y = x - 5$, if $x = 0$, $y = -5$.

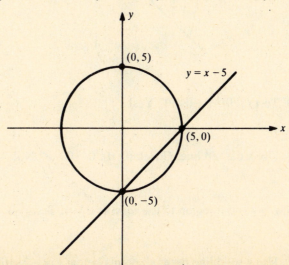

Fig. 5.20

5.4 SYSTEMS OF INEQUALITIES

For Probs. 5.209 to 5.216, graph the given inequality.

5.209 $2x - 3y < 6$

▮ See Fig. 5.21. We graph the line $2x - 3y = 6$. Then we find which "side" of the plane satisfies the inequality by testing a point. Dash the line $2x - 3y = 6$ since the inequality is $<$, not $\leq$. Test $(0, 0)$: $2(0) - 3(0) < 6$. Thus, the $(0, 0)$ side of the plane is shaded.

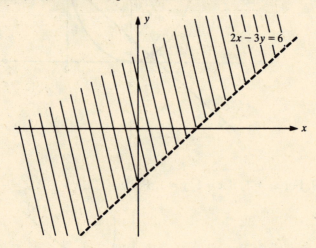

Fig. 5.21

5.210 $x \leq y$

▮ See Fig. 5.22. Test $(1, 0)$: $1 \leq 0$? No! Also use a solid line since it is $\leq$.

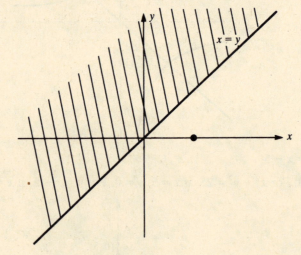

Fig. 5.22

5.211 $x + y \geq 0$

▮ See Fig. 5.23. Test $(1, 0)$: $1 + 0 \geq 0$? Yes!

5.212 $4x - y > 8$

▮ See Fig. 5.24. Use a dashed line. Check $(0, 0)$: $0 - 0 > 8$? No!

5.213 $x > 4$

▮ See Fig. 5.25. $x > 4$ is the region to the right of $x = 4$. Test $(-1, 0)$.

5.214 $y \leq -2$

▮ See Fig. 5.26. Use a solid line here; all points below $y = -2$ satisfy $y \leq -2$.

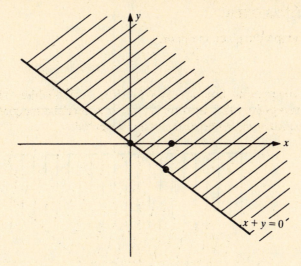

Fig. 5.23

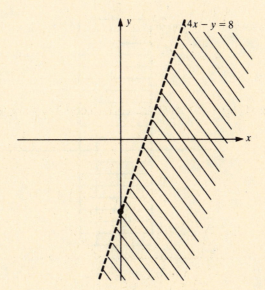

Fig. 5.24

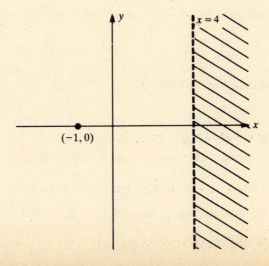

Fig. 5.25

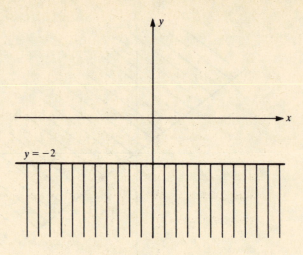

Fig. 5.26

5.215 $-5 < x \le 1$

▮ See Fig. 5.27. Be careful! One line is dashed, and the other is not.

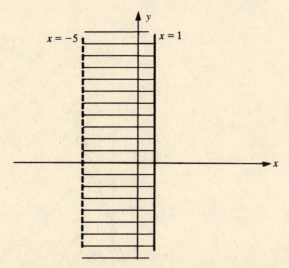

Fig. 5.27

5.216 $1 \le y < 0$

▮ There is no region satisfying this inequality. If $1 \le y$, then $y \ge 1$; thus, y is not less than zero. No solution.

For Probs. 5.217 to 5.231, find the solution set of each system graphically.

5.217 $-2 \le x < 2,\ -1 < y \le 6$

▮ See Fig. 5.28. We graph $2 \le x < 2$ and $-1 < y \le 6$ on the same set of axes and find where they intersect. Notice the cross hatched region: That is the solution set.

5.218 $-4 \le x < -1,\ -2 < y \le 5$

▮ See Fig. 5.29. We sketch the regions $-4 \le x < -1$ and $-2 < y \le 5$ on the same axes and find the intersection of these regions. The crosshatched region is the solution set.

5.219 $x < 5,\ y > 2$

▮ See Fig. 5.30. The crosshatched quadrant is the solution set.

5.220 $2x + y \le 8,\ 0 \le x \le 3,\ 0 \le y \le 5$

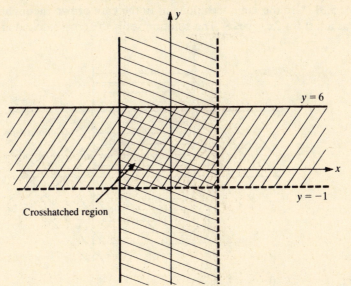

Fig. 5.28

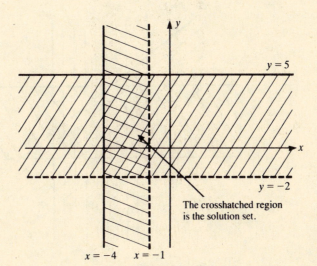

Fig. 5.29

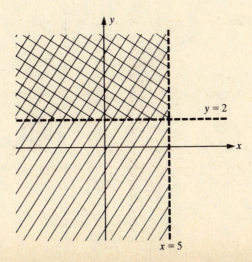

Fig. 5.30

❚ See Fig. 5.31. Use the same technique as in the case of two inequalities. Use all solid lines, and use $(0, 0)$ as a test for $2x + y \le 8$. The solution set is the crosshatched area.

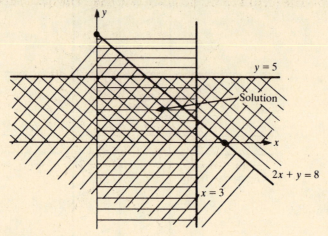

Fig. 5.31

5.221 $x + 3y \le 12$, $0 \le x \le 8$, $0 \le y \le 3$

❚ See Fig. 5.32. All lines are solid. Use $(0, 0)$ as a test for $x + 3y \le 12$.

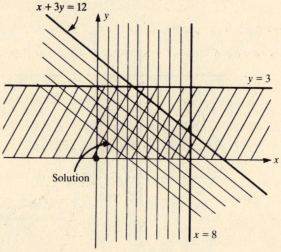

Fig. 5.32

5.222 $2x + y \le 8$, $x + 3y \le 12$, $x \ge 0$, $y \ge 0$

❚ See Fig. 5.33. The solution set is in quadrant I since $x, y \ge 0$. Use $(0, 0)$ as a test point for both lines.

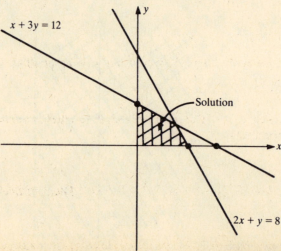

Fig. 5.33

5.223 $x + y \geq 1, 2x - 3y \leq 6, x \geq 0, y \geq 0$

▌ See Fig. 5.34. Use $(0, 0)$ as a test point for both lines. Use quadrant I only!

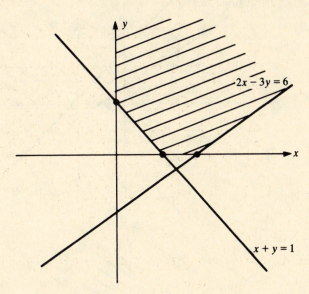

Fig. 5.34

5.224 $x + 2y \leq 10, 3x + y \leq 15, x \geq 0, y \geq 0$

▌ See Fig. 5.35. Once again use solid lines and $(0, 0)$ as a test point.

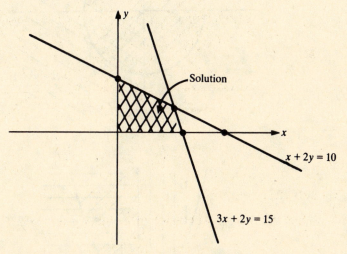

Fig. 5.35

5.225 $3x + 4y \geq 8, 4x + 3y \geq 24, x, y \geq 0$

▌ See Fig. 5.36. Be careful here! The solution set is in quadrant I.

5.226 $x^2 + y^2 \leq 1, x \geq 0$

▌ See Fig. 5.37. Test $(0, 0)$. The inside of the circle satisfies the first inequality. The crosshatched region is the solution set.

5.227 $y \geq x^2, x \geq 0$

▌ See Fig. 5.38. Use $(0, 1)$ as a test. Is $1 > 0^2$? Yes! The crosshatched region is the solution set.

5.228 $y \geq x^2, y < 1$

▌ See Fig. 5.39 and Prob. 5.227. The crosshatched region is the solution set.

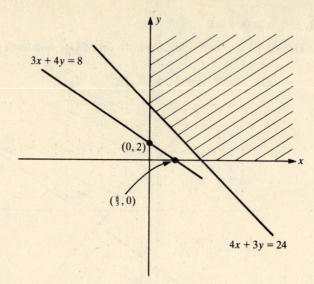

$3x + 4y = 8$

$(0, 2)$

$(\frac{8}{3}, 0)$

$4x + 3y = 24$

Fig. 5.36

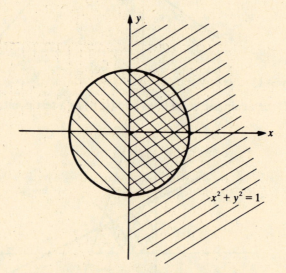

$x^2 + y^2 = 1$

Fig. 5.37

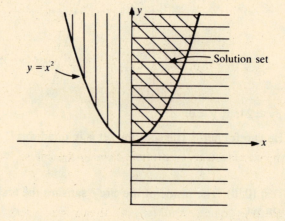

Solution set

$y = x^2$

Fig. 5.37

5.229 $y > x^2$, $y \leq 1$

┃ See Fig. 5.40 and Prob. 5.228. The crosshatched region is the solution set.

5.230 $y \leq x^2$, $y \leq 0$

┃ See Fig. 5.41. The solution set here is the half plane below $y = 0$.

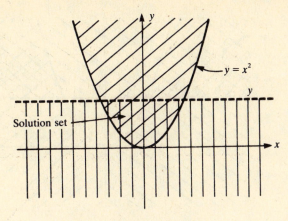

Fig. 5.39

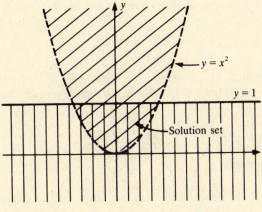

Fig. 5.40

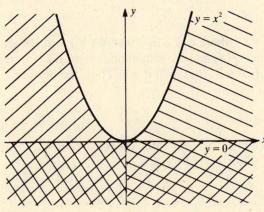

Fig. 5.41

5.231 $y \le x^2,\ y > 1$

    ❚ See Fig. 5.42.

For Probs. 5.232 to 5.235, find a parametric representation for the line segment $\overline{P_1P_2}$.

5.232 $P_1(2, 3),\ P_2(5, 8)$

    ❚ If the line l connects $P_1(x_1, y_1)$ and $P_2(x_2, y_2)$, then $x = x_1 + t(x_2 - x_1)$ and $y = y_1 + t(y_2 - y_1)$ for any $t \in \mathcal{R}$ represents (x, y) on l. In this case, $x = 2 + t(5 - 2)$ and $y = 3 + t(8 - 3)$, or $x = 2 + 3t$ and $y = 3 + 5t$. These last two equations represent (x, y) on line $\overleftrightarrow{P_1P_2}$. If $0 \le t \le 1$, the segment $\overline{P_1P_2}$ is represented.

5.233 $P_1(3, 5),\ P_2(-1, 2)$

    ❚ See Prob. 5.232. $x_1 = 3,\ x_2 = -1,\ y_1 = 5,\ y_2 = 2$. Then $x = 3 + t(-1 - 3)$ and $y = 5 + t(2 - 5)$. Then

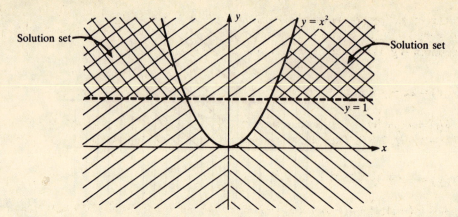

Fig. 5.42

$x = 3 - 4t$ and $y = 5 - 3t$. These equations represent the line. If $0 \le t \le 1$, they represent the line segment.

5.234 $P_1(-4, 0)$, $P_2(2, -5)$

▌ Here $x_1 = -4$, $y_1 = 0$, $x_2 = 2$, $y_2 = -5$. Then $x = -4 + t[2 - (-4)]$ and $y = 0 + t(-5 - 0)$, or $x = -4 + 6t$ and $y = -5t$. If $0 \le t \le 1$, these equations represent the line segment.

5.235 $P_1(7, 3)$, $P_2(-2, -3)$

▌ $x = 7 + t(-2 - 7)$ and $y = 3 + t(-3 - 3)$, or $x = 7 - 9t$ and $y = 3 - 6t$. If $0 \le t \le 1$, these equations represent the line segment.

For Probs. 5.236 to 5.240, find the maximum and minimum values for $f(t)$ in the given range.

5.236 $f(t) = 2t + 5$, $0 \le t \le 4$

▌ If $f(t) = at + b$, $a \ne 0$ (where $a, b \in \mathcal{R}$), then if $c \le t \le d$, the extrema for f occur at c and d. Also if $a > 0$, the maximum is $f(d)$ and the minimum is $f(c)$. If $a < 0$, the maximum is $f(c)$ and the minimum is $f(d)$. Here $a = 2 > 0$; $c = 0$, $d = 4$. Thus, the maximum is $f(4) = 4(2) + 5 = 13$, and the minimum is $f(0) = 5$.

5.237 $f(t) = 3t - 2$, $-2 \le t \le 3$

▌ Here $a = 3 > 0$ (see Prob. 5.236). Thus, the maximum $= f(d) = f(3) = 9 - 2 = 7$, and the minimum $= f(c) = f(-2) = -6 - 2 = -8$.

5.238 $f(t) = -3t + 2$, $1 \le t \le 6$

▌ Then $a = -3 < 0$. The maximum $= f(c) = f(1) = -1$, and the minimum $= f(d) = f(6) = -18 + 2 = -16$.

5.239 $f(t) = -t - 4$, $-5 \le t \le 2$

▌ $a = -1 < 0$. The maximum $= f(c) = f(-5) = 1$, and the minimum $= f(d) = f(2) = -6$.

5.240 $f(t) = 7$, $0 \le t \le 5$

▌ The above theorem does not apply since $a = 0$. However, if $f(t) = 7$, then $f = 7$ for all t, and f is constant.

For Probs. 5.241 to 5.243, express $f(x, y)$ as $g(t)$ for $\overline{P_1P_2}$, and then find the extrema.

5.241 $f(x, y) = 3x + 2y - 5$; $P_1 = (2, 1)$, $P_2(8, 6)$

▌ $f(x, y) = f[x_1 + t(x_2 - x_1), y_1 + t(y_2 - y_1)] = ax_1 + by_1 + c + [a(x_2 - x_1) + b(y_2 - y_1)]t = g(t)$, where

$P_1 = (x_1, y_1)$, $P_2 = (x_2, y_2)$, and $f(x, y) = ax + by + c$. Then $g(t) = ax_1 + by_1 + c + [a(x_2 - x_1) + b(y_2 - y_1)]t = g(t) = 3(2) + 2(1) + (-5) + [3(8 - 2) + 2(6 - 1)]t = 3 + (18 + 10)t = 3 + 28t$. Since $28 > 0$, the maximum is $g(1) = 31$ and the minimum is $g(0) = 3$.

5.242 $f(x, y) = 2x - y + 3$; $P_1 = (-3, 0)$, $P_2 = (2, 3)$

▌ $g(t) = ax_1 + by_1 + c + t[a(x_2 - x_1) + b(y_2 - y_1)] = 2(-3) + (-1)(0) + 3 + t[2(2 + 3) + (-1)(3 - 0)] = -3 + 7t$; $7 > 0$. Thus, the maximum is $g(1) = 4$, and the minimum is $g(10) = -3$.

5.243 $f(x, y) = -x + 4y + 2$; $P_1 = (4, -3)$, $P_2 = (-1, 4)$

▌ $g(t) = (-1)(4) + 4(-3) + 2 + t[-1(-1 - 4) + 4(4 + 3)] = -14 - 33t$. The maximum $= 19 = g(1)$, and the minimum $= -14 = g(0)$.

For Probs. 5.244 and 5.245, find the extrema for f on $\overline{P_1 P_2}$.

5.244 $f(x, y) = 5x + 2y - 3$; $P_1(2, 3)$, $P_2(5, -1)$

▌ The extrema must occur at P_1 and P_2, the endpoints of $\overline{P_1 P_2}$. Thus, the maximum is $f(P_1)$ or $f(P_2)$, and the minimum is $f(P_1)$ or $f(P_2)$. $f(P_1) = f(2, 3) = 5(2) + 2(3) - 3 = 13$, and $f(P_2) = f(5, -1) = 5(5) + 2(-1) - 3 = 20$. Since $20 > 13$, the maximum $= 20$ and the minimum $= 13$.

5.245 $f(x, y) = -4x - 2y + 2$; $P_1(3, 2)$, $P_2(-2, -4)$

▌ $f(P_1) = -4(3) - 2(2) + 2 = -14$, and $f(P_2) = -4(-2) - 2(-4) + 2 = 18$. Since $18 > -14$, $f(P_1) =$ the minimum $= -14$ and $f(P_2) =$ the maximum $= 18$.

For Probs. 5.246 to 5.251, find the maximum and minimum values of f on the set S determined by the given linear inequalities.

5.246 $-2x + 3y - 6 \le 0$, $y \ge 0$, $x \le 0$; $f(x, y) = 2x + y - 1$

▌ See Fig. 5.43. We notice that S is convex. Thus, the maximum and minimum of f on S occur at the vertices. $f(0, 0) = 0 + 0 - 1 = -1$. $f(-3, 0) = -6 + 0 - 1 = -7 =$ minimum. $f(0, 2) = 0 + 2 - 1 = 1 =$ maximum.

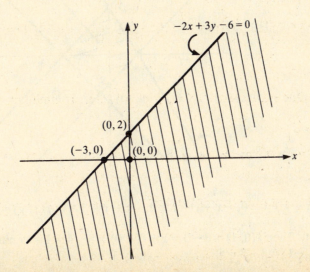

Fig. 5.43

5.247 $2x - y + 2 \geq 0$, $x + y - 2 \leq 0$, $y \geq 0$, $f(x, y) = -x + 3y - 5$

▮ See Fig. 5.44. S has vertices $(-1, 0)$, $(0, 2)$, $(2, 0)$ and is convex. Then $f(-1, 0) = 1 - 5 = -4$. $f(2, 0) = -2 - 5 = -7 =$ minimum. $f(0, 2) = 6 - 5 = 1 =$ maximum.

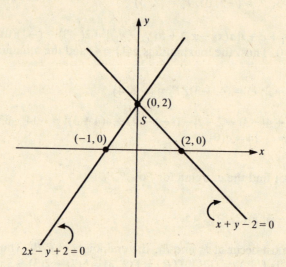

Fig. 5.44

5.248 $2x - y + 2 \geq 0$, $x + y - 2 \leq 0$, $y \geq 0$; $f(x, y) = 2x - 6y + 8$

▮ See Prob. 5.247. This is the same region. Then $f(-1, 0) = -2 + 8 = 6$. $f(2, 0) = 4 + 8 = 12 =$ maximum. $f(0, 2) = -12 + 8 = -4 =$ minimum.

5.249 $3x - 2y + 6 \geq 0$, $x + y + 2 \geq 0$, $x - y - 3 \leq 0$, $x + y - 3 \leq 0$; $f(x, y) = x + y - 8$

▮ See Fig. 5.45. Then $f(0, 3) = 0 + 0 - 8 = -8$. $f(-2, 0) = -2 + 0 - 8 = -10$. $f(3, 0) = 3 + 0 - 8 = -5 =$ maximum. $f(\frac{1}{2}, -2\frac{1}{2}) = \frac{1}{2} - 2\frac{1}{2} - 8 = -10 =$ minimum.

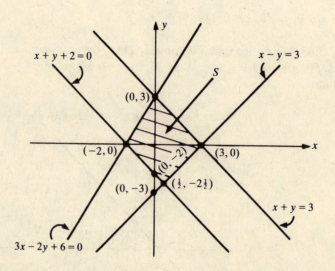

Fig. 5.45

5.250 $x - y + 1 \geq 0$; $x + y + 1 \geq 0$, $-x + y + 1 \geq 0$; $-x - y + 1 \geq 0$; $f(x, y) = -4x + 3y + 8$

▮ See Fig. 5.46. Then $f(0, 1) = 8 + 3 = 11$. $f(1, 0) = 8 - 4 = 4 =$ minimum. $f(-1, 0) = 8 + 4 = 12 =$ maximum. $f(0, -1) = 8 - 3 = 5$.

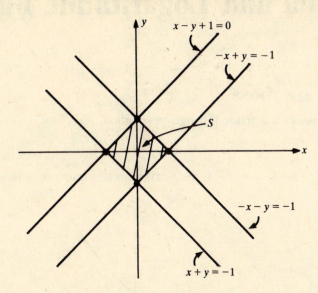

$x - y + 1 = 0$

$-x + y = -1$

S

$-x - y = -1$

$x + y = -1$

Fig. 5.46

5.251 $x - y + 1 \geq 0$, $x + y + 1 \geq 0$, $-x + y + 1 \geq 0$, $-x - y + 1 \geq 0$; $f(x, y) = x$

▮ Here S is the same region as in Prob. 5.250 above. Then $f(0, 1) = x = 0$. $f(1, 0) = x = 1 =$ maximum. $f(1, 0) = x = -1 =$ minimum. $f(0, -1) = x = 0$.

CHAPTER 6
Exponential and Logarithmic Functions

6.1 EXPONENTIAL FUNCTIONS

For Probs. 6.1 to 6.20, sketch the graph of the given equation.

6.1 $y = 2^x$

▌ See Fig. 6.1. If $x = 0$, $y = 2^0 = 1$; the x axis is an asymptote; as x increases, y increases.

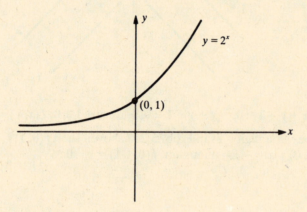

Fig. 6.1

6.2 $y = -2^x$

▌ See Fig. 6.2. If $x = 0$, $y = -2^0 = -1$; $-2^x \neq 0$ implies that y is never zero (the x axis is an asymptote); as x gets large, y decreases.

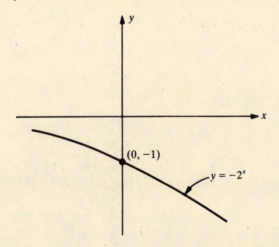

Fig. 6.2

6.3 $y = 3^x$, $y = 4^x$ on the same axes.

▌ See Fig. 6.3. $y = 4^x$ is steeper; both equations increase as x increases, but 4^x increases more dramatically.

6.4 $y = -2^x$, $y = -3^x$ on the same axes.

▌ See Fig. 6.4. The reasoning is the same as in Prob. 6.3. $y = -3^x$ will decrease more dramatically.

6.5 $y = 3^{-x}$

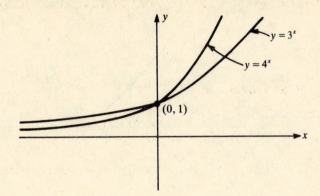

Fig. 6.3

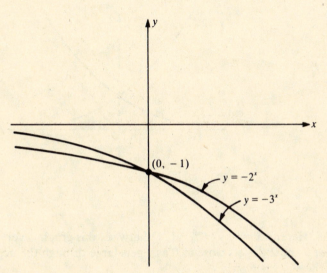

Fig. 6.4

▌ See Fig. 6.5. If $y = 3^{-x}$, then $y = 1/3^x$. If $x = 0$, $y = 1$. The x axis is an asymptote; and as x increases, y decreases.

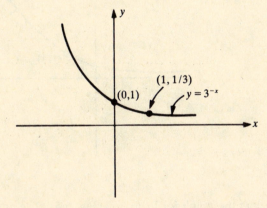

Fig. 6.5

6.6 $y = 4^{-x}$

▌ See Fig. 6.6. See Prob. 6.5; this is a similar function. Which is steeper?

6.7 $y = 2^x$ and $y = 2^{-x}$ on the same axes.

▌ See Fig. 6.7. Since $2^x \cdot 2^{-x} = 1$, we suspect that something interesting will occur. Notice that one curve is the image of the other in the y axis.

6.8 $y = 2^{x-1}$

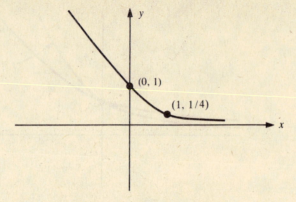

Fig. 6.6

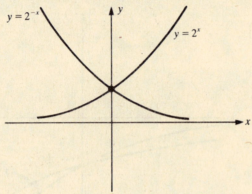

Fig. 6.7

▌ See Fig. 6.8. Here, if $x = 1$, $y = 2^0 = 1$. Otherwise, this graph is very similar to that of $y = 2^x$. Since y is never 0 and 2^{x-1} gets close to 0 as x gets large through the negative numbers, the x axis is an asymptote.

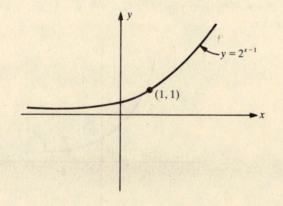

Fig. 6.8

6.9 $y = -2^{-x}$

▌ See Fig. 6.9. Look at the graph for Prob. 6.5 (Fig. 6.5). The graph of $y = 2^{-x}$ is similar. The graph of $y = -2^{-x}$ is the ordinate of each point in $y = 2^{-x}$ negated.

6.10 $y = 3^{-0.5x}$

▌ See Fig. 6.10. $3^{-0.5x} = 1/3^{0.5x}$. If $x = 0$, $y = 1$. The x axis is again an asymptote, since y approaches 0 as x grows large.

6.11 $y = 2 + 2^x$

▌ See Fig. 6.11. We take the graph of 2^x and add 2 to each ordinate.

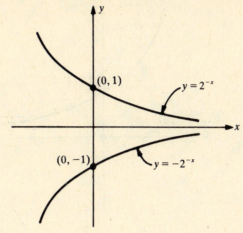

Fig. 6.9

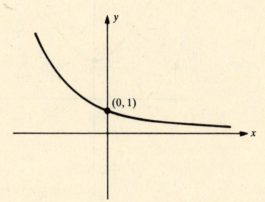

Fig. 6.10

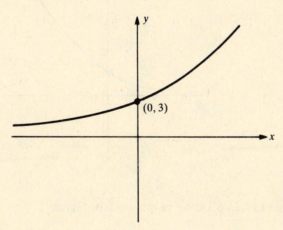

Fig. 6.11

6.12 $y = 3^{1-x}$

▮ See Fig. 6.12. If $x = 1$, $y = 3^0 = 1$. We notice that the x axis is again an asymptote. Also as x takes on large negative values, y gets large.

6.13 $y = 2^{|x|}$

▮ See Fig. 6.13. If $x = 0$, $y = 1$. For $x > 0$, $2^{|x|} = 2^x$; for $x < 0$, $2^{|x|} = 2^{-x}$. We combine these two graphs.

6.14 $y = 2^{2x}$

▮ See Fig. 6.14. When $x = 0$, $y = 1$; besides that, as x increases, y increases dramatically.

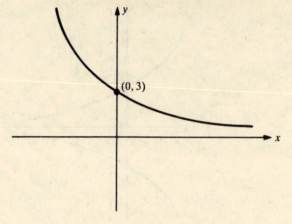

Fig. 6.12

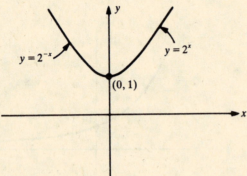

Fig. 6.13

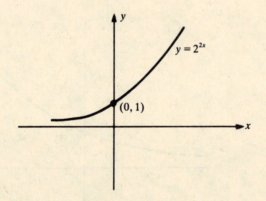

Fig. 6.14

6.15 $y = -2^{2x}$

▮ See Fig. 6.15 and Prob. 6.14; we negate each ordinate.

6.16 $y = e^x$

▮ See Fig. 6.16. Recall that e is the base for natural logarithms. In the calculus you will learn that e is the number which $(1 + 1/b)^n$ approaches as n gets arbitrarily large. Numerically, $e \approx 2.718$; it is an irrational number. Thus, when $x = 0$, $e^0 = 1$; as x increases, so does e^x.

6.17 $y = 2e^{-x}$

▮ See Fig. 6.17. $2e^{-x} = 2(1/e^x)$. If $x = 0$, $1/e^x = 1$; then $2(1/e^x) = 2(1) = 2$. As x gets large, $2/e^x$ decreases

6.18 $y = 2^{x+3}$

▮ See Fig. 6.18. There are two ways (at least!) in which we can obtain the graph. Notice that $2^{x+3} = 2^x 2^3 = 8 \cdot 2^x$. We can get the same graph by noting directly that when $x = 0$, $y = 2^3 = 8$.

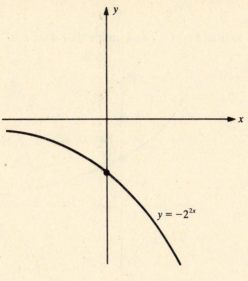

$y = -2^{2x}$

Fig. 6.15

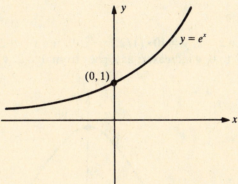

$y = e^x$

$(0, 1)$

Fig. 6.16

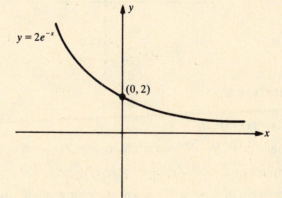

$y = 2e^{-x}$

$(0, 2)$

Fig. 6.17

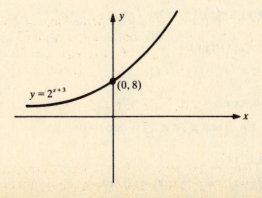

$y = 2^{x+3}$

$(0, 8)$

Fig. 6.18

6.19 $y = 2e^{-x} + 5$

▌ See Fig. 6.19. We look at Fig. 6.17 and notice that if we add 5 to each ordinate, we get our function.

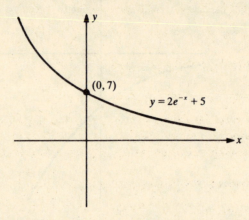

$(0, 7)$

$y = 2e^{-x} + 5$

Fig. 6.19

6.20 $y = 10 \cdot 2^{-x^2}$, $-2 \leq x \leq 2$

▌ See Fig. 6.20. When $x = 2$, $y = 10 \cdot (1/2^4) = \frac{10}{16}$. When $x = -2$, $y = \frac{10}{16}$. When $x = 0$, $y = 10 \cdot 1/2^0 = 10$. As x goes from -2 to 0, y increases; as x goes from 0 to 2, y decreases.

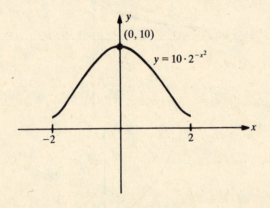

$(0, 10)$

$y = 10 \cdot 2^{-x^2}$

-2 2

Fig. 6.20

For Probs. 6.21 to 6.27, let $f(x) = a^x$.

6.21   Prove that $f(x + 2) = f(x) \cdot f(2)$.

▌ If $f(x) = a^x$, then $f(x + 2) = a^{x+2} = a^x a^2$. Since $a^x = f(x)$ and $a^2 = f(2)$, $f(x + 2) = a^x a^2 = f(x) \cdot f(2)$.

6.22   Generalize the statement in Prob. 6.21, and prove your generalization.

▌ Prove that $f(x + y) = f(x)f(y)$. *Proof*: $f(x + y) = a^{x+y} = a^x a^y = f(x)f(y)$.

6.23   Prove that $f(x - y) = f(x)/f(y)$.

▌ $f(x - y) = a^{x-y} = a^x/a^y = f(x)/f(y)$.

6.24   Prove that f is one-to-one.

▌ Suppose that $m \neq n$. Then $a^m \neq a^n$; f is one-to-one.

6.25   Prove that $f(-x) = 1/f(x)$.

▌ $f(x) = a^x$; then $f(-x) = a^{-x} = 1/a^x = 1/f(x)$.

6.26 Prove that $f(2+x) = a^2 f(x)$.

▮ $f(2+x) = a^{2+x} = a^2 a^x = a^2 a^x$.

6.27 Generalize the statement in Prob. 6.26, and prove your statement.

▮ Prove that $f(b+x) = a^b f(x)$. *Proof:* $f(b+x) = a^{b+x} = a^b a^x = a^b f(x)$.

For Probs. 6.28 to 6.30, compute the compound amount. Use a calculator to perform the arithmetic.

6.28 $2000 at 12 percent compounded semiannually for 3 years

▮ $A(n) = P(1+i)^n$, where $A(n) =$ compound amount at the end of n interest periods, $i =$ rate per interest period, $P =$ amount invested. Here $n = 6$, $i = 0.12/2 = 0.06$. Then $A(6) = \$2000(1+0.06)^6 = \$2000(1.06)^6 \approx \$2000(1.42)$ (rounded to two places on the calculator) $= \$2840$.

6.29 $6000 at 15 percent compounded quarterly for 4 years, 6 months

▮ $A(n) = P(1+i)^n$, where $n = 18$, $P = \$6000$, $i = 0.15/4 = 0.0375$. Then $A(18) = \$6000(1 + 0.0375)^{18} = \$6000(1.0375)^{18} \approx \$6000(1.94)$ (rounded to the hundreds place) $= \$11,640$.

6.30 $5050 at $11\frac{3}{4}$ percent compounded daily for 2 years

▮ $A(n) = P(1+i)^n$, where $n = 730$, $P = \$5050$, $i = 0.1175/365 \approx 0.00032$ (rounded to five places). Then $A(730) = \$5050(1.00032)^{730} = \$6379$.

For Probs. 6.31 to 6.33, compute the principal P invested to yield the following compound amounts A.

6.31 $5000 at 10 percent compounded annually for 5 years

▮ Since $A = P(1+i)^n$, $P = A(1+i)^{-n}$; here $A = \$5000$, $i = 0.1/1 = 0.1$, $n = 5.0$. Then $P = \$5000(1+0.1)^{-5} = \$5000(1.1)^{-5} \approx \$5000(0.62)$ (rounded to the hundreds place) $= \$3100$.

6.32 $6750 at $12\frac{1}{2}$ percent compounded semiannually for 3 years

▮ $P = \$6750(1+0.0625)^{-6} = \$6750(1.0625)^{-6} \approx \$6750(0.69)$ (rounded) $= \$4657.50$.

6.33 $10,000 at $14\frac{3}{4}$ percent compounded quarterly for $6\frac{1}{2}$ years

▮ $P = \$10,000(1+0.036875)^{-26} = \$10,000(1.036875)^{-26} \approx \$10,000(0.39)$ (rounded) $= \$3900$.

For Probs. 6.34 to 6.36, compute the amount due in each, given that the interest is compounded continuously.

6.34 $3000 at 10 percent for 5 years

▮ $A = Pe^{rt} = \$3000(e^{5(0.1)}) = \$3000(e^{0.5}) \approx \$4946$ (rounded to the nearest dollar).

6.35 $4550 at $12\frac{1}{2}$ percent for 3 years

▮ $A = \$4550(e^{(0.125)(3)}) = \$4550(e^{0.375}) \approx \$6620$.

6.36 $7500 at $16\frac{1}{4}$ percent for $4\frac{1}{2}$ years

▮ $A = \$7500(e^{0.73125}) \approx \$15,581$. (Does that amount shock you?)

For Probs. 6.37 to 6.40, answer true or false, and explain your answer.

6.37 If $a > b$ and $n \in \mathscr{Z}$, then $a^n > b^n$.

▮ False. $4 > 2$, but $4^{-1} = \frac{1}{4}$, $2^{-1} = \frac{1}{2}$, and $\frac{1}{4} < \frac{1}{2}$.

6.38 If $a > b$ and $n \in \mathcal{N}$, then $a^{n^2} > b^{n^2}$.

    ∎ True. $a^n > b^n \ \forall n \in \mathcal{Z}$, but $n^2 \in \mathcal{N}$. Thus, $a^{n^2} > b^{n^2}$.

6.39 The graphs of $y = a^x$ and $y = b^x$ (where $a \neq b$; $a, b \in \mathcal{N}$) *must* intersect $\forall a, b$ so chosen.

    ∎ True. If $x = 0$, $a^0 = b^0 = 1$; they intersect at $(0, 1)$.

6.40 The graphs of $y = a^x$ and $y = 2a^x$ have no points of intersection.

    ∎ True. Suppose that $a^x = 2a^x$; $a^x \neq 0$, so we can divide by a^x. Then $1 = 2$.

6.2 LOGARITHMIC FUNCTIONS

For Probs. 6.41 to 6.45, rewrite in an equivalent exponential form.

6.41 $\log_{10} 100 = 2$

    ∎ Remember that if $\log_a b = x$, then $a^x = b$, and conversely. Thus, if we let $a = 10$, $b = 100$, $x = 2$, then $\log_{10} 100 = 2$ and $10^2 = 100$. However, by convention the base 10 is understood and therefore dropped. So $\log_{10} 100$ is written simply $\log 100$.

6.42 $\log 10,000 = 4$

    ∎ If $\log 10,000 = 4$, then $10^4 = 10,000$. Remember: A logarithm is an exponent.

6.43 $\log_2 8 = 3$

    ∎ See Prob. 6.41. Here, $a = 2$, $b = 8$, $x = 2$, and $2^3 = 8$.

6.44 $\log_4 64 = 3$

    ∎ Using the formula in Prob. 6.41, where $a = 4$, $b = 64$, $x = 3$, we have $4^3 = 64$.

6.45 $\log_{14} 1 = 0$

    ∎ $a = 14$, $b = 1$, $x = 0$, and $14^0 = 1$.

For Probs. 6.46 to 6.51, rewrite in an equivalent logarithmic form.

6.46 $7^2 = 49$

    ∎ Again, we use the formula $a^x = b \Leftrightarrow \log_a b = x$. Thus, $7^2 = 49 \Leftrightarrow \log_7 49 = 2$.

6.47 $4^c = 256$

    ∎ Here, $a = 4$, $b = 256$, $x = c$, and $\log_4 256 = c$.

6.48 $u = v^x$

    ∎ If $u = v^x$, then $a = v$, $b = u$, $x = x$, and $\log_v u = x$.

6.49 $9 = 27^{2/3}$

    ∎ Then $\log_{27} 9 = \frac{2}{3}$.

6.50 $625^{0.25} = 5$

    ∎ Do not be fooled or misled by decimals (or anything else!). The logarithm-exponential conversion still holds; $\log_{625} 5 = 0.25$.

6.51 $729^{1/6} = 3$

▎ $\log_{729} 3 = \frac{1}{6}$.

For Probs. 6.52 to 6.66, evaluate the given expression.

6.52 $\log_2 8$

▎ We are looking for x such that $\log_2 8 = x$. Then $2^x = 8$, but $x = 3$, so $\log_2 8 = 3$.

6.53 $\log 10^3$

▎ Remember that $\log a$ means $\log_{10} a$, and $\ln a$ means $\log_e a$. We let $\log 10^3 = x$ which means $10^x = 10^3$. Then $x = 3$ and $\log 10^3 = 3$. Also, remember that $\log_q r = s$ means "s is the power to which q is raised to yield r." To what power must 10 be raised to give 10^3?

6.54 $\log_2 125$

▎ If $t = \log_5 125$, then $5^t = 125$; $t = 3$.

6.55 $\log_a a^2$

▎ To yield a^2, raise a to the second power. $\log_a a^2 = 2$.

6.56 $\log_2 \frac{1}{16}$

▎ If $2^x = \frac{1}{16}$, then $2^{-x} = 16$, and $-x = 4$, or $x = -4$. See Prob. 6.57.

6.57 $\log_3 \frac{1}{27}$

▎ Compare this to Prob. 6.56. If $\log_3 \frac{1}{27} = x$, then $3^x = \frac{1}{27}$, which means $(\frac{1}{3})^{-x} = (\frac{1}{3})^3$, and $-x = 3$, or $x = -3$.

6.58 $\log_3 1$

▎ If $\log_3 1 = y$, then $3^y = 1$, or $y = 0$.

6.59 $\log_{4000} 1$

▎ If $4000^x = 1$, $x = 0$.

6.60 $\ln e^{-2}$

▎ $\ln e^{-2} = \log_e e^{-2}$. Then observe directly that $\ln e^{-2} = -2$, or that $\log_e e^{-2} = x$ means $e^x = e^{-2}$, or $x = -2$.

6.61 $\log_{32} 2$

▎ If $\log_{32} 2 = p$, then $32^p = 2$, or $(2^5)^p = 2$. Then $5p = 1$, and $p = \frac{1}{5}$.

6.62 $\log_2 2^{-4}$

▎ Observe directly that $\log_2 2^{-4} = -4$, or that if $\log_2 2^{-4} = s$, then $2^s = 2^{-4}$, or $s = -4$.

6.63 $\log_b b^u$

▎ If $\log_b b^u = k$, then $b^k = b^u$, or $k = u$. Then $\log_b b^u = u$. Think about it! Doesn't this make sense?

6.64 $\log_b b^{uv}$

▎ If $\log_b b^{uv} = k$, then $b^k = b^{uv}$, or $k = uv$. Notice that $k = uv$ also follows directly from the result in Prob. 6.63.

6.65 $\log_2 \sqrt{8}$

   ▌ If $\log_2 \sqrt{8} = x$, then $2^x = \sqrt{8} = \sqrt{2^3} = 2^{3/2}$, and $x = \frac{3}{2}$.

6.66 $\log_5 \sqrt[3]{5}$

   ▌ If $\log_5 \sqrt[3]{5} = x$, then $5^x = \sqrt[3]{5} = 5^{1/3}$, and $x = \frac{1}{3}$.

For Probs. 6.67 to 6.82, solve the equation.

6.67 $\log 100 = x$

   ▌ If $\log 100 = x$, then $10^x = 100 = 10^2$ and $x = 2$..

6.68 $\log x = 2$

   ▌ If $\log x = 2$, then $10^2 = x$ and $x = 100$.

6.69 $\log_2 x = 1$

   ▌ If $\log_2 x = 1$, then $2^1 = x$ and $x = 2$.

6.70 $\log_3 3 = x$

   ▌ $3^x = 3$ (Ask yourself, Which 3 in the equation is the base?) Then $3^x = 3^1$, and $x = 1$.

6.71 $\log x = 0$

   ▌ If $\log x = 0$, then $10^0 = x$, or $x = 1$.

6.72 $\log_x 81 = 4$

   ▌ $\log_x 81 = 4$. Then $x^4 = 81$, or $x = 3$ ($3^4 = 81$).

6.73 $\log_x \frac{1}{27} = -3$

   ▌ $\log_x \frac{1}{27} = -3$. Then $x^{-3} = \frac{1}{27} = 1/3^3 = 3^{-3}$, or $x = 3$.

6.74 $\ln e^x = 5$

   ▌ If $\ln e^x = 5$, then $\log_e e^x = 5$. Thus, $e^5 = e^x$, or $x = 5$.

6.75 $\ln e^{x+2} = 7$

   ▌ Then $\log_e e^{x+2} = 7$. Thus, $e^7 = e^{x+2}$, $x + 2 = 7$, or $x = 5$.

6.76 $\log_x 27x = 4$

   ▌ Then $x^4 = 27x$, $x^4 - 27x = 0$, $x(x^3 - 27) = 0$, and $x = 0$, $x = 3$. But $x = 0$ is extraneous (0 is not a logarithmic base), so $x = 3$.

6.77 $\log_{49} \frac{1}{7} = y$

   ▌ Then $49^y = \frac{1}{7}$, $(7^2)^y = \frac{1}{7}$, $7^{2y} = \frac{1}{7} = 7^{-1}$, $2y = -1$, or $y = -\frac{1}{2}$.

6.78 $\log_b 1000 = \frac{3}{2}$

   ▌ Then $b^{3/2} = 1000$, but $1000 = 10^3$. Thus $1000 = 100^{3/2}$, $b^{3/2} = 100^{3/2}$, or $b = 100$.

6.79 $\log_b 4 = \frac{2}{3}$

   ▌ Then $b^{2/3} = 4$. Raise each side to the power $\frac{3}{2}$. Then $(b^{2/3})^{3/2} = 4^{3/2}$ and $b = 8$. Compare this with Prob. 6.77, which can be done by using this technique as well.

6.80 $\log_b b = 1$

▌ If $\log_b b = 1$, then $b^1 = b$; b can be any positive real except 1. Thus, $b \neq 1$, $b > 0$ since b is the logarithm base.

6.81 $\log_b 1 = 0$

▌ Then $b^0 = 1$. This is true for all b. However, b must be positive and not 1 since b is the logarithm base.

6.82 $\log_{e^2} x = 10$

▌ Here, the base is e^2. Then $(e^2)^{10} = x$, or $x = e^{20}$.

For Probs. 6.83 to 6.87, evaluate the given expression where $f(x) = \log x$, $g(x) = 10^x$, $h(x) = \ln x$, $k(x) = e^x$, $l(x) = x^2$.

6.83 $f \circ g(x)$

▌ $f \circ g(x) = f(g(x)) = \log 10^x$. But $\log 10^x = x$. $f \circ g(x) = x$.

6.84 $g \circ f(x)$

▌ $g \circ f(x) = 10^{\log x} = x \ (x > 0)$.

6.85 $h \circ k(x)$

▌ $h \circ k(x) = \ln e^x = \log_e e^x = x$. Thus, $h \circ k(x) = x$.

6.86 $f \circ l(10)$

▌ $l(x) = x^2$; thus, $l(10) = 10^2 = 100$. Then $f \circ l(10) = f(l(10)) = \log 10^2 = 2$.

6.87 $l \circ h(3)$

▌ $h(x) = \ln x$; thus, $h(3) = \ln 3$. Then $l \circ h(3) = l(h(3)) = l(\ln 3) = (\ln 3)^2$.

For Probs. 6.88 to 6.97, sketch the given relation.

6.88 $y = \log_2 x$

▌ See Fig. 6.21. When $x = 1$, $y = \log_2 1 = 0$; when $x = 2$, $y = \log_2 2 = 1$. Also as x takes on values between 0 and 1, y decreases as x approaches 0; $\log_2 \frac{1}{2} = -1$, $\log_2 \frac{1}{4} = -2$, $\log_2 \frac{1}{8} = -3$, etc. The y axis is a vertical asymptote.

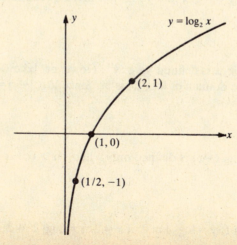

Fig. 6.21

6.89 $y = \log_4 x$

❚ See Fig. 6.22 and Prob. 6.88. $\log_4 1 = 0$, $\log_4 4 = 1$, $\log_4 \frac{1}{4} = -1$. The y axis is a vertical asymptote.

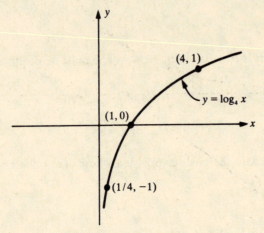

Fig. 6.22

6.90 $y = \log_2 x$, $y = \log_4 x$ on the same axes.

❚ See Fig. 6.23.

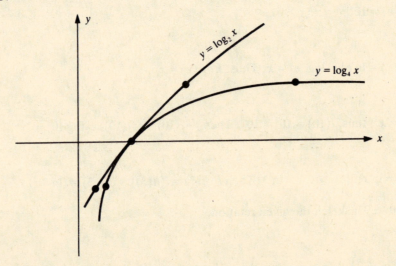

Fig. 6.23

6.91 $y = \log_2 x$, $y = 2^x$ on the same axes.

❚ See Fig. 6.24. Since $y = \log_2 x$ and $y = 2^x$ are inverse functions, their graphs are mirror images about $y = x$.

6.92 $y = \log_3 |x|$

❚ See Fig. 6.25. $y = \log_3 x$ is defined $\forall x \in \mathcal{R}^+$. However, $|x| \geq 0 \ \forall x \in \mathcal{R}$. Thus, while the domain of $\log_3 x$ is $\{x | x \in \mathcal{R}^+\}$, the domain of $\log_3 |x|$ is $\mathcal{R}$. Since $\log_x |-x| = \log_3 |x| = y$, $f(x) = \log_x |x|$ is symmetric about the y axis.

6.93 $y = \log_5 (-x)$

❚ See Fig. 6.26. $y = \log_5 (-x)$ is defined only when $-x > 0$; $-x > 0$ implies that $x < 0$. The domain of $f(x) = \log_5 (-x)$ is $\mathcal{R}^-$.

6.94 $y = \log_2 x^2$

❚ See Fig. 6.27. If $x = 1$, $y = \log_2 1 = 0$; if $x = 2$, $y = \log_2 2^2 = 2$; if $x = 4$, $y = \log_2 4^2 = \log_2 2^4 = 4$,

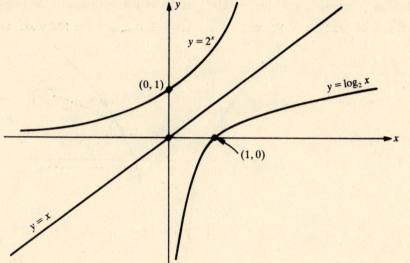

Fig. 6.24

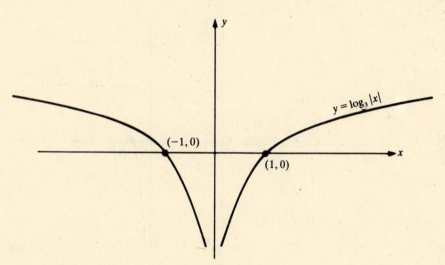

Fig. 6.25

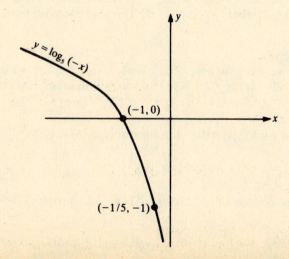

Fig. 6.26

etc. Also, as x decreases, y decreases. The y axis is an asymptote, and the graph is symmetric about the y axis.

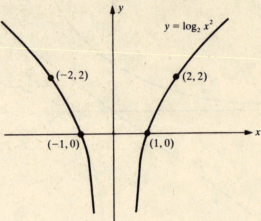

Fig. 6.27

6.95 $y = \log_2 (x - 2)$

▌ See Fig. 6.28. Compare this to $y = \log_2 x$. Here, the asymptote is the line $x = 2$. Also if $x = 3$, $y = \log_2 1 = 0$. If $x = 6$, $y = \log_2 4 = 2$.

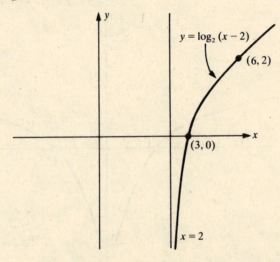

Fig. 6.28

6.96 $y = \log_2 2x$

▌ See Fig. 6.29. Again, we look to $y = \log_2 x$ for help. Here, we double each x value. If $x = \frac{1}{2}$, then $y = \log_2 2x = \log_2 1 = 0$; if $x = 1$, $y = \log_2 2 = 1$. The asymptote remains the same.

6.97 $y = \log_{1/2} x$

▌ See Fig. 6.30. Be very careful here. The base $b < 1$. If $x = 1$, $y = \log_{1/2} 1 = 0$; if $x = \frac{1}{2}$, $y = \log_{1/2} \frac{1}{2} = 1$; if $x = 2$, $y = \log_{1/2} 2 = -1$. The y axis is an asymptote since as x decreases, y increases.

6.98   Rewrite the equation $y = \log |x|$ without using absolute value signs.

▌
$$|x| = \begin{cases} x & x \geq 0 \\ -x & x < 0 \end{cases}$$

But $x = 0$ is not in the domain of this function. Thus, an equivalent definition is

$$y = \begin{cases} \log x & x > 0 \\ \log (-x) & x < 0 \end{cases}$$

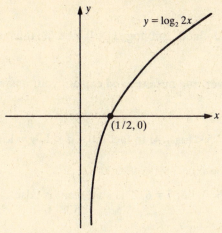

Fig. 6.29

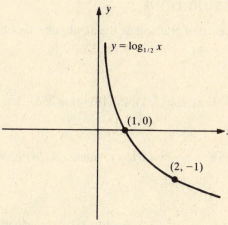

Fig. 6.30

6.99 Prove that $g(x) = \log_a x$ is one-to-one.

▮ To prove this, we use the definition of one-to-one. If $\log_a x = \log_a y$ and $\log_a x = p$, then $\log_a y = p$, and $a^p = x$, $a^p = y$. Thus, $x = y$. $g(x)$ is one-to-one.

6.100 Prove that the graphs of $y = \log_a x$ and $y = \log_{1/a} x$ are mirror images of each other about the x axis.

▮ We need to show that, for all x, $\log_a x = -\log_{1/a} x$. Then the y values are the same, and the graphs will be mirror images. Let $y = \log_a x$; then $a^y = x$, $1/a^y = 1/x$, $a^{-y} = 1/x$, $1/a^{-y} = x$, $(1/a)^{-y} = x$, $-y = \log_{1/a} x$, and the proof is complete.

6.101 Prove that $\log 3$ is irrational.

▮ This will be a proof by contradiction. Suppose that $\log 3 = a/b$ with $b, a \in \mathscr{Z}$, $b \neq 0$, $(a, b) = 1$. Then $10^{a/b} = 3$, which implies that $10^a = 3^b$. Then, since $5 \mid 10^a$, $5 \mid 3^b$. But 5 is *not* a factor of 3^b.

For Probs. 6.102 to 6.105, find the domain and range of the given function.

6.102 $y = \log_3 (x + 1)$

▮ We know $x + 1 \neq 0$; thus, the domain $= \{x \in \mathscr{R} \mid x > -1\}$, and the range $= \mathscr{R}$.

6.103 $y = \log_3 (2x - 5)$

▮ We must have $2x - 5 > 0$; thus, the domain $= \{x \in \mathscr{R} \mid x > \frac{5}{2}\}$, and the range $= \mathscr{R}$.

6.104 $y = \log_5 (x^2 + 1)$

▮ Since $x^2 + 1 > 0 \; \forall x \in \mathscr{R}$, the domain $= \mathscr{R}$, and the range $= \mathscr{R}$.

6.105 $y = |\log_6 x|$

▌ $x > 0$ for $\log_6 x$ to be defined. But $|\log_6 x| \geq 0 \ \forall x \in \mathcal{R}$; thus, the domain $= \{x \in \mathcal{R} \mid x > 0\}$, and the range $= \{y \in \mathcal{R} \mid y \geq 0\}$.

For Probs. 6.106 and 6.107, answer true or false, and explain your answer.

6.106   If $x > y$, then $\log_a x > \log_a y$.

▌ False; $5 > 4$, but $\log_{1/2} 5 < \log_{1/2} 4$. In fact if $0 < a < 1$, $y = \log_a x$ is a decreasing function.

6.107   If $\log_a x = \log_b x$, then $a = b$.

▌ $\log_5 1 = 0$, and $\log_6 1 = 0$; but $5 \neq 6$. The statement is false.

6.3   PROPERTIES OF LOG FUNCTIONS

For Probs. 6.108 to 6.116, write each expression as the algebraic sum of logarithms. The base is any positive real number except 1.

6.108 $\log (251)(46)(18)$

▌ $\log xy = \log x + \log y$; thus, $\log (251)(46)(18) = \log 251 + \log (46)(18) = \log 251 + \log 46 + \log 18$.

6.109 $\log (34)^2(2.7)$

▌ $\log xy = \log x + \log y$, and $\log z^2 = 2 \log z$; thus, $\log (34)^2(2.7) = \log (34)^2 + \log (2.7) = 2 \log 34 + \log 2.7$.

6.110 $\log (24)^{1/2}(35)^3$

▌ In general, $\log a^x = x \log a$; thus, $\log (24)^{1/2}(35)^3 = \log (24)^{1/2} + \log (35)^3 = \frac{1}{2} \log 24 + 3 \log 35$.

6.111 $\log \dfrac{(83)(41)}{29}$

▌ $\log \dfrac{x}{y} = \log x - \log y$; thus, $\log \dfrac{(83)(41)}{29} = \log (83)(41) - \log 29 = (\log 83 + \log 41) - \log 29$.

6.112 $\log \dfrac{(49)(65)}{(71)(86)}$

▌ $\log \dfrac{(49)(65)}{(71)(86)} = \log (49)(65) - \log (71)(86) = (\log 49 + \log 65) - (\log 71 + \log 86) =$ $\log 49 + \log 65 - \log 71 - \log 86$.

6.113 $\log \dfrac{(2.7)^2(58)^{1/3}}{(75)(89)^2}$

▌ $\log \dfrac{(2.7)^2(58)^{1/3}}{(75)(89)^2} = \log (2.7)^2(58)^{1/3} - \log (75)(89)^2 = (2 \log 2.7 + \frac{1}{3} \log 58) -$ $(\log 75 + 2 \log 89) = 2 \log 2.7 + \frac{1}{3} \log 58 - \log 75 - 2 \log 89$.

6.114 $\log \sqrt{\dfrac{(87)(28)}{15}}$

▌ $\log \sqrt{\dfrac{(87)(28)}{15}} = \log \left[\dfrac{(87)(28)}{15} \right]^{1/2} = \frac{1}{2} \log \dfrac{(87)(28)}{15} = \frac{1}{2}(\log 87 + \log 28 - \log 15)$.

6.115 $\log a^n b^m$

▮ $\log a^n b^m = \log a^n + \log b^m = n \log a + m \log b$.

6.116 $\log \sqrt[n]{a^{n-1}p}$

▮ $\log \sqrt[n]{a^{n-1}p} = \log (a^{n-1}p)^{1/n} = (1/n) \log a^{n-1}p = (1/n)(\log a^{n-1} + \log p) = (1/n)[(n-1) \log a + \log p] = \dfrac{n-1}{n} \log a + \dfrac{\log p}{n}$.

For Probs. 6.117 to 6.122, express each as a single logarithm.

6.117 $\log a + \log b + \log c$

▮ $\log a + \log b = \log ab$. Thus, $\log a + \log b + \log c = \log ab + \log c = \log (ab)(c) = \log abc$.

6.118 $\log a + \log b + \log c + \log d$

▮ $\log a + \log b = \log ab$; $\log c + \log d = \log cd$. Thus, $\log a + \log b + \log c + \log d = \log ab + \log cd = \log abcd$.

6.119 $\log a - \log b - \log c + \log d$

▮ $\log a - \log b - \log c = \log a/b - \log c = \log \dfrac{a/b}{c} = \log \dfrac{a}{bc}$. Thus,

$\log a - \log b - \log c + \log d = \log \left(\dfrac{a}{bc} \cdot d \right) = \log \dfrac{ad}{bc}$.

6.120 $2 \log x - 3 \log y + \log z$

▮ $2 \log x = \log x^2$; $3 \log y = \log y^3$. Thus, $2 \log x - 3 \log y + \log z = \log x^2 - \log y^3 + \log z = \log \left(\dfrac{x^2}{y^3} \cdot z \right) = \log \dfrac{x^2 z}{y^3}$.

6.121 $3 \log x - \log (x - 2)$

▮ $3 \log x = \log x^3$; thus $3 \log x - \log (x - 2) = \log \dfrac{x^3}{x - 2}$.

6.122 $\log 1 + \log 5376$

▮ For any base, $\log 1 = 0$. Thus, $\log 1 + \log 5376 = \log 5376$.

For Probs. 6.123 to 6.133, evaluate the given expression given that $\log 2 = 0.3010$ and $\log 3 = 0.4771$.

6.123 $\log 8$

▮ $\log 8 = \log 2^3 = 3 \log 2 = 3(0.3010) = 0.9030$.

6.124 $\log 32$

▮ $32 = 2^5$; thus, $\log 32 = \log 2^5 = 5 \log 2 = 5(0.3010) = 1.5050$.

6.125 $\log 60$

▮ $\log 60 = \log (6 \cdot 10) = \log 6 + \log 10 = (\log 3 + \log 2) + \log 10 = 0.4771 + 0.3010 + 1 = 1.7781$.

6.126 $\log 600$

▌ $\log 600 = \log(100 \cdot 6) = \log 100 + \log 6 = 2 + \log 3 + \log 2 = 2 + 0.4771 + 0.3010 = 2.7781.$

6.127 $\log 60\underbrace{\cdots 000}_{n \text{ zeros}}$

▌ See Probs. 6.125 and 6.126 above. Then for each extra zero, we will add 1 to the result. $\log 6000 = 1 + \log 600 = 3.7781,$ etc. $\log 60\underbrace{\cdots 000}_{n \text{ zeros}} = n.7781.$

6.128 $\log 54$

▌ $\log 54 = \log(9 \cdot 6) = \log(3^2 \cdot 3 \cdot 2) = \log(3^3 \cdot 2) = 3\log 3 + \log 2 = 3(0.4771) + 0.3010 = 1.7323.$

6.129 $\log 540$

▌ Since $540 = 54 \cdot 10,$ $\log 540 = 1 + \log 54 = 1.7323 + 1 = 2.7323.$

6.130 $\log \sqrt{12}$

▌ $\log \sqrt{12} = \log(12)^{1/2} = \frac{1}{2}\log 12 = \frac{1}{2}\log(2^2 \cdot 3) = \frac{1}{2}\log 2^2 + \frac{1}{2}\log 3 = 2 \cdot \frac{1}{2}\log 2 + \frac{1}{2}\log 3 = 0.3010 + 0.4771/2 = 0.53955.$

6.131 $\log_{10} \sqrt[3]{18}$

▌ $18 = 3^2 \cdot 2;$ thus, $\log(18)^{1/3} = \frac{1}{3}\log 18 = \frac{1}{3}\log(3^2 \cdot 2) = \frac{1}{3}(2\log 3 + \log 2) = \frac{2}{3}\log 3 + \frac{1}{3}\log 2 = 0.3181 + 0.1003$ (rounding off) $= 0.4184.$

6.132 $\log \frac{64}{9}$

▌ $\log \frac{64}{9} = \log(2^6/3^2) = \log 2^6 - \log 3^2 = 6\log 2 - 2\log 3 = 1.806 - 0.9542 = 0.8518.$

6.133 $\log_{10} \sqrt[4]{72}$

▌ $\log(72)^{1/4} = \frac{1}{4}\log 72 = \frac{1}{4}\log(3^2 \cdot 2^3) = \frac{1}{4}\log 3^2 + \frac{1}{4}\log 2^3 = \frac{1}{2}\log 3 + \frac{3}{4}\log 2 = \frac{1}{2}(0.4771) + \frac{3}{4}(0.3010) = 0.23855 + 0.22575 = 0.4643.$

For Probs. 6.134 to 6.139, obtain the required logarithm.

6.134 $\log_2 (16)(1024)$

▌ $\log_2 (16)(1024) = \log_2 16 + \log_2 1024 = 4 + 10 = 14.$ (*Note*: $2^{10} = 1024.$)

6.135 $\log_2 (8)(16{,}384)$

▌ $\log_2 (8)(16{,}384) = \log_2 8 + \log_2 16{,}384.$ But $2^{14} = 16{,}384.$ Thus, $\log_2 8 + \log_2 16{,}384 = 3 + 14 = 17.$

6.136 $\log_2 (256)(4096)$

▌ $\log_2 (256)(4096) = \log_2 256 + \log_2 4096 = 8 + 12 = 20.$

6.137 $\log_2 (1024)^4$

▌ $\log_2 (1024)^3 = 3\log_2 1024 = 3 \cdot 10 = 30.$

6.138 $\log_2 (16{,}384)^{-2}$

▌ $\log_2 (16{,}384)^{-2} = -2\log 16{,}384 = -2 \cdot 14 = -28.$

6.139 $\log_2 \sqrt[4]{65{,}536}$

▌ $2^{16} = 65{,}536.$ Thus, $\log_2 (65{,}536)^{1/4} = \frac{1}{4}\log_2 65{,}536 = \frac{1}{4} \cdot 16 = 4.$

For Probs. 6.140 to 6.146, write each expression in terms of a single logarithm with a coefficient of 1.

6.140 $2 \log_b x - \log_b y$

▮ $2 \log_b x = \log_b x^2$; thus, $2 \log_b x - \log_b y = \log_b x^2 - \log_b y = \log_b (x^2/y)$.

6.141 $\log_b m - \frac{1}{2} \log_b n$

▮ $\log_b m - \frac{1}{2} \log_b n = \log_b m - \log_b n^{1/2} = \log_b (m/n^{1/2})$.

6.142 $3 \log_b x + 2 \log_b y - 4 \log_b z$

▮ $3 \log_b x + 2 \log_b y - 4 \log_b z = \log_b x^3 + \log_b y^2 - \log_b z^4 = \log_b (x^3 y^2 / z^4)$.

6.143 $\frac{1}{3} \log_b w - 3 \log_b x - 5 \log_b y$

▮ $\frac{1}{3} \log_b w - 3 \log_b x - 5 \log_b y = \log_b w^{1/3} - \log_b x^3 - \log_b y^5 = \log_b \dfrac{w^{1/3}}{x^{1/3}} - \log_b y^5 = \log_b \dfrac{w^{1/3}/x^{1/3}}{y^5} = \log_b \dfrac{w^{1/3}}{x^{1/3}y^5}$.

6.144 $\frac{1}{5}(2 \log_b x + 3 \log_b y)$

▮ $\frac{1}{5}(2 \log_b x + 3 \log_b y) = \frac{2}{5} \log_b x + \frac{3}{5} \log_b y = \log_b x^{2/5} + \log_b y^{3/5} = \log_b x^{2/5}y^{3/5}$. See Prob. 6.145.

6.145 $\frac{1}{7}(3 \log_b x + 4 \log_b y)$

▮ $3 \log_b x + 4 \log_b y = \log_b x^3 + \log_b y^4 = \log_b x^3 y^4$; thus, $\frac{1}{7}(3 \log_b x + 4 \log_b y) = \frac{1}{7}(\log_b x^3 y^4) = \log_b (x^3 y^4)^{1/7}$. Compare this to the method used in Prob. 6.144.

6.146 $\frac{1}{3}(\log_b x - \log_b y)$

▮ $\log_b x - \log_b y = \log_b (x/y)$; thus $\frac{1}{3} \log_b (x/y) = \log_b (x/y)^{1/3}$.

For Probs. 6.147 to 6.149, prove the stated logarithm property.

6.147 $\log_b xy = \log_b x + \log_b y$

▮ Assume that $x = b^t$, $y = b^s$; then $\log_b xy = \log_b (b^t b^s) = \log_b b^{t+s}$; but $\log_b b^z = z$ for all b, z; thus, $\log_b b^{t+s} = t + s = \log_b x + \log_b y$. The proof is complete.

6.148 $\log_b (x/y) = \log_b x - \log_b y$

▮ Assume that $x = b^y$, $y = b^s$; then $\log_b (x/y) = \log_b (b^t/b^s) = \log_b b^{t-s} = t - s = \log_b x - \log_b y$.

6.149 $\log_b x^s = s \log_b x$

▮ Assume that $x = b^t$. Then $\log_b x^s = \log_b (b^t)^s = \log_b b^{ts} = ts = st = s \log_b x$.

For Probs. 6.150 to 6.156, rewrite the given expression in terms of common logarithms.

6.150 $\log_6 7$

▮ Using the property $\log_b x = \log_a x/\log_a b$, we want $b = 6$, $x = 7$, $a = 10$ (this is to be in terms of common logarithms). Then $\log_6 7 = \log 7/\log 6$ ($\log 7 = \log_{10} 7$).

6.151 $\log_{14} 15$

▮ $\log_4 15 = \log_b x$ where $b = 4$, $x = 15$. Then $\log_b x = \log_4 15 = \log_a x/\log_a b = \log 15/\log 4$.

6.152 $\log_{1/2} 10$

▮ $\log_{1/2} 10 = \log_b x$ where $b = \frac{1}{2}$, $x = 10$. Then $\log_b x = \log_{1/2} 10 = \log_a x / \log_a b = \log 10 / \log \frac{1}{2} = 1 / \log \frac{1}{2} = -1 / \log 2$.

6.153 $\log_{1/3} 30$

▮ $\log_{1/3} 30 = \log_b x$ where $b = \frac{1}{3}$, $x = 30$. Then $\log_b x = \log_{1/3} 30 = \log_a x / \log_a b = \log 30 / \log \frac{1}{3} = \log (10 \cdot 3)/(-\log 3) = (\log 10 + \log 3)/(-\log 3) = (1 + \log 3)/(-\log 3)$.

6.154 $\log_{100} 10$

▮ $\log_{100} 10 = \log 10 / \log 100 = \frac{1}{2}$. Alternately, $\log_{100} 10 = \log_{100} (100)^{1/2} = \frac{1}{2} \log_{100} 100 = \frac{1}{2} \cdot 1 = \frac{1}{2}$.

6.155 $\log_{20} e$

▮ Here, e is *not* the base; 20 is. $\log_{20} e = \dfrac{\log e}{\log 20} = \dfrac{\log e}{\log (10 \cdot 2)} = \dfrac{\log e}{\log 10 + \log 2} = \dfrac{\log e}{1 + \log 2}$.

6.156 $\log_{20} e^3$

▮ See Prob. 6.155, $\log_{20} e^3 = 3 \log_{20} e = 3 \dfrac{\log e}{1 + \log 2} = \dfrac{3 \log e}{1 + \log 2}$.

For Probs. 6.157 to 6.161, rewrite the given expression in terms of logarithms with a base of 13.

6.157 $\log_4 8$

▮ $\log_4 8 = \log_b x$ where $b = 4$, $x = 8$. Here $a = 13$; then $\log_b x = \log_4 8 = \log_a x / \log_a b = \log_{13} 8 / \log_{13} 4$.

6.158 $\log_{12} 15$

▮ $\log_{12} 15 = \log_b x$ where $b = 12$, $x = 15$. Then $\log_{12} 15 = \log_{13} 15 / \log_{13} 12$. Which is larger, $\log_{12} 15$ or $\log_{13} 15$?

6.159 $\ln 10$

▮ $\ln 10 = \log_e 10 = \log_{13} 10 / \log_{13} e$.

6.160 $\ln e^3$

▮ $\ln e^3 = \log_e e^3 = \log_{13} e^3 / \log_{13} e$. But $\ln e^3 = \log_e e^3 = 3$. Thus, $\log_{13} e^3 / \log_{13} e = 3$.

6.161 $\log_{11} 121$

▮ $\log_{11} 121 = \log_{13} 121 / \log_{13} 11$. But $\log_{11} 121 = \log_{11} (11)^2 = 2$. Thus, $\log_{13} 121 / \log_{13} 11 = 2$. Compare this with Prob. 6.159. Do you notice a pattern?

For Probs. 6.162 to 6.174, tell whether the given statement is true or false, and explain your answer.

6.162 $(\log 15)^0 = 1$

▮ This is true; $\log 15 \in \mathcal{R}$, and $a^0 = 1 \ \forall a \in \mathcal{R}$. Do not be tricked by the fact that the real number here is a logarithm. *All* real numbers can be written in logarithm form!

6.163 If $f(x) = \log_8 x$, then the range of f is $\mathcal{R}$.

▮ True. Look at their graphs, or recall that the logarithm is the exponent. In $\log_a b = q$, a and b are the numbers which are restricted, not q!

6.164 The domain of $y = \log_x b$ is $\mathcal{R}$.

▐ False. For example, if $x = 1$, then 1 is the logarithmic base, which is impossible.

6.165 $\log_x 100 = -\log_{1/x} \frac{1}{100}$

▐ Let $x = 10$; then $\log 100 = 2$, but $\log_{1/10} \frac{1}{100} = 2$. The statement is false.

6.166 $\log_x a = \log_{1/x} (1/a)$

▐ If $\log_x a = b$, then $x^b = a$, $1/x^b = 1/a$, $(1/x)^b = 1/a$, and $\log_{1/x} 1/a = b$. The statement is true.

6.167 $f(x) = \log_b x$ is an always-decreasing function.

▐ False. $f(x) = \log_b x$ is decreasing for $0 < b < 1$. If $b = 10$, for example, $\log x_1 > \log x_2$ when $x_1 > x_2$.

6.168 $\log_k ab = (\log_k a)(\log_k b)$

▐ Clearly, this is a false statement. For all k, a, b, $\log_k ab = \log_k a + \log_k b$. When $k = e$, $a = e$, $b = e$, for example, $\ln (e \cdot e) = \ln e^2 = 2$, but $\ln e \ln e = 1 \cdot 1 = 1$; $2 \neq 1$.

6.169 $g(x) = \ln |x|$ has domain $= \mathcal{R}$.

▐ False; $0 \notin$ domain of g. However, all other real numbers are in the domain.

6.170 Let $g(x) = \log_a x$ and $h(x) = \log_b x$; then $g \circ h(ab) = \log_a (1 + \log a)$.

▐ $h(ab) = \log_b ab = \log_b a + \log_b b = 1 + \log_b a$. $g(1 + \log_b a) = \log_a (1 + \log_b a)$. The statement is true.

6.171 Using functions g and h in Prob. 6.170, $g \circ h(ab) = \log_a 1 + \log_a (\log_b a)$.

▐ False. For this to be true, we would be saying that "the log of a sum is the sum of the logs." But this is false since, for example, $\log_2 8 = \log_2 2^3 = 3$, thus $\log_2 (4 + 4) = 3$, not $2 + 2$.

6.172 $\log_a (1/x) = -\log_a x$

▐ $\log_a (1/x) = \log_a (x^{-1}) = -\log_a x$. True.

6.173 $\log_b a = 1/\log_a b$

▐ $\log_b a = \log_k a/\log_k b$ for all log bases k. Let $k = a$; $\log_b a = \log_a a/\log_a b = 1/\log_a b$. True.

6.174 The graph of $h(x) = \ln e^x$ is a straight line.

▐ $h(x) = \ln e^x = x$; thus $h(x)$ is the straight line with slope 1 passing through $(0, 0)$. True.

6.175 Find x so that $\frac{3}{2} \log_b 4 - \frac{2}{3} \log_b 8 + 2 \log_b 2 = \log_b x$.

▐ Then $\log_b x = \log_b 4^{3/2} - \log_b 8^{2/3} + \log_b 2^2 = \log_b 8 - \log_b 4 + \log_b 4 = \log_b (\frac{8}{4} \cdot 4)$, or $x = 8$.

6.176 Find x so that $3 \log_b 2 + \frac{1}{2} \log_b 25 - \log_b 20 = \log_b x$.

▐ $\log_b x = \log_b 8 + \log_b 5 - \log_b 20 = \log_b (8 \cdot 5/20) = \log_b x$, or $x = 2$.

6.177 Write $\log_b y - \log_b c + kt = 0$ in exponential form free of logarithms.

▐ Then $\log_b y - \log_b c = -kt$, and $b^{(\log_b y - \log_b c)} = b^{-kt}$, $b^{\log_b y}/b^{\log_b c} = b^{-kt}$, $y/c = b^{-kt}$, or $y = cb^{-kt}$.

6.178 Prove that $\log_a xyz = \log_a x + \log_a y + \log_a z$.

▐ $\log_a xy + \log_a z = (\log_a x + \log_a y) + \log_a z = \log_a x + \log_a y + \log_a z$, and the proof is complete.

6.179 Let $f(x) = 7^x$. Find the domain and range of $f^{-1}(x)$.

▮ If $f(x) = 7^x$, then $f^{-1}(x) = \log_7 x$. Then the domain of f is $\mathcal{R}^+$, and the range (i.e., the f^{-1} values) is $\mathcal{R}$.

6.180 Solve the equation $A = Pe^{rt}$ for r.

▮ If $A = Pe^{rt}$, then $\ln A = \ln Pe^{rt} = \ln P + \ln e^{rt} = \ln P + rt \ln e = \ln P + rt$ (since $\ln e = 1$). Thus, $rt = \ln A - \ln P$, and $r = (\ln A - \ln P)/t$.

6.181 Solve the equation $\ln(\log x) = 1$ for x.

▮ If $\ln(\log x) = 1$, then $e^{\ln(\log x)} = e^1$. But $e^{\ln P} = P$ for all P. Thus, $e^{\ln(\log x)} = \log x$. Then $\log x = e$, $10^{\log x} = 10^e$, and $x = 10^e$.

6.182 Check the result in Prob. 6.181.

▮ $x = 10^e$ where $\ln(\log x) \overset{?}{=} 1$. $\ln(\log 10^e) = \ln(e \log 10) = \ln(e \cdot 1) = \ln e = 1$. The result checks.

6.4 NATURAL AND COMMON LOGARITHMS

Note: Many problems in this section involve the use of a calculator. If you do not know how to use your calculator for logarithm computations, check the instruction manual.

For Probs. 6.183 to 6.187, express the given number in scientific notation.

6.183 137.65

▮ Scientific notation is of the numerical form $N \times 10^M$, $1 \le N \le 10$, $M \in \mathcal{Z}$. Thus, for 137.65 we move the decimal point over two spaces, so $137.65 = 1.3765 \times 10^2$. Here, $M = 2$, $N = 1.3765$.

6.184 14,632.60004000

▮ Let $M = 4$, $N = 1.463260004000$. Then $N \times 10^M = 1.463260004000 \times 10^4 = 14,632.60004000$. (We have moved the decimal point over four spaces.)

6.185 1.6843

▮ $1.6843 = 1.6843 \times 10^0$. Notice that, in essence, the number given was in scientific notation. The 0 exponent comes about since we moved the decimal over "0" spaces.

6.186 0.00684

▮ $0.00684 = 6.84 \times 10^{-3}$. We can easily see the -3 exponent by observing that we have moved the decimal point three spaces to the right.

6.187 $4.26 \cdot 9^2$

▮ $4.26 \times 9^2 = 345.06 = 3.4506 \times 10^2$.

For Probs. 6.188 and 6.189, rewrite the numbers by using ordinary notation.

6.188 $4.63 \cdot 10^{-4}$

▮ $4.63 \times 10^{-4} = 4.63 \times 0.0001 = 0.000463$. Another, and easier, way to do this is to observe that a 0.0001 multiplier moves the decimal point four spaces to the left ($M < 0$).

6.189 $4.5686 \cdot 10^3$

▮ Simply move the decimal three places to the right ($M > 0$). Then $4.5686 \times 10^3 = 4568.6$.

For Probs. 6.190 to 6.193, find the characteristic of the logarithm (base 10) of the given number.

6.190   146

▌ To determine the characteristic of a number P, put P into scientific notation: $P = N \times 10^M$, $1 \le N < 10$, $M \in \mathcal{Z}$. Then $M =$ the characteristic. Thus, $146 = 1.46 \times 10^2$, and $2 =$ characteristic.

6.191   14,673.1

▌ $14,673.1 = 1.46731 \times 10^4$. $4 =$ characteristic.

6.192   0.003634

▌ $0.003634 = 3.634 \times 10^{-3}$, and the characteristic $= -3$.

6.193   4.638

▌ $4.638 = 4.638 \times 10^0$; and the characteristic is 0.

For Probs. 6.194 to 6.199, use a logarithm table (base 10) to find the common logarithm of each number.

6.194   log 5.73

▌ First notice that the characteristic is 0, since $5.73 = 5.73 \times 10^0$. Looking up 573 in the table, we find 7582. Thus, $\log 5.73 = 0.7582$ where 0 is the characteristic and 7582 is the mantissa.

6.195   log 57.3

▌ $57.3 = 5.73 \times 10^1$. We find 7582 in the table for 573. Thus, $\log 57.3 = 1.7582$. The 1 in 1.7582 is the characteristic found in scientific notation.

6.196   0.00573

▌ We again use the 7582 found in the table for 573. Then, $\log 0.00573 = \log (5.73 \times 10^{-3}) = \log 5.73 + \log 10^{-3} = 0.7582 + (-3) = 0.7582 - 3$. You can, if you like, rewrite $0.7582 - 3$ as -2.2418, but $0.7582 - 3$ is a perfectly acceptable form.

6.197   0.0101

▌ $0.0101 = 1.01 \times 10^{-2}$. Look up 101 in your table, and find 0043. Thus, $\log (0.0101) = \log (1.01 \times 10^{-2}) = \log 1.01 + \log 10^{-2} = 0.0043 - 2 \, (= -1.9957)$.

6.198 5.13×10^{-4}

▌ $\log (5.13 \times 10^{-4}) = \log 5.13 + \log 10^{-4} = 0.7101 - 4$.

6.199 51.3×10^{-7}

▌ $\log (51.3 \times 10^{-7}) = \log (5.13 \times 10^{-6}) = 0.7101 - 6$.

For Probs. 6.200 to 6.202, rewrite the given log values in standard logarithmic form.

6.200 $5.1983 - 8$

▌ Standard log form is $\log n =$ mantissa + characteristic where the mantissa is nonnegative and less than 1. Thus, $5.1983 - 8$ has a "mantissa" of 5.1983: we subtract 5 from 5.1983, and add 5 to -8 to obtain $(5.1983 - 5) + [(-8) + 5] = 0.1983 - 3$. Do you see the power of this method? We now know that, in scientific notation, n is of the form $a.bc \times 10^{-3}$.

6.201 -1.1776

▌ -1.1776 is not in standard log form. $-1.1776 = (-1.1776 + 2)$ (adding 2 is the least you can add to -1.1776 to get it to be nonnegative) $- 2$. Then, $(-1.1776 + 2) - 2 = 0.8224 - 2$.

6.202 −3.2076

┃ −3.2076 = (−3.2076 + 4) − 4 = 0.7924 − 4.

For Probs. 6.203 to 6.206, use a common logarithm table to find the antilogarithm of the given number.

6.203 0.7993

┃ Look for 0.7993 in your table. It is going to correspond to 630. Since 0.7993 has a 0 characteristic, antilog $0.7993 = 6.30 \times 10^0 = 6.30$.

6.204 2.5729

┃ We find in our table that 0.5729 corresponds to 374. Since the characteristic is 2, antilog $2.5729 = 3.74 \times 10^2 = 374$.

6.205 0.9460 − 2

┃ Then antilog $(0.9460 − 2) = 8.83 \times 10^{-2} = 0.0883$.

6.206 −4.0052

┃ $-4.0052 = (-4.0052 + 5) - 5 = 0.9948 - 5$. Since 0.9948 corresponds to 988 in the table, antilog $(-4.0052) = $ antilog $(0.9948 - 5) = 9.88 \times 10^{-5}$.

For Probs. 6.207 to 6.210, use a calculator to find the indicated quantity, rounded to six decimal places.

6.207 **(a)** log 82,734, **(b)** ln 82,734

┃ **(a)** If you enter 82,734 and press the log key, you will get log 82,734. What could be easier? *Press*: $\boxed{8}\,\boxed{2}\,\boxed{7}\,\boxed{3}\,\boxed{4}\,\boxed{\log}$. *On screen*: 4.917684.
(b) We repeat the same procedure, but use the ln key instead. *Press*: $\boxed{8}\,\boxed{2}\,\boxed{7}\,\boxed{3}\,\boxed{4}\,\boxed{\ln}$. *On screen*: 11.323386.

6.208 **(a)** log 843,250, **(b)** ln 843,250

┃ **(a)** *Press*: $\boxed{8}\,\boxed{4}\,\boxed{3}\,\boxed{2}\,\boxed{5}\,\boxed{0}\,\boxed{\log}$. *On screen*: 5.9259563. Rounding, we get 5.925956.
(b) *Press*: $\boxed{8}\,\boxed{4}\,\boxed{3}\,\boxed{2}\,\boxed{5}\,\boxed{0}\,\boxed{\ln}$. *On screen*: 13.645019.

6.209 log 0.0103

┃ *Press*: $\boxed{\cdot}\,\boxed{0}\,\boxed{1}\,\boxed{0}\,\boxed{3}\,\boxed{\log}$. *On screen*: −1.9871628. Rounding, we get −1.987163. Then −1.987163 = (−1.987163 + 2) − 2 = 0.012837 − 2.

6.210 ln 0.081043

┃ *Press*: $\boxed{\cdot}\,\boxed{0}\,\boxed{8}\,\boxed{1}\,\boxed{0}\,\boxed{4}\,\boxed{3}\,\boxed{\ln}$. *On screen*: −2.5127754. Rounding, we get −2.512775. Then (−2.512775 + 3) − 3 = 0.487225 − 3.

For Probs. 6.211 to 6.216, use a calculator to find x to four significant digits.

6.211 log $x = 5.3027$

┃ Recall that $f(x) = \log x$ and $g(x) = 10^x$ are inverse functions. Thus, to go from log x to x, we find $10^{\log x}$. *Press*: 5.3027 $\boxed{10^x}$. *On screen*: 200,770.55. Then $x = 200,770.55 \approx 200,800$ to four significant digits.

6.212 log $x = 1.9168$

┃ *Press*: 1.9168 $\boxed{10^x}$. *On screen*: 82.565763. Then $x = 82.565763 \approx 82.57$ to four significant digits.

6.213 $\log x = -3.1773$

▮ *Press*: 3.1773 $\boxed{\pm}$ $\boxed{10^x}$. *On screen*: 0.0006648. Then $x = 0.0006648$ (already is to four significant digits).

6.214 $\log x = -2.0411$

▮ *Press*: 2.0411 $\boxed{\pm}$ $\boxed{10^x}$. *On screen*: 0.009097. Then $x = 0.009097 \approx 0.0091$ to four significant digits.

6.215 $\ln x = 3.8655$

▮ *Press*: 3.8655 $\boxed{e^x}$. *On screen*: 47.72713. Then $x = 47.72713 \approx 47.73$ to four significant digits.

6.216 $\ln x = -0.3916$

▮ *Press*: 0.3916 $\boxed{\pm}$ $\boxed{e^x}$. *On screen*: 0.6759745. Then $x = 0.6759745 \approx 0.6760$ to four significant digits.

For Probs. 6.217 to 6.220, find x to five significant digits, using a calculator. Your calculator may operate slightly differently from that exhibited here.

6.217 $x = \log (5.3147 \times 10^{12})$

▮ $\log (5.3147 \times 10^{12}) = \log 5.3147 + \log 10^{12} = \log 5.3147 + 12$. Find $\log 5.3147$ on the calculator. *Press*: 5.3147 $\boxed{\log}$. *On screen*: 0.7254788. Then $x = 12.7254788 \approx 12.725$ to five significant digits. (*Note*: We did not enter 5.3147×10^{12} on the calculator. Why?)

6.218 $x = \log (2.0991 \times 10^{17})$

▮ $\log (2.0991 \times 10^{17}) = \log 2.0991 + \log 10^{17} = 0.322033 + 17$ (0.322033 is obtained from the calculator) $= 17.322031 \approx 17.322$.

6.219 $x = \ln (6.7917 \times 10^{-12})$

▮ $x = \ln (6.7917 \times 10^{-12}) = \ln 6.7917 + \ln 10^{-12} = \ln 6.7917 + \dfrac{\log_{10} 10^{-12}}{\log_{10} e} =$ $1.9157013 + \dfrac{-12}{0.4342974} \approx -25.715$.

6.220 $\log x = 32.068523$

▮ Enter 32.068523 into your calculator, and press $\boxed{10^x}$. Then $x = 1.1709 \times 10^{32}$ is what you will see on the screen (actually, you will see $1.1709 - 32$). Alternatively, if $\log x = 32.068523$, $x = 10^{32} \times$ antilog $0.068523 = 10^{32} \times 1.1709086$. (To find the antilog 0.068523, enter 0.068523 $\boxed{10^x}$ into your calculator to obtain $\approx 1.1709 \times 10^{32}$.)

For Probs. 6.221 to 6.223, use the change-of-base formula and a calculator to find the indicated quantity to four significant places.

6.221 $\log_5 372$

▮ Recall that $\log_b M = \log_k M / \log_k b$. Then $\log_5 372 = \ln 372 / \ln 5$. *Press*: 372 $\boxed{\ln}$ $\div$ 5 $\boxed{\ln}$. *On screen*: 3.67765155. Rounding, we get 3.6776.

6.222 $\log_8 0.352$

▮ $\log_8 0.0352 = \ln 0.0352 / \ln 8$. *Press*: 0.0352 $\boxed{\ln}$ $\div$ 8 $\boxed{\ln}$. *On screen*: -1.6094269. Rounding, we get -1.6094.

6.223 $\log_3 0.1483$

▌ $\log_3 0.1483 = \ln 0.1483/\ln 3$. Using the key steps in Probs. 6.221 and 6.222, we get $-1.737208 \approx -1.7372$.

For Probs. 6.224 to 6.231, perform the indicated computations. Obtain answers to three significant digits. Use a common logarithm table.

6.224 $(3.86)(59.1)$

▌ $\log 3.86 = 0.5866$ and $\log 59.1 = 1.7716$. Then since $\log (3.86)(59.1) = 2.3582$, $(3.86)(59.1) = 228$. (We find 2.28 in the table, and we use the characteristic of 2 to get 228. We do not interpolate since we only want the result to three significant digits. 2.28 corresponds to 0.3579 in the table, which is closest to 0.3582.)

6.225 $(807)(0.276)$

▌ Let $P = (807)(0.276)$. Then $\log P = \log 807 + \log 0.276 = 2.9069 + (-0.5591) = 2.3478$. Thus, $P = 223$ (0.3483 is closest to 0.3478). We use $10^2 \times 2.23$ since the characteristic is 2.

6.226 $(6.23)(4.71)(32.8)$

▌ Let Q = the product. Then $\log Q = \log 6.23 + \log 4.71 + \log 32.8 = 0.7945 + 0.6730 + 1.5159 = 2.9834$. Then since 0.9834 is closest to 0.9836 which corresponds to 9.63, $\log Q = 2.9834$ and $Q = 963$.

6.227 $984 \div 237$

▌ Let $S = \frac{948}{237}$. Then $\log S = \log 948 - \log 237 = 2.9768 - 2.3747 = 0.6021$, and $S = 4.00$. (Actually, 0.6021 is in the table!)

6.228 $\dfrac{(24.7)(3.28)}{47.6}$

▌ Let T represent the solution. Then $\log T = \log 24.7 + \log 3.28 - \log 47.6 = 1.3927 + 0.5159 - 1.6776 = 0.231$, which is closest to 0.2304, which corresponds to 1.70. $T = 1.70$.

6.229 $(7.25)^2$

▌ If $P = (7.25)^2$, then $\log P = 2 \log 7.25 = 2 \times 0.8603 = 1.7206$. Then since 0.7206 corresponds to 5.25 in the table (that is the closest), $P = 52.5$.

6.230 $\sqrt[3]{9.84}$

▌ Let $R = \sqrt[3]{9.84}$. Then $R = (9.84)^{1/3}$, and $\log R = \log (9.84)^{1/3} = \frac{1}{3} \log 9.84 = \frac{1}{3} \times 0.9930 = 0.331$. Thus, $R = 2.14$ (0.331 is closest to 0.3304 in the table).

6.231 $\sqrt{27.3} \sqrt[3]{0.629}$

▌ Let T = the product. Then $\log T = \frac{1}{2} \log 27.3 + \frac{1}{3} \log 0.629 = \frac{1}{2} \times 1.4362 + \frac{1}{3} \times 0.7987 = 0.7181 + 0.2662 = 0.9843$. Thus, $T = 9.64$ (closest value from table).

For Probs. 6.232 and 6.233, find N to four significant digits if $\log N$ is the number given. Use the common log table.

6.232 $\log N = 0.5801$

▮ We locate two numbers in the table r and s such that $r < 0.5801 < s$; they are 0.5798 and 0.5809:

$$0.0011\begin{bmatrix}0.5798 \\ 0.5801 \\ 0.5809\end{bmatrix}0.0008 \begin{matrix}3.80 \\ x \\ 3.81\end{matrix}\Big]3.81-x\Bigg]0.01$$

Then $0.0011/0.0008 = 0.01/(3.81-x)$, $(0.0008)(0.01) = [(0.0011)(3.81)-x]$, $8(0.01) = 11(3.81-x)$, $0.08 = 41.91 - 11x$, $11x = 41.83$, and $x \approx 3.802727 \approx 3.803$ (to four significant places).

6.233 $\log N = 0.4063$

▮ From the table, $0.4048 < 0.4063 < 0.4065$.

$$0.0017\begin{bmatrix}0.4065 \\ 0.4063 \\ 0.4048\end{bmatrix}0.0015 \begin{matrix}2.55 \\ x \\ 2.54\end{matrix}\Big]x-2.54\Bigg]0.01$$

Then $0.0017/0.0015 = 0.01/(x-2.54)$, $\frac{17}{15} = 0.01/(x-2.54)$, $17x = (15)(0.01)+(2.54)(17)$, $17x = 43.33$, and $x \approx 2.548823 \approx 2.549$.

For Probs. 6.234 and 6.235, find $\log N$ by using interpolation.

6.234 $N = 53.62$

▮ $\log 53.6 = 1.7292$ and $\log 53.7 = 1.7300$. Interpolating, we get

$$0.1\begin{bmatrix}53.7 & 1.7300 \\ 0.02\begin{bmatrix}53.62 & x \\ 53.6 & 1.7292\end{bmatrix}x-1.7292\end{bmatrix}0.0008$$

Then $0.1/0.02 = 0.0008/(x-1.7292)$, $10(x-1.7292) = 0.0008(2)$, $10x - 17.292 = 0.0016$, $10x = 0.0016 + 17.292$, and $x = 1.72936 \approx 1.7294$. (The accuracy matches the table's accuracy.)

6.235 $N = 141.6$

▮ $\log 141 = 2.1492$ and $\log 142 = 2.1523$. Interpolating, we get

$$0.6\begin{bmatrix}142 & 2.1523 \\ 141.6 & 1 & x \\ 141 & 2.1492\end{bmatrix}x-2.1492\end{bmatrix}0.0031$$

Then $1/0.6 = 0.0031/(x-2.1492)$, $x - 2.1492 = 0.00186$, and $x = 2.15106 \approx 2.1511$.

6.5 LOGARITHMIC AND EXPONENTIAL EQUATIONS

For Probs. 6.236 to 6.242, solve the given equation for x in terms of y.

6.236 $y = 10^x$

▮ If $y = 10^x$, then $x = \log y$. You can get that answer by remembering that $\log_a x$ and a^x are inverse functions, or by taking the log of both sides of the equation $y = 10^x$. Then $\log y = \log 10^x$ and $\log y = x$.

6.237 $y = 10^{-x}$

▮ If $y = 10^{-x}$, then $\log y = \log 10^{-x}$ and $\log y = -x$. $x = -\log y = \log y^{-1} = \log(1/y)$.

6.238 $y = 5(10^{-2x})$

▮ $\log y = \log 5(10^{-2x}) = \log 5 + \log 10^{-2x} = \log 5 + (-2x)$. Then $x = (\log 5 - \log y)/2 = \frac{1}{2}\log(5/y) = \log\sqrt{5/y}$.

6.239 $y = 3(10^{2x})$

▮ $\log y = \log 3(10^{2x}) = \log 3 + \log 10^{2x} = \log 3 + 2x$. Then $2x = \log y - \log 3 = \log (y/3)$ and $x = \frac{1}{2} \log (y/3) = \log (y/3)^{1/2}$.

6.240 $y = \dfrac{e^x + e^{-x}}{2}$

▮ Let $u = e^x$. (This is a common trick.) Then $e^{-x} = 1/u$, so $2y = u + 1/u = (u^2 + 1)/u$ and $u^2 - 2uy + 1 = 0$, a quadratic in u. Then $u = (2y + \sqrt{4y^2 - 4})/2 = y + \sqrt{y^2 - 1}$, and $e^x = y \pm \sqrt{y^2 - 1}$. Thus, $x = \ln (y \pm \sqrt{y^2 - 1})$.

6.241 $y = \dfrac{e^x - e^{-x}}{2}$

▮ Using the procedure in Prob. 6.240, let $u = e^x$. Then $1/u = e^{-x}$ and $2y = u - 1/u = (u^2 - 1)/u$. Then $u^2 - 1 = 2yu$, $u^2 - 2yu - 1 = 0$, $u = y + \sqrt{y^2 + 1}$, and we obtain $x = \ln (y + \sqrt{y^2 + 1})$ by solving the quadratic.

6.242 $y = \ln (\sqrt{x^2 + 1} + x)$

▮ This is the solution from Prob. 6.241; thus $x = (e^y - e^{-y})/2$.

For Probs. 6.243 to 6.264, solve the given equation; when appropriate, give answers to three decimal places.

6.243 $5^x = 625$

▮ $5^x = 5^3$, or $x = 3$. (If $5^x = 5^3$, then the exponents must be equal; equivalently, if $\log_5 5^x = \log_5 5^3$, $x = 3$.)

6.244 $3^x = 27$

▮ $3^x = 3^3$, $\log_3 3^x = \log_3 3^3$, or $x = 3$.

6.245 $3^{x^2-1} = 27$

▮ $3^{x^2-1} = 3^3$, $\log_3 3^{x^2-1} = \log_3 3^3$, $x^2 - 1 = 3$, $x^2 = 4$, or $x = \pm 2$. (Check both results!)

6.246 $5^{x^2-3x} = 625$

▮ $5^{x^2-3x} = 5^4$, $x^2 - 3x = 4$, $x^2 - 3x - 4 = 0$, $(x - 4)(x + 1) = 0$, or $x = 4, -1$. (Check these!)

6.247 $5^{x^2-4x+3} = 1$

▮ $5^{x^2-4x+3} = 5^0$, $\log_5 5^{x^2-4x+3} = \log_5 5^0$, $x^2 - 4x + 3 = 0$, $(x - 1)(x - 3) = 0$, or $x = 1, 3$.

6.248 $4^{x^2+3x} = 256$

▮ $4^{x^2+3x} = 4^4$, $x^2 + 3x = 4$, $x^2 + 3x - 4 = 0$, $(x + 4)(x - 1) = 0$, or $x = -4, 1$.

6.249 $2^x = 41$

▮ $\log 2^x = \log 41$, $x \log 2 = \log 41$, or $x = \log 41/\log 2 = 1.6128/0.3010 \approx 5.358$.

6.250 $5^{x+1} = 3.02$

▮ $\log 5^{x+1} = \log 3.02$, $(x + 1) \log 5 = \log 3.02$, $x + 1 = \log 3.02/\log 5$, or $x = \log 3.02/\log 5 - 1 = 0.4800/0.6990 - 1 \approx -0.310$.

6.251 $195^x = 2.68$

▮ $\log 195^x = \log 2.68$, $x \log 195 = \log 2.68$, or $x = \log 2.68/\log 195 = 0.4281/2.2900 \approx 0.187$.

6.252 $\log_5 (x - 1) + \log_5 (x + 3) = 1$

▮ Then $\log_5 (x - 1)(x + 3) = 1$, $5^{\log_5 (x-1)(x+3)} = 5^1$, $(x - 1)(x + 3) = 5$, $x^2 + 2x - 3 = 5$, $x^2 + 2x - 8 = 0$, $(x + 4)(x - 2) = 0$, or $x = -4, 2$. It is crucial to check these answers since the domain of $\log x$ is so restricted. If $x = -4$, $\log_5 (-5) + \log_5 (-1)$ is undefined and $x = -4$ is extraneous. If $x = 2$, $\log_5 1 + \log_5 5 = 0 + 1 = 1$. Thus, the solution is $x = 2$.

6.253 $\log_5 (2x + 1) + \log_5 (3x - 1) = 2$

▮ Then $\log_5(2x + 1)(3x - 1) = 2$, $(2x + 1)(3x - 1) = 5^2$, $6x^2 + x - 1 = 25$, $6x^2 + x - 26 = 0$, $(6x + 13)(x - 2) = 0$, $6x = -13$, or $x = -\frac{13}{6}, 2$. But $x = -\frac{13}{6}$ is extraneous, since $3x - 1 < 0$; thus $x = 2$ is the solution.

6.254 $\log_6 3 + \log_6 (x + 6) = 2$

▮ $\log_6 3(x + 6) = 2$, $3x + 18 = 36$, $3x = 18$, or $x = 6$. Since $\log_6 3 + \log_6 12 = \log_6 36 = 2$, the answer checks.

6.255 $\log_3 (x + 2) + \log_3 (x + 4) = 1$

▮ $\log_3 (x + 2)(x + 4) = 1$, $(x + 2)(x + 4) = 3$, $x^2 + 6x + 8 = 3$, $x^2 + 6x + 5 = 0$, $(x + 1)(x + 5) = 0$, or $x = -1, -5$. But $x = -5$ is extraneous. If $x = -1$, $\log_3 1 + \log_3 3 = 0 + 1 = 1$. Thus, $x = -1$.

6.256 $\log_5 (2x + 4) - \log_5 (x - 1) = 1$

▮ $\log_5 \dfrac{2x + 4}{x - 1} = 1$, $\dfrac{2x + 4}{x - 1} = 5$, $2x + 4 = 5x - 5$, $3x = -9$, or $x = -3$. Since $x = -3$, $x - 1 < 0$; thus there are no solutions.

6.257 $\log_2 (3x + 1) - \log_2(x - 3) = 3$

▮ $\log_2 \dfrac{3x + 1}{x - 3} = 3$, $3x + 1 = 8x - 24$ $(2^3 = 8)$, $5x = 25$, or $x = 5$. Since $\log_2 16 - \log_2 2 = 4 - 1 = 3$, the answer checks.

6.258 $\log_3 (x + 11) + \log_3 (x + 3) = 2$

▮ $\log_3 (x + 11)(x + 3) = 2$, $x^2 + 14x + 33 = 9$, $x^2 + 14x + 24 = 0$, $(x + 12)(x + 2) = 0$, or $x = -2, -12$. But $x = -12$ is extraneous. If $x = -2$, $\log_3 9 - \log_3 1 = 2 - 0 = 2$. Thus, $x = -2$.

6.259 $\log_5 (3x + 7) + \log_5 (x - 5) = 2$

▮ $\log_5 (3x + 7)(x - 5) = 2$, $3x^2 - 8x - 35 = 25$, $3x^2 - 8x - 60 = 0$, $(3x + 10)(x - 6) = 0$, or $x = -\frac{10}{3}, 6$. $x = 6$ is the solution.

6.260 $\log 1/(2x) = -\log 6$

▮ $\log 1/(2x) = \log 6^{-1}$, $1/(2x) = 6^{-1} = \frac{1}{6}$, $2x = 6$, or $x = 3$. Since $\log \frac{1}{6} = \log 6^{-1} = -\log 6$, the answer checks.

6.261 $\log (3x + 4) = \log (5x - 6)$

▮ Then $3x + 4 = 5x - 6$, $2x = 10$, or $x = 5$. Check this solution.

6.262 $2 \ln 5x = 3 \ln x$

▮ If $2 \ln 5x = 3 \ln x$, then $\ln (5x)^2 = \ln x^3$. Then $(5x)^2 = x^3$, $25x^2 = x^3$, $x^3 - 25x^2 = 0$, $x^2(x - 25) = 0$, or $x = 0, 25$. But $x = 0$ is extraneous, since $5x = 0$ and $\ln 0$ is undefined.

6.263 $\log x = \ln e$

▌ If $\log x = \ln e$, then $10^{\log x} = 10^{\ln e}$, or $x = 10^{\ln e}$. But $\ln e = 1$, so $x = 10$.

6.264 $\ln 25x - \ln 5x = 2 \ln 5x - \ln 25x$

▌ $\ln \dfrac{25x}{5x} = \ln \dfrac{(5x)^2}{25x}$. $\ln 5 = \ln \dfrac{25x^2}{25x}$, $\ln 5 = \ln x$, or $x = 5$.

CHAPTER 7

Trigonometric Functions

7.1 ANGLE MEASUREMENT

For Probs. 7.1 to 7.8, determine the quadrant containing the given angle. Do *not* sketch the angle to determine the quadrant.

7.1   120°

     ▮ Since 90° < 120° < 180°, 120° is in quadrant II. List a few other quadrant II angles before you go on to another problem.

7.2   239.5°

     ▮ Since 180° < 239.5° < 270°, 239.5° is in quadrant III.

7.3   −70°

     ▮ We note that −90° < −70° < 0°, and so −70° is in quadrant IV. We also might note that −70° + 360° = 290°. Thus, 70° is coterminal with 290°. Since 270° < 290° < 360°, 290 (or −70°) is in quadrant IV.

7.4   −420°

     ▮ −420° + 360° = −60° and −60° + 360° = 300°. Thus, −420° and 300° are coterminal; 300° is in quadrant IV, and therefore so is −420°.

7.5   −46.5π

     ▮ $2\pi \times 24 = 48\pi$, and $48\pi + (-46.5\pi) = 1.5\pi = 3\pi/2$. This is a quadrantal angle.

7.6   −π/10

     ▮ Since $-\pi/2 < -\pi/10 < 0$, $-\pi/10$ is in quadrant IV. Also, $2\pi + (-\pi/10) = 19\pi/10$, and $19\pi/10$ is in quadrant IV. Since $-\pi/10$ and $19\pi/10$ are coterminal, $-\pi/10$ is in quadrant IV.

7.7   π/17

     ▮ Since $0 < \pi/17 < \pi/2$, $\pi/17$ is in quadrant I.

7.8   −1654°

     ▮ $360° \times 5 = 1800°$, and $1800° + (-1654°) = 146°$. Since 146° is in quadrant II and 146° and −1654° are coterminal, −1654° is in quadrant II.

For Probs. 7.9 to 7.15, sketch the angle of each of the following measures. In your sketch, place the angle in standard position.

7.9   225°

     ▮ See Fig. 7.1. 180° + 45° = 225°

7.10   420°

     ▮ See Fig. 7.2. 420° = 360° + 60°.

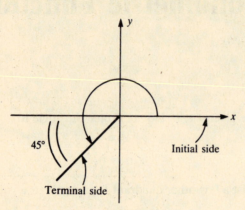

Fig. 7.1

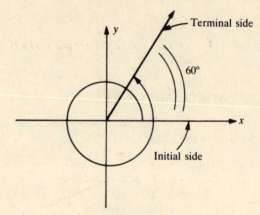

Fig. 7.2

7.11   621°

▌ See Fig. 7.3. When we divide 621° by 360°, the quotient is 1. 360° × 1 = 360°, and 621° − 360° = 261° = 180° + 81°.

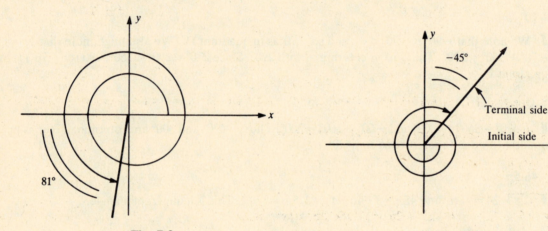

Fig. 7.3

Fig. 7.4

7.12   −675°

▌ See Fig. 7.4. −675° = −360° + (−315°) = −360° + (−270°) + (−45°).

7.13   −π/3

▌ See Fig. 7.5. −π/2 < −π/3 < 0.

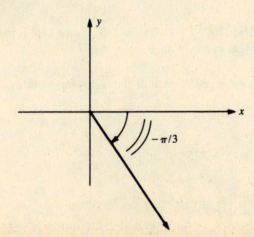

Fig. 7.5

7.14 $23\pi/6$

▌ See Fig. 7.6. $23\pi/6 = 2\pi + 11\pi/6 = 2\pi + (2\pi - \pi/6)$.

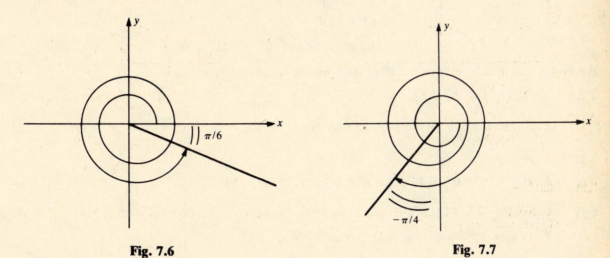

Fig. 7.6   Fig. 7.7

7.15 $-19\pi/4$

▌ See Fig. 7.7. $-19\pi/4 = -16\pi/4 + (-3\pi/4) = -4\pi + (-3\pi/4) = -4\pi + (-\pi/2) + (-\pi/4)$.

For Probs. 7.16 to 7.20, perform the indicated operation.

7.16 $65°13'$
         $+71°26'$

▌ $65° + 71° = 136°$; $13' + 26' = 39'$. So the answer is $136°39'$.

7.17 $114°29'46''$
         $-81°\ 4'11''$

▌ $114° - 81° = 33°$; $29' - 4' = 25'$; $46'' - 11'' = 35''$. So the answer is $33°25'35''$.

7.18 $127°10'14''$
         $-85°\ 5'53''$

▌ $10'14'' = 9'74''$ $(1' = 60'')$. Then

$$127°9'74''$$
$$-85°5'53''$$
$$\overline{42°4'21''}$$

7.19 $11'11''$
         $+58'59''$

▌

$$11'11''$$
$$58'59''$$
$$\overline{69'70''}$$

Since $1' = 60''$, $69'70'' = 70'10''$. Then since $1° = 60'$, $70'10'' = 1°10'10''$.

7.20   49°   57″
             +61°59′53″

┃
$$49° \ 0′ \ 57″$$
$$+61°59′ \ 53″$$

$$110°59′110″ = 110°60′50″ = 111°0′50″$$

For Probs. 7.21 to 7.24, express the given angle in degrees, minutes, and seconds.

7.21   40.25°

┃ $40.25° = 40° + 0.25° = 40° + 0.25 \times 60'$ (since $60' = 1°$) $= 40° + 15' = 40°15' = 40°15'0''$.

7.22   75.2°

┃ $75.2° = 75° + 0.2° = 75° + 0.2 \times 60' = 75° + 12' = 75°12'0''$.

7.23   17.45°

┃ $17.45° = 17° + 0.45 \times 60' = 17° + 27' = 17°27'0''$.

7.24   14.458°

┃ $14.458° = 14° + 0.458 \times 60' = 14° + 27.48' = 14° + 27' + 0.48 \times 60'' = 14° + 27' + 28.8'' = 14°27'28.8''$.

For Probs. 7.25 to 7.27, rewrite the given angle in decimal form.

7.25   10°10′

┃ $60' = 1°$; thus, $10' = 1° \times \frac{10}{60} = (\frac{1}{6})° = 0.1\bar{6}°$. Then $10°10' = 10.1\bar{6}°$.

7.26   27°15′25″

┃ $27°15'25'' = 27° + (\frac{15}{60})° + [25/(60)^2]° = 27° + 0.25° + 0.0069\bar{4}° = 27.2569\bar{4}°$.

7.27   45°45′45″

┃ $45°45'45'' = 45° + (\frac{45}{60})° + [45/(60)^2]° = 45° + 0.75° + 0.0125° = 45.7625°$.

For Probs. 7.28 to 7.37, convert the given angle to degrees if it is in radians and to radians if it is in degrees.

7.28 $\pi/4$

┃ *(a)* 2π rad $= 360°$. Then $\pi/4$ rad $= 360° \times (\pi/4)/(2\pi) = 360° \times \frac{1}{8} = 45°$. *(b)*   Alternatively, $\pi = 180°$ and $\pi/4 = 180°/4 = 45°$.

7.29 $6\pi/9$

┃ *(a)* $6\pi/9 = 180° \times \frac{6}{9} = 120°$; or *(b)* $6\pi/9 = 2\pi/3 = \frac{2}{3}(180°) = 120°$.

7.30 $-\pi/6$

┃ $-\pi/6 = 180° \times (-\frac{1}{6}) = -30°$.

7.31 -7π

┃ $-7\pi = -7(180°) = -1260°$.

7.32 $-17\pi/6$

┃ $-17\pi/6 = -\frac{17}{6}(180°) = -510°$.

7.33 2

    ❚ π rad $= 180°$. Thus 1 rad $= (180/\pi)°$ and 2 rad $= 2(180/\pi)° = (360/\pi)°$.

7.34 45°

    ❚ $180° = \pi$ rad. $45° = \pi \times \frac{45}{180} = \pi \times \frac{1}{4} = \pi/4$ rad.

7.35 410°

    ❚ $410° = \pi \times \frac{410}{180} = \pi \times 2.2\overline{7} = 2.2\overline{7}\pi$ rad.

7.36 −135°

    ❚ $-135° = \pi \times (-\frac{135}{180}) = \pi \times (-0.75) = -0.75\pi = -3\pi/4$ rad.

7.37 7°30′

    ❚ $7°30' = 7° + (\frac{30}{60})° = 7.5° = \pi \times 7.5/180 = \pi \times 0.041\overline{6} = 0.041\overline{6}\,\pi$ rad.

For Probs. 7.38 and 7.39, find the radian measure of a central angle θ subtended by an arc s in a circle of radius R, where R and s are given in each of the following.

7.38 $R = 4$ cm, $s = 24$ cm

    ❚ $\theta = \dfrac{s}{R}$ (in radians) $= \dfrac{24 \text{ cm}}{4 \text{ cm}} = 6$ rad.

7.39 $R = 10$ cm, $s = 10\pi$ cm

    ❚ $\theta = \dfrac{s}{R} = \dfrac{10\pi \text{ cm}}{10 \text{ cm}} = \pi$.

7.40 Find five angles coterminal with −155°.

    ❚ See Fig. 7.8. (1) $360° + (-155°) = 205$. (2) $2 \times 360° + (-155°) = 565°$.
(3) $3 \times 360° + (-155°) = 925°$. (4) $15 \times 360° + (-155°) = 5245°$.
(5) $5245° - (35 \times 360°) = 5245° - 12,600° = -7355°$.

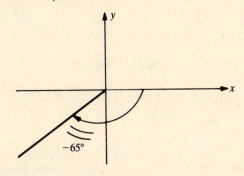

Fig. 7.8

For Probs. 7.41 to 7.43, answer true or false, and justify your answer.

7.41 If $\theta_1 = \theta_2$ in degree measure, then $\theta_1 = \theta_2$ in radian measure.

    ❚ True. Radians and degrees are simply *units* of measure.

7.42 An angle measuring 4π rad is congruent to an angle measuring 2π rad.

    ❚ False. Be very careful here. These two angles are coterminal, but not congruent. The same is true, for example, for 120° and (360 + 120)°.

7.43   If x is the supplement of y, then $x/2$ is the complement of y.

▮ False. Here $x + y = 180°$. Then $(x + y)/2 = 90°$, but $x/2 + y \neq 90°$.

For Probs. 7.44 to 7.46, a wheel is turning at the rate of 48 revolutions per minute (r/min). Express this angular speed in:

7.44   r/s

▮ $48 \text{ r/min} = \frac{48}{60} \text{ r/s} = \frac{4}{5} \text{ r/s}.$

7.45   rad/min

▮ Since $1 \text{ r} = 2\pi \text{ rad}$, $48 \text{ r/min} = 48(2\pi) \text{ rad/min} = 301.6 \text{ rad/min}.$

7.46   rad/s

▮ $48 \text{ r/min} = \frac{4}{5} \text{ r/s} = \frac{4}{5}(2\pi) \text{ rad/s} = 5.03 \text{ rad/s}$ or $48 \text{ r/min} = 96\pi \text{ rad/min} = 9\pi/60 \text{ rad/s} = 5.03 \text{ rad/s}.$

7.47   The minute hand of a clock is 12 in long. How far does the tip of the hand move during 20 min?

▮ During 20 min, the hand moves through an angle $\theta = 120° = 2\pi/3 \text{ rad}$, and the tip of the hand moves a distance $s = r\theta = 12(2\pi/3) = 8\pi \text{ in} = 25.1 \text{ in}.$

7.48   A central angle of a circle of radius 30 in intercepts an arc of 6 in. Express the central angle θ in radians and in degrees.

▮ $\theta = s/r = \frac{6}{30} = 1 \text{ rad} = 11°27'33''.$

7.49   A railroad curve is to be laid out on a circle. What radius should be used if the track is to change direction by 25° in a distance of 120 ft?

▮ We are required to find the radius of a circle on which a central angle $\theta = 25° = 5\pi/36 \text{ rad}$ intercepts an arc of 120 ft. Then

$$r = \frac{s}{\theta} = \frac{120}{5\pi/36} = \frac{864}{\pi} \text{ ft} = 275 \text{ ft}$$

7.50   Assuming the earth to be a sphere of radius 3960 mi, find the distance of a point in latitude 36°N from the equator.

▮ Since $36° = \pi/5 \text{ rad}$, $s = r\theta = 3960(\pi/5) = 2488 \text{ mi}.$

7.51   Two cities 270 mi apart lie on the same meridian. Find their difference in latitude.

▮ $\theta = s/r = 270/3960 = \frac{3}{44} \text{ rad}$ or $3°54.4'.$

7.52   A wheel 4 ft in diameter is rotating at 80 r/min. Find the distance (in feet) traveled by a point on the rim in 1 s, that is, the linear speed of the point (in feet per second).

▮
$$80 \text{ r/min} = 80\left(\frac{2\pi}{60}\right) \text{ rad/s} = \frac{8\pi}{3} \text{ rad/s}$$

Then in 1 s the wheel turns through an angle $\theta = 8\pi/3 \text{ rad}$, and a point on the wheel will travel a distance $s = r\theta = 2(8\pi/3) \text{ ft} = 16.8 \text{ ft}$. The linear velocity is 16.8 ft/s.

7.53   Find the diameter of a pulley which is driven at 360 r/min by a belt moving at 40 ft/s.

▮
$$360 \text{ r/min} = 360\left(\frac{2\pi}{60}\right) \text{ rad/s} = 12\pi \text{ rad/s}$$

Then in 1 s the pulley turns through an angle $\theta = 12\pi$ rad, and a point on the rim travels a distance $s = 40$ ft.

$$d = 2r = 2\left(\frac{s}{\theta}\right) = 2\left(\frac{40}{12\pi}\right) \text{ ft} = \frac{20}{3\pi} \text{ ft} = 2.12 \text{ ft}$$

7.54 A point on the rim of a turbine wheel of 10-ft diameter moves with a linear speed of 45 ft/s. Find the rate at which the wheel turns (angular speed) in radians per second and in revolutions per second.

❚ In 1 s a point on the rim travels a distance $s = 45$ ft. Then in 1 s the wheel turns through an angle $\theta = s/r = 45/5 = 9$ rad, and its angular speed is 9 rad/s.
  Since $1 \text{ r} = 2\pi$ rad or 1 rad $= 1/(2\pi)$ r, 9 rad/s $= 9[1/(2\pi)]$ r/s $= 1.43$ r/s.

7.55 Express each of the following angles in mils: (**a**) 18°, (**b**) 16°20′, (**c**) 0.22 rad, (**d**) 1.6 rad.

❚ Since $1° = \frac{160}{9}$ mils and 1 rad $= 1000$ mils;

(**a**) $18° = 18\left(\frac{160}{9}\right)$ mils $= 320$ mils.

(**b**) $16°20′ = \frac{49}{3}\left(\frac{160}{9}\right)$ mils $= 290$ mils.

(**c**) 0.22 rad $= 0.22(1000)$ mils $= 220$ mils.
(**d**) 1.6 rad $= 1.6(1000)$ mils $= 1600$ mils.

7.56 Express each of the following angles in degrees and in radians: (**a**) 40 mils, (**b**) 100 mils.

❚ (**a**) 40 mils $= 40(\frac{9}{160})° = 2°15′$, and 40 mils $= 40(0.001)$ rad $= 0.04$ rad.
(**b**) 100 mils $= 100(\frac{9}{160})° = 5°37.5′$, and 100 mils $= 100(0.001)$ rad $= 0.1$ rad.

7.57 At 5000-yd range, a battery subtends an angle of 15 mils. Find the width of the battery.

❚ $R = \frac{5000}{1000} = 5$, $m = 15$, and $W = Rm = 5(15) = 75$ yd.

7.58 A ship 360 ft long is found to subtend an angle of 40 mils at an observation post onshore. Find the distance from shore to ship.

❚ $W = 360$ ft $= 120$ yd, $m = 40$, and $R = W/m = 120/40 = 3$. The required distance is 3000 yd.

7.59 A shell is observed to burst 200 yd to the left of the target. What angular correction should be made in aiming the gun if the range is (**a**) 5000 yd and (**b**) 7500 yd?

❚ (**a**) The correction is $m = W/R = 200/5 = 40$ mils, to the right.
(**b**) The correction is $m = W/R = 200/7.5 = 27$ mils, to the right.

For Probs. 7.60 to 7.63, in which quadrant will θ terminate?

7.60 $\sin \theta$ and $\cos \theta$ are both negative.

❚ Since $\sin \theta = y/r$ and $\cos \theta = x/r$, both x and y are negative. (Recall that r is always positive.) Thus, θ is a third-quadrant angle.

7.61 $\sin \theta$ and $\tan \theta$ are both positive.

❚ Since $\sin \theta$ is positive, y is positive; since $\tan \theta = y/x$ is positive, x is also positive. Thus, θ is a first-quadrant angle.

7.62 $\sin \theta$ is positive, and secant θ is negative.

> Since sin θ is positive, y is positive; since sec θ is negative, x is negative. Thus, θ is a second-quadrant angle.

7.63 sec θ is negative, and tan θ is negative.

> Since sec θ is negative, x is negative; since tan θ is negative, y is positive. Thus, θ is a second-quadrant angle.

For Probs. 7.64 to 7.67, in which quadrants may θ terminate?

7.64 sin θ is positive.

> Since sin θ is positive, y is positive. Then x may be positive or negative, and θ is a first- or second-quadrant angle.

7.65 cos θ is negative.

> Since cos θ is negative, x is negative. Then y may be positive or negative, and θ is a second- or third-quadrant angle.

7.66 tan θ is negative.

> Since tan θ is negative, either y is positive and x is negative or y is negative and x is positive. Thus, θ may be a second- or fourth-quadrant angle.

7.67 sec θ is positive.

> Since sec θ is positive, x is positive. Thus, θ may be a first- or fourth-quadrant angle.

7.68 Find the values of cos θ and tan θ, given sin $\theta = \frac{8}{17}$ and θ in quadrant I.

> Let P be a point on the terminal line of θ. Since sin $\theta = y/r = \frac{8}{17}$, we take $y = 8$ and $r = 17$. Since θ is in quadrant I, x is positive; thus

$$x = \sqrt{r^2 - y^2} = \sqrt{17^2 - 8^2} = 15$$

To draw the figure, locate the point $P(15, 8)$, join it to the origin, and indicate angle θ. Then

$$\cos \theta = \frac{x}{r} = \frac{15}{17} \quad \text{and} \quad \tan \theta = \frac{y}{x} = \frac{8}{15}$$

The choice of $y = 8$, $r = 17$ is one of convenience. Note that $\frac{8}{17} = \frac{16}{34}$ and we might have taken $y = 16$, $r = 34$. Then $x = 30$, cos $\theta = \frac{30}{34} = \frac{15}{17}$, and tan $\theta = \frac{16}{30} = \frac{8}{15}$.

7.69 Determine the values of cos θ and tan θ if sin $\theta = m/n$, a negative fraction.

> Since sin θ is negative, θ is in quadrant III or IV.
> **(a)** In quadrant III: Take $y = m$, $r = n$, $x = -\sqrt{n^2 - m^2}$; then

$$\cos \theta = \frac{x}{r} = \frac{-\sqrt{n^2 - m^2}}{n} \quad \text{and} \quad \tan \theta = \frac{y}{x} = \frac{-m}{\sqrt{n^2 - m^2}}$$

> **(b)** In quadrant IV: Take $y = m$, $r = n$, $x = +\sqrt{n^2 - m^2}$; then

$$\cos \theta = \frac{x}{r} = \frac{\sqrt{n^2 - m^2}}{n} \quad \text{and} \quad \tan \theta = \frac{y}{x} = \frac{m}{\sqrt{n^2 - m^2}}$$

7.70 Evaluate sin $0°$ + 2 cos $0°$ + 3 sin $90°$ + 4 cos $90°$ + 5 sec $0°$ + 6 csc $90°$.

> $0 + 2(1) + 3(1) + 4(0) + 5(1) + 6(1) = 16$.

7.71   Evaluate $\sin 180° + 2 \cos 180° + 3 \sin 270° + 4 \cos 270° - 5 \sec 180° - 6 \csc 270°$.

▮ $0 + 2(-1) + 3(-1) + 4(0) - 5(-1) - 6(-1) = 6$.

7.2 TRIGONOMETRIC FUNCTIONS

For Probs. 7.72 to 7.96, find the value of the trigonometric function at the given angle. Do *not* use a calculator.

7.72 $\sin 90°$

▮ See Fig. 7.9. Here the point $(0, 1)$ is chosen on the terminal ray of the angle. From the equation $\sin \theta = b/R$ for any angle θ where b is the ordinate of (a, b) on the terminal ray, we have $\sin 90° = 1/1 = 1$. [Clearly, $R = 1$, the distance from $(0, 0)$ to $(0, 1)$.]

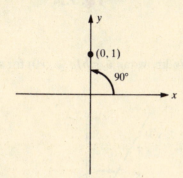

Fig. 7.9

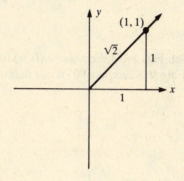

Fig. 7.10

7.73 $\sin 45°$

▮ See Fig. 7.10. for convenience, choose $(1, 1)$ on the terminal ray. Then $\sin \theta = b/R = 1/\sqrt{2} = \sqrt{2}/2$ [$\sqrt{2}$ = distance from $(0, 0)$ to $(1, 1)$].

7.74 $\cos 60°$

▮ See Fig. 7.11. $\cos \theta = a/R = \frac{1}{2}$. We choose $(1, \sqrt{3})$ on the 60° ray for convenience.

7.75 $\sin 30°$

▮ See Fig. 7.12. Then $\sin 30° = b/R = \frac{1}{2}$.

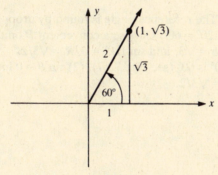

Fig. 7.11

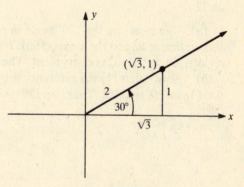

Fig. 7.12

7.76 $\cot 30°$

▮ See Fig. 7.13. $\cot \theta = a/b$ for any angle θ. Then $\cot \theta = \sqrt{3}/1 = \sqrt{3}$ when $\theta = 30°$, so $\cot 30° = \sqrt{3}$.

7.77 $\csc 45°$

▮ See Fig. 7.14. For any angle θ, $\csc \theta = 1/\sin \theta = R/b$. Thus, $\csc 45° = \sqrt{2}/1 = \sqrt{2}$.

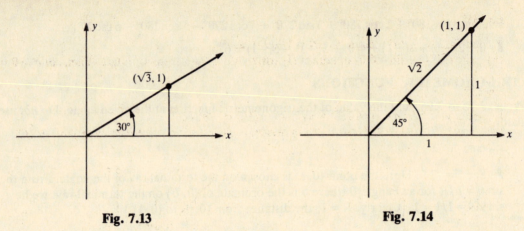

Fig. 7.13   **Fig. 7.14**

7.78   tan 90°

▮ See Fig. 7.15. We choose $(0, 1)$ on the terminal ray. We know $\tan \theta = b/a$ $(a \neq 0)$ for any θ, but here $a = 0$. Thus, tan 90° is *not* defined.

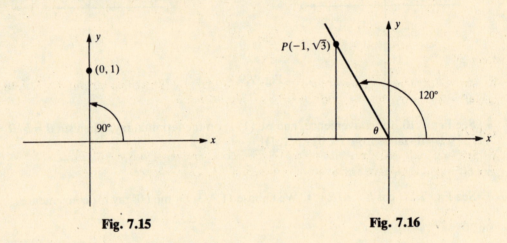

Fig. 7.15   **Fig. 7.16**

7.79   sin 120°

▮ *(a)*   We illustrate the 120° angle in Fig. 7.16. The reference angle is found by dropping a perpendicular line to the x axis. Here $\theta = 180° - 120° = 60°$. Choose a convenient P on the terminal ray of 120°; $(-1, \sqrt{3})$ is convenient. Then $R = 2$, $b = \sqrt{3}$, and $\sin 120° = b/R = \sqrt{3}/2$.

(b)   Notice that (1) the reference angle for 120° is 60° (see Fig. 7.17), (2) $\sin \theta > 0$ in quadrant II, and (3) $\sin 60° = \sqrt{3}/2$. Thus, $\sin 120° = +\sin 60° = \sqrt{3}/2$.

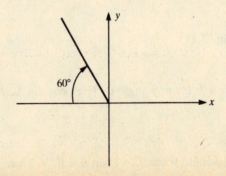

Fig. 7.17

7.80   sin 135°

▮ See Fig. 7.18. (*a*) $\sin \theta = b/R = 1/\sqrt{2} = \sqrt{2}/2$.
(*b*)   sin 135° = +sin 45° in quadrant II, where 45° is the reference angle. Then $\sin \theta > 0 = \sqrt{2}/2$.

7.81   cos 135°

▮ See Fig. 7.19. (*a*) $\cos 135° = a/R = -1/\sqrt{2} = -\sqrt{2}/2$.
(*b*) $\cos 135° = -\cos 45°$ (cos θ < 0 in quadrant II and 45° = reference angle) $= -\sqrt{2}/2$.

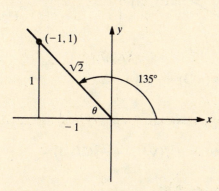

Fig. 7.18

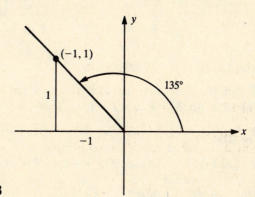

Fig. 7.19

7.82   tan 120°

▮ $\tan 120° = -\tan 60°$ (quadrant II) $= -\sqrt{3}$. *Note*: If you forget that $\tan 60° = \sqrt{3}$, then recall that $\sin 60° = \sqrt{3}/2$, $\cos 60° = \frac{1}{2}$, and $\tan 60° = (\sqrt{3}/2) \cdot 2 = \sqrt{3}$.

7.83   cot 120°

▮ See Fig. 7.20. (*a*) $\cot 120° = -\cot 60° = -1/\tan 60° = -1/\sqrt{3} = -\sqrt{3}/3$.
(*b*)   Use the reference angle: $\cot 120° = a/b = -1/\sqrt{3} = -\sqrt{3}/3$.

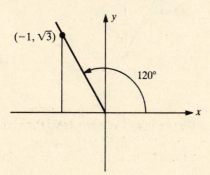

Fig. 7.20

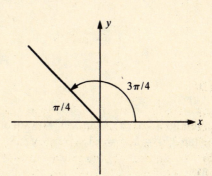

Fig. 7.21

7.84   sin (3π/4)

▮ See Fig. 7.21. $\sin (3\pi/4) = \sin (\pi/4) = \sqrt{2}/2$. [Recall: $\sin (\pi/4) = b/R = 1/\sqrt{2} = \sqrt{2}/2$.]

7.85   cos (5π/6)

▮ See Fig. 7.22.   (*a*) $\cos (5\pi/6) = -\cos (\pi/6)$ (cos θ < 0 in quadrant II and π/6 = reference angle) $= -\sqrt{3}/2$.
(*b*) $\cos (5\pi/6) = a/R = -\sqrt{3}/2$.

7.86   csc 210°.

▮ See Fig. 7.23. (*a*) $\csc 210° = -\csc 30° = -1/\sin 30° = -1/\frac{1}{2} = -2$.
(*b*) $\csc 210° = R/b = 2/-1 = -2$.

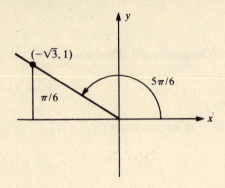

Fig. 7.22

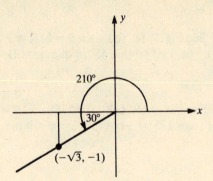

Fig. 7.23

7.87 sec (4π/3)

▌ See Fig. 7.24. (*a*) sec (4π/3) = −sec (π/3) = −1/cos (π/3) = −1/½ = −2.
(*b*) sec (4π/3) = R/a = 2/−1 = −2.

7.88 cot 300°

▌ See Fig. 7.25. (*a*) cot 300° = −cot 60° = −cos 60°/sin 60° = −½/(√3/2) = −½(2/√3) =
−1/√3 = −√3/3.
(*b*) cot 300° = a/b = 1/−√3 = −√3/3.

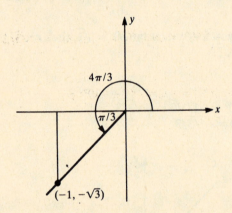

Fig. 7.24

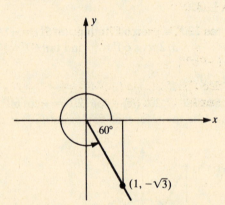

Fig. 7.25

7.89 tan 330°

▌ See Fig. 7.26. (*a*) tan 330° = −tan 30° = −sin 30°/cos 30° = −½/(√3/2) = −½(2/√3) = −√3/3.
(*b*) tan 330° = b/a = −1/√3 = −√3/3.

7.90 cos 315°

▌ See Fig. 7.27. cos 315° = cos 45° = √2/2 (cos θ > 0 in quadrant IV).

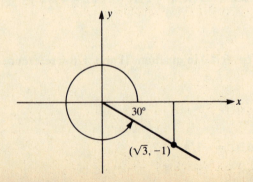

Fig. 7.26

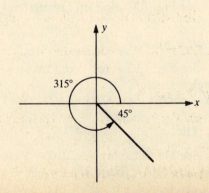

Fig. 7.27

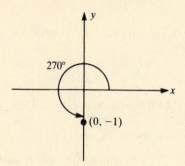

Fig. 7.28

7.91 cos 270°

▮ See Fig. 7.28. $\cos 270° = a/R = 0/1 = 0$.

7.92 sin 480°

▮ $\sin 480° = \sin (360° + 120°) = \sin 120° = \sin 60° = \sqrt{3}/2$.

7.93 cos 870°

▮ $\cos 870° = \cos (720° + 150°) = \cos 150° = -\cos 30° = -\sqrt{3}/2$.

7.94 cos (−870°)

▮ $\cos (-870°) = \cos 870° \ [\cos (-\theta) = \cos \theta] = \cos (720° + 150°) = \cos 150° = -\cos 30° = -\sqrt{3}/2$.

7.95 sin (−960°)

▮ See Fig. 7.29. $\sin (-960°) = -\sin 960° \ [\sin (-\theta) = -\sin \theta] = -\sin (720° + 240°) = -(-\sin 60°) = \sin 60° = \sqrt{3}/2$.

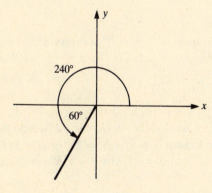

Fig. 7.29

7.96 cot (−1035°)

▮ $\cot (-1035°) = -\cot 1035° = -\cot (720° + 315°) = -\cot 315° = -\cot 45° = -1$.

For Probs. 7.97 to 7.105, evaluate the given expression.

7.97 $\sin^2 0° + \cos^2 0°$

▮ $\sin 0° = 0$, and $\cos 0° = 1$. Then $\sin^2 0° + \cos^2 0° = 0^2 + 1^2 = 1$.

7.98 $\sin^2 (\pi/2) + \cos^2 (\pi/2)$

▮ $\sin (\pi/2) = 1$, and $\cos (\pi/2) = 0$. Then $\sin^2 (\pi/2) + \cos^2 (\pi/2) = 1^2 + 0^2 = 1$.

7.99 $\sin^2 315° + \cos^2 315°$

▮ $\sin 315° = -\sin 45° = -\sqrt{2}/2$, and $\cos 315° = \cos 45° = \sqrt{2}/2$. Then $\sin^2 315° + \cos^2 315° = (-\sqrt{2}/2)^2 + (\sqrt{2}/2)^2 = \frac{1}{2} + \frac{1}{2} = 1$.

7.100 $1 + \tan^2 (\pi/2)$

▮ $\tan (\pi/2)$ is undefined ($\tan \theta = b/a$, and $a = 0$). Thus, $1 + \tan^2 (\pi/2)$ is undefined.

7.101 $\cos (\pi/4) \cos (\pi/2) - \sin (\pi/4) \sin (\pi/2)$

▮ $\cos (\pi/4) = \sqrt{2}/2$, $\cos (\pi/2) = 0$, $\sin (\pi/4) = \sqrt{2}/2$, and $\sin (\pi/2) = 1$. Thus, $\cos (\pi/4) \cos (\pi/2) - \sin (\pi/4) \sin (\pi/2) = \sqrt{2}/2 \cdot 0 - \sqrt{2}/2 \cdot 1 = -\sqrt{2}/2$.

7.102 $\sin 0 + \sin (\pi/2) + \sin \pi$

▮ $\sin 0 = 0$, $\sin (\pi/2) = 1$, and $\sin \pi = 0$. Then $\sin 0 + \sin (\pi/2) + \sin \pi = 0 + 1 + 0 = 1$.

7.103 $2 \sin (\pi/2) \cos (\pi/2)$

▮ $\sin (\pi/2) = 1$, and $\cos (\pi/2) = 0$. Then $2 \sin (\pi/2) \cos (\pi/2) = 2 \cdot 1 \cdot 0 = 0$.

7.104 $\cos (\pi/4) + \cos (\pi/2) + \cos \pi$

▮ $\cos (\pi/4) = \sqrt{2}/2$, $\cos (\pi/2) = 0$, and $\cos \pi = -1$. Thus, $\cos (\pi/4) + \cos (\pi/2) + \cos \pi = \sqrt{2}/2 + 0 - 1 = \sqrt{2}/2 - 1$.

7.105 $\sin e^x$ where $x = \ln \pi$

▮ $\sin e^{\ln \pi} = \sin \pi = 0$.

For Probs. 7.106 to 7.111, find the value of the other five trigonometric functions for the indicated θ. Do *not* find θ.

7.106 $\sin \theta = \frac{3}{5}$, $\cos \theta < 0$

▮ $\sin \theta = \frac{3}{5} = b/R$. Then from $a^2 + b^2 = R^2$ we have $a^2 + 9 = 25$, or $a = \pm 4$. But $\cos \theta < 0$, so $a = -4$. Then $\cos \theta = a/R = -\frac{4}{5}$; $\tan \theta = b/a = -\frac{3}{4}$; $\cot \theta = -\frac{4}{3}$; $\csc \theta = 1/\sin \theta = \frac{5}{3}$; and $\sec \theta = 1/\cos \theta = -\frac{5}{4}$.

7.107 $\tan \theta = -\frac{4}{3}$, $\sin \theta < 0$

▮ $\tan \theta = -\frac{4}{3} = b/a$. Then $b = 4$, $a = -3$ or $b = -4$, $a = 3$. But $\sin \theta = b/R$ and $\sin \theta < 0$. Thus, $b < 0$. Then $b = -4$, $a = 3$, and $R^2 = a^2 + b^2 = 25$, or $R = 5$ ($R > 0$ always). Then $\cot \theta = -\frac{3}{4}$; $\sin \theta = b/R = -\frac{4}{5}$; $\csc \theta = 1/\sin \theta = -\frac{5}{4}$; $\cos \theta = a/R = -\frac{3}{5}$; and $\sec \theta = -\frac{5}{3}$.

7.108 $\cos \theta = -\sqrt{5}/3$, $\cot \theta > 0$

▮ If $\cos \theta = -\sqrt{5}/3$, then $R = 3$, $a = -\sqrt{5}$. From $a^2 + b^2 = R^2$, $b^2 = R^2 - a^2 = 9 - 5 = 4$. Then $b = \pm 2$. But $\cot \theta > 0$, so $a/b > 0$, and $a < 0$. Thus, $b < 0$, and $b = -2$. Then $\sec \theta = -3/\sqrt{5}$; $\tan \theta = b/a = -2/-\sqrt{5} = 2/\sqrt{5}$; $\cot \theta = \sqrt{5}/2$; $\sin \theta = b/R = -\frac{2}{3}$; and $\csc \theta = -\frac{3}{2}$.

7.109 $\cos \theta = -\sqrt{5}/3$, $\tan \theta > 0$

▮ Note that $\tan \theta = b/a$ and $\cot \theta = a/b$. Then $\tan \theta > 0$ whenever $\cot \theta > 0$. Thus, this problem is equivalent to Prob. 7.108, and the answers are identical.

7.110 $\tan \theta = -\sqrt{2}$, $\sin \theta < 0$

▮ $\tan \theta = b/a = -\sqrt{2} = -\sqrt{2}/1$, so $b = -\sqrt{2}$, $a = 1$ or $b = \sqrt{2}$, $a = -1$. But $\sin \theta < 0$, so $b < 0$ and $b = -\sqrt{2}$, $a = 1$. Then $R = \sqrt{1^2 + (-\sqrt{2})^2} = \sqrt{3}$. Thus, $\cot \theta = -1/\sqrt{2}$; $\sin \theta = b/R = -\sqrt{2}/\sqrt{3} = -\sqrt{2/3}$; $\csc \theta = -\sqrt{3}/2$; $\cos \theta = 1/\sqrt{3}$; and $\sec \theta = \sqrt{3}$.

7.111 $\tan \theta = -\sqrt{2}$, $\cos \theta > 0$

▮ $\tan \theta = -\sqrt{2}$, so $b = -\sqrt{2}$, $a = 1$ or $b = \sqrt{2}$, $a = -1$. But $\cos \theta > 0$, so $a > 0$ and $b = -\sqrt{2}$, $a = 1$. The results are identical to Prob. 7.110. Do you see why?

For Probs. 7.112 to 7.115, answer true or false, and justify your answer.

7.112 If x is an angle measured in radians, then $\cos(360 + x) = \cos x$.

▮ False. In radian measure, x and $360 + x$ are not coterminal. However, $\cos(2\pi + x) = \cos x$.

7.113 If $\sin x = \cos x$, then x must be $45°$.

▮ False. $x = 45°$ is one solution. Others are $45° + 360°$, $45° + 2(360°)$, etc.

7.114 $\sin[\sin(\pi/2)] > 0$

▮ $\sin(\pi/2) = 1$. Since $0 < 1 < \pi/2$, the angle α measuring 1 rad is in quadrant I and $\sin \alpha > 0$. True.

7.115 $\cos[\sin(3\pi/2)] > 0$

▮ $\sin(3\pi/2) = -1$, but $-\pi/2 < -1 < 0$; thus the angle β measuring -1 rad is in quadrant IV and $\cos \beta > 0$. True.

7.3 INVERSE TRIGONOMETRIC FUNCTIONS

For Probs. 7.116 to 7.146, find the value of the given expression. Do *not* use a calculator or a table of values.

7.116 $\mathrm{Sin}^{-1} \frac{1}{2}$

▮ If $y = \mathrm{Sin}^{-1} \frac{1}{2}$, then $\sin y = \frac{1}{2}$ (by definition of the $\mathrm{Sin}^{-1} x$ function). Then $y = \pi/6$ (30°). *Remember*: If $y = \mathrm{Sin}^{-1} x$, then $-\pi/2 \le y \le \pi/2$. If $y = \mathrm{Cos}^{-1} x$, then $0 \le y \le \pi$. If $y = \mathrm{Tan}^{-1} x$, then $-\pi/2 < y < \pi/2$.

7.117 $\mathrm{Sin}^{-1}(\sqrt{3}/2)$

▮ If $y = \mathrm{Sin}^{-1}(\sqrt{3}/2)$, then $\sin y = \sqrt{3}/2$ $(-\pi/2 \le y \le \pi/2)$. Thus, $y = 60°$ $(\pi/3)$.

7.118 $\mathrm{Arccos} \frac{1}{2}$

▮ "Arc" notation is just another means of expressing inverse trigonometric functions. $\mathrm{Arccos} \frac{1}{2} = \mathrm{Cos}^{-1} \frac{1}{2}$. Then if $\mathrm{Arccos} \frac{1}{2} = y$, $\cos y = \frac{1}{2}$ $(0 \le y \le \pi)$. Thus, $y = 60°$.

7.119 $\mathrm{Cos}^{-1}(1/\sqrt{2})$

▮ If $y = \mathrm{Cos}^{-1}(1/\sqrt{2})$, then $\cos y = 1/\sqrt{2}$ $(= \sqrt{2}/2)$ $(0 \le y \le \pi)$; $y = 45°$.

7.120 $\mathrm{Sin}^{-1} 1 + \mathrm{Sin}^{-1} 0$

▮ If $y = \mathrm{Sin}^{-1} 1$, then $\sin y = 1$ $(-\pi/2 \le y \le \pi/2)$, and $y = \pi/2$. $\mathrm{Sin}^{-1} 0 = y$, so $\sin y = 0$, and $y = 0$. Thus, $\mathrm{Sin}^{-1} 1 + \mathrm{Sin}^{-1} 0 = \pi/2 + 0 = \pi/2$.

7.121 $\mathrm{Sin}^{-1} 1 + \mathrm{Sin}^{-1}(-1)$

▮ $\mathrm{Sin}^{-1} 1 = \pi/2$ (see Prob. 7.120). If $y = \mathrm{Sin}^{-1}(-1)$, then $\sin y = -1$ $(-\pi/2 \le y \le \pi/2)$ and $y = -\pi/2$. (*Note*: $y = 3\pi/2$ is incorrect. Look at the range!) Thus, $\mathrm{Sin}^{-1} 1 + \mathrm{Sin}^{-1}(-1) = \pi/2 + (-\pi/2) = 0$.

7.122 $\mathrm{Arctan} \sqrt{3}$

▮ If $y = \mathrm{Arctan} \sqrt{3}$, then $\tan y = \sqrt{3}$ $(-\pi/2 < y < \pi/2)$ and $y = 60°$. (Did you forget that $\tan 60° = \sqrt{3}$? If so, recall that $\sin 60° = \sqrt{3}/2$, $\cos 60° = \frac{1}{2}$, and $\tan 60° = \sin 60°/\cos 60°$.)

7.123 $Tan^{-1}(-1/\sqrt{3})$

▮ If $y = Tan^{-1}(-1/\sqrt{3})$, then $\tan y = -1/\sqrt{3}$ $(-\pi/2 < y < \pi/2)$. Thus, $y = -30°$. (Note that 150° is wrong!)

7.124 $Arccot \sqrt{3}$

▮ If $y = Arccot \sqrt{3}$, then $\cot y = \sqrt{3}$ $(0 < y < \pi)$ and $\tan y = 1/\sqrt{3}$. Thus, $y = 30°$.

7.125 $Arcsec(-\sqrt{2})$

▮ See Fig. 7.30. If $y = Arcsec(-\sqrt{2})$, then $\sec y = -\sqrt{2}$ $(0 \le y \le \pi,$ but $y \ne \pi/2)$. Then $\cos y = -1/\sqrt{2}$, so $y = 135°$.

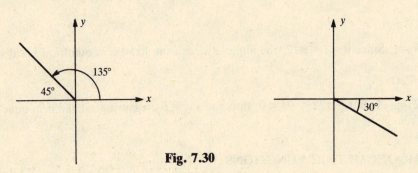

Fig. 7.30   Fig. 7.31

7.126 $Arccsc(-2)$

▮ See Fig. 7.31. If $y = Arccsc(-2)$, then $\csc y = -2$ $(-\pi/2 \le y \le \pi/2,$ but $y \ne 0)$. Then, $\sin y = -\frac{1}{2}$, and $y = -30°$.

7.127 $Tan^{-1}(-1) + Cot^{-1}(-1)$

▮ $Tan^{-1}(-1) = y$, so $\tan y = -1$ $(-\pi/2 < y < \pi/2)$ and $y = -\pi/4$. $Cot^{-1}(-1) = y$, so $\cot y = -1$ $(0 < y < \pi)$ and $y = 3\pi/4$. Then $Tan^{-1}(-1) + Cot^{-1}(-1) = -\pi/4 + 3\pi/4 = \pi/2$.

7.128 $\sin(Arcsin \frac{1}{2})$

▮ This really says, "The sine of the angle whose sine is $\frac{1}{2}$." Thus, $\sin(Arcsin \frac{1}{2}) = \frac{1}{2}$.

7.129 $\cos(Cos^{-1} 0.72134)$

▮ See Prob. 7.128. This is simply 0.72134. In fact, what is $\cos(Cos^{-1} l)$, where $-1 \le l \le 1$? It must be l itself.

7.130 $Arcsin[\sin(\pi/3)]$

▮ $\sin(\pi/3) = \sqrt{3}/2$. $Arcsin[\sin(\pi/3)] = Arcsin(\sqrt{3}/2) =$ the angle whose sin is $\sqrt{3}/2$ (between $-\pi/2$ and $\pi/2) = \pi/3$.

7.131 $\cos[Arccos(\sqrt{3}/2)]$

▮ See Prob. 7.128. $\cos[Arccos(\sqrt{3}/2)] = \sqrt{3}/2$.

7.132 $\tan(Arccos 0.5)$

▮ $Arccos 0.5 = y$ means $\cos y = \frac{1}{2}$, or $y = 60°$. Then $\tan(Arccos 0.5) = \tan 60° = \dfrac{\sin 60°}{\cos 60°} = \dfrac{\sqrt{3}/2}{\frac{1}{2}} = \sqrt{3}$.

7.133 Arccos $[\cos{(\pi/2)}]$

▮ $\cos{(\pi/2)} = 0$. Arccos $0 = y$ means $\cos{y} = 0 \ (0 \leq y \leq \pi)$, and $y = \pi/2$. Thus, Arccos $[\cos{(\pi/2)}] = \pi/2$. Look at Prob. 7.130. Do you see a pattern?

7.134 $\sin{[\text{Arccos}\,(-\sqrt{2}/2)]}$

▮ Arccos $(-\sqrt{2}/2) = y = $ means $\cos{y} = -\sqrt{2}/2 \ (0 \leq y \leq \pi)$, and $y = 135°$. Thus, $\sin{[\text{Arccos}\,(-\sqrt{2}/2)]} = \sin{135°} = \sqrt{2}/2$.

7.135 $\cos{[\text{Arctan}\,(-1)]}$

▮ Arctan $(-1) = -\pi/4 \ (-\pi/2 < y < \pi/2)$. Then $\cos{[\text{Arctan}\,(-1)]} = \cos{(-\pi/4)} = \sqrt{2}/2$.

7.136 Arccos $[\sin{(3\pi/4)}]$

▮ See Fig. 7.32. $\sin{(3\pi/4)} = \sin{(\pi/4)} = \sqrt{2}/2$. Then Arccos $(\sqrt{2}/2) = \pi/4$ [since $\cos{(\pi/4)} = \sqrt{2}/2$].

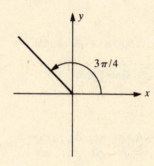

Fig. 7.32

7.137 $\sin{[\text{Arccos}\,(-\sqrt{3}/2)]}$

▮ Arccos $(-\sqrt{3}/2) = y$, so $\cos{y} = -\sqrt{3}/2 \ (0 \leq y \leq \pi)$, and $y = 150°$. Then $\sin{150°} = \sin{30°} = 0.5$.

7.138 $\cos{(2\,\text{Tan}^{-1}\,1)}$

▮ $\text{Tan}^{-1}\,1 = \pi/4$. Then $2\,\text{Tan}^{-1}\,1 = \pi/2$ and $\cos{(\pi/2)} = 0$. Thus, $\cos{(2\,\text{Tan}^{-1}\,1)} = 0$.

7.139 Sec $[\text{Arccos}\,(-\sqrt{3}/2)]$

▮ Arccos $\left(-\dfrac{\sqrt{3}}{2}\right) = 150°$ (since $0 \leq y \leq \pi$). Then $\sec{150°} = \dfrac{1}{\cos{150°}} = \dfrac{1}{-\sqrt{3}/2} = \dfrac{-2}{\sqrt{3}} = \dfrac{-2\sqrt{3}}{3}$.

7.140 $\csc{[\text{Arcsin}\,(\sqrt{2}/2)]}$

▮ Arcsin $\dfrac{\sqrt{2}}{2} = \dfrac{\pi}{4}$. Then $\csc{\dfrac{\pi}{4}} = \dfrac{1}{\sin{(\pi/4)}} = \dfrac{1}{\sqrt{2}/2} = \dfrac{2}{\sqrt{2}} = \dfrac{2\sqrt{2}}{2} = \sqrt{2}$.

7.141 $\sec{[\text{Arccos}\,(\sqrt{3}/2)]}$

▮ Arccos $\dfrac{\sqrt{3}}{2} = \dfrac{\pi}{6}$. Then $\sec{\dfrac{\pi}{6}} = \dfrac{1}{\cos{(\pi/6)}} = \dfrac{1}{\sqrt{3}/2} = \dfrac{2}{\sqrt{3}} = \dfrac{2\sqrt{3}}{3}$.

7.142 $\tan{[\text{Arcsin}\,(\sqrt{3}/2)]}$

▮ Arcsin $\dfrac{\sqrt{3}}{2} = 60°$. Then $\tan{60°} = \dfrac{\sin{60°}}{\cos{60°}} = \dfrac{\sqrt{3}/2}{\frac{1}{2}} = \sqrt{3}$.

7.143 $\cot{[\text{Arccos}\,(-\sqrt{2}/2)]}$

■ Arccos $(-\sqrt{2}/2) = y$, so $\cos y = -\sqrt{2}/2$ $(0 \le y \le \pi)$. Thus, $y = 135°$, and then

$$\cot 135° = \frac{\cos 135°}{\sin 135°} = \frac{-\sqrt{2}/2}{\sqrt{2}/2} = -1.$$

7.144   csc [Arctan $(-\sqrt{3})$]

■ Arctan $(-\sqrt{3}) = y$ means that $\tan y = -\sqrt{3}$ $(-\pi/2 < y < \pi/2)$, so $y = -60°$. Then csc $(-60°) =$

$$\frac{1}{\sin (-60°)} = \frac{1}{-\sin 60°} = \frac{1}{-\sqrt{3}/2} = \frac{-2}{\sqrt{3}}.$$

7.145 $\text{Tan}^{-1}\,[\tan\,(\text{Cos}^{-1}\tfrac{1}{2})]$

■ $\text{Cos}^{-1}\tfrac{1}{2} = \pi/3$ $(0 \le y \le \pi)$. Then $\tan (\pi/3) = \sqrt{3}$. Finally, $\text{Tan}^{-1}\sqrt{3} = \pi/3$. Another method would be to note that $\text{Tan}^{-1} (\tan \theta) = \theta$ $(-\pi/2 < \theta < \pi/2)$. Then $\text{Tan}^{-1}\,[\tan\,(\text{Cos}^{-1}\tfrac{1}{2})] = \text{Cos}^{-1}\tfrac{1}{2} = \pi/3$.

7.146   Arccos (Arcsin 0)

■ Arcsin $0 = 0$. Then Arccos (Arcsin 0) = Arccos $0 = \pi/2$ $(0 \le y \le \pi)$.

For Probs. 7.147 to 7.153, prove the given statement.

7.147   Arcsin $(-x) = -$Arcsin x

■ Let $y = $ Arcsin $(-x)$. Then $\sin y = -x$, and $\sin (-y) = -(-x) = x$. Then $-y = $ Arcsin x, and $y = -$Arcsin x. Thus, $-$Arcsin $x = y = $ Arcsin $(-x)$.

7.148   Arccos $(-x) = \pi - $ Arccos x

■ Let $y = \pi - $ Arccos x. Then Arccos $x = \pi - y$, and $x = \cos (\pi - y) = \cos \pi \cos y + \sin \pi \sin y$ [using the cos of a sum formula (see Chap. 8)] $= -\cos y$. Thus, $y = $ Arccos $(-x)$, and Arccos $(-x) = y = \pi - $ Arccos x.

7.149   Arctan $(-x) = -$Arctan x

■ Let $y = $ Arctan $(-x)$; then $\tan y = -x$, and $x = -\tan y = \tan (-y)$. Thus, $-y = $ Arctan x, $y = -$Arctan x, and Arctan $(-x) = y = -$Arctan x.

7.150   Arctan $x + $ Arctan $y = $ Arctan $\dfrac{x + y}{1 - xy}$, where $|x| < 1$, $|y| < 1$

■ Let $a = $ Arctan x, $b = $ Arctan y. Then $x = \tan a$ and $y = \tan b$. Thus,

Arctan $\dfrac{x + y}{1 - xy} = $ Arctan $\dfrac{\tan a + \tan b}{1 - \tan a \tan b} = $ Arctan $[\tan (a + b)]$ (for the tangent of a sum see Chap. 8) $= a + b = $ Arctan $x + $ Arctan y, and the two sides of the identity are equal.

7.151   Sec (Arccos x) $= 1/x$

■ Let $y = $ Arccos x. Then $\cos y = x$, and $\sec y = 1/x$; but $y = $ Arccos x, so Sec (Arccos x) $= 1/x$.

7.152   csc (Arcsin x) $= 1/x$

■ Let $y = $ Arcsin x. Then $\sin y = x$, and $\csc y = 1/x$. Thus csc (Arcsin x) $= 1/x$.

7.153   2 Arctan $x = $ Arctan $\dfrac{2x}{1 - x^2}$, $|x| < 1$

∎ $2 \operatorname{Arctan} x = \operatorname{Arctan} \dfrac{x+x}{1-x \cdot x} = \operatorname{Arctan} \dfrac{2x}{1-x^2}$; see Prob. 7.150 and let $x = y$.

For Probs. 7.154 to 7.156 verify the given statement.

7.154 $\operatorname{Arctan} \frac{1}{2} + \operatorname{Arctan} \frac{1}{3} = \pi/4$

∎ $\operatorname{Arctan} x + \operatorname{Arctan} y = \operatorname{Arctan} \dfrac{x+y}{1-xy}$, so $\operatorname{Arctan} \frac{1}{2} + \operatorname{Arctan} \frac{1}{3} = \operatorname{Arctan} \dfrac{\frac{1}{2}+\frac{1}{3}}{1-\frac{1}{6}} =$
$\operatorname{Arctan} \dfrac{\frac{5}{6}}{\frac{5}{6}} = \operatorname{Arctan} 1 = \pi/4$.

7.155 $2 \operatorname{Arctan} \frac{1}{3} + \operatorname{Arctan} \frac{1}{7} = \pi/4$

∎ $2 \operatorname{Arctan} \frac{1}{3} = \operatorname{Arctan} \dfrac{\frac{2}{3}}{1-\frac{1}{9}} = \operatorname{Arctan} \frac{3}{4}$. Then $\operatorname{Arctan} \frac{3}{4} + \operatorname{Arctan} \frac{1}{7} = \operatorname{Arctan} \dfrac{\frac{3}{4}+\frac{1}{7}}{1-\frac{3}{28}} = \operatorname{Arctan} \dfrac{\frac{25}{28}}{\frac{25}{28}} = \dfrac{\pi}{4}$.

7.156 $\operatorname{Arctan} \frac{1}{2} + \operatorname{Arctan} \frac{1}{5} + \operatorname{Arctan} \frac{1}{8} = \pi/4$

∎ $\operatorname{Arctan} \frac{1}{2} + \operatorname{Arctan} \frac{1}{5} = \operatorname{Arctan} \dfrac{\frac{1}{2}+\frac{1}{5}}{1-\frac{1}{10}} = \operatorname{Arctan} \frac{7}{9}$. Then $\operatorname{Arctan} \frac{7}{9} + \operatorname{Arctan} \frac{1}{8} =$
$\operatorname{Arctan} \dfrac{\frac{7}{9}+\frac{1}{8}}{1-\frac{7}{72}} = \operatorname{Arctan} \dfrac{\frac{65}{72}}{\frac{65}{72}} = \dfrac{\pi}{4}$.

7.157 Write $\sin (\operatorname{Cos}^{-1} x)$ as an algebraic expression in x free of all trigonometric functions.

∎ Let $\theta = \operatorname{Cos}^{-1} x$. Then, $0 \le \theta \le \pi$ and if θ' is θ's reference angle, Fig. 7.33 describes θ. Then $\overline{AB} = \sqrt{1-x^2}$ and $\sin \theta = \sqrt{1-x^2}/1 = \sqrt{1-x^2}$. Thus, $\sin (\operatorname{Cos}^{-1} x) = \sin \theta = \sqrt{1-x^2}$. Note that whether θ is a quadrant I or II angle does not alter this.

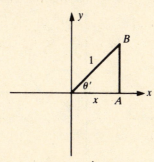

Fig. 7.33

7.4 GRAPHING THE TRIGONOMETRIC FUNCTIONS

For Probs. 7.158 to 7.167, find the period of the function.

7.158 $y = \csc x$

∎ We note that $\csc (x + 2\pi) = \csc x \; \forall x$ in the domain; also, no number p smaller than 2π satisfies the equation $\csc (x + p) = \csc x$ for all x. Thus, period $= 2\pi$.

7.159 $y = \cot x$

∎ $\cot (x + \pi) = \cot x$ for all x in the domain. No number p smaller than π satisfies $\cot (x + p) = \cot x \; \forall x$. Period $= \pi$.

7.160 $f(x) = \sec x$

∎ $\sec (x + 2\pi) = \sec x \; \forall x \in$ domain; thus, $2\pi =$ period since no smaller p satisfies $\sec (x + p) = \sec x \; \forall x$.

7.161 $f(x) = -\sec x$

  ▌ If $\sec(x + 2\pi) = \sec x \; \forall x$, then $-\sec(x + 2\pi) = -\sec x$. Period $= 2\pi$.

7.162 $f(x) = 4 \sin x$

  ▌ The period is not affected by the number 4. It is only affected by that which we take the sine of. Here period $= 2\pi$.

7.163 $y = 7 \sin 3x$

  ▌ The period of $y = q \sin rx$ is $2\pi/|r|$. Here, $r = 3$. The period $= 2\pi/3$.

7.164 $y = -6 \cos(2x/7)$

  ▌ The period of $y = q \cos rx$ is $2\pi/|r|$. Here, $r = \frac{2}{7}$; thus, the period $= 2\pi/\frac{2}{7} = 2\pi \cdot \frac{7}{2} = 7\pi$.

7.165 $y = 4 \sin(2x + \pi)$

  ▌ $y = 4 \sin(2x + \pi)$ which is of the form $y = A \sin(Bx + C)$ where $A = 4$, $B = 2$. Then the period $= 2\pi/B = 2\pi/2 = \pi$.

7.166 $f(x) = 3 \tan(2x + \pi)$

  ▌ Here, $f(x) = A \tan(Bx + C)$ where $B = 3$. Then the period $= \pi/B = \pi/3$.

7.167 $f(x) = |\sin x|$

  ▌ See Fig. 7.34. The period is π (not 2π) since $|\sin x| \geq 0$ for all x; thus, between π and 2π, we get a repeat of the graph from 0 to π.

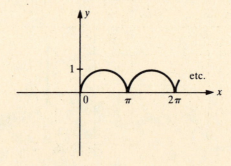

Fig. 7.34

For Probs. 7.168 to 7.170, determine the phase shift.

7.168 $y = \sin(x + \pi)$

  ▌ If $y = A \sin(Bx + C)$, $B > 0$, then the phase shift $= |C/B|$ to the right if $C/B < 0$; C/B to the left if $C/B > 0$. Here $A = 1$, $B = 1$, $C = \pi$, and $C/B = \pi/1 > 0$; thus, the phase shift $= \pi$ units to the left.

7.169 $f(x) = 2 \cos(2x + \pi/3)$

  ▌ Here $B = 2$, and $C = \pi/3$. Then $C/B = (\pi/3)/2 = \pi/6 > 0$. Phase shift $= C/B = \pi/6$ units to the left.

7.170 $f(x) = \frac{1}{3} \sin(2x - \pi/7)$

  ▌ Here $B = 2$, and $C = -\pi/7$. Then $C/B = -\pi/14 < 0$. Phase shift $= |C/B| = \pi/14$ to the *right*.

For Probs. 7.171 to 7.173, use the trigonometric graphs to solve the given equation.

7.171 $\sin x = 0$

▮ See Fig. 7.35. Clearly, $\sin x = 0$ when $x = \pm n\pi$ for all whole numbers n.

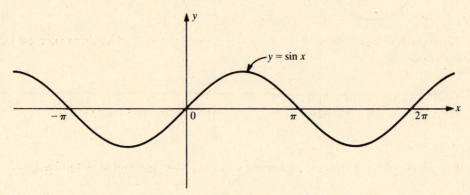

Fig. 7.35

7.172 $\tan x = 0$

▮ See Fig. 7.36. Clearly, $\tan x = 0$ when $x = n\pi \ \forall n \in Z$.

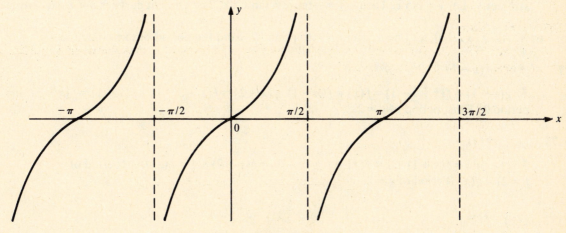

Fig. 7.36

7.173 $\csc x = 0$

▮ See Fig. 7.37. Clearly, $\csc x$ is never zero. Notice why! $\csc x = 1/\sin x$ and $1 \neq 0$.

For Probs. 7.174 to 7.176, answer true or false and justify your answer.

7.174 The secant function is even.

▮ Does $f(-x) = f(x) \ \forall x$? $\sec(-x) = 1/\cos(-x) = 1/\cos x = \sec x$. Thus, $f(-x) = f(x)$ and $\sec x$ is even. True.

7.175 The cosecant function is odd.

▮ $\csc(-x) = \dfrac{1}{\sin(-x)} = \dfrac{1}{-\sin x} = -\csc x$. Thus, $\csc(-x) = -\csc x$, and $\csc x$ is odd. True.

7.176 The tangent curve is symmetric about the x axis.

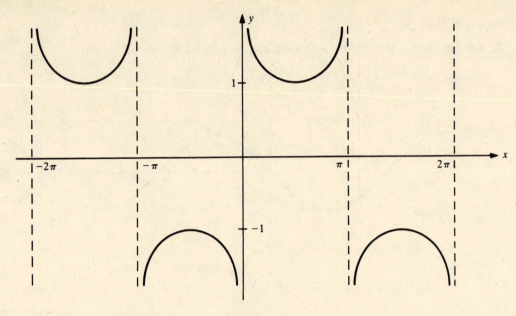

Fig. 7.37

▮ Recall that if $y = f(x)$ and replacing y by $-y$ does not alter the function, then f exhibits x-axis symmetry. Let $y = \tan x$. Then $-y = -\tan x \neq \tan x \; \forall x$. The tangent curve is *not* x-axis symmetric.

For Probs. 7.177 to 7.180, let f be a function with period 2. Find the period of each function.

7.177 $g(x) = f(x - 1)$

▮ $g(x + 1) = f(x + 1 - 1) = f(x) \neq f(x - 1)$. $g(x + 2) = f(x + 2 - 1) = f(x + 1) = f(x - 1)$ since f has period 2. Thus, period of $g = 2$.

7.178 $g(x) = f(x) + 1$

▮ $g(x + 1) = f(x + 1) + 1 \neq f(x) + 1$. $g(x + 2) = f(x + 2) + 1 = f(x) + 1$ (period of $f = 2$) $= g(x)$. Period of $g = 2$.

7.179 $g(x) = f(2x)$

▮ $g(x + 1) = f[2(x + 1)] = f(2x + 2) = f(2x) = g(x)$. Period of $g = 1$.

7.180 $g(x) = f(0.4x)$

▮ $g(x + 5) = f[0.4(x + 5)] = f(0.4x + 2) = f(0.4x) = g(x)$. No smaller number works. Period of $g = 5$.

For Probs. 7.181 to 7.186, sketch, on the same axes, one complete period of each function.

7.181 $y = \sin x, \; y = \sin 2x$

▮ See Fig. 7.38. Period of $\sin 2x = 2\pi/2 = \pi$. Both functions have an amplitude $= 1$.

7.182 $y = \sin x, \; y = 2 \sin x$

▮ See Fig. 7.39. Both functions have a period $= 2\pi$; $2 \sin x$ has amplitude $= 2$.

7.183 $y = \cos x, \; y = \cos 2x$

▮ See Fig. 7.40. Both functions have amplitude $= 1$; period of $\cos 2x = 2\pi/2 = \pi$.

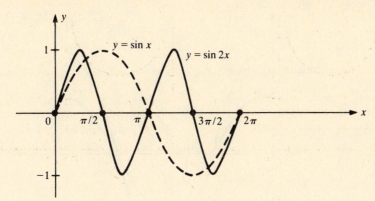

Fig. 7.38

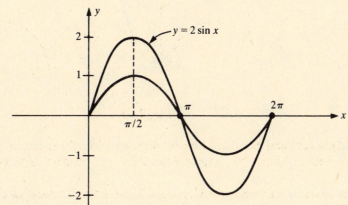

Fig. 7.39

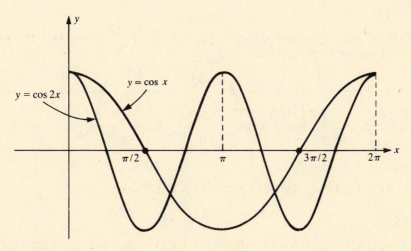

Fig. 7.40

7.184 $y = \cos x$, $y = 2 \cos x$

▮ See Fig. 7.41. Amplitude of $2 \cos x = 2$.

7.185 $y = \sin x$, $y = \sin 3x$.

▮ See Fig. 7.42. Period of $\sin 3x = 2\pi/3$. *Sketching hint*: Sketch three copies of $\sin x$ for $\sin 3x$, and *then* label the points on the x axis. Finally, sketch $\sin x$.

7.186 $y = \sin x$, $y = \sin \frac{1}{2}x$.

▮ See Fig. 7.43. Period of $\sin (x/2) = 2\pi/\frac{1}{2} = 4\pi$. *Sketching hint*: Sketch $\sin x$ for $\sin (x/2)$, and then label the x axis from 0 to 4π.

For Probs. 7.187 to 7.198, sketch the graph of the given equation.

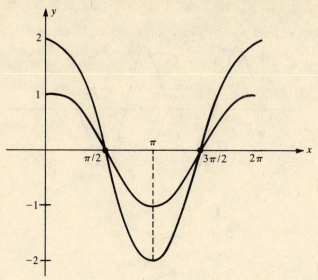

Fig. 7.41

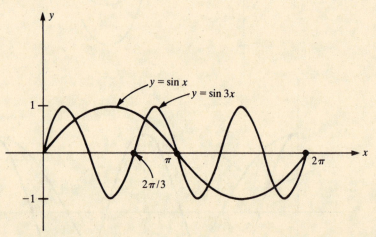

Fig. 7.42

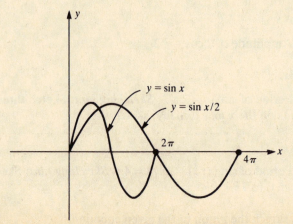

Fig. 7.43

7.187 $y = -\sin x$

▮   See Fig. 7.44. $-\sin (\pi/2) = -1$; $-\sin (3\pi/2) = -(-1) = 1$.

7.188 $f(x) = -2 \sin x$

▮   See Fig. 7.45 and Prob. 7.187. Multiply each y value by 2.

7.189 $y = -3 \sin 2x$

▮   See Fig. 7.46. Amplitude $= 3 \ (= |-3|)$; period $= 2\pi/2 = \pi$. *Sketching hint*: Sketch a copy of $\sin x$, but observe the period. Label the x axis appropriately. Observe the amplitude, and label the y axis.

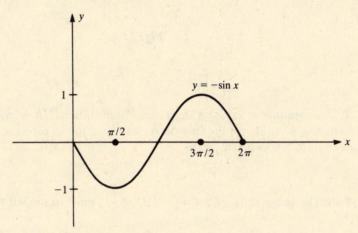

Fig. 7.44

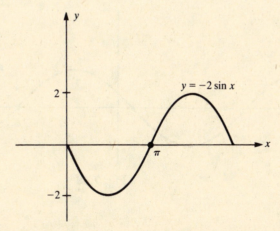

Fig. 7.45

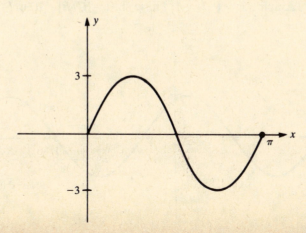

Fig. 7.46

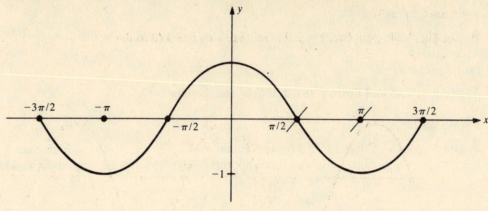

Fig. 7.47

7.190 $y = \sin(x + \pi/2)$

▮ See Fig. 7.47. Amplitude = 1; $C = \pi/2$, $B = 1$. Phase shift = $C/B = (\pi/2)/1 = \pi/2$ units to the left ($C/B > 0$). *Sketching hint*: Sketch the sine curve, and then put in the axes. Notice the sine curve between the two slash marks on the graph.

7.191 $y = \sin(x + \pi/3)$

▮ See Fig. 7.48. The phase shift is $C/B = (\pi/3)/1 = \pi/3$ units to the left ($C/B > 0$).

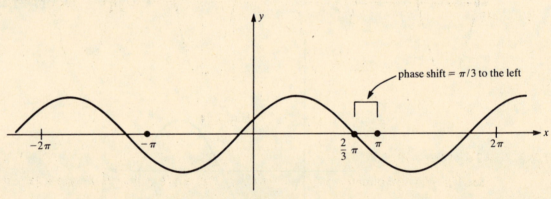

Fig. 7.48

7.192 $y = -\sin(x - \pi/4)$

▮ See Fig. 7.49. Amplitude = $|-1| = 1$. Phase shift = $|C/B| = |(-\pi/4)/1| = \pi/4$ units to the right ($C/B < 0$).

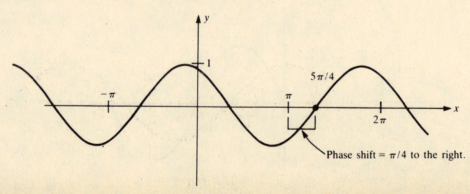

Fig. 7.49

7.193 $y = 2\cos(x + \pi/3)$

▍ See Fig. 7.50. Amplitude = 2. Phase shift = $C/B = \pi/3$ to the left.

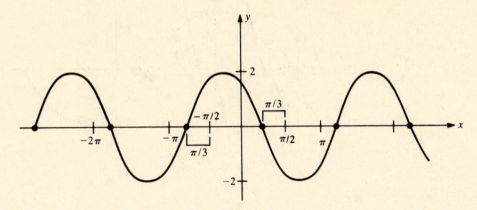

Fig. 7.50

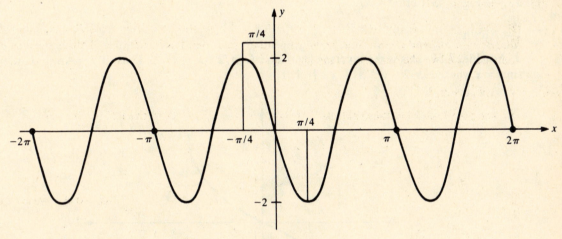

Fig. 7.51

7.194 $y = 2\cos(2x + \pi/2)$.

▍ See Fig. 7.51. Amplitude = 2. Phase shift = $(\pi/2)/2 = \pi/4$ to the left. Period = $2\pi/2 = \pi$.

7.195 $y = \frac{1}{2}\cos(x - \pi/4)$ $(-\pi \le x \le 3\pi)$

▍ See Fig. 7.52. Amplitude = $\frac{1}{2}$. Period = 2π ($= 2\pi/1$). Phase shift = $|-\pi/4| = \pi/4$ units to the right.

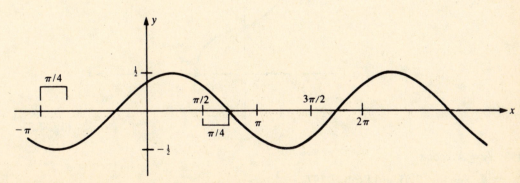

Fig. 7.52

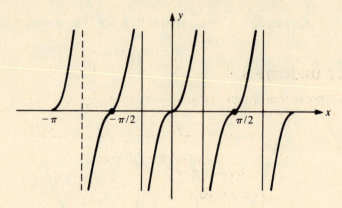

Fig. 7.53

7.196 $y = 3 \tan 2x \; (-\pi \le x \le \pi)$

┃ See Fig. 7.53. Period $= \pi/2$.

7.197 $y = \text{Arcsin } x$

┃ See Fig. 7.54. This is the inverse of $y = \sin x \; (-\pi/2 \le x \le \pi/2)$. Thus, if $y = \text{Arcsin } x$, then $\sin y = x$ where $-\pi/2 \le y \le \pi/2$, $x \in [-1, 1]$.

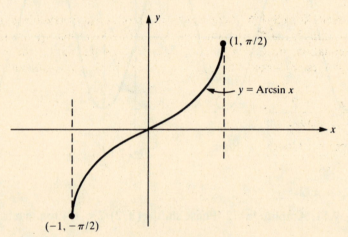

Fig. 7.54

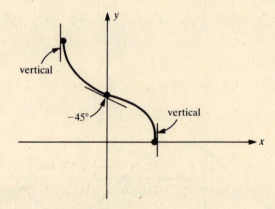

Fig. 7.55

7.198 $y = \text{Arccos } x$

┃ See Fig. 7.55 and Prob. 7.197.

CHAPTER 8
Trigonometric Equations and Inequalities

8.1 ELEMENTARY IDENTITIES

For Probs. 8.1 to 8.53, prove that the given equation is an identity. Refer to Fig. 8.1 when necessary.

$$\boxed{\begin{array}{c} \textit{Fundamental Relations} \\[4pt] \begin{array}{ll} \sin\theta\csc\theta = 1 & \tan\theta = \dfrac{\sin\theta}{\cos\theta} \\[10pt] \cos\theta\sec\theta = 1 & \\[6pt] \tan\theta\cot\theta = 1 & \cot\theta = \dfrac{\cos\theta}{\sin\theta} \end{array} \\[20pt] \sin^2\theta + \cos^2\theta = 1 \\[6pt] 1 + \cot^2\theta = \csc^2\theta \\[6pt] 1 + \tan^2\theta = \sec^2\theta \end{array}}$$

Fig. 8.1

8.1 $\sin x(\csc x - \sin x) = \cos^2 x$

▮ Ask yourself, Which side of the equation is more complicated? Begin working there. Continue to simplify that side of the equation until it looks exactly like the other side of the equation. $\sin x(\csc x - \sin x) = \sin x\csc x - \sin^2 x = 1 - \sin^2 x$ (since $\sin x = 1/\csc x$) $= \cos^2 x$ (since $\sin^2 x + \cos^2 x = 1$), and the identity is established.

8.2 $\dfrac{\sec x}{\sin x} + \dfrac{\csc x}{\cos x} = \dfrac{2}{\sin x \cos x}$

▮ $\dfrac{\sec x}{\sin x} + \dfrac{\csc x}{\cos x} = \dfrac{1/\cos x}{\sin x} + \dfrac{1/\sin x}{\cos x}$ since $\sec x = \dfrac{1}{\cos x}$, $\csc x = \dfrac{1}{\sin x}$

$= \dfrac{1}{\sin x \cos x} + \dfrac{1}{\sin x \cos x} = \dfrac{2}{\sin x \cos x}$

8.3 $\cos x(\sec x - \cos x) = \sin^2 x$

▮ $\cos x(\sec x - \cos x) = \cos x \sec x - \cos^2 x = 1 - \cos^2 x = \sin^2 x$ (since $\sec x \cos x = 1$, $1 - \cos^2 x = \sin^2 x$).

8.4 $\csc x(\csc x - \sin x) = \cot^2 x$

▮ $\csc x(\csc x - \sin x) = \csc^2 x - \csc x \sin x = \cot^2 x$ (since $\csc x \sin x = 1$, $\csc^2 x - 1 = \cot^2 x$).

8.5 $(\sin x + \cos x)\sin x = 1 - \cos x(\cos x - \sin x)$

▮ $(\sin x + \cos x)\sin x = \sin^2 x + \cos x \sin x$

$= 1 - \cos^2 x + \cos x \sin x$ (since $\sin^2 x + \cos^2 x = 1$)

$= 1 - (\cos^2 x - \cos x \sin x) = 1 - \cos x(\cos x - \sin x)$

8.6 $\dfrac{\sin x + \cos x}{\tan x} = \cos x + \dfrac{\cos^2 x}{\sin x}$

$$\blacksquare \quad \frac{\sin x + \cos x}{\tan x} = \frac{\sin x}{\tan x} + \frac{\cos x}{\tan x}$$

$$= \frac{\sin x}{\sin x/\cos x} + \frac{\cos x}{\sin x/\cos x} \quad \left(\text{since } \tan x = \frac{\sin x}{\cos x}\right)$$

$$= \frac{\sin x \cos x}{\sin x} + \frac{\cos x \cos x}{\sin x} = \cos x + \frac{\cos^2 x}{\sin x}$$

8.7 $(\sin x + \cos x)^2 = 1 + 2 \sin x \cos x$

$\blacksquare \quad (\sin x + \cos x)^2 = \sin^2 x + 2 \sin x \cos x + \cos^2 x = (\sin^2 x + \cos^2 x) + 2 \sin x \cos x = 1 + 2 \sin x \cos x.$

8.8 $(\sin x - \cos x)^2 = 1 - 2 \sin x \cos x$

$\blacksquare \quad (\sin x - \cos x)^2 = \sin^2 x - 2 \sin x \cos x + \cos^2 x = 1 - 2 \sin x \cos x.$

8.9 $(\sin x - \cos x)^2 + (\sin x + \cos x)^2 = 2$

$\blacksquare \quad (\sin x - \cos x)^2 + (\sin x + \cos x)^2 = \sin^2 x + \cos^2 x - 2 \sin x \cos x + \sin^2 x + \cos^2 x + 2 \sin x \cos x = 1 - 2 \sin x \cos x + 1 + 2 \sin x \cos x = 2 + 0 = 2.$

8.10 $\sin^4 x = 1 - (\cos^2 x + \sin^2 x \cos^2 x)$

$\blacksquare \quad \sin^4 x = \sin^2 x \sin^2 x = \sin^2 x (1 - \cos^2 x) \quad (\text{since } \sin^2 x + \cos^2 x = 1)$

$$= \sin^2 x - \sin^2 x \cos^2 x$$

$$= 1 - \cos^2 x - \sin^2 x \cos^2 x = 1 - (\cos^2 x + \sin^2 x \cos^2 x)$$

8.11 $\dfrac{\sin^2 x}{1 - \sin^2 x} + \dfrac{\cos^2 x}{1 - \cos^2 x} = \dfrac{\tan^4 x + 1}{\tan^2 x}$

$\blacksquare \quad \dfrac{\sin^2 x}{1 - \sin^2 x} + \dfrac{\cos^2 x}{1 - \cos^2 x} = \dfrac{\sin^2 x}{\cos^2 x} + \dfrac{\cos^2 x}{\sin^2 x}$

$$= \tan^2 x + \cot^2 x \left(\text{since } \frac{\sin x}{\cos x} = \tan x\right) = \tan^2 x + \frac{1}{\tan^2 x} = \frac{\tan^4 x + 1}{\tan^2 x}$$

8.12 $(\sin x + \tan x)^2 + 1 - \sec^2 x = \sin^2 x + \dfrac{2 \sin^2 x}{\cos x}$

$\blacksquare \quad (\sin x + \tan x)^2 + 1 - \sec^2 x = \sin^2 x + 2 \sin x \tan x + \tan^2 x + 1 - \sec^2 x$

$$= \sin^2 x + 2 \sin x \tan x + (\sec^2 x - 1) + 1 - \sec^2 x$$

$$(\text{since } \tan^2 x = \sec^2 x - 1)$$

$$= \sin^2 x + 2 \sin x \tan x = \sin^2 x + 2 \sin x \frac{\sin x}{\cos x}$$

$$= \sin^2 x + \frac{2 \sin^2 x}{\cos x}$$

8.13 $\dfrac{1}{\sin x + 1} - \dfrac{1}{\sin x - 1} = 2 \sec^2 x$

$\blacksquare \quad \dfrac{1}{\sin x + 1} - \dfrac{1}{\sin x - 1} = \dfrac{(\sin x - 1) - (\sin x + 1)}{(\sin x + 1)(\sin x - 1)} = \dfrac{-2}{\sin^2 x - 1}$

$$= \frac{+2}{1 - \sin^2 x} = \frac{2}{\cos^2 x} = 2 \sec^2 x$$

8.14 $(1 - \sin x)(1 + \sin x) + \sin^2 x = 1$

$\blacksquare$ $(1 - \sin x)(1 + \sin x) + \sin^2 x = (1 - \sin^2 x) + \sin^2 x = \cos^2 x + \sin^2 x = 1$ or, more simply, $(1 - \sin^2 x) + \sin^2 x = 1 + 0 = 1.$

8.15 $\dfrac{\sin x}{1 - \sin^2 x} + \dfrac{\cos x}{1 - \cos^2 x} = \dfrac{\sin^3 x + \cos^3 x}{\sin^2 x \cos^2 x}$

$\blacksquare$ $\dfrac{\sin x}{1 - \sin^2 x} + \dfrac{\cos x}{1 - \cos^2 x} = \dfrac{\sin x}{\cos^2 x} + \dfrac{\cos x}{\sin^2 x}$ (since $\sin^2 x + \cos^2 x = 1$)

$$= \dfrac{\sin^3 x + \cos^3 x}{\cos^2 x \sin^2 x}$$

8.16 $\dfrac{1}{\sin x + \cot x} - \dfrac{1}{\sin x - \cot x} = \dfrac{-2 \sin x \cos x}{\sin^4 x - \cos^2 x}$

$\blacksquare$ $\dfrac{1}{\sin x + \cot x} - \dfrac{1}{\sin x - \cot x} = \dfrac{\sin x - \cot x - (\sin x + \cot x)}{(\sin x + \cot x)(\sin x - \cot x)}$

$$= \dfrac{-2 \cot x}{\sin^2 x - \cot^2 x} = \dfrac{(-2 \cos x)/\sin x}{(\sin^2 x)/1 - (\cos^2 x)/\sin^2 x}$$

$$= \dfrac{-2 \sin x \cos x}{\sin^4 x - \cos^2 x} \qquad \text{(here we multiplied by } \sin^2 x / \sin^2 x)$$

8.17 $\dfrac{1 - \cos^2 x}{\sin^2 x \cos x} = \sec x$

$\blacksquare$ $\dfrac{1 - \cos^2 x}{\sin^2 x \cos x} = \dfrac{\sin^2 x}{\sin^2 x \cos x} = \dfrac{1}{\cos x} = \sec x$

8.18 $(\sin x + \cot x)^2 = \dfrac{\sin^4 x \cos^2 x}{\sin^2 x} + 2 \cos x$

$\blacksquare$ $(\sin x + \cot x)^2 = \sin^2 x + \cot^2 x + 2 \sin x \cot x$

$$= \sin^2 x + \dfrac{\cos^2 x}{\sin^2 x} + 2 \sin x \dfrac{\cos x}{\sin x} \qquad \left(\text{since } \cot x = \dfrac{\cos x}{\sin x}\right)$$

$$= \sin^2 x + \dfrac{\cos^2 x}{\sin^2 x} + 2 \cos x = \dfrac{\sin^4 x + \cos^2 x}{\sin^2 x} + 2 \cos x$$

8.19 $\dfrac{1 - \cos x}{\sin x} + \dfrac{\sin x}{1 - \cos x} = 2 \csc x$

$\blacksquare$ $\dfrac{1 - \cos x}{\sin x} + \dfrac{\sin x}{1 - \cos x} = \dfrac{(1 - \cos x)^2 + \sin^2 x}{\sin x(1 - \cos x)} = \dfrac{1 - 2 \cos x + \cos^2 x + \sin^2 x}{\sin x(1 - \cos x)}$

$$= \dfrac{2 - 2 \cos x}{\sin x(1 - \cos x)} \qquad (\text{since } \sin^2 x + \cos^2 x = 1) = \dfrac{2(1 - \cos x)}{\sin x(1 - \cos x)}$$

$$= \dfrac{2}{\sin x} = 2 \csc x \qquad \left(\text{since } \sin x = \dfrac{1}{\csc x}\right)$$

8.20 $\dfrac{1 + \sec x}{\tan x} - \dfrac{\tan x}{\sec x} = \dfrac{1 + \sec x}{\sec x \tan x}$

❚ $\dfrac{1+\sec x}{\tan x}-\dfrac{\tan x}{\sec x}=\dfrac{(\sec x)(1+\sec x)-\tan^2 x}{\sec x \tan x}=\dfrac{\sec x+\sec^2 x-\tan^2 x}{\sec x \tan x}$

$$=\dfrac{\sec x+1}{\sec x \tan x} \qquad (\text{since } \sec^2 x-\tan^2 x=1)$$

8.21 $\dfrac{\cos x}{\sin x}+\dfrac{\sin x}{\cos x}=\sec x \csc x$

❚ $\dfrac{\cos x}{\sin x}+\dfrac{\sin x}{\cos x}=\dfrac{\cos^2 x+\sin^2 x}{\sin x \cos x}=\dfrac{1}{\sin x \cos x}=\dfrac{1}{\sin x}\dfrac{1}{\cos x}$

$$=\csc x \sec x \qquad (\text{since } \sin x \csc x=1 \text{ and } \cos x \sec x=1)$$

8.22 $\sec^2 x \tan^2 x-\tan^2 x=\tan^4 x$

❚ $\sec^2 x \tan^2 x-\tan^2 x=\tan^2 x(\sec^2 x-1)$

$$=\tan^2 x \tan^2 x \qquad (\text{since } \tan^2 x=\sec^2 x-1)=\tan^4 x$$

8.23 $2\sin^2 x-\cos^2 x+1=3\sin^2 x$

❚ $2\sin^2 x-\cos^2 x+1=2\sin^2 x+(1-\cos^2 x)=2\sin^2 x+\sin^2 x=3\sin^2 x.$

8.24 $\sec^4 x-\tan^4 x=\sec^2 x+\tan^2 x$

❚ $\sec^4 x-\tan^4 x=(\sec^2 x+\tan^2 x)(\sec^2 x-\tan^2 x) \qquad (\text{difference of two perfect squares})$

$$=(\sec^2 x+\tan^2 x)(1) \qquad (\text{since } \sec^2 x-\tan^2 x=1)$$

$$=\sec^2 x+\tan^2 x$$

8.25 $\dfrac{1+\cot x}{\cot x}=\tan x+\csc^2 x-\cot^2 x$

❚ This is the same as Prob. 8.26. Here we begin with the right-hand side. Sometimes both sides look fairly complicated, and it is hard to decide where to begin.

$$\tan x+\csc^2 x-\cot^2 x=\tan x+1=\dfrac{1}{\cot x}+\dfrac{\cot x}{\cot x}=\dfrac{1+\cot x}{\cot x}$$

8.26 $\dfrac{1+\cot x}{\cot x}=\tan x+\csc^2 x-\cot^2 x$

❚ $\dfrac{1+\cot x}{\cot x}=\dfrac{1}{\cot x}+\dfrac{\cot x}{\cot x}=\tan x+1$

$$=\tan x+(\csc^2 x-\cot^2 x) \qquad (\text{since } \csc^2 x-\cot^2 x=1)$$

8.27 $(\sin x-\cot x)(\sin x+\cot x)=\dfrac{1-\cos^2 x(\sin^2 x+1)}{\sin^2 x}$

❚ $(\sin x-\cot x)(\sin x+\cot x)=\sin^2 x-\cot^2 x=1-\cos^2 x-\cot^2 x=1-(\cos^2 x+\cot^2 x)$

$$=1-\left(\dfrac{\cos^2 x}{1}+\dfrac{\cos^2 x}{\sin^2 x}\right)=1-\dfrac{\cos^2 x \sin^2 x+\cos^2 x}{\sin^2 x}$$

$$=1-\dfrac{\cos^2 x(\sin^2 x+1)}{\sin^2 x}$$

8.28 $(\sin x + \cos x)^4 = 1 + 4 \sin x \cos x + 4(\sin x \cos x)^2$

▌ $(\sin x + \cos x)^4 = (\sin x + \cos x)^2 (\sin x + \cos x)^2$

$$= (\sin^2 x + 2 \sin x \cos x + \cos^2 x)(\sin^2 x + 2 \sin x \cos x + \cos^2 x)$$

$$= (1 + 2 \sin x \cos x)(1 + 2 \sin x \cos x) = 1 + 4 \sin x \cos x + 4(\sin x \cos x)^2$$

8.29 $\dfrac{1 - \cos x}{\csc x} = \dfrac{\sin^3 x}{1 + \cos x}$

▌ $\dfrac{1 - \cos x}{\csc x} = \dfrac{1 - \cos x}{1/\sin x} = (\sin x)(1 - \cos x)$

$$= (\sin x)(1 - \cos x)\left(\frac{\sin^2 x}{\sin^2 x}\right) = \frac{(\sin^3 x)(1 - \cos x)}{\sin^2 x}$$

$$= \frac{(\sin^3 x)(1 - \cos x)}{1 - \cos^2 x}$$

$$= \frac{(\sin^3 x)(1 - \cos x)}{(1 - \cos x)(1 + \cos x)} = \frac{\sin^3 x}{1 + \cos x} \qquad \text{(see Prob. 8.30)}$$

8.30 $\dfrac{1 - \cos x}{\csc x} = \dfrac{\sin^3 x}{1 + \cos x}$

▌ $\dfrac{\sin^3 x}{1 + \cos x} = \dfrac{\sin x \sin^2 x}{1 + \cos x} = \dfrac{(\sin x)(1 - \cos^2 x)}{1 + \cos x}$

$$= \frac{(\sin x)(1 - \cos x)(1 + \cos x)}{1 + \cos x} = \sin x(1 - \cos x) = \frac{1 - \cos x}{\csc x}$$

Which method do you think is preferable, that of Prob. 8.29 or of Prob. 8.30? Why?

8.31 $\dfrac{\cos^2 \theta - \sin^2 \theta}{\sin \theta \cos \theta} = \cot \theta - \tan \theta$

▌ $\dfrac{\cos^2 \theta - \sin^2 \theta}{\sin \theta \cos \theta} = \dfrac{\cos^2 \theta}{\sin \theta \cos \theta} - \dfrac{\sin^2 \theta}{\sin \theta \cos \theta}$

$$= \frac{\cos \theta \cos \theta}{\sin \theta \cos \theta} - \frac{\sin \theta \sin \theta}{\sin \theta \cos \theta} = \cot \theta - \tan \theta$$

Note the first line of the equations. There were many algebraic techniques that seemed applicable here. The minus sign on both sides convinced us to use this technique. Do not be afraid to abort an attempt that appears futile!

8.32 $\dfrac{\cos \theta}{\sec \theta} + \dfrac{\sin \theta}{\csc \theta} = 1$

▌ $\dfrac{\cos \theta}{\sec \theta} + \dfrac{\sin \theta}{\csc \theta} = \dfrac{\cos \theta}{1/\cos \theta} + \dfrac{\sin \theta}{1/\sin \theta} = \cos^2 \theta + \sin^2 \theta = 1$

8.33 $\dfrac{\csc \theta}{\sec \theta} + \dfrac{\cos \theta}{\sin \theta} = 2 \cot \theta$

▌ $\dfrac{\csc \theta}{\sec \theta} + \dfrac{\cos \theta}{\sin \theta} = \dfrac{1/\sin \theta}{1/\cos \theta} + \dfrac{\cos \theta}{\sin \theta} = \dfrac{\cos \theta}{\sin \theta} + \dfrac{\cos \theta}{\sin \theta} = \cot \theta + \cot \theta = 2 \cot \theta$

8.34 $(1 - \cos \theta)(1 + \sec \theta) = \sec \theta - \cos \theta$

▮ $(1 - \cos \theta)(1 + \sec \theta) = 1 - \cos \theta + \sec \theta - \sec \theta \cos \theta = 1 - \cos \theta + \sec \theta - 1 = \sec \theta - \cos \theta$.

8.35 $\sec^2 \theta + \csc^2 \theta = \sec^2 \theta \csc^2 \theta$

▮ $\sec^2 \theta + \csc^2 \theta = \dfrac{1}{\cos^2 \theta} + \dfrac{1}{\sin^2 \theta} = \dfrac{\sin^2 \theta + \cos^2 \theta}{\sin^2 \theta \cos^2 \theta}$

$$= \dfrac{1}{\sin^2 \theta \cos^2 \theta} = \dfrac{1}{\sin^2 \theta} \cdot \dfrac{1}{\cos^2 \theta} = \sec^2 \theta \csc^2 \theta$$

Note that at the beginning we went from $\sec \theta$ and $\csc \theta$ to $\sin \theta$ and $\cos \theta$, and back, then at the end we went to $\sec \theta$ and $\csc \theta$. This is a common technique: We tend to feel more comfortable with the sine and cosine functions.

8.36 $\tan^2 \theta - \sin^2 \theta = \tan^2 \theta \sin^2 \theta$

▮ $\tan^2 \theta - \sin^2 \theta = \dfrac{\sin^2 \theta}{\cos^2 \theta} - \sin^2 \theta = \dfrac{\sin^2 \theta - \sin^2 \theta \cos^2 \theta}{\cos^2 \theta}$

$$= \dfrac{\sin^2 \theta (1 - \cos^2 \theta)}{\cos^2 \theta} = \dfrac{\sin^2 \theta}{\cos^2 \theta} \cdot \dfrac{1 - \cos^2 \theta}{1} = \tan^2 \theta \sin^2 \theta$$

8.37 $\csc \theta - \cot \theta \cos \theta = \sin \theta$

▮ (See the discussion for Prob. 8.35.)

$$\csc \theta - \cot \theta \cos \theta = \dfrac{1}{\sin \theta} - \dfrac{\cos \theta}{\sin \theta} \cos \theta = \dfrac{1 - \cos^2 \theta}{\sin \theta} = \dfrac{\sin^2 \theta}{\sin \theta} = \sin \theta$$

8.38 $\sec^2 \theta + \cot^2 \theta = \csc^2 \theta + \tan^2 \theta$

▮ $\sec^2 \theta + \cot^2 \theta \overset{?}{=} \csc^2 \theta + \tan^2 \theta$

$$\dfrac{1}{\cos^2 \theta} + \dfrac{\cos^2 \theta}{\sin^2 \theta} \overset{?}{=} \dfrac{1}{\sin^2 \theta} + \dfrac{\sin^2 \theta}{\cos^2 \theta}$$

$$\dfrac{\sin^2 \theta + \cos^4 \theta}{\cos^2 \theta \sin^2 \theta} \overset{?}{=} \dfrac{\cos^2 \theta + \sin^4 \theta}{\sin^2 \theta \cos^2 \theta}$$

This seems to be futile. Try this instead:

$$\sec^2 \theta + \cot^2 \theta \overset{?}{=} \csc^2 \theta + \tan^2 \theta$$

$$(\tan^2 \theta + 1) + \cot^2 \theta \overset{?}{=} (1 + \cot^2 \theta) + \tan^2 \theta$$

$$\tan^2 \theta + 1 + \cot^2 \theta = \tan^2 \theta + 1 + \cot^2 \theta$$

8.39 $\dfrac{\sin \theta}{\csc \theta} + \dfrac{\cos \theta}{\sec \theta} = \sin \theta \csc \theta$

▮ See Prob. 8.40. We can work on both sides of the equation as long as we do *not* make use of the equals sign.

$$\dfrac{\sin \theta}{\csc \theta} + \dfrac{\cos \theta}{\sec \theta} \overset{?}{=} \sin \theta \csc \theta$$

$$\dfrac{\sin \theta}{1/\sin \theta} + \dfrac{\cos \theta}{1/\cos \theta} \overset{?}{=} \sin \theta \csc \theta$$

$$\sin^2 \theta + \cos^2 \theta \overset{?}{=} 1$$

$$1 = 1$$

8.40 $\dfrac{\sin \theta}{\csc \theta} + \dfrac{\cos \theta}{\sec \theta} = \sin \theta \csc \theta$

▮ $\dfrac{\sin \theta}{\csc \theta} + \dfrac{\cos \theta}{\sec \theta} = \dfrac{\sin \theta}{1/\sin \theta} + \dfrac{\cos \theta}{1/\cos \theta} = \sin^2 \theta + \cos^2 \theta$

$$= 1 = \sin \theta \csc \theta \qquad \left(\text{since } \sin \theta = \dfrac{1}{\csc \theta}\right)$$

8.41 $\dfrac{\cot \theta - \tan \theta}{\sin \theta \cos \theta} = \csc^2 \theta - \sec^2 \theta$

▮ $\dfrac{\cot \theta - \tan \theta}{\sin \theta \cos \theta} = \dfrac{\cot \theta}{\sin \theta \cos \theta} - \dfrac{\tan \theta}{\sin \theta \cos \theta} = \dfrac{(\cos \theta)/\sin \theta}{\sin \theta \cos \theta} - \dfrac{(\sin \theta)/\cos \theta}{\sin \theta \cos \theta}$

$$= \dfrac{\cos \theta}{\sin^2 \theta \cos \theta} - \dfrac{\sin \theta}{\sin \theta \cos^2 \theta} = \dfrac{1}{\sin^2 \theta} - \dfrac{1}{\cos^2 \theta}$$

$$= \csc^2 \theta - \sec^2 \theta \qquad \left(\text{since } \csc \theta = \dfrac{1}{\sin \theta} \text{ and } \sec \theta = \dfrac{1}{\cos \theta}\right)$$

8.42 $\dfrac{\tan^2 \theta - \sin^2 \theta}{\sin^2 \theta} = \tan^2 \theta$

▮ $\dfrac{\tan^2 \theta - \sin^2 \theta}{\sin^2 \theta} = \dfrac{\tan^2 \theta}{\sin^2 \theta} - \dfrac{\sin^2 \theta}{\sin^2 \theta} = \dfrac{(\sin^2 \theta)/\cos^2 \theta}{\sin^2 \theta} - 1 \qquad \left(\text{since } \tan^2 \theta = \dfrac{\sin^2 \theta}{\cos^2 \theta}\right)$

$$= \dfrac{1}{\cos^2 \theta} - 1 = \sec^2 \theta - 1 = \tan^2 \theta \qquad (\text{since } \sec^2 \theta - 1 = \tan^2 \theta)$$

8.43 $\dfrac{\sin \theta \cos \theta}{\cos^2 \theta - \sin^2 \theta} = \dfrac{\tan \theta}{1 - \tan^2 \theta}$

▮ $\dfrac{\tan \theta}{1 - \tan^2 \theta} = \dfrac{\sin \theta/\cos \theta}{1 - \sin^2 \theta/\cos^2 \theta} \cdot \dfrac{\cos \theta}{\cos \theta}$

$$= \dfrac{\sin \theta}{\cos \theta - \sin^2 \theta/\cos \theta} \cdot \dfrac{\cos \theta}{\cos \theta} = \dfrac{\sin \theta \cos \theta}{\cos^2 \theta - \sin^2 \theta}$$

8.44 $\cot \theta + \dfrac{\sin \theta}{1 + \cos \theta} = \csc \theta$

▮ $\cot \theta + \dfrac{\sin \theta}{1 + \cos \theta} = \dfrac{\cos \theta}{\sin \theta} + \dfrac{\sin \theta}{1 + \cos \theta} = \dfrac{\cos \theta(1 + \cos \theta) + \sin^2 \theta}{\sin \theta(1 + \cos \theta)}$

$$= \dfrac{\cos \theta + \cos^2 \theta + \sin^2 \theta}{\sin \theta(1 + \cos \theta)} = \dfrac{\cos \theta + 1}{\sin \theta(1 + \cos \theta)} = \dfrac{1}{\sin \theta} = \csc \theta$$

8.45 $\dfrac{\sin x + \tan x}{\cot x + \csc x} = \sin x \tan x$

▮ $\dfrac{\sin x + \tan x}{\cot x + \csc x} = \dfrac{\sin x + \sin x/\cos x}{\cos x/\sin x + 1/\sin x} = \dfrac{(\sin x \cos x + \sin x)/\cos x}{(\cos x + 1)/\sin x}$

$$= \dfrac{(\sin x \cos x + \sin x) \sin x}{\cos x(\cos x + 1)} = \dfrac{\sin^2 x \cos x + \sin^2 x}{\cos^2 x + \cos x}$$

$$= \dfrac{\sin^2 x(\cos x + 1)}{\cos x(\cos x + 1)} = \dfrac{\sin^2 x}{\cos x} = \sin x \cdot \dfrac{\sin x}{\cos x}$$

$$= \sin x \tan x$$

8.46 $\dfrac{1}{\sec\theta + \tan\theta} = \sec\theta - \tan\theta$

▌ We begin by multiplying by the conjugate.

$$\frac{1}{\sec\theta + \tan\theta} = \frac{1}{\sec\theta + \tan\theta} \cdot \frac{\sec\theta - \tan\theta}{\sec\theta - \tan\theta} = \frac{\sec\theta - \tan\theta}{(\sec\theta + \tan\theta)(\sec\theta - \tan\theta)}$$

$$= \frac{\sec\theta - \tan\theta}{\sec^2\theta - \tan^2\theta} = \frac{\sec\theta - \tan\theta}{1} = \sec\theta - \tan\theta$$

8.47 $\tan^2 x \csc^2 x \cot^2 x \sin^2 x = 1$

▌ $\tan^2 x \csc^2 x \cot^2 x \sin^2 x = \dfrac{\sin^2 x}{\cos^2 x} \cdot \dfrac{1}{\sin^2 x} \cdot \dfrac{\cos^2 x}{\sin^2 x}\sin^2 x = \dfrac{1}{1} = 1$

8.48 $\dfrac{\sec^2\theta}{\sec^2\theta - 1} = \csc^2\theta$

▌ $\dfrac{\sec^2\theta}{\sec^2\theta - 1} = \dfrac{\sec^2\theta}{\tan^2\theta} = \dfrac{1/\cos^2\theta}{\sin^2\theta/\cos^2\theta} = \dfrac{1}{\sin^2\theta} = \csc^2\theta$

8.49 $\dfrac{\sin^4\theta - \cos^4\theta}{\sin^2\theta - \cos^2\theta} = 1$

▌ $\dfrac{\sin^4\theta - \cos^4\theta}{\sin^2\theta - \cos^2\theta} = \dfrac{(\sin^2\theta - \cos^2\theta)(\sin^2\theta + \cos^2\theta)}{\sin^2 - \cos^2\theta} = \sin^2\theta + \cos^2\theta = 1$

8.50 $\dfrac{\cos\theta - \sin\theta}{\cos\theta + \sin\theta} = \dfrac{\cot\theta - 1}{\cot\theta + 1}$

▌ Think! What will change $\cos\theta$ into $\cot\theta$?

$$\frac{\cos\theta - \sin\theta}{\cos\theta + \sin\theta} = \frac{\cos\theta - \sin\theta}{\cos\theta + \sin\theta} \cdot \frac{1/\sin\theta}{1/\sin\theta} = \frac{\cot\theta - 1}{\cot\theta + 1}$$

8.51 $\dfrac{\sin x}{1 - \cos x} = \dfrac{1 + \cos x}{\sin x}$

▌ $\dfrac{\sin x}{1 - \cos x} = \dfrac{\sin x}{1 - \cos x} \cdot \dfrac{1 + \cos x}{1 + \cos x} = \dfrac{(\sin x)(1 + \cos x)}{1 - \cos^2 x}$

$$= \frac{(\sin x)(1 + \cos x)}{\sin^2 x} = \frac{1 + \cos x}{\sin x}$$

8.52 $\ln\tan x = \ln\sin x - \ln\cos x$

▌ $\ln\tan x = \ln\dfrac{\sin x}{\cos x} \left(\text{since } \tan x = \dfrac{\sin x}{\cos x}\right) = \ln\sin x - \ln\cos x$ (property of logs)

8.53 $(x\sin\theta - y\cos\theta)^2 + (x\cos\theta + y\sin\theta)^2 = x^2 + y^2$

▌ $(x\sin\theta - y\cos\theta)^2 + (x\cos\theta + y\sin\theta)^2$

$$= x^2\sin^2\theta - 2xy\sin\theta\cos\theta + y^2\cos^2\theta + x^2\cos^2\theta + 2xy\sin\theta\cos\theta + y^2\sin^2\theta$$

$$= x^2\sin^2\theta + x^2\cos^2\theta + y^2\sin^2\theta + y^2\cos^2\theta$$

$$= x^2(\sin^2\theta + \cos^2\theta) + y^2(\sin^2\theta + \cos^2\theta)$$

$$= x^2 \cdot 1 + y^2 \cdot 1 = x^2 + y^2$$

8.54 Prove the quotient relations:
$$\tan \theta = \frac{\sin \theta}{\cos \theta} \qquad \cot \theta = \frac{\cos \theta}{\sin \theta}$$

▮ For any angle θ, $\sin \theta = y/r$, $\cos \theta = x/r$, $\tan \theta = y/x$, and $\cot \theta = x/y$, where $P(x, y)$ is any point on the terminal side of θ at a distance r from the origin. Then $\tan \theta = \dfrac{y}{x} = \dfrac{y/r}{x/r} = \dfrac{\sin \theta}{\cos \theta}$ and $\cot \theta =$

$\dfrac{x}{y} = \dfrac{x/r}{y/r} = \dfrac{\cos \theta}{\sin \theta}$. $\left(\text{Also } \cot \theta = \dfrac{1}{\tan \theta} = \dfrac{\cos \theta}{\sin \theta}. \right)$

8.55 Prove the pythagorean relations: (**a**) $\sin^2 \theta + \cos^2 \theta = 1$, (**b**) $1 + \tan^2 \theta = \sec^2 \theta$, (**c**) $1 + \cot^2 \theta = \csc^2 \theta$.

▮ For $P(x, y)$ defined as in Prob. 8.54, we have
$$(1) \qquad x^2 + y^2 = r^2$$

(**a**) Dividing (1) by r^2, we get $(x/r)^2 + (y/r)^2 = 1$ and $\sin^2 \theta + \cos^2 \theta = 1$.

(**b**) Dividing (1) by x^2 gives $1 + (y/x)^2 = (r/x)^2$ and $1 + \tan^2 \theta = \sec^2 \theta$. Also dividing $\sin^2 \theta + \cos^2 \theta = 1$ by $\cos^2 \theta$, we get $\left(\dfrac{\sin \theta}{\cos \theta} \right)^2 + 1 = \left(\dfrac{1}{\cos \theta} \right)^2$ or $\tan^2 \theta + 1 = \sec^2 \theta$.

(**c**) Dividing (1) by y^2, we get $(x/y)^2 + 1 = (r/y)^2$ and $\cot^2 \theta + 1 = \csc^2 \theta$. Also dividing $\sin^2 \theta + \cos^2 \theta = 1$ by $\sin^2 \theta$ gives $1 + \left(\dfrac{\cos \theta}{\sin \theta} \right)^2 = \left(\dfrac{1}{\sin \theta} \right)^2$ or $1 + \cot^2 \theta = \csc^2 \theta$.

8.56 Express each of the other functions of θ in terms of $\sin \theta$.

▮ $\cos^2 \theta = 1 - \sin^2 \theta$ and $\cos \theta = \pm\sqrt{1 - \sin^2 \theta}$

$$\tan \theta = \frac{\sin \theta}{\cos \theta} = \frac{\sin \theta}{\pm\sqrt{1 - \sin^2 \theta}} \qquad \cot \theta = \frac{1}{\tan \theta} = \frac{\pm\sqrt{1 - \sin^2 \theta}}{\sin \theta}$$

$$\sec \theta = \frac{1}{\cos \theta} = \frac{1}{\pm\sqrt{1 - \sin^2 \theta}} \qquad \csc \theta = \frac{1}{\sin \theta}$$

Note that $\cos \theta = \pm\sqrt{1 - \sin^2 \theta}$. Writing $\cos \theta = \sqrt{1 - \sin^2 \theta}$ limits angle θ to those quadrants (first and fourth) in which the cosine is positive.

8.57 Express each of the other functions of θ in terms of $\tan \theta$.

▮ $\sec^2 \theta = 1 + \tan^2 \theta$ and $\sec \theta = \pm\sqrt{1 + \tan^2 \theta}$, $\cos \theta = \dfrac{1}{\sec \theta} = \dfrac{1}{\pm\sqrt{1 + \tan^2 \theta}}$, $\dfrac{\sin \theta}{\cos \theta} = \tan \theta$

and $\sin \theta = \tan \theta \cos \theta = \tan \theta \dfrac{1}{\pm\sqrt{1 + \tan^2 \theta}} = \dfrac{\tan \theta}{\pm\sqrt{1 + \tan^2 \theta}}$, $\csc \theta = \dfrac{1}{\sin \theta} = \dfrac{\pm\sqrt{1 + \tan^2 \theta}}{\tan \theta}$,

$\cot \theta = \dfrac{1}{\tan \theta}$.

8.58 Using the fundamental relations, find the values of the functions of θ, given $\sin \theta = \frac{3}{5}$.

▮ From $\cos^2 \theta = 1 - \sin^2 \theta$, $\cos \theta = \pm\sqrt{1 - \sin^2 \theta} = \pm\sqrt{1 - (\frac{3}{5})^2} = \pm\sqrt{\frac{16}{25}} = \pm\frac{4}{5}$.

Now $\sin \theta$ and $\cos \theta$ are both positive when θ is a first-quadrant angle while $\sin \theta = +$ and $\cos \theta = -$ when θ is a second-quadrant angle. Thus,

first quadrant		second quadrant	
$\sin \theta = \frac{3}{5}$	$\cot \theta = \frac{4}{3}$	$\sin \theta = \frac{3}{5}$	$\cot \theta = -\frac{4}{3}$
$\cos \theta = \frac{4}{5}$	$\sec \theta = \frac{5}{4}$	$\cos \theta = -\frac{4}{5}$	$\sec \theta = -\frac{5}{4}$
		$\tan \theta = -\frac{3}{4}$	$\csc \theta = \frac{5}{3}$
$\tan \theta = \dfrac{\frac{3}{5}}{\frac{4}{5}} = \frac{3}{4}$	$\csc \theta = \frac{5}{3}$		

8.59   Using the fundamental relations, find the values of the functions of θ, given $\tan \theta = -\frac{5}{12}$.

   ❙   Since $\tan \theta = -$, θ is either a second- or fourth-quadrant angle.

<table>
<tr><td colspan="2" align="center">**second quadrant**</td><td colspan="2" align="center">**fourth quadrant**</td></tr>
<tr><td colspan="2">$\tan \theta = -\frac{5}{12}$</td><td colspan="2">$\tan \theta = -\frac{5}{12}$</td></tr>
<tr><td colspan="2">$\cot \theta = 1/\tan \theta = -\frac{12}{5}$</td><td colspan="2">$\cot \theta = -\frac{12}{5}$</td></tr>
<tr><td colspan="2">$\sec \theta = -\sqrt{1+\tan^2 \theta} = -\frac{13}{12}$</td><td colspan="2">$\sec \theta = \frac{13}{12}$</td></tr>
<tr><td colspan="2">$\cos \theta = 1/\sec \theta = -\frac{12}{13}$</td><td colspan="2">$\cos \theta = \frac{12}{13}$</td></tr>
<tr><td colspan="2">$\csc \theta = \sqrt{1+\cot^2 \theta} = \frac{13}{5}$</td><td colspan="2">$\csc \theta = -\frac{13}{5}$</td></tr>
<tr><td colspan="2">$\sin \theta = 1/\csc \theta = \frac{5}{13}$</td><td colspan="2">$\sin \theta = -\frac{5}{13}$</td></tr>
</table>

8.60   Perform the indicated operations.

   ❙   (a) $(\sin \theta - \cos \theta)(\sin \theta + \cos \theta) = \sin^2 \theta - \cos^2 \theta$
   (b) $(\sin A + \cos A)^2 = \sin^2 A + 2 \sin A \cos A + \cos^2 A$
   (c) $(\sin x + \cos y)(\sin y - \cos x) = \sin x \sin y - \sin x \cos x + \sin y \cos y - \cos x \cos y$
   (d) $(\tan^2 A - \cot A)^2 = \tan^4 A - 2 \tan^2 A \cot A + \cot^2 A$

   (e) $1 + \dfrac{\cos \theta}{\sin \theta} = \dfrac{\sin \theta + \cos \theta}{\sin \theta}$

   (f) $1 - \dfrac{\sin \theta}{\cos \theta} + \dfrac{2}{\cos^2 \theta} = \dfrac{\cos^2 \theta - \sin \theta \cos \theta + 2}{\cos^2 \theta}$

8.61   Factor.

   ❙   (a) $\sin^2 \theta - \sin \theta \cos \theta = \sin \theta (\sin \theta - \cos \theta)$
   (b) $\sin^2 \theta + \sin^2 \theta \cos^2 \theta = \sin^2 \theta (1 + \cos^2 \theta)$
   (c) $\sin^2 \theta + \sin \theta \sec \theta - 6 \sec^2 \theta = (\sin \theta + 3 \sec \theta)(\sin \theta - 2 \sec \theta)$
   (d) $\sin^3 \theta \cos^2 \theta - \sin^2 \theta \cos^3 \theta + \sin \theta \cos^2 \theta = \sin \theta \cos^2 \theta (\sin^2 \theta - \sin \theta \cos \theta + 1)$
   (e) $\sin^4 \theta - \cos^4 \theta = (\sin^2 \theta + \cos^2 \theta)(\sin^2 \theta - \cos^2 \theta)$
        $= (\sin^2 \theta + \cos^2 \theta)(\sin \theta - \cos \theta)(\sin \theta + \cos \theta)$

8.62   Simplify each of the following.

   ❙   (a) $\sec \theta - \sec \theta \sin^2 \theta = \sec \theta(1 - \sin^2 \theta) = \sec \theta \cos^2 \theta = \dfrac{1}{\cos \theta} \cos^2 \theta = \cos \theta$

   (b) $\sin \theta \sec \theta \cot \theta = \sin \theta \cdot \dfrac{1}{\cos \theta} \cdot \dfrac{\cos \theta}{\sin \theta} = \dfrac{\sin \theta \cos \theta}{\cos \theta \sin \theta} = 1$

   (c) $\sin^2 \theta(1 + \cot^2 \theta) = \sin^2 \theta \csc^2 \theta = \sin^2 \theta \cdot \dfrac{1}{\sin^2 \theta} = 1$

   (d) $\sin^2 \theta \sec^2 \theta - \sec^2 \theta = (\sin^2 \theta - 1) \sec^2 \theta = -\cos^2 \theta \sec^2 \theta = -\cos^2 \theta \cdot \dfrac{1}{\cos^2 \theta} = -1$

   (e) $(\sin \theta + \cos \theta)^2 + (\sin \theta - \cos \theta)^2 = \sin^2 \theta + 2 \sin \theta \cos \theta + \cos^2 \theta$
        $\qquad\qquad\qquad\qquad\qquad\qquad + \sin^2 \theta - 2 \sin \theta \cos \theta + \cos^2 \theta$
        $\qquad\qquad\qquad\qquad\qquad = 2(\sin^2 \theta + \cos^2 \theta) = 2$

   (f) $\tan^2 \theta \cos^2 \theta + \cot^2 \theta \sin^2 \theta = \dfrac{\sin^2 \theta}{\cos^2 \theta} \cos^2 \theta + \dfrac{\cos^2 \theta}{\sin^2 \theta} \sin^2 \theta$

        $\qquad\qquad\qquad\qquad\qquad = \sin^2 \theta + \cos^2 \theta = 1$

(g) $\quad \tan \theta + \dfrac{\cos \theta}{1 + \sin \theta} = \dfrac{\sin \theta}{\cos \theta} + \dfrac{\cos \theta}{1 + \sin \theta} = \dfrac{\sin \theta(1 + \sin \theta) + \cos^2 \theta}{\cos \theta(1 + \sin \theta)}$

$$= \dfrac{\sin \theta + \sin^2 \theta + \cos^2 \theta}{\cos \theta(1 + \sin \theta)} = \dfrac{\sin \theta + 1}{\cos \theta(1 + \sin \theta)}$$

$$= \dfrac{1}{\cos \theta} = \sec \theta$$

Verify the identities in Probs. 8.63 to 8.70.

8.63 $\quad \sec^2 \theta - \csc^2 \theta = \dfrac{\sin^2 \theta - \cos^2 \theta}{\sin^2 \theta \cos^2 \theta}$

∎ $\quad \sec^2 \theta - \csc^2 \theta = \dfrac{1}{\cos^2 \theta} - \dfrac{1}{\sin^2 \theta} = \dfrac{\sin^2 \theta - \cos^2 \theta}{\sin^2 \theta \cos^2 \theta}$

8.64 $\quad \sec^4 \theta - \sec^2 \theta = \tan^4 \theta + \tan^2 \theta$

∎ $\quad \tan^4 \theta + \tan^2 \theta = \tan^2 \theta(\tan^2 \theta + 1) = \tan^2 \theta \sec^2 \theta = (\sec^2 \theta - 1) \sec^2 \theta = \sec^4 \theta - \sec^2 \theta$
or $\sec^4 \theta - \sec^2 \theta = \sec^2 \theta(\sec^2 \theta - 1) = \sec^2 \theta \tan^2 \theta = (1 + \tan^2 \theta) \tan^2 \theta = \tan^2 \theta + \tan^4 \theta$

8.65 $\quad 2 \csc x = \dfrac{\sin x}{1 + \cos x} + \dfrac{1 + \cos x}{\sin x}$

∎ $\quad \dfrac{\sin x}{1 + \cos x} + \dfrac{1 + \cos x}{\sin x} = \dfrac{\sin^2 x + (1 + \cos x)^2}{\sin x(1 + \cos x)}$

$$= \dfrac{\sin^2 x + 1 + 2 \cos x + \cos^2 x}{\sin x(1 + \cos x)} = \dfrac{2 + 2 \cos x}{\sin x(1 + \cos x)}$$

$$= \dfrac{2(1 + \cos x)}{\sin x(1 + \cos x)} = \dfrac{2}{\sin x} = 2 \csc x$$

8.66 $\quad \dfrac{1 - \sin x}{\cos x} = \dfrac{\cos x}{1 + \sin x}$

∎ $\quad \dfrac{\cos x}{1 + \sin x} = \dfrac{\cos^2 x}{\cos x(1 + \sin x)} = \dfrac{1 - \sin^2 x}{\cos x(1 + \sin x)}$

$$= \dfrac{(1 - \sin x)(1 + \sin x)}{\cos x(1 + \sin x)} = \dfrac{1 - \sin x}{\cos x}$$

8.67 $\quad \dfrac{\sec A - \csc A}{\sec A + \csc A} = \dfrac{\tan A - 1}{\tan A + 1}$

∎ $\quad \dfrac{\sec A - \csc A}{\sec A + \csc A} = \dfrac{\dfrac{1}{\cos A} - \dfrac{1}{\sin A}}{\dfrac{1}{\cos A} + \dfrac{1}{\sin A}} = \dfrac{\dfrac{\sin A}{\cos A} - 1}{\dfrac{\sin A}{\cos A} + 1} = \dfrac{\tan A - 1}{\tan A + 1}$

8.68 $\quad \dfrac{\tan x - \sin x}{\sin^3 x} = \dfrac{\sec x}{1 + \cos x}$

▮ $\dfrac{\tan x - \sin x}{\sin^3 x} = \dfrac{\sin x/\cos x - \sin x}{\sin^3 x} = \dfrac{\sin x - \sin x \cos x}{\cos x \sin^3 x} = \dfrac{\sin x(1 - \cos x)}{\cos x \sin^3 x}$

$\qquad = \dfrac{1 - \cos x}{\cos x \sin^2 x} = \dfrac{1 - \cos x}{\cos x(1 - \cos^2 x)}$

$\qquad = \dfrac{1}{\cos x(1 + \cos x)} = \dfrac{\sec x}{1 + \cos x}$

8.69 $\quad \dfrac{\cos A \cot A - \sin A \tan A}{\csc A - \sec A} = 1 + \sin A \cos A$

▮ $\dfrac{\cos A \cot A - \sin A \tan A}{\csc A - \sec A} = \dfrac{\cos A \dfrac{\cos A}{\sin A} - \sin A \dfrac{\sin A}{\cos A}}{\dfrac{1}{\sin A} - \dfrac{1}{\cos A}} = \dfrac{\cos^3 A - \sin^3 A}{\cos A - \sin A}$

$\qquad = \dfrac{(\cos A - \sin A)(\cos^2 A + \cos A \sin A + \sin^2 A)}{\cos A - \sin A}$

$\qquad = \cos^2 A + \cos A \sin A + \sin^2 A$

$\qquad = 1 + \cos A \sin A$

8.70 $\quad \dfrac{\sin \theta - \cos \theta + 1}{\sin \theta + \cos \theta - 1} = \dfrac{\sin \theta + 1}{\cos \theta}$

▮ $\dfrac{\sin \theta + 1}{\cos \theta} = \dfrac{(\sin \theta + 1)(\sin \theta + \cos \theta - 1)}{\cos \theta(\sin \theta + \cos \theta - 1)} = \dfrac{\sin^2 \theta + \sin \theta \cos \theta + \cos \theta - 1}{\cos \theta(\sin \theta + \cos \theta - 1)}$

$\qquad = \dfrac{-\cos^2 \theta + \sin \theta \cos \theta + \cos \theta}{\cos \theta(\sin \theta + \cos \theta - 1)} = \dfrac{\cos \theta(\sin \theta - \cos \theta + 1)}{\cos \theta(\sin \theta + \cos \theta - 1)}$

$\qquad = \dfrac{\sin \theta - \cos \theta + 1}{\sin \theta + \cos \theta - 1}$

8.2 ADDITION AND SUBTRACTION IDENTITIES

(Refer to Fig. 8.2 when necessary for problems in this section.)

> **Addition and Subtraction Identities**
>
> $\sin(\theta \pm \alpha) = \sin \theta \cos \alpha \pm \cos \theta \sin \alpha$
>
> $\cos(\theta \pm \alpha) = \cos \theta \cos \alpha \pm \sin \theta \sin \alpha$
>
> $\tan(\theta \pm \alpha) = \dfrac{\tan \theta \pm \tan \alpha}{1 \mp \tan \theta \tan \alpha}$

Fig. 8.2

For Probs. 8.71 to 8.80, find the value of the given expression. Do *not* use tables or a calculator.

8.71 $\quad \sin 15°$

▮ Here we use the identities $\sin(x + y) = \sin x \cos y + \cos x \sin y$ and $\sin(x - y) = \sin x \cos y - \cos x \sin y$. $15° = 45° - 30°$. Thus, $\sin 15° = \sin(45° - 30°) = \sin 45° \cos 30° - \cos 45° \sin 30° = \dfrac{\sqrt{2}}{2} \cdot \dfrac{\sqrt{3}}{2} - \dfrac{\sqrt{2}}{2} \cdot \dfrac{1}{2} = \dfrac{\sqrt{6}}{4} - \dfrac{\sqrt{2}}{4} = \dfrac{\sqrt{6} - \sqrt{2}}{4}$.

8.72   cos 15°

▮ Compare this to Prob. 8.71. Here we use the identities $\cos(x+y) = \cos x \cos y - \sin x \sin y$ and $\cos(x-y) = \cos x \cos y + \sin x \sin y$. Thus, $\cos 15° = \cos(60° - 45°) = \cos 60° \cos 45° + \sin 60° \sin 45° = \frac{1}{2} \cdot \frac{\sqrt{2}}{2} + \frac{\sqrt{3}}{2} \cdot \frac{\sqrt{2}}{2} = \frac{\sqrt{2} + \sqrt{6}}{4}$.

8.73   tan 75°

▮ Here we use the identity $\tan(x \pm y) = \dfrac{\tan x \pm \tan y}{1 \mp \tan x \tan y}$. Thus, $\tan 75° = \tan(45° + 30°) = \dfrac{\tan 45° + \tan 30°}{1 - \tan 45° \tan 30°} = \dfrac{1 + 1/\sqrt{3}}{1 - 1(1/\sqrt{3})} = \dfrac{(\sqrt{3}+1)/\sqrt{3}}{(\sqrt{3}-1)/\sqrt{3}} = \dfrac{\sqrt{3}+1}{\sqrt{3}-1}$. If you forgot what tan 30° is, use $\sin 30°/\cos 30°$.

8.74   cot 75°

▮ $\cot 75° = \dfrac{1}{\tan 75°} = \dfrac{1}{(\sqrt{3}+1)/(\sqrt{3}-1)}$ (see Prob. 8.73) $= \dfrac{\sqrt{3}-1}{\sqrt{3}+1}$.

8.75   csc ($\pi/12$)

▮ $\pi/12 = \pi/3 - \pi/4 \ (= 4\pi/12 - 3\pi/12)$. Thus,

$$\csc \frac{\pi}{12} = \csc \left(\frac{\pi}{3} - \frac{\pi}{4} \right) = \frac{1}{\sin(\pi/3 - \pi/4)}.$$

But
$$\sin \left(\frac{\pi}{3} - \frac{\pi}{4} \right) = \sin \frac{\pi}{3} \cos \frac{\pi}{4} - \cos \frac{\pi}{3} \sin \frac{\pi}{4}$$

$$= \frac{\sqrt{3}}{2} \cdot \frac{\sqrt{2}}{2} - \frac{1}{2} \cdot \frac{\sqrt{2}}{2} = \frac{\sqrt{6} - \sqrt{2}}{4}$$

Thus, $\csc \dfrac{\pi}{12} = \dfrac{4}{\sqrt{6} - \sqrt{2}}$.

8.76 $\tan \dfrac{7\pi}{12}$

▮ $\tan \dfrac{7\pi}{12} = \tan \left(\dfrac{\pi}{3} + \dfrac{\pi}{4} \right) = \dfrac{\tan(\pi/3) + \tan(\pi/4)}{1 - \tan(\pi/3)\tan(\pi/4)} = \dfrac{\sqrt{3}+1}{1 - \sqrt{3} \cdot 1} = \dfrac{\sqrt{3}+1}{1 - \sqrt{3}}$.

8.77   sin 22° cos 38° + cos 22° sin 38°

▮ $\sin 22° \cos 38° + \cos 22° \sin 38° = \sin(22° + 38°) = \sin 60° = \sqrt{3}/2$.

8.78   cos 14° cos 76° − sin 14° sin 76°

▮ $\cos 14° \cos 76° - \sin 14° \sin 76° = \cos(14° + 76°) = \cos 90° = 0$.

8.79 $\dfrac{\tan 80° + \tan 40°}{1 - \tan 80° \tan 40°}$

▮ $\dfrac{\tan 80° + \tan 40°}{1 - \tan 80° \tan 40°} = \tan(80° + 40°) = \tan 120° = -\tan 60°$ (quadrant II) $= -3$.

8.80   cos 260° cos 70° − sin 260° sin 70°

▮ $\cos 260° \cos 70° - \sin 260° \sin 70° = \cos(260° + 70°) = \cos 330° = \cos 30°$ (since $x > 0$ in quadrant IV) $= \frac{3}{2}$.

For Probs. 8.81 to 8.99, prove that the given equation is an identity.

8.81 $\dfrac{\tan 4\theta - \tan 3\theta}{1 + \tan 4\theta \tan 3\theta} = \tan \theta$

▮ $\dfrac{\tan 4\theta - \tan 3\theta}{1 + \tan 4\theta \tan 3\theta}$ is the formula for $\tan(x - y)$ and therefore is equal to $\tan(4\theta - 3\theta) = \tan \theta$.

8.82 $\dfrac{\tan(45° - x)}{\tan(45° + x)} = \dfrac{(1 - \tan x)^2}{(1 + \tan x)^2}$

▮ $\tan(45° - x) = \dfrac{\tan 45° - \tan x}{1 + \tan 45° \tan x} = \dfrac{1 - \tan x}{1 + \tan x}$

$\tan(45° + x) = \dfrac{\tan 45° + \tan x}{1 - \tan 45° \tan x} = \dfrac{1 + \tan x}{1 - \tan x}$

Thus,

$$\frac{\tan(45° - x)}{\tan(45° + x)} = \frac{\dfrac{1 - \tan x}{1 + \tan x}}{\dfrac{1 + \tan x}{1 - \tan x}} = \frac{(1 - \tan x)^2}{(1 + \tan x)^2}$$

8.83 $\sec(\pi/2 - x) = \csc x$

▮ $\sec(\pi/2 - x) = \dfrac{1}{\cos(\pi/2 - x)} = \dfrac{1}{\cos(\pi/2)\cos x + \sin(\pi/2)\sin x}$

[from the identity for $\cos(x - y)$]

$= \dfrac{1}{0 \cos x + \sin x} = \dfrac{1}{\sin x} = \csc x$ (since $\sin x \csc x = 1$)

8.84 $\cot(\pi/2 - x) = \tan x$

▮ $\cot(\pi/2 - x) = \dfrac{1}{\tan(\pi/2 - x)} = \dfrac{1}{\dfrac{\tan(\pi/2) - \tan x}{1 + \tan(\pi/2)\tan x}}$. But $\tan(\pi/2)$ is undefined. Abort this

attempt. (This happens frequently.) Try using $\tan x = \dfrac{\sin x}{\cos x}$ and $\cot x = \dfrac{\cos x}{\sin x}$ instead:

$\cot(\pi/2 - x) = \dfrac{\cos(\pi/2 - x)}{\sin(\pi/2 - x)}$. But $\cos(\pi/2 - x) = \sin x$ and $\sin(\pi/2 - x) = \cos x$. Thus,

$\cot(\pi/2 - x) = \dfrac{\cos(\pi/2 - x)}{\sin(\pi/2 - x)} = \dfrac{\sin x}{\cos x} = \tan x.$

8.85 $\sin(\pi - x) = \sin x$

▮ $\sin(\pi - x) = \sin \pi \cos x - \cos \pi \sin x = 0 \cos x - (-1)\sin x = 0 + \sin x = \sin x.$

8.86 $\dfrac{\sin(\pi - x)}{\sin(\pi + x)} = -1$ $(\sin A \neq 0)$

▮ $\sin(\pi - x) = \sin x$ (see Prob. 8.85). Then $\sin(\pi + x) = \sin \pi \cos x + \cos \pi \sin x = 0 + (-\sin x) = -\sin x.$ Thus, $\dfrac{\sin(\pi - x)}{\sin(\pi + x)} = \dfrac{\sin x}{-\sin x} = -1$ $(\sin x \neq 0).$

8.87 $\cos 2x = \cos^2 x - \sin^2 x$

$\blacksquare$ $\cos 2x = \cos (x + x) = \cos x \cos x - \sin x \sin x = \cos^2 x - \sin^2 x$.

8.88 $\cos 2x = 2 \cos^2 x - 1 = 1 - 2 \sin^2 x$

$\blacksquare$ **(a)** $\cos 2x = \cos^2 x - \sin^2 x$ (see Prob. 8.87) $= \cos^2 x - (1 - \cos^2 x) = 2 \cos^2 x - 1$.
(b) $\cos 2x = \cos^2 x - \sin^2 x = (1 - \sin^2 x) - \sin^2 x = 1 - 2 \sin^2 x$. Thus,
$\cos 2x = 1 - 2 \sin^2 x = 2 \cos^2 x - 1$.

8.89 $\tan 2x = \dfrac{2 \tan x}{1 - \tan^2 x}$

$\blacksquare$ $\tan 2x = \tan (x + x) = \dfrac{\tan x + \tan x}{1 - \tan x \tan x} = \dfrac{2 \tan x}{1 - \tan^2 x}$.

8.90 $\sin 2x = 2 \sin x \cos x$

$\blacksquare$ $\sin 2x = \sin (x + x) = \sin x \cos x + \cos x \sin x = 2 \sin x \cos x$.

8.91 $\sin 3x = \sin 2x \cos x + \cos 2x \sin x$

$\blacksquare$ $3x = 2x + x$. Then $\sin 3x = \sin (2x + x) = \sin 2x \cos x + \cos 2x \sin x$.

8.92 $\sin 3x = 2 \sin x \cos^2 x + \cos^2 x \sin x - \sin^3 x$

$\blacksquare$ $\sin 3x = \sin 2x \cos x + \cos 2x \sin x$ (see Prob. 8.91)

$\qquad = (2 \sin x \cos x) \cos x + (\cos^2 x - \sin^2 x) \sin x$

$\qquad = 2 \sin x \cos^2 x + \cos^2 x \sin x - \sin^3 x$

8.93 $2 \sin (\pi/6 + x) = \sqrt{3} \sin x + \cos x$

$\blacksquare$ $2 \sin \left(\dfrac{\pi}{6} + x \right) = 2 \left(\sin \dfrac{\pi}{6} \cos x + \cos \dfrac{\pi}{6} \sin x \right) = 2 \left(\dfrac{1}{2} \cos x + \dfrac{\sqrt{3}}{2} \sin x \right)$

$\qquad = \cos x + \dfrac{\sqrt{3}}{2} \cdot 2 \sin x = \sqrt{3} \sin x + \cos x$

8.94 $\sqrt{2} \cos (\pi/4 - x) = \sin x + \cos x$

$\blacksquare$ $\sqrt{2} \cos \left(\dfrac{\pi}{4} - x \right) = \sqrt{2} \left(\cos \dfrac{\pi}{4} \cos x + \sin \dfrac{\pi}{4} \sin x \right) = \sqrt{2} \left(\dfrac{\sqrt{2}}{2} \cos x + \dfrac{\sqrt{2}}{2} \sin x \right)$

$\qquad = \tfrac{2}{2} \cos x + \tfrac{2}{2} \sin x = \cos x + \sin x$

8.95 $\cot (x + y) = \dfrac{\cot x \cot y - 1}{\cot x + \cot y}$

$\blacksquare$ $\cot (x + y) = \dfrac{1}{\tan (x + y)} = \dfrac{1}{\dfrac{\tan x + \tan y}{1 - \tan x \tan y}} = \dfrac{1 - \tan x \tan y}{\tan x + \tan y}$

$\qquad = \dfrac{\dfrac{1}{1} - \dfrac{1}{\cot x} \dfrac{1}{\cot y}}{\dfrac{1}{\cot x} + \dfrac{1}{\cot y}} \cdot \dfrac{\cot x \cot y}{\cot x \cot y} = \dfrac{\cot x \cot y - 1}{\cot x + \cot y}$

8.96 $\dfrac{\sin(\theta+h)-\sin\theta}{h}=(\cos\theta)\dfrac{\sin h}{h}-(\sin\theta)\dfrac{1-\cos h}{h}$

$\blacksquare$ $\dfrac{\sin(\theta+h)-\sin\theta}{h}=\dfrac{\sin\theta\cos h+\cos\theta\sin h-\sin\theta}{h}$

$$=\dfrac{\sin\theta\cos h}{h}+\dfrac{\cos\theta\sin h}{h}-\dfrac{\sin\theta}{h}$$

$$=\dfrac{\sin\theta\cos h}{h}-\dfrac{\sin\theta}{h}+\dfrac{\cos\theta\sin h}{h}$$

$$=\dfrac{\sin\theta(\cos h-1)}{h}+\dfrac{\cos\theta\sin h}{h}$$

$$=\dfrac{-\sin\theta(1-\cos h)}{h}+\dfrac{\cos\theta\sin h}{h}$$

8.97 $\tan x-\tan y=\dfrac{\sin(x-y)}{\cos x\cos y}$

$\blacksquare$ $\dfrac{\sin(x-y)}{\cos x\cos y}=\dfrac{\sin x\cos y-\cos x\sin y}{\cos x\cos y}=\dfrac{\sin x\,\cancel{\cos y}}{\cos x\,\cancel{\cos y}}-\dfrac{\cancel{\cos x}\sin y}{\cancel{\cos x}\cos y}$

$$=\dfrac{\sin x}{\cos x}-\dfrac{\sin y}{\cos y}=\tan x-\tan y$$

8.98 $\cot x-\tan y=\dfrac{\cos(x+y)}{\sin x\cos y}$

$\blacksquare$ $\dfrac{\cos(x+y)}{\sin x\cos y}=\dfrac{\cos x\cos y-\sin x\sin y}{\sin x\cos y}=\dfrac{\cos x\,\cancel{\cos y}}{\sin x\,\cancel{\cos y}}-\dfrac{\cancel{\sin x}\sin y}{\cancel{\sin x}\cos y}$

$$=\dfrac{\cos x}{\sin x}-\dfrac{\sin y}{\cos y}=\cot x-\tan y$$

8.99 $\cot 2x=\dfrac{\cot^2 x-1}{2\cot x}$

$\blacksquare$ $\cot 2x=\dfrac{1}{\tan 2x}=\dfrac{1}{(2\tan x)/(1-\tan^2 x)}$ (see Prob. 8.89)

$$=\dfrac{1-(1/\cot^2 x)}{2/\cot x}\cdot\dfrac{\cot^2 x}{\cot^2 x}=\dfrac{\cot^2 x-1}{2\cot x}$$

8.100 Evaluate $\sin[\mathrm{Cos}^{-1}(-\tfrac{4}{5})+\mathrm{Sin}^{-1}(-\tfrac{3}{5})]$

$\blacksquare$ See Fig. 8.3. $\sin(x+y)=\sin x\cos y+\cos x\sin y$. Let $x=\mathrm{Cos}^{-1}(-\tfrac{4}{5})$ and $y=\mathrm{Sin}^{-1}(-\tfrac{3}{5})$.

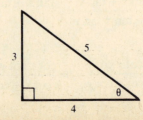

Fig. 8.3

We get

$$\underbrace{\sin \text{Cos}^{-1}(-\tfrac{4}{5})}_{+\tfrac{3}{5} \text{ (quad. II)}} \underbrace{\cos \text{Sin}^{-1}(-\tfrac{3}{5})}_{+\tfrac{4}{5} \text{ (quad. IV)}} + \underbrace{\cos \text{Cos}^{-1}(-\tfrac{4}{5})}_{\text{quad. II}} \underbrace{\sin \text{Sin}^{-1}(-\tfrac{3}{5})}_{\text{quad. IV}} = (+\tfrac{3}{5}) \cdot (+\tfrac{4}{5}) + (-\tfrac{4}{5})(-\tfrac{3}{5})$$

$$= +\tfrac{12}{25} + \tfrac{12}{25} = \tfrac{24}{25}$$

8.101 Evaluate $\sin (x + y + \pi/2)$

▮ $x + y + \pi/2 = (x + y) + \pi/2$. Then $\sin [(x + y) + \pi/2] = \sin (x + y) \cos (\pi/2) + \cos (x + y) \sin (\pi/2) = \sin (x + y) \cdot 0 + \cos (x + y) \cdot 1 = \cos (x + y)$.

8.102 Simplify: $\sin (x + y) \cdot \dfrac{\sqrt{3}}{2} + \cos (x + y) \cdot \tfrac{1}{2}$

▮ This is of the form $\sin A \cos B + \cos A \sin B$, where $A = x + y$ and $B = \pi/6$. But $\sin A \cos B + \cos A \sin B = \sin (A + B)$. Thus, the solution is $\sin (x + y + \pi/6)$.

8.3 DOUBLE- AND HALF-ANGLE IDENTITIES

For Probs. 8.103 to 8.108, prove that the given formula is true for all x. Use Fig. 8.4 when necessary.

Double- and Half-Angle Identities

$$\sin 2x = 2 \sin x \cos x$$

$$\cos 2x = \cos^2 x - \sin^2 x = 1 - 2 \sin^2 x = 2 \cos^2 x - 1$$

$$\tan 2x = \frac{2}{\cot x - \tan x}$$

$$\sin \frac{x}{2} = \pm \sqrt{\frac{1 - \cos x}{2}}$$

$$\cos \frac{x}{2} = \pm \sqrt{\frac{1 + \cos x}{2}}$$

$$\tan \frac{x}{2} = \frac{\sin x}{1 + \cos x} = \frac{1 - \cos x}{\sin x}$$

Fig. 8.4

8.103 $\sin 2x = 2 \sin x \cos x$

▮ $\sin (x + y) = \sin x \cos y + \cos x \sin y$. Thus, letting $x = y$ in the above equation gives $\sin (x + x) = \sin x \cos x + \cos x \sin x = 2 \sin x \cos x$, and the proof is complete.

8.104 $\tan 2x = \dfrac{2}{\cot x - \tan x}$

▮ $\tan (x + y) = \dfrac{\tan x + \tan y}{1 - \tan x \tan y}$. Thus,

$$\tan 2x = \tan (x + x) = \frac{2 \tan x}{1 - \tan^2 x} = \frac{2/\cot x}{1/1 - 1/\cot^2 x} \cdot \frac{\cot^2 x}{\cot^2 x} = \frac{2 \cot x}{\cot^2 x - 1} \cdot \frac{1/\cot x}{1/\cot x} = \frac{2}{\cot x - \tan x},$$

and the proof is complete.

8.105 $\cos 2x = \cos^2 x - \sin^2 x = 1 - 2 \sin^2 x = 2 \cos^2 x - 1$

▮ Letting $x = y$, $\cos (x + y) = \cos x \cos y - \sin x \sin y = \cos x \cos x - \sin x \sin x = \cos^2 x - \sin^2 x = (1 - \sin^2 x) - \sin^2 x = 1 - 2 \sin^2 x$. From $\cos^2 x - \sin^2 x$ we also have $= \cos^2 x - (1 - \cos^2 x) = 2 \cos^2 x - 1$, and the proof is complete.

8.106 $\cos\dfrac{x}{2} = \pm\sqrt{\dfrac{1+\cos x}{2}}$

▌ $\cos 2t = 2\cos^2 t - 1$. Let $t = x/2$. Then $\cos x = 2\cos^2(x/2) - 1$, $2\cos^2(x/2) = \cos x + 1$,

$\cos(x/2) = \pm\sqrt{\dfrac{1+\cos x}{2}}$, and the proof is complete.

8.107 $\sin\dfrac{x}{2} = \pm\sqrt{\dfrac{1-\cos x}{2}}$

▌ $\cos 2t = 1 - 2\sin^2 t$. Let $t = \dfrac{x}{2}$. Then $\cos x = 1 - 2\sin^2\dfrac{x}{2}$, $\sin^2\dfrac{x}{2} = \dfrac{1-\cos x}{2}$,

$\sin\dfrac{x}{2} = \pm\sqrt{\dfrac{1-\cos x}{2}}$, and the proof is complete.

8.108 $\tan\dfrac{x}{2} = \dfrac{\sin x}{1+\cos x} = \dfrac{1-\cos x}{\sin x}$

▌ $\tan\dfrac{x}{2} = \dfrac{\sin x/2}{\cos x/2} = \dfrac{\pm\sqrt{(1-\cos x)/2}}{\pm\sqrt{(1+\cos x)/2}} = \pm\sqrt{\dfrac{1-\cos x}{1+\cos x}}$

$= \pm\sqrt{\dfrac{1-\cos x}{1+\cos x}\cdot\dfrac{1+\cos x}{1+\cos x}}$ (multiplying by 1)

$= \pm\sqrt{\dfrac{1-\cos^2 x}{(1+\cos x)^2}} = \pm\sqrt{\dfrac{\sin^2 x}{(1+\cos x)^2}}$.

Thus, $\left|\tan\dfrac{x}{2}\right| = \dfrac{|\sin x|}{|1+\cos x|} = \dfrac{|\sin x|}{1+\cos x}$ since $|\cos x| \le 1$ for all x. Actually, $\tan(x/2)$ and $\sin x$ always

agree in sign (check this!), so $\tan\dfrac{x}{2} = \dfrac{\sin x}{1+\cos x} = \dfrac{\sin x}{1+\cos x}\cdot\dfrac{1-\cos x}{1-\cos x}$

$= \dfrac{\sin x(1-\cos x)}{1-\cos^2 x} = \dfrac{\sin x(1-\cos x)}{\sin^2 x} = \dfrac{1-\cos x}{\sin x}$.

Thus, $\tan\dfrac{x}{2} = \dfrac{\sin x}{1+\cos x} = \dfrac{1-\cos x}{\sin x}$, and the proof is complete.

For Probs. 8.109 to 8.116, find the exact value of the given expression without using a table or calculator.

8.109 $\sin 22.5°$

▌ $\sin 22.5° = \sin(45°/2)$. Since $\sin\dfrac{x}{2} = \pm\sqrt{\dfrac{1-\cos x}{2}}$, $\sin\dfrac{45°}{2} = \pm\sqrt{\dfrac{1-\cos 45°}{2}}$. We reject

the minus answer since this is a quadrant I angle. Thus, the answer is $= \sqrt{\dfrac{1-1/\sqrt{2}}{2}}$.

8.110 $\sin 75°$

▌ $75° = \dfrac{150°}{2}$. Then $\sin 75° = \sqrt{\dfrac{1-\cos 150°}{2}} = \sqrt{\dfrac{1-(-\sqrt{3}/2)}{2}} = \sqrt{\dfrac{1+\sqrt{3}/2}{2}\cdot\dfrac{2}{2}} = \sqrt{\dfrac{2+\sqrt{3}}{4}} =$

$\dfrac{\sqrt{2+\sqrt{3}}}{2}$.

8.111 $\cos 15°$

▌ $\cos 15° = \cos\dfrac{30°}{2} = \sqrt{\dfrac{1+\cos 30°}{2}} = \sqrt{\dfrac{1+\sqrt{3}/2}{2}} = \dfrac{\sqrt{2+\sqrt{3}}}{2}$. Look at Prob. 8.110; does the fact

that these two results are the same surprise you? It shouldn't!

8.112 cos 165°

$\blacksquare$ $\cos 165° = -\sqrt{\dfrac{1 + \cos 330°}{2}}$ (Be careful! We have a minus sign here because $\cos x < 0$ in quadrant II.)

$= -\sqrt{\dfrac{1 + \sqrt{3}/2}{2} \cdot \dfrac{2}{2}} = -\dfrac{\sqrt{2 + \sqrt{3}}}{2}$

8.113 tan 165°

$\blacksquare$ $\tan \dfrac{x}{2} = \dfrac{1 - \cos x}{\sin x}$. Then $\tan 165° = \dfrac{1 - \cos 330°}{\sin 330°} = \dfrac{1 - \sqrt{3}/2}{-\frac{1}{2}} \cdot \dfrac{2}{2} = \dfrac{2 - \sqrt{3}}{-1} = \sqrt{3} - 2.$

8.114 tan 22.5°

$\blacksquare$ $\tan \dfrac{x}{2} = \dfrac{\sin x}{1 + \cos x}$. Then $\tan 22.5° = \dfrac{\sin 45°}{1 + \cos 45°} = \dfrac{\sqrt{2}/2}{1 + \sqrt{2}/2} \cdot \dfrac{2}{2} = \dfrac{\sqrt{2}}{2 + \sqrt{2}}.$

8.115 sin 165°

$\blacksquare$ $\sin 165° = \sin \dfrac{330°}{2} = \pm\sqrt{\dfrac{1 - \cos 330°}{2}} = \sqrt{\dfrac{1 - \cos 330°}{2}}$ ($\sin x > 0$ in quadrant II) $=$

$\sqrt{\dfrac{1 - \sqrt{3}/2}{2} \cdot \dfrac{2}{2}} = \dfrac{\sqrt{2 - \sqrt{3}}}{2}.$

8.116 cos 375°

$\blacksquare$ $\cos 375° = \cos \dfrac{750°}{2} = \sqrt{\dfrac{1 + \cos 750°}{2}}$ (Here we have a positive answer since 375° is in quadrant I, and thus $\cos x > 0$.)

$= \sqrt{\dfrac{1 + \cos (750° - 720°)}{2}} = \sqrt{\dfrac{1 + \cos 30°}{2}} = \sqrt{\dfrac{1 + \sqrt{3}/2}{2}} = \dfrac{\sqrt{2 + \sqrt{3}}}{2}.$

For Probs. 8.117 to 8.131, establish that the given statement is an identity.

8.117 $\csc 2\theta = \dfrac{\sec \theta \csc \theta}{2}$

$\blacksquare$ $\csc 2\theta = \dfrac{1}{\sin 2\theta} = \dfrac{1}{2 \sin \theta \cos \theta}$ (since $\sin 2\theta = 2 \sin \theta \cos \theta$) $= \dfrac{1}{2} \cdot \dfrac{1}{\sin \theta} \cdot \dfrac{1}{\cos \theta} =$

$\dfrac{\sec \theta \csc \theta}{2}$ (since $\sin \theta \csc \theta = 1$ and $\cos \theta \sec \theta = 1$).

8.118 $\sec 2\theta = \dfrac{1}{2 \cos^2 \theta - 1}$

$\blacksquare$ $\sec 2\theta = \dfrac{1}{\cos 2\theta} = \dfrac{1}{2 \cos^2 \theta - 1}$ (since $\cos 2\theta = 2 \cos^2 \theta - 1$).

8.119 $\cot 2\theta = \dfrac{\cot^2 \theta - 1}{2 \cot \theta}$

$\blacksquare$ $\tan 2\theta = \dfrac{2 \tan \theta}{1 - \tan^2 \theta}$. Thus, $\cot 2\theta = \dfrac{1}{\tan 2\theta} = \dfrac{1}{(2 \tan \theta)/(1 - \tan^2 \theta)} = \dfrac{1 - \tan^2 \theta}{2 \tan \theta} =$

$\dfrac{1 - 1/\cot^2 \theta}{2/\cot \theta} \cdot \dfrac{\cot^2 \theta}{\cot^2 \theta}$ $(\tan \theta \cot \theta = 1)$ $= \dfrac{\cot^2 \theta - 1}{2 \cot \theta}.$

8.120 $\sec 2\theta = \dfrac{\sec^2 \theta}{2 - \sec^2 \theta}$

▮ See Prob. 8.118. $\sec 2\theta = \dfrac{1}{2\cos^2 \theta - 1} = \dfrac{1}{2/\sec^2 \theta - 1} \cdot \dfrac{\sec^2 \theta}{\sec^2 \theta}$ (multiplying by 1) $= \dfrac{\sec^2 \theta}{2 - \sec^2 \theta}$.

8.121 $\dfrac{\sin 2x}{1 + \cos 2x} = \tan x$

▮ We want the 1 to drop out, so we let $\cos 2x = 2\cos^2 x - 1$. Then

$$\frac{\sin 2x}{1 + \cos 2x} = \frac{2 \sin x \cos x}{1 + (2\cos^2 x - 1)} = \frac{2 \sin x \cos x}{2\cos^2 x} = \frac{\sin x}{\cos x} = \tan x.$$

8.122 $\cot 4x = \dfrac{1 - \tan^2 2x}{2 \tan 2x}$

▮ $\tan 4x = \tan 2(2x)$, so $\cot 4x = \dfrac{1}{\tan 2(2x)} = \dfrac{1}{(2 \tan 2x)/(1 - \tan^2 2x)} = \dfrac{1 - \tan^2 2x}{2 \tan 2x}$.

8.123 $\cos 2x + 2\sin^2 x = 1$

▮ $2\sin^2 x = 2(1 - \cos^2 x) = 2 - 2\cos^2 x$. Thus, $\cos 2x + 2\sin^2 x = (2\cos^2 x - 1) + (2 - 2\cos^2 x) = -1 + 2 = 1$.

8.124 $4\sin^2 \theta \cos^2 \theta = 1 - \cos^2 2\theta$

▮ $\sin 2\theta = 2 \sin \theta \cos \theta$. Thus, $(2 \sin \theta \cos \theta)^2 = (\sin 2\theta)^2$, or $4\sin^2 \theta \cos^2 \theta = \sin^2 2\theta = 1 - \cos^2 2\theta$.

8.125 $\tan (\theta - 45°) + \tan (\theta + 45°) = 2 \tan 2\theta$

▮ $\tan (\theta - 45°) + \tan (\theta + 45°) = \dfrac{\tan \theta - \tan 45°}{1 + \tan \theta \tan 45°} + \dfrac{\tan \theta + \tan 45°}{1 - \tan \theta \tan 45°}$

$$= \frac{\tan \theta - 1}{1 + \tan \theta} + \frac{\tan \theta + 1}{1 - \tan \theta}$$
$$= \frac{(\tan \theta - 1)(1 - \tan \theta) + (\tan \theta + 1)^2}{1 - \tan^2 \theta}$$
$$= \frac{\tan \theta - 1 - \tan^2 \theta + \tan \theta + \tan^2 \theta + 2 \tan \theta + 1}{1 - \tan^2 \theta}$$
$$= \frac{4 \tan \theta}{1 - \tan^2 \theta} = 2 \frac{2 \tan \theta}{1 - \tan^2 \theta} = 2 \tan 2\theta$$

8.126 $\cot \dfrac{x}{2} = \dfrac{1 + \cos x}{\sin x}$

▮ $\cot \dfrac{x}{2} = \dfrac{1}{\tan (x/2)} = \dfrac{1}{(1 - \cos x)/\sin x} = \dfrac{\sin x}{1 - \cos x} \cdot \dfrac{1 + \cos x}{1 + \cos x}$

$$= \frac{\sin x(1 + \cos x)}{1 - \cos^2 x} = \frac{\sin x(1 + \cos x)}{\sin^2 x} = \frac{1 + \cos x}{\sin x}.$$

8.127 $\sec^2 \dfrac{\theta}{2} = \dfrac{2}{1 + \cos \theta}$

▮ $\cos^2 \dfrac{\theta}{2} = \dfrac{1 + \cos \theta}{2}$. Then $\sec^2 \dfrac{\theta}{2} = \dfrac{1}{\cos^2 (\theta/2)} = \dfrac{1}{(1 + \cos \theta)/2} = \dfrac{2}{1 + \cos \theta}$.

8.128 $\sin^2 \dfrac{\theta}{2} = \dfrac{\sec\theta - 1}{2\sec\theta}$

▮ $\sin^2 \dfrac{\theta}{2} = \dfrac{1 - \cos\theta}{2} = \dfrac{1 - 1/\sec\theta}{2} \cdot \dfrac{\sec\theta}{\sec\theta}$ $\left(\text{since } \cos\theta = \dfrac{1}{\sec\theta}\right) = \dfrac{\sec\theta - 1}{2\sec\theta}$.

(When would this formula be useful?)

8.129 $\csc^2 \dfrac{\theta}{2} = \dfrac{2}{1 - \cos\theta}$

▮ $\csc^2 \dfrac{\theta}{2} = \dfrac{1}{\sin^2(\theta/2)} = \dfrac{1}{(1 - \cos\theta)/2} = \dfrac{2}{1 - \cos\theta}$.

8.130 $\cot^2 \dfrac{\theta}{2} = \dfrac{1 + \cos\theta}{1 - \cos\theta}$

▮ $\cot^2 \dfrac{\theta}{2} = \dfrac{\cos^2(\theta/2)}{\sin^2(\theta/2)} = \dfrac{(1 + \cos\theta)/2}{(1 - \cos\theta)/2} = \dfrac{1 + \cos\theta}{1 - \cos\theta}$

8.131 $\tan \dfrac{x}{2} \sin x = \dfrac{\tan x - \sin x}{\sin x \sec x}$

▮   Here we simplify both sides.

(1) $\qquad\qquad \tan\dfrac{x}{2}\sin x = \dfrac{1 - \cos x}{\sin x}\sin x = 1 - \cos x$ (LHS).

(2) $\qquad\qquad \dfrac{\tan x - \sin x}{\sin x \sec x} = \dfrac{\sin x/\cos x - \sin x}{\sin x/\cos x}$

Then multiplying by $\dfrac{\cos x}{\cos x}$, we have

$$\dfrac{\sin x - \sin x \cos x}{\sin x} = \dfrac{(\sin x)(1 - \cos x)}{\sin x} = 1 - \cos x \qquad \text{(RHS)}$$

Since LHS = RHS, the identity is established.

For Probs. 8.132 to 8.137, find the exact value without using a table or calculator.

8.132 $\sin(2\,\mathrm{Cos}^{-1}\tfrac{3}{5})$

▮   See Fig. 8.5. $\sin 2x = 2\sin x \cos x$. Then $\sin(2\,\mathrm{Cos}^{-1}\tfrac{3}{5}) = 2\sin(\mathrm{Cos}^{-1}\tfrac{3}{5})\cos(\mathrm{Cos}^{-1}\tfrac{3}{5}) = 2\cdot\tfrac{4}{5}\cdot\tfrac{3}{5} = \tfrac{24}{25}$ (since $\sin\mathrm{Cos}^{-1}\tfrac{3}{5} = \tfrac{4}{5}$).

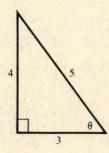

Fig. 8.5

8.133 $\cos(2\,\mathrm{Cos}^{-1}\tfrac{3}{5})$

▮   Let $x = \mathrm{Cos}^{-1}\tfrac{3}{5}$. Then $\cos 2x = 2\cos^2 x - 1$. Therefore, $\cos(2\,\mathrm{Cos}^{-1}\tfrac{3}{5}) = 2\cos^2(\mathrm{Cos}^{-1}\tfrac{3}{5}) - 1 = 2(\tfrac{3}{5})^2 - 1 = 2\cdot\tfrac{9}{25} - 1 = \tfrac{18}{25} - 1 = -\tfrac{7}{25}$.

8.134 $\tan \left[2 \cos^{-1} \left(-\frac{4}{5}\right)\right]$

▮ See Fig. 8.6. $\tan 2x = \dfrac{2 \tan x}{1 - \tan^2 x}$. Then $\tan \left[2 \cos^{-1} \left(-\frac{4}{5}\right)\right] = \dfrac{2 \tan \cos^{-1} \left(-\frac{4}{5}\right)}{1 - \tan^2 \left[\cos^{-1} \left(-\frac{4}{5}\right)\right]}$.

Since $\tan \cos^{-1} \left(\frac{4}{5}\right) = \frac{3}{4}$ and $\tan \cos^{-1} \left(-\frac{4}{5}\right) = -\frac{3}{4}$ ($\tan x < 0$ in quadrant II),

$$\dfrac{2 \tan \cos^{-1} \left(-\frac{4}{5}\right)}{1 - \tan^2 \left[\cos^{-1} \left(-\frac{4}{5}\right)\right]} = \dfrac{2\left(-\frac{3}{4}\right)}{1 - \left(-\frac{3}{4}\right)^2} = \dfrac{-\frac{6}{4}}{1 - \frac{9}{16}} = -\frac{6}{4} \cdot \frac{16}{7} = -\frac{24}{7}.$$

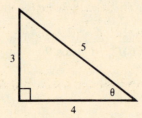

Fig. 8.6

8.135 $\sin (x/2)$ if $\cos x = \frac{1}{3}$, $0° < x < 90°$

▮ $\sin \dfrac{x}{2} = \sqrt{\dfrac{1 - \cos x}{2}}$ (quadrant I) $= \sqrt{\dfrac{1 - \frac{1}{3}}{2}} = \sqrt{\dfrac{\frac{2}{3}}{2}} = \sqrt{\dfrac{1}{3}}$.

8.136 $\tan 2x$ if $\cot x = -\frac{12}{5}$, $-\pi/2 < x < 0$

▮ $\tan 2x = \dfrac{2 \tan x}{1 - \tan^2 x} = \dfrac{2(1/\cot x)}{1 - (1/\cot x)^2} = \dfrac{2\left(-\frac{5}{12}\right)}{1 - \left(-\frac{5}{12}\right)^2} = \dfrac{-\frac{10}{12}}{1 - \frac{25}{144}} = -\frac{10}{12} \cdot \frac{144}{119} = -\frac{120}{119}.$

8.137 $\cos (x/2)$ if $\sin x = -\frac{1}{3}$, $\pi < x < 3\pi/2$

▮ See Fig. 8.7. $\cos \dfrac{x}{2} = -\sqrt{\dfrac{1 + \cos x}{2}}$ (quadrant III) $= -\sqrt{\dfrac{1 + \sqrt{8}/3}{2}} = -\sqrt{\dfrac{1 + 2\sqrt{2}/3}{2}} = -\sqrt{\dfrac{3 + 2\sqrt{2}}{6}}.$

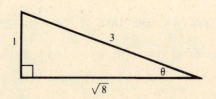

Fig. 8.7

8.4 PRODUCT AND SUM IDENTITIES

8.138 State the four basic product identities.

▮ $2 \sin A \cos B = \sin (A + B) + \sin (A - B)$ $\qquad$ $2 \cos A \sin B = \sin (A + B) - \sin (A - B)$

$2 \cos A \cos B = \cos (A + B) + \cos (A - B)$ $\qquad$ $2 \sin A \sin B = -\cos (A + B) + \cos (A - B)$

8.139 State the four basic sum identities.

▮ $\sin A + \sin B = 2 \sin \dfrac{A + B}{2} \cos \dfrac{A - B}{2}$ $\qquad$ $\sin A - \sin B = 2 \cos \dfrac{A + B}{2} \sin \dfrac{A - B}{2}$

$\cos A + \cos B = 2 \cos \dfrac{A + B}{2} \cos \dfrac{A - B}{2}$ $\qquad$ $\cos B - \cos A = 2 \sin \dfrac{A + B}{2} \sin \dfrac{A - B}{2}$

Be careful with the $\cos B - \cos A$ identity. Look carefully at A and B.

8.140 Derive the identity $2 \sin A \cos B = \sin (A + B) + \sin (A - B)$.

▌ $\sin (A + B) = \sin A \cos B + \cos A \sin B$ and $\sin (A - B) = \sin A \cos B - \cos A \sin B$. Adding these, we get $\sin (A + B) + \sin (A - B) = 2 \sin A \cos B$. (You should try to establish the other three; use this derivation as a model!)

8.141 Derive the identity $\sin X + \sin Y = 2 \sin \dfrac{X + Y}{2} \cos \dfrac{X - Y}{2}$.

▌ Let $X = A + B$ and $Y = A - B$. Then, solving for A and B gives $X + Y = 2A$, and $X - Y = 2B$, so $A = \dfrac{X + Y}{2}$ and $B = \dfrac{X - Y}{2}$. Thus, $2 \sin \dfrac{X + Y}{2} \cos \dfrac{X - Y}{2} = \sin X + \sin Y$ by substituting directly into the identity in Prob. 8.140.

For Probs. 8.142 to 8.146, write each expression as a sum or difference.

8.142 $2 \sin 45° \cos 73°$

▌ $2 \sin A \cos B = \sin (A + B) + \sin (A - B)$. Then letting $A = 45°$ and $B = 73°$, we get $\sin (45° + 73°) + \sin (45° - 73°) = \sin 118° + \sin (-28°) = \sin 118° - \sin 28°$. See Prob. 8.143.

8.143 $2 \cos 73° \sin 45°$

▌ This is the same as Prob. 8.142, but here we find the answer in a different way. $2 \cos A \sin B = \sin (A + B) - \sin (A - B)$. Then letting $A = 73°$ and $B = 45°$, we have $\sin 118° - \sin 28°$. Notice that either formula could be used here. Obviously, the answers are the same either way.

8.144 $2 \cos 28° \cos 16°$

▌ $2 \cos A \cos B = \cos (A + B) + \cos (A - B)$. Then letting $A = 28°$ and $B = 16°$ gives $\cos (28° + 16°) + \cos (28° - 16°) = \cos 44° + \cos 12°$.

8.145 $2 \cos 15° \cos 17°$

▌ $2 \cos 15° \cos 17° = \cos (A + B) + \cos (A - B)$. Letting $A = 15°$ and $B = 17°$, we get $\cos (15° + 17°) + \cos (15° - 17°) = \cos 32° + \cos (-2°) = \cos 32° + \cos 2°$ ($\cos x > 0$ in quadrant IV).

8.146 $\sin 5p \sin 6p$

▌ $2 \sin A \sin B = -\cos (A + B) + \cos (A - B)$, and $\sin A \sin B = \dfrac{-\cos (A + B) + \cos (A - B)}{2}$. Letting $A = 5p$ and $B = 6p$, we get $\dfrac{-\cos (5p + 6p) + \cos (5p - 6p)}{2} = \dfrac{-\cos 11p + \cos (-p)}{2} = \dfrac{-\cos 11p}{2} + \dfrac{\cos (-p)}{2}$.

8.147 Express each of the following as a sum or difference: (**a**) $\sin 40° \cos 30°$, (**b**) $\cos 110° \sin 55°$, (**c**) $\cos 50° \cos 35°$, (**d**) $\sin 55° \sin 40°$.

▌   (**a**) $\sin 40° \cos 30° = \frac{1}{2}[\sin (40° + 30°) + \sin (40° - 30°)] = \frac{1}{2}(\sin 70° + \sin 10°)$.
  (**b**) $\cos 110° \sin 55° = \frac{1}{2}[\sin (110° + 55°) - \sin (110° - 55°)] = \frac{1}{2}(\sin 165° - \sin 55°)$.
  (**c**) $\cos 50° \cos 35° = \frac{1}{2}[\cos (50° + 35°) + \cos (50° - 35°)] = \frac{1}{2}(\cos 85° + \cos 15°)$.
  (**d**) $\sin 55° \sin 40° = -\frac{1}{2}[\cos (55° + 40°) - \cos (55° - 40°)] = -\frac{1}{2}(\cos 95° - \cos 15°)$.

For Probs. 8.148 to 8.152, rewrite each expression as a product.

8.148 $\sin 20° + \sin 15°$

▌ $\sin A + \sin B = 2 \sin \dfrac{A+B}{2} \cos \dfrac{A-B}{2}$. Letting $A = 20°$ and $B = 15°$, we get

$\sin 20° + \sin 15° = 2 \sin \dfrac{20° + 15°}{2} \cos \dfrac{20° - 15°}{2} = 2 \sin 17.5° \cos 2.5°.$

8.149 $\sin 51° - \sin 61°$

▌ $\sin A - \sin B = 2 \cos \dfrac{A+B}{2} \sin \dfrac{A-B}{2}$. Letting $A = 51°$ and $B = 61°$, we find

$\sin 51° - \sin 61° = 2 \cos \dfrac{51° + 61°}{2} \sin \dfrac{51° - 61°}{2} = 2 \cos 56° \sin (-5°) = -2 \cos 56° \sin 5°$

$(\sin x < 0$ in quadrant IV$)$.

8.150 $\cos \dfrac{\pi}{12} - \cos \dfrac{\pi}{6}$

▌ $\cos \dfrac{\pi}{12} - \cos \dfrac{\pi}{6} = 2 \sin \dfrac{\pi/6 + \pi/12}{2} \sin \dfrac{\pi/6 - \pi/12}{2} = 2 \sin \dfrac{3\pi/12}{2} \sin \dfrac{\pi/12}{2} = 2 \sin \dfrac{\pi}{8} \sin \dfrac{\pi}{24}.$

8.151 $\cos 5x + \cos 7x$

▌ $\cos 5x + \cos 7x = 2 \cos \dfrac{5x + 7x}{2} \cos \dfrac{5x - 7x}{2} = 2 \cos \dfrac{12x}{2} \cos \left(\dfrac{-2x}{2}\right) = 2 \cos 6x \cos (-x) =$

$2 \cos 6x \cos x.$

8.152 $\sin 620° - \sin 720°$

▌ $\sin 620° - \sin 720° = 2 \cos \dfrac{620° + 720°}{2} \sin \dfrac{620° - 720°}{2} = 2 \cos \dfrac{1340°}{2} \sin \left(\dfrac{-100°}{2}\right) =$

$2 \cos 670° \sin (-50°) = -2 \cos 310° \sin 50°.$

8.153 Express each of the following as a product:
(**a**) $\sin 50° + \sin 40°$, (**b**) $\sin 70° - \sin 20°$, (**c**) $\cos 55° + \cos 25°$, (**d**) $\cos 35° - \cos 75°$.

▌ (**a**) $\sin 50° + \sin 40° = 2 \sin \frac{1}{2}(50° + 40°) \cos \frac{1}{2}(50° - 40°) = 2 \sin 45° \cos 5°.$
(**b**) $\sin 70° - \sin 20° = 2 \cos \frac{1}{2}(70° + 20°) \sin \frac{1}{2}(70° - 20°) = 2 \cos 45° \sin 25°.$
(**c**) $\cos 55° + \cos 25° = 2 \cos \frac{1}{2}(55° + 25°) \cos \frac{1}{2}(55° - 25°) = 2 \cos 40° \cos 15°.$
(**d**) $\cos 35° - \cos 75° = -2 \sin \frac{1}{2}(35° + 75°) \sin \frac{1}{2}(35° - 75°) = -2 \sin 55° \sin (-20°) = 2 \sin 55° \sin 20°.$

For Probs. 8.154 to 8.164, use a product or sum identity to evaluate the given expression.

8.154 $2 \sin 75° \cos 15°$

▌ $2 \sin 75° \cos 15° = \sin (75° + 15°) + \sin (75° - 15°) = \sin 90° + \sin 60° = 1 + \dfrac{\sqrt{3}}{2} = \dfrac{2 + \sqrt{3}}{2}.$

8.155 $\cos 75° \cos 15°$

▌ $\cos 75° \cos 15° = \dfrac{\cos (75° + 15°) + \cos (75° - 15°)}{2} = \dfrac{\cos 90° + \cos 60°}{2} = \dfrac{0 + \frac{1}{2}}{2} = \dfrac{1}{4}.$

8.156 $\sin 15° \sin 45°$

▌ $\sin 15° \sin 45° = \dfrac{-\cos (15° + 45°) + \cos (15° - 45°)}{2} = \dfrac{-\cos 60° + \cos (-30°)}{2}$

$= \dfrac{-\frac{1}{2} + \sqrt{3}/2}{2} = \dfrac{-1 + \sqrt{3}}{4}.$

8.157 sin 45° sin 15° (See Prob. 8.156.)

▮ $\sin 45° \sin 15° = \dfrac{-\cos(45° + 15°) + \cos(45° - 15°)}{2} = \dfrac{-\cos 60° + \cos 30°}{2} = \dfrac{-\frac{1}{2} + \sqrt{3}/2}{2} = \dfrac{-1 + \sqrt{3}}{4}.$

You should not be surprised that these answers are the same. What law governs this situation?

8.158 cos 52½° cos 7½°

▮ $\cos 52\frac{1}{2}° \cos 7\frac{1}{2}° = \dfrac{\cos(52\frac{1}{2}° + 7\frac{1}{2}°) + \cos(52\frac{1}{2}° - 7\frac{1}{2}°)}{2} = \dfrac{\cos 60° + \cos 45°}{2} = \dfrac{\frac{1}{2} + \sqrt{2}/2}{2} = \dfrac{\sqrt{2} + 1}{4}.$

8.159 sin 75° + sin 15°

▮ $\sin 75° + \sin 15° = 2 \sin \dfrac{75° + 15°}{2} \cos \dfrac{75° - 15°}{2} = 2 \sin 45° \cos 30° = 2 \dfrac{\sqrt{2}}{2} \cdot \dfrac{\sqrt{3}}{2} = \dfrac{2\sqrt{6}}{4} = \dfrac{\sqrt{6}}{2}.$

8.160 cos 75° − cos 15°

▮ $\cos 75° - \cos 15° = 2 \sin \dfrac{15° + 75°}{2} \sin \dfrac{15° - 75°}{2} = 2 \sin 45° \sin(-30°) = (2\sqrt{2}/2) \cdot -\frac{1}{2} = -\dfrac{\sqrt{2}}{2}.$

Note that $\cos 75° - \cos 15° = -(\sin 75° - \sin 15°)$.

8.161 sin 75° − sin 15°

▮ $\sin 75° - \sin 15° = 2 \cos \dfrac{75° + 15°}{2} \sin \dfrac{75° - 15°}{2} = 2 \cos 45° \sin 30° = (2\sqrt{2}/2) \cdot \frac{1}{2} = \sqrt{2}/2.$

Note that $\cos 75° - \cos 15° = -(\sin 75° - \sin 15°)$.

8.162 cos 300° + cos 120°

▮ $\cos 300° + \cos 120° = 2 \cos \dfrac{300° + 120°}{2} \cos \dfrac{300° - 120°}{2} = 2 \cos 210° \cos 90° = 0$. *Note*: You can

check this answer by noting that $\cos 300° = \cos 60° = \frac{1}{2}$, $\cos 120° = -\cos 60° = -\frac{1}{2}$, and $\cos 300° + \cos 120° = 0$.

8.163 cos 75° sin 45°

▮ $\cos 75° \sin 45° = \dfrac{\sin(75° + 45°) - \sin(75° - 45°)}{2} = \dfrac{\sin 120° - \sin 30°}{2}$

$= \dfrac{\sin 60° - \sin 30°}{2} = \dfrac{\sqrt{3}/2 - \frac{1}{2}}{2} = \dfrac{\sqrt{3} - 1}{4}.$

8.164 cos 75° cos 45°

▮ $\cos 75° \cos 45° = \dfrac{\cos(75° + 45°) + \cos(75° - 45°)}{2} = \dfrac{\cos 120° + \cos 30°}{2}$

$= \dfrac{-\cos 60° + \cos 30°}{2} = \dfrac{-\frac{1}{2} + \sqrt{3}/2}{2} = \dfrac{\sqrt{3} - 1}{4}.$

Note that the answer here is the same as in Prob. 8.163.

8.165 Prove: $\dfrac{\sin 4A + \sin 2A}{\cos 4A + \cos 2A} = \tan 3A$

▮ $\dfrac{\sin 4A + \sin 2A}{\cos 4A + \cos 2A} = \dfrac{2 \sin \frac{1}{2}(4A + 2A) \cos \frac{1}{2}(4A - 2A)}{2 \cos \frac{1}{2}(4A + 2A) \cos \frac{1}{2}(4A - 2A)} = \dfrac{\sin 3A}{\cos 3A} = \tan 3A.$

8.166 Prove: $\dfrac{\sin A - \sin B}{\sin A + \sin B} = \dfrac{\tan \frac{1}{2}(A - B)}{\tan \frac{1}{2}(A + B)}$

❚ $\dfrac{\sin A - \sin B}{\sin A + \sin B} = \dfrac{2 \cos \frac{1}{2}(A + B) \sin \frac{1}{2}(A - B)}{2 \sin \frac{1}{2}(A + B) \cos \frac{1}{2}(A - B)} = \cot \frac{1}{2}(A + B) \tan \frac{1}{2}(A - B) = \dfrac{\tan \frac{1}{2}(A - B)}{\tan \frac{1}{2}(A + B)}$.

8.167 Prove: $\cos^3 x \sin^2 x = \frac{1}{16}(2 \cos x - \cos 3x - \cos 5x)$

❚ $\cos^3 x \sin^2 x = (\sin x \cos x)^2 \cos x = \frac{1}{4} \sin^2 2x \cos x = \frac{1}{4}(\sin 2x)(\sin 2x \cos x)$

$\qquad = \frac{1}{4}(\sin 2x)[\frac{1}{2}(\sin 3x + \sin x)] = \frac{1}{8}(\sin 3x \sin 2x + \sin 2x \sin x)$

$\qquad = \frac{1}{8}\{-\frac{1}{2}(\cos 5x - \cos x) + [-\frac{1}{2}(\cos 3x - \cos x)]\}$

$\qquad = \frac{1}{16}(2 \cos x - \cos 3x - \cos 5x)$

8.168 Prove: $1 + \cos 2x + \cos 4x + \cos 6x = 2 \cos 3x (\cos 3x + \cos x)$

❚ $1 + (\cos 2x + \cos 4x) + \cos 6x = 1 + 2 \cos 3x \cos x + \cos 6x$

$\qquad\qquad\qquad\qquad = (1 + \cos 6x) + 2 \cos 3x \cos x$

$\qquad\qquad\qquad\qquad = 2 \cos^2 3x + 2 \cos 3x \cos x$

$\qquad\qquad\qquad\qquad = 2 \cos 3x (\cos 3x + \cos x)$

8.169 Transform $4 \cos x + 3 \sin x$ to the form $c \cos (x - a)$.

❚ Since $c \cos (x - a) = c(\cos x \cos a + \sin x \sin a)$, set $c \cos a = 4$ and $c \sin a = 3$. Then $\cos a = 4/c$ and $\sin a = 3/c$. Since $\sin^2 a + \cos^2 a = 1$, $c = 5$ and -5. Using $c = 5$, $\cos a = \frac{4}{5}$, $\sin a = \frac{3}{5}$, and $a = 36°52'$. Thus, $4 \cos x + 3 \sin x = 5 \cos (x - 36°52')$. Using $c = -5$, we get $a = 216°52'$ and $4 \cos x + 3 \sin x = -5 \cos (x - 216°52')$.

8.170 Find the maximum and minimum values of $4 \cos x + 3 \sin x$ on the interval $0 \le x \le 2\pi$.

❚ From Prob. 8.169, $4 \cos x + 3 \sin x = 5 \cos (x - 36°52')$.
Now, on the prescribed interval, $\cos \theta$ attains its maximum value 1 when $\theta = 0$ and its minimum value -1 when $\theta = 0$ and $\theta = \pi$. Thus, the maximum value of $4 \cos x + 3 \sin x$ is 5 which occurs when $x - 36°52' = 0$ or when $x = 36°52'$, while the minimum value is -5 which occurs when $x - 36°52' = \pi$ or when $x = 216°52'$.

For Probs. 8.171 to 8.181, establish that the given equation is an identity.

8.171 $\dfrac{\cos 3x + \cos x}{\sin 3x + \sin x} = \cot 2x$

❚ Using the sum formulas, we have $A = 3x$ and $B = x$,

$$\frac{\cos 3x + \cos x}{\sin 3x + \sin x} = \frac{2 \cos \dfrac{3x + x}{2} \cos \dfrac{3x - x}{2}}{2 \sin \dfrac{3x + x}{2} \cos \dfrac{3x - x}{2}} = \frac{\cos 2x}{\sin 2x} = \cot 2x.$$

8.172 $\dfrac{\sin x + \sin 3x}{\cos x - \cos 3x} = 1$

$$\blacksquare \quad \frac{\sin x + \sin 3x}{\cos x - \cos 3x} = \frac{2 \sin \dfrac{x + 3x}{2} \cos \dfrac{x - 3x}{2}}{2 \sin \dfrac{3x + x}{2} \cos \dfrac{3x - x}{2}} = \frac{\cos (-x)}{\cos x} = \frac{\cos x}{\cos x} = 1.$$

8.173 $\dfrac{\cos 5x - \cos 7x}{\sin 7x - \sin 5x} = \tan 6x$

$$\blacksquare \quad \frac{\cos 5x - \cos 7x}{\sin 7x - \sin 5x} = \frac{2 \sin \dfrac{7x + 5x}{2} \sin \dfrac{7x - 5x}{2}}{2 \cos \dfrac{7x + 5x}{2} \sin \dfrac{7x - 5x}{2}} = \frac{2 \sin 6x \sin x}{2 \cos 6x \sin x} = \frac{\sin 6x}{\cos 6x} = \tan 6x.$$

8.174 $\dfrac{\sin 5x - \sin 3x}{\cos 5x + \cos 3x} = \tan x$

$$\blacksquare \quad \frac{\sin 5x - \sin 3x}{\cos 5x + \cos 3x} = \frac{2 \cos 4x \sin x}{2 \cos 4x \cos x} = \frac{\sin x}{\cos x} = \tan x.$$

8.175 $\dfrac{\sin B + \sin A}{\cos B - \cos A} = \cot \dfrac{A - B}{2}$

$$\blacksquare \quad \frac{\sin B + \sin A}{\cos B - \cos A} = \frac{2 \sin \dfrac{B + A}{2} \cos \dfrac{B - A}{2}}{2 \sin \dfrac{A + B}{2} \sin \dfrac{A - B}{2}}$$

$$= \frac{\cos \dfrac{B - A}{2}}{\sin \dfrac{A - B}{2}} = \frac{\cos \left[-\left(\dfrac{A - B}{2} \right) \right]}{\sin \dfrac{A - B}{2}}$$

$$= \frac{\cos \dfrac{A - B}{2}}{\sin \dfrac{A - B}{2}} \quad [\text{since } \cos (-x) = \cos x] = \cot \frac{A - B}{2}$$

8.176 $\dfrac{\sin A + \sin B}{\cos A + \cos B} = \tan \dfrac{A + B}{2}$

$$\blacksquare \quad \frac{\sin A + \sin B}{\cos A + \cos B} = \frac{2 \sin \dfrac{A + B}{2} \cos \dfrac{A - B}{2}}{2 \cos \dfrac{A + B}{2} \cos \dfrac{A - B}{2}} = \frac{\sin \dfrac{A + B}{2}}{\cos \dfrac{A + B}{2}} = \tan \frac{A + B}{2}.$$

8.177 $\dfrac{\sin 5x - \sin 9x}{\cos 5x - \cos 9x} = -\cot 7x$

$$\blacksquare \quad \frac{\sin 5x - \sin 9x}{\cos 5x - \cos 9x} = \frac{2 \cos \dfrac{5x + 9x}{2} \sin \dfrac{5x - 9x}{2}}{2 \sin \dfrac{9x + 5x}{2} \sin \dfrac{9x - 5x}{2}} = \frac{2 \cos 7x \sin (-2x)}{2 \sin 7x \sin 2x}$$

$$= \frac{-\cos 7x \sin 2x}{\sin 7x \sin 2x} \quad [\text{since } \sin (-x) = -\sin x] = -\cot 7x.$$

8.178 $(\cos x - \cos 5x)(\cos x + \cos 5x) = \sin 4x \sin 6x$

▎ $\cos x - \cos 5x = 2 \sin \dfrac{5x + x}{2} \sin \dfrac{5x - x}{2}$ and $\cos x + \cos 5x = 2 \cos \dfrac{5x + x}{2} \cos \dfrac{5x - x}{2}$.

Thus, $(\cos 5x - \cos 5x)(\cos x + \cos 5x) = 4 \sin 3x \sin 2x \cos 3x \cos 2x = 4 \sin 3x \cos 3x \sin 2x \cos 2x = (2 \sin 3x \cos 3x)(2 \sin 2x \cos 2x)$. Letting $A = 3x$, we get
$2 \sin 3x \cos 3x = 2 \sin A \cos A = \sin 2A = \sin 6x$. Letting $B = 2x$ gives $2 \sin 2x \cos 2x = \sin 2B = \sin 4x$. Therefore, $(\cos x - \cos 5x)(\cos x + \cos 5x) = \sin 4x \sin 6x$.

8.179 $\dfrac{\cos 2\theta - \sin 2\theta}{\sin \theta \cos \theta} = \cot \theta - \tan \theta - 2$

▎ $\dfrac{\cos 2\theta - \sin 2\theta}{\sin \theta \cos \theta} = \dfrac{\cos^2 \theta - \sin^2 \theta - 2 \sin \theta \cos \theta}{\sin \theta \cos \theta}$

$= \dfrac{\cos^2 \theta}{\sin \theta \cos \theta} - \dfrac{\sin^2 \theta}{\sin \theta \cos \theta} - \dfrac{2 \sin \theta \cos \theta}{\sin \theta \cos \theta} = \cot \theta - \tan \theta - 2.$

Note that we did not use one of the four sum formulas here. Why?

8.180 $\dfrac{4}{\cos \theta - \cos 3\theta} = 2 \csc 2\theta \csc \theta$

▎ $\dfrac{4}{\cos \theta - \cos 3\theta} = \dfrac{4}{2 \sin \dfrac{3\theta + \theta}{2} \sin \dfrac{3\theta - \theta}{2}} = \dfrac{4}{2 \sin 2\theta \sin \theta} = \dfrac{2}{\sin 2\theta \sin \theta} = 2 \csc 2\theta \csc \theta.$

8.181 Assume that $A + B + C = \pi$, and prove that $\sin (A + B) = \sin C$ is an identity.

▎ $\sin (A + B) = \sin (\pi - C)$ (since $A + B = \pi - C$) $= \sin \pi \cos C - \cos \pi \sin C$ [from the $\sin (A + B)$ formula] $= 0 + \sin C = \sin C.$

8.5 MISCELLANEOUS IDENTITIES

For Probs. 8.182 to 8.201, verify the given identity. Problems in this section combine the techniques of the first four sections of this chapter as well as earlier chapters.

8.182 $\tan (\pi + x) = \tan x$

▎ $\tan (\pi + x) = \dfrac{\tan \pi + \tan x}{1 - \tan \pi \tan x} = \dfrac{0 + \tan x}{1 - 0} = \tan x.$ (*Note*: You may also obtain this result by using the technique of Prob. 8.187.)

8.183 $\sec (\pi + x) = -\sec x$

▎ $\sec (\pi + x) = \dfrac{1}{\cos (\pi + x)} = \dfrac{1}{\cos \pi \cos x - \sin \pi \sin x} = \dfrac{1}{-\cos x - 0} = -\dfrac{1}{\cos x} = -\sec x.$

8.184 $\tan (\pi - x) = -\tan x$

▎ $\tan (\pi - x) = \dfrac{\tan \pi - \tan x}{1 + \tan \pi \tan x} = \dfrac{0 - \tan x}{1 + 0} = -\tan x.$

8.185 $\cot(\pi - x) = -\cot x$

▎ $\cot(\pi - x) = \dfrac{1}{\tan(\pi - x)} = \dfrac{1}{-\tan x}$ (see Prob. 8.184) $= -\cot x.$

8.186 $\sec(\pi - x) = -\sec x$

$\blacksquare$ $\sec(\pi - x) = \dfrac{1}{\cos(\pi - x)} = \dfrac{1}{\cos\pi\cos x + \sin\pi\sin x} = \dfrac{1}{-\cos x + 0} = -\dfrac{1}{\cos x} = -\sec x.$

8.187 $\tan\left(\dfrac{\pi}{2} + u\right) = -\cot u$

$\blacksquare$ $\tan\left(\dfrac{\pi}{2} + u\right) = \dfrac{\sin(\pi/2 + u)}{\cos(\pi/2 + u)} = \dfrac{\sin(\pi/2)\cos u + \cos(\pi/2)\sin u}{\cos(\pi/2)\cos u - \sin(\pi/2)\sin u} = \dfrac{\cos u + 0}{0 - \sin u} = -\dfrac{\cos u}{\sin u} = -\cot u.$

8.188 $\csc(\pi - u) = \csc u$

$\blacksquare$ $\csc(\pi - u) = \dfrac{1}{\sin(\pi - u)} = \dfrac{1}{\sin\pi\cos u - \cos\pi\sin u} = \dfrac{1}{0 - (-\sin u)} = \dfrac{1}{\sin u} = \csc u.$

8.189 $\tan(2\pi - x) = \tan(-x)$

$\blacksquare$ $\tan(2\pi - x) = \dfrac{\tan 2\pi - \tan x}{1 + \tan 2\pi \tan x} = \dfrac{0 - \tan x}{1 + 0} = -\tan x = \tan(-x).$

8.190 $\cot(2\pi - u) = \cot(-u)$

$\blacksquare$ $\cot(2\pi - u) = \dfrac{1}{\tan(2\pi - u)} = \dfrac{1}{\tan(-u)}$ (see Prob. 8.189) $= \cot(-u).$

8.191 $\dfrac{\cot x}{\csc x - 1} = \dfrac{\csc x + 1}{\cot x}$

$\blacksquare$ $\dfrac{\cot x}{\csc x - 1} = \dfrac{\cot x}{\csc x - 1}\cdot\dfrac{\csc x + 1}{\csc x + 1} = \dfrac{(\cot x)(\csc x - 1)}{\csc^2 x - 1} = \dfrac{(\cot x)(\csc x + 1)}{\cot^2 x} = \dfrac{\csc x + 1}{\cot x}.$

8.192 $\dfrac{\tan x}{\sec x - 1} - \dfrac{\tan x}{\sec x + 1} = 2\cot x$

$\blacksquare$ $\dfrac{\tan x}{\sec x - 1} - \dfrac{\tan x}{\sec x + 1} = \dfrac{(\tan x)(\sec x + 1) - (\tan x)(\sec x - 1)}{\sec^2 x - 1} = \dfrac{2\tan x}{\tan^2 x} = \dfrac{2}{\tan x} = 2\cot x.$

8.193 $\ln|\cos x| + \ln|\sec x| = 0$

$\blacksquare$ $\ln|\cos x| = \ln\left|\dfrac{1}{\sec x}\right|.$ Thus, $\ln|\cos x| + \ln|\sec x| = \ln\left|\dfrac{1}{\sec x}\right| + \ln|\sec x| =$

$\ln\left|\dfrac{1}{\sec x}\sec x\right| = \ln 1 = 0.$

8.194 $\tan\theta\sin 2\theta = 2\sin^2\theta$

$\blacksquare$ $\tan\theta\sin 2\theta = (\tan\theta)(2\sin\theta\cos\theta) = \dfrac{\sin\theta}{\cos\theta}2\sin\theta\cos\theta = 2\sin^2\theta.$

8.195 $\cot\theta\sin 2\theta = 1 + \cos 2\theta$

$\blacksquare$ $\cot\theta\sin 2\theta = \dfrac{\cos\theta}{\sin\theta}(2\sin\theta\cos\theta) = 2\cos^2\theta = 1 + (2\cos^2\theta - 1) = 1 + \cos 2\theta.$

8.196 $\dfrac{1 + \cos 2\theta}{\sin 2\theta} = \cot\theta$

$$\frac{1 + \cos 2\theta}{\sin 2\theta} = \frac{1 + (2\cos^2 \theta - 1)}{2 \sin \theta \cos \theta} = \frac{2 \cos^2 \theta}{2 \sin \theta \cos \theta} = \frac{\cos \theta}{\sin \theta} = \cot \theta.$$

8.197 $\sin x = 2 \sin \frac{1}{2}x \cos \frac{1}{2}x$

▮ $\sin 2\theta = 2 \sin \theta \cos \theta$. Let $\theta = \frac{1}{2}x$. Then $\sin (2 \cdot \frac{1}{2}x) = 2 \sin \frac{1}{2}x \cos \frac{1}{2}x$, or $\sin x = 2 \sin \frac{1}{2}x \cos \frac{1}{2}x$.

8.198 $\cos (x - 3\pi/2) = -\sin x$

▮ $\cos (x - 3\pi/2) = \cos x \cos (3\pi/2) + \sin x \sin (3\pi/2) = 0 + (\sin x)(-1) = -\sin x.$

8.199 $(1 - \cos x)(\csc x + \cot x) = \sin x$

▮ $(1 - \cos x)(\csc x + \cot x) = \csc x - \cos x \csc x + \cot x - \cos x \cot x$

$$= \csc x - \cos x \cdot \frac{1}{\sin x} + \frac{\cos x}{\sin x} - \cos x \cdot \frac{\cos x}{\sin x}$$

$$= \csc x - \frac{\cos^2 x}{\sin x} = \frac{1}{\sin x} - \frac{\cos^2 x}{\sin x}$$

$$= \frac{1 - \cos^2 x}{\sin x} = \frac{\sin^2 x}{\sin x} = \sin x$$

8.200 $\left(\dfrac{1 - \cot x}{\csc x}\right)^2 = 1 - \sin 2x$

▮ $\left(\dfrac{1 - \cot x}{\csc x}\right)^2 = \dfrac{1 - 2 \cot x + \cot^2 x}{\csc^2 x} = \dfrac{\csc^2 x - 2 \cot x}{\csc^2 x}$ (since $\csc^2 x = 1 + \cot^2 x$)

$$= 1 - \frac{2 \cot x}{\csc^2 x} = 1 - \frac{2 \cos x}{\sin x} \cdot \sin^2 x = 1 - 2 \cos x \sin x = 1 - \sin 2x.$$

8.201 $\sin (x + 9\pi/2) = \cos x$

▮ $\sin (x + 9\pi/2) = \sin x \cos (9\pi/2) + \cos x \sin (9\pi/2) = \sin x \cos (\pi/2) + \cos x \sin (\pi/2) = 0 + \cos x = \cos x.$

8.202 A and B are acute; find $A + B$, given that $\tan A = \frac{1}{4}$ and $\tan B = \frac{3}{5}$.

▮ $\tan (A + B) = \dfrac{\tan A + \tan B}{1 - \tan A \tan B} = \dfrac{\frac{1}{4} + \frac{3}{5}}{1 - \frac{1}{4} \cdot \frac{3}{5}} = \dfrac{(5 + 12)/20}{1 - \frac{3}{20}} = \dfrac{\frac{17}{20}}{\frac{17}{20}} = 1.$

If $\tan (A + B) = 1$ and A and B are acute, then $A + B = 45°$.

8.203 If $\tan (x + y) = 33$ and $\tan x = 3$, show that $\tan y = 0.3$.

▮ $\tan (x + y) = 33 = \dfrac{\tan x + \tan y}{1 - \tan x \tan y} = \dfrac{3 + \tan y}{1 - 3 \tan y}$. Then $3 + \tan y = 33 - 99 \tan y$, $100 \tan y = 30$, and $\tan y = 0.3$.

8.204 Prove that $\tan 50° - \tan 40° = 2 \tan 10°$.

▮ $\tan (50° - 40°) = \dfrac{\tan 50° - \tan 40°}{1 + \tan 50° \tan 40°}$. Then $\tan 50° - \tan 40° = (1 + \tan 50° \tan 40°) \tan 10° = 2 \tan 10°$.

8.6 TRIGONOMETRIC EQUATIONS

For Probs. 8.205 to 8.215, determine whether the given number is a solution of the given equation.

8.205 $3\pi/4$, $1 + \tan x = 0$

▌ Letting $x = 3\pi/4$, we get $1 + \tan(3\pi/4) = 1 + [-\tan(\pi/4)] = 1 - 1 = 0$. Yes, it is a solution.

8.206 $\pi/2$, $1 + \tan x = 0$

▌ $\tan(\pi/2)$ is undefined. No, it is not a solution.

8.207 $\pi/3$, $2\sin x = \sqrt{3}$

▌ Letting $x = \pi/3$ gives $2\sin(\pi/3) = 2 \cdot \sqrt{3}/2 = \sqrt{3}$. Yes, it is a solution.

8.208 $\pi/6$, $2\sec x = \tan x + \cot x$

▌ $2\sec \pi/6 = \dfrac{2}{\cos(\pi/6)} = \dfrac{2}{\sqrt{3}/2} = \dfrac{4}{\sqrt{3}}$, and $\tan\dfrac{\pi}{6} + \cot\dfrac{\pi}{6} = \dfrac{1}{\sqrt{3}} + \sqrt{3} = \dfrac{1 + (\sqrt{3})^2}{\sqrt{3}} = \dfrac{4}{\sqrt{3}}$.
Yes, it is a solution.

8.209 $\pi/3$, $2\sin 2x = 1$

▌ $2\sin(2 \cdot \pi/3) = 2\sin(2\pi/3) = 2\sin(\pi/3) = 2\sqrt{3}/2 = \sqrt{3} \neq 1$. No, it is not a solution.

8.210 $\pi/4$, $2\tan^2 x + \tan x - 3 = 0$

▌ $2\tan^2(\pi/4) + \tan(\pi/4) - 3 = 2 \cdot 1^2 + 1 - 3 = 3 - 3 = 0$. Yes, it is a solution.

8.211 π, $\sin x = \cos x$

▌ $\sin \pi = 0$, and $\cos \pi = -1$, but $0 \neq -1$. No, it is not a solution.

8.212 $\pi/3$, $\csc x + \cot x = \sqrt{3}$

▌ $\csc\dfrac{\pi}{3} + \cot\dfrac{\pi}{3} = \dfrac{1}{\sin(\pi/3)} + \dfrac{\cos(\pi/3)}{\sin(\pi/3)} = \dfrac{1}{\sqrt{3}/2} + \dfrac{\frac{1}{2}}{\sqrt{3}/2} = \dfrac{2}{\sqrt{3}} + \dfrac{1}{\sqrt{3}} = \dfrac{3}{\sqrt{3}} = \dfrac{3}{\sqrt{3}}\dfrac{\sqrt{3}}{\sqrt{3}} = \dfrac{3\sqrt{3}}{3} = \sqrt{3}$.
Yes, it is a solution.

8.213 $11\pi/6$, $2\sin x - \csc x = 1$

▌ $2\sin(11\pi/6) - \csc(11\pi/6) = -2\sin(\pi/6) - [-1/\sin(\pi/6)] = -2 \cdot \frac{1}{2} + 1/\frac{1}{2} = -1 + 2 = 1$. Yes, it is a solution.

8.214 $5\pi/4$, $\tan x + 3\cot x = 4$

▌ $\tan(5\pi/4) + 3\cot(5\pi/4) = \tan(\pi/4) + 3\cot(\pi/4) = 1 + 3 = 4$. Yes, it is a solution.

8.215 $2\pi/3$, $\cos x - \sqrt{3}\sin x = 1$

▌ $\cos(2\pi/3) - \sqrt{3}\sin(2\pi/3) = -\cos(\pi/3) - \sqrt{3}\sin(\pi/3) = -\frac{1}{2} - \sqrt{3} \cdot \sqrt{3}/2 = -\frac{1}{2} - \frac{3}{2} = -\frac{4}{2} \neq 1$. No, it is not a solution.

For Probs. 8.216 to 8.240, solve the given equation, finding all solutions which lie in the interval $[0, 2\pi)$.

8.216 $1 + \cos x = 0$

▌ If $1 + \cos x = 0$, then $\cos x = -1$, or $x = \pi$. There are no other angles in $[0, 2\pi)$ whose cosines are -1.

8.217 $1 - \sin x = 0$

▮ If $1 - \sin x = 0$, then $\sin x = 1$, or $x = \pi/2$. There are no others in $[0, 2\pi)$; for example, $\sin (3\pi/2) = -1$.

8.218 $1 + \sqrt{2} \sin \theta = 0$

▮ If $1 + \sqrt{2} \sin \theta = 0$, then $\sqrt{2} \sin \theta = -1$, or $\sin \theta = -1/\sqrt{2} \, (= -\sqrt{2}/2)$. Then $\theta =$ the angle in quadrants III and IV with a $\pi/4$ reference angle ($\sin \theta < 0$), so $\theta = 5\pi/4$ or $\theta = 7\pi/4$.

8.219 $1 - \sqrt{2} \cos \theta = 0$

▮ If $1 - \sqrt{2} \cos \theta = 0$, then $-\sqrt{2} \cos \theta = -1$, $\sqrt{2} \cos \theta = 1$, or $\cos \theta = 1/\sqrt{2} = \sqrt{2}/2$. We know $\cos \theta > 0$ in quadrants I and IV. Since $\cos \theta = \sqrt{2}/2$, the reference angle is $\pi/4$, so $\theta = \pi/4$ or $\theta = 7\pi/4$.

8.220 $4 \cos^2 x - 3 = 0$

▮ $4 \cos^2 x = 3$, $\cos^2 x = \frac{3}{4}$, and $\cos x = \pm\sqrt{3}/4 = \sqrt{3}/2$. If $\cos x = -\sqrt{3}/2$, then x has a $\pi/6$ reference angle. And x must be a quadrant I or IV angle since $\cos x > 0$. $x = \pi/6$ or $x = 11\pi/6$. If $\cos x = \sqrt{3}/2$, x is in quadrant I or II. $x = 5\pi/6$ or $7\pi/6$. *Check*: $4 \cos^2 (7\pi/6) \stackrel{?}{=} 3$, $4(-\sqrt{3}/2)^2 \stackrel{?}{=} 3$, $3 = 3$. The other three solutions check as well.

8.221 $2 \sin^2 x - 1 = 0$

▮ $2 \sin^2 x = 1$, $\sin^2 x = \frac{1}{2}$, or $\sin x = \pm\sqrt{\frac{1}{2}} = \pm 1\sqrt{2} = \pm\sqrt{2}/2$. Then $\sin x = \sqrt{2}/2$ or $\sin x = -\sqrt{2}/2$. If $\sin x = \sqrt{2}/2$, x has a $\pi/4$ reference angle and is in quadrant I or II. If $\sin x = -\sqrt{2}/2$, x has the same reference angle but is in quadrant III or IV. Thus, $x = \pi/4, 3\pi/4, 7\pi/4, \pi/4$. *Check*: $2 \sin^2 (3\pi/4) \stackrel{?}{=} 1$, $2(\sqrt{2}/2)^2 \stackrel{?}{=} 1$, $1 = 1$. The others check as well.

8.222 $\sin^2 \theta = \sin \theta$

▮ $\sin^2 \theta - \sin \theta = 0$, $(\sin \theta)(\sin \theta - 1) = 0$, and $\sin \theta = 0$ or $\sin \theta = 1$. If $\sin \theta = 0$, then $\theta = 0$ or $\theta = \pi$. If $\sin \theta = 1$, then $\theta = \pi/2$ (not $3\pi/2$). Thus, $\theta = 0, \pi/2, \pi$. *Check*: $\sin^2 (\pi/2) - \sin (\pi/2) = 1 - 1 = 0$. Check the others.

8.223 $\cos^2 \theta = \cos \theta$

▮ $\cos^2 \theta - \cos \theta = 0$, $(\cos \theta)(\cos \theta - 1) = 0$, and $\cos \theta = 0$ or $\cos \theta = 1$. If $\cos \theta = 0$, then $\theta = \pi/2$ or $\theta = 3\pi/2$. If $\cos \theta = 0$, then $\theta = 0$. Thus, $\theta = 0, \pi/2, 3\pi/2$.

8.224 $2 \sin x \cos x - \cos x$

▮ $2 \sin x \cos x - \cos x = 0$, and $\cos x (2 \sin x - 1) = 0$. Then $\cos x = 0$, and $x = \pi/2$ or $3\pi/2$; or $2 \sin x = 1$, $\sin x = \frac{1}{2}$, and $x = \pi/6$ or $5\pi/6$. Thus, $x = \pi/6, \pi/2, 3\pi/2, 5\pi/6$. Check these in the original equation.

8.225 $2 \sin x \cos x = \sin x$

▮ $2 \sin x \cos x - \sin x = 0$, $\sin x (2 \cos - 1) = 0$, and $\sin x = 0$ or $2 \cos x - 1 = 0$. If $\sin x = 0$, then $x = 0$. If $\cos x = \frac{1}{2}$, $x = \pi/3$ or $5\pi/3$. Check these answers.

8.226 $1 + \cos (x/2) = 0$

▮ $\cos (x/2) = -1$. Then $x/2 = \pi$, so $x = 2\pi$. *Check*: $1 + \cos (2\pi/2) \stackrel{?}{=} 0$, $1 + \cos \pi \stackrel{?}{=} 0$, $1 + (-1) = 0$.

8.227 $1 - \sin (x/2) = 0$

▮ $\sin (x/2) = 1$, $x/2 = \pi/2$, and $x = \pi$.

8.228 $1 - \sin(x/8) = 0$

▮ $\sin(x/8) = 1$, $x/8 = \pi/2$, $2x = 8\pi$, and $x = 4\pi$. There are no solutions in the interval $[0, 2\pi)$.

8.229 $\sin 2x = 1$

▮ $\sin 2x = 1$, $2x = \pi/2$, or $x = \pi/4$.

8.230 $\sin x \cos x = 0$

▮ $\sin x \cos x = 0$, so $\sin x = 0$ or $\cos x = 0$. If $\sin x = 0$, then $x = 0$ or π. If $\cos x = 0$, then $x = \pi/2$ or $3\pi/2$. See Prob. 8.231.

8.231 $8 \sin x \cos x = 0$

▮ $2 \sin x \cos x = \sin 2x$. Then multiplying by 4, we get $8 \sin x \cos x = 4 \sin 2x$. Thus, $4 \sin 2x = 0$, $\sin 2x = 0$, and $2x = 0$, π, 2π, or 3π. If $2x = 0$, $x = 0$. If $2x = \pi$, $x = \pi/2$. If $2x = 2\pi$, $x = \pi$. If $2x = 3\pi$, $x = 3\pi/2$. See Prob. 8.230.

8.232 $\tan(x/2) = \sqrt{3}$

▮ If $\tan(x/2) = \sqrt{3}$, then $x/2 = \pi/3$ or $4\pi/3$. If $x/2 = \pi/3$, $3x = 2\pi$ and $x = 2\pi/3$. If $x/2 = 4\pi/3$, $3x = 8\pi$ and $x = 8\pi/3$. But this is extraneous since it is not in the interval $[0, 2\pi)$. Hence, the solution is $x = 2\pi/3$.

8.233 $1 + \sin x = 1 - \sin x$

▮ Then $2 \sin x = 0$, $\sin x = 0$, and $x = 0$ or π.

8.234 $\tan x + 1 = 0$

▮ $\tan x = -1$. Then $x =$ quadrant II and IV angles with a $\pi/4$ reference angle. Thus, $x = 3\pi/4$ or $7\pi/4$.

8.235 $\cos^2 x + \cos x = 0$

▮ $\cos x (\cos x + 1) = 0$. If $\cos x = 0$, then $x = \pi/2$ or $3\pi/2$. If $\cos x = -1$, then $x = \pi$.

8.236 $\sin x \tan x = \sin x$

▮ $\sin x \tan x - \sin x = 0$; $\sin x (\tan x - 1) = 0$. If $\sin x = 0$, then $x = 0$ or π. If $\tan x = 1$, then $x = \pi/4$ or $5\pi/4$.

8.237 $(2 \sin x + 1)(\cos x + 1) = 0$

▮ If $2 \sin x + 1 = 0$, then $2 \sin x = -1$ or $\sin x = -\frac{1}{2}$. In this case $x = 7\pi/6$ or $11\pi/6$. If $\cos x + 1 = 0$, then $\cos x = -1$, and in this case $x = \pi$. Then $2 \sin x = -1$, and $\sin x = -\frac{1}{2}$. Thus, $x = 7\pi/6$ or $11\pi/6$ or $\cos x + 1 = 0$, $\cos x = -1$, $x = \pi$.

8.238 $\sin x + \cos x = 0$

▮ If $\sin x + \cos x = 0$, then $\sin x = -\cos x$, $\sin x/\cos x = -1$ ($\cos x \neq 0$), and $\tan x = -1$. Thus, $x = 3\pi/4$ or $7\pi/4$. (Note that $3\pi/4 \neq 0$, and thus division by $\cos x$ was okay; check the results.)

8.239 $1 - \tan^2 x = 0$

▮ $1 - \tan^2 x = 0$ is the difference of two squares, so we have $(1 - \tan x)(1 + \tan x) = 0$. If $\tan x = 1$, then $x = \pi/4$ or $5\pi/4$. If $\tan x = -1$, then $x = 3\pi/4$ or $7\pi/4$.

8.240 $2 \cos^2 x = 1 + \sin x$

    ▌ $2\cos^2 x - 1 = \sin x$, or $\cos 2x = \sin x$. (***a***) x has a $\pi/6$ reference angle $[\cos(\pi/3) = \frac{1}{2} = \sin(\pi/6)]$. But if the signs of $\cos 2x$ and $\sin x$ must agree, then $x = \pi/6, 5\pi/6$.
(***b***) Otherwise, x is $3\pi/2$, since $\sin(3\pi/2) = -1$ and $\cos(6\pi/2) = -1$. The solutions are therefore $\pi/6, 5\pi/6, 3\pi/2$.

For Probs. 8.241 to 8.262, solve for all x such that $0 \le x < 2\pi$.

8.241 $2\sin x - 1 = 0$

    ▌ Here $\sin x = \frac{1}{2}$ and $x = \pi/6, 5\pi/6$.

8.242 $\sin x \cos x = 0$

    ▌ From $\sin x = 0$, $x = 0, \pi$; from $\cos x = 0$, $x = \pi/2, 3\pi/2$. The required solutions are $x = 0, \pi/2, \pi, 3\pi/2$.

8.243 $(\tan x - 1)(4\sin^2 x - 3) = 0$

    ▌ From $\tan x - 1 = 0$, $\tan x = 1$ and $x = \pi/4, 5\pi/4$; from $4\sin^2 x - 3 = 0$, $\sin x = \pm\sqrt{3}/2$ and $x = \pi/3, 2\pi/3, 4\pi/3, 5\pi/3$. The required solutions are $x = \pi/4, \pi/3, 2\pi/3, 5\pi/4, 4\pi/3, 5\pi/3$.

8.244 $\sin^2 x + \sin x - 2 = 0$

    ▌ Factoring, we get $(\sin x + 2)(\sin x - 1) = 0$. From $\sin x + 2 = 0$, $\sin x = -2$ and there is no solution; from $\sin x - 1 = 0$, $\sin x = 1$ and $x = \pi/2$. The required solution is $x = \pi/2$.

8.245 $3\cos^2 x = \sin^2 x$

    ▌ First solution: Replacing $\sin^2 x$ by $1 - \cos^2 x$, we have $3\cos^2 x = 1 - \cos^2 x$ or $4\cos^2 x = 1$. Then $\cos x = \pm\frac{1}{2}$, and the required solutions are $x = \pi/3, 2\pi/3, 4\pi/3, 5\pi/3$.
    Second solution: Dividing the equation by $\cos^2 x$, we have $3 = \tan^2 x$. Then $\tan x = \pm\sqrt{3}$, and the solutions above are obtained.

8.246 $2\sin x - \csc x = 1$

    ▌ Multiplying the equation by $\sin x$ gives $2\sin^2 x - 1 = \sin x$, and rearranging, we have $2\sin^2 x - \sin x - 1 = (2\sin x + 1)(\sin x - 1) = 0$. From $2\sin x + 1 = 0$, $\sin x = -\frac{1}{2}$ and $x = 7\pi/6, 11\pi/6$; from $\sin x = 1$, $x = \pi/2$.
    Check. For $x = \pi/2$, $2\sin x - \csc x = 2(1) - 1 = 1$; for $x = 7\pi/6$ and $11\pi/6$, $2\sin x - \csc x = 2(-\frac{1}{2}) - (-2) = 1$. The solutions are $x = \pi/2, 7\pi/6, 11\pi/6$.

8.247 $2\sec x = \tan x + \cot x$

    ▌ Transforming to sines and cosines and clearing fractions, we have

$$\frac{2}{\cos x} = \frac{\sin x}{\cos x} + \frac{\cos x}{\sin x} \qquad \text{or} \qquad 2\sin x = \sin^2 x + \cos^2 x = 1$$

Then $\sin x = \frac{1}{2}$ and $x = \pi/6, 5\pi/6$.

8.248 $\tan x + 3\cot x = 4$

    ▌ Multiplying by $\tan x$ and rearranging, we get $\tan^2 x - 4\tan x + 3 = (\tan x - 1)(\tan x - 3) = 0$. From $\tan x - 1 = 0$, $\tan x = 1$ and $x = \pi/4, 5\pi/4$; from $\tan x - 3 = 0$, $\tan x = 3$ and $x = 71°34'$, $251°34'$.
    Check: For $x = \pi/4$ and $5\pi/4$, $\tan x + 3\cot x = 1 + 3(1) = 4$; for $x = 71°34'$ and $251°34'$, $\tan x + 3\cot x = 3 + 3(1/3) = 4$. The solutions are $45°, 71°34', 225°, 251°34'$.

8.249 $\csc x + \cot x = \sqrt{3}$

    ▌ First solution: Writing the equation in the form $\csc x = \sqrt{3} - \cot x$ and squaring, we have $\csc^2 x = 3 - 2\sqrt{3}\cot x + \cot^2 x$.

Replacing $\csc^2 x$ by $1 + \cot^2 x$ and combining, this becomes $2\sqrt{3} \cot x - 2 = 0$. Then $\cot x = 1/\sqrt{3}$ and $x = \pi/3, 4\pi/3$.

Check: For $x = \pi/3$, $\csc x + \cot x = 2/\sqrt{3} + 1/\sqrt{3} = \sqrt{3}$; for $x = 4\pi/3$, $\csc x + \cot x = -2/\sqrt{3} + 1/\sqrt{3} \neq \sqrt{3}$. The required solution is $x = \pi/3$.

Second solution: Upon making the indicated replacement, the equation becomes $\dfrac{1}{\sin x} + \dfrac{\cos x}{\sin x} = \sqrt{3}$, and clearing fractions, we get $1 + \cos x = \sqrt{3} \sin x$. Squaring both members, we have $1 + 2 \cos x + \cos^2 x = 3 \sin^2 x = 3(1 - \cos^2 x)$ or

$$4 \cos^2 x + 2 \cos x - 2 = 2(2 \cos x - 1)(\cos x + 1) = 0$$

From $2 \cos x - 1 = 0$, $\cos x = \frac{1}{2}$ and $x = \pi/3, 5\pi/3$; from $\cos x + 1 = 0$, $\cos x = -1$ and $x = \pi$.

Now $x = \pi/3$ is the solution. The values $x = \pi$ and $5\pi/3$ are to be excluded since $\csc \pi$ is not defined while $\csc (5\pi/3)$ and $\cot (5\pi/3)$ are both negative.

8.250 $\cos x - \sqrt{3} \sin x = 1$

▌ *First solution*: Putting the equation in the form $\cos x - 1 = \sqrt{3} \sin x$ and squaring, we have $\cos^2 x - 2 \cos x + 1 = 3 \sin^2 x = 3(1 - \cos^2 x)$; then by combining and factoring, $4 \cos^2 x - 2 \cos x - 2 = 2(2 \cos x + 1)(\cos x - 1) = 0$. From $2 \cos x + 1 = 0$, $\cos x = -\frac{1}{2}$ and $x = 2\pi/3, 4\pi/3$; from $\cos x - 1 = 0$, $\cos x = 1$ and $x = 0$.

Check: For $x = 0$, $\cos x - \sqrt{3} \sin x = 1 - \sqrt{3}(0) = 1$; for $x = 2\pi/3$, $\cos x - \sqrt{3} \sin x = -\frac{1}{2} - \sqrt{3}(\sqrt{3}/2) \neq 1$; for $x = 4\pi/3$, $\cos x - \sqrt{3} \sin x = -\frac{1}{2} - \sqrt{3}(-\sqrt{3}/2) = 1$. The required solutions are $x = 0, 4\pi/3$.

Second solution: The left member of the given equation may be put in the form

$$\sin \theta \cos x + \cos \theta \sin x = \sin (\theta + x)$$

in which θ is a known angle, by dividing the given equation by $r > 0$, $\dfrac{1}{r} \cos x + \left(\dfrac{-\sqrt{3}}{r}\right) \sin x = \dfrac{1}{r}$, and setting $\sin \theta = \dfrac{1}{r}$ and $\cos \theta = \dfrac{-\sqrt{3}}{r}$. Since $\sin^2 \theta + \cos^2 \theta = 1$, $(1/r)^2 + (-\sqrt{3}/r)^2 = 1$ and $r = 2$. Now $\sin \theta = \frac{1}{2}$ and $\cos \theta = -\sqrt{3}/2$ so that the given equation may be written as $\sin (\theta + x) = \frac{1}{2}$ with $\theta = 5\pi/6$. Then $\theta + x = 5\pi/6 + x = \arcsin \frac{1}{2} = \pi/6, 5\pi/6, 13\pi/6, 17\pi/6, \ldots$ and $x = -2\pi/3, 0, 4\pi/3, 2\pi, \ldots$. As before, the required solutions are $x = 0, 4\pi/3$.

Note that r is the positive square root of the sum of the squares of the coefficients of $\cos x$ and $\sin x$ when the equation is written in the form $a \cos x + b \sin x = 1$, that is,

$$r = \sqrt{a^2 + b^2}$$

The equation will have no solution if $a/\sqrt{a^2 + b^2}$ is greater than 1 or less than -1.

8.251 $2 \cos x = 1 - \sin x$

▌ *First solution*: As in Prob. 8.250, we obtain

$$4 \cos^2 x = 1 - 2 \sin x + \sin^2 x$$

$$4(1 - \sin^2 x) = 1 - 2 \sin x + \sin^2 x$$

$$5 \sin^2 x - 2 \sin x - 3 = (5 \sin x + 3)(\sin x - 1) = 0$$

From $5 \sin x + 3 = 0$, $\sin x = -\frac{3}{5} = -0.6000$ and $x = 216°52', 323°8'$; from $\sin x - 1 = 0$, $\sin x = 1$ and $x = \pi/2$.

Check: For $x = \pi/2$, $2(0) = 1 - 1$; for $x = 216°52'$, $2(-\frac{4}{5}) \neq 1 - (-\frac{3}{5})$; for $x = 323°8'$, $2(\frac{4}{5}) = 1 - (-\frac{3}{5})$. The required solutions are $x = 90°, 323°8'$.

Second solution: Writing the equation as $2 \cos x + \sin x = 1$ and dividing by $r = \sqrt{2^2 + 1^2} = \sqrt{5}$, we have

(1) $$\dfrac{2}{\sqrt{5}} \cos x + \dfrac{1}{\sqrt{5}} \sin x = \dfrac{1}{\sqrt{5}}$$

Let $\sin \theta = 2/\sqrt{5}$ and $\cos \theta = 1/\sqrt{5}$; then (1) becomes

$$\sin \theta \cos x + \cos \theta \sin x = \sin (\theta + x) = \frac{1}{\sqrt{5}}$$

with $\theta = 63°26'$. Now $\theta + x = 63°26' + x = \arcsin (1/\sqrt{5}) = \arcsin 0.4472 = 26°34'$, $153°26'$, $386°34', \ldots$ and $x = 90°, 323°8'$ as before.

8.252 $\sin 3x = -\frac{1}{2}\sqrt{2}$

▌ Since we require x such that $0 \le x < 2\pi$, $3x$ must be such that $9 \le 3x < 6\pi$. Then $3x = 5\pi/4, 7\pi/4$, $13\pi/4, 15\pi/4, 21\pi/4, 23\pi/4$ and $x = 5\pi/12, 7\pi/12, 13\pi/12, 5\pi/4, 7\pi/4, 23\pi/4$. Each of these values is a solution.

8.253 $\cos \frac{1}{2}x = \frac{1}{2}$

▌ Since we require x such that $0 \le x < 2\pi$, $\frac{1}{2}x$ must be such that $0 \le \frac{1}{2}x < \pi$. Then $\frac{1}{2}x = \pi/3$ and $x = 2\pi/3$.

8.254 $\sin 2x + \cos x = 0$

▌ Substituting for $\sin 2x$, we have $2 \sin x \cos x + \cos x = \cos x (2 \sin x + 1) = 0$. From $\cos x = 0$, $x = \pi/2, 3\pi/2$; from $\sin x = -\frac{1}{2}$, $x = 7\pi/6, 11\pi/6$. The required solutions are $x = \pi/2, 7\pi/6$, $3\pi/2, 11\pi/6$.

8.255 $2 \cos^2 \frac{1}{2}x = \cos^2 x$

▌ Substituting $1 + \cos x$ for $2 \cos^2 \frac{1}{2}x$, the equation becomes $\cos^2 x - \cos x - 1 = 0$; then $\cos x = (1 \pm \sqrt{5})/2 = 1.6180, -0.6180$. Since $\cos x$ cannot exceed 1, we consider $\cos x = -0.6180$ and obtain the solutions $x = 128°10', 231°50'$.

Note: To solve $\sqrt{2} \cos \frac{1}{2}x = \cos x$ and $\sqrt{2} \cos \frac{1}{2}x = -\cos x$, we square and obtain the equation of this problem. The solution of the first of these equations is $231°50'$, and the solution of the second is $128°10'$.

8.256 $\cos 2x + \cos x + 1 = 0$

▌ Substituting $2 \cos^2 x - 1$ for $\cos 2x$, we have $2 \cos^2 x + \cos x = \cos x (2 \cos x + 1) = 0$. From $\cos x = 0$, $x = \pi/2, 3\pi/2$; from $\cos x = -\frac{1}{2}$, $x = 2\pi/3, 4\pi/3$. The required solutions are $x = \pi/2, 2\pi/3, 3\pi/2, 4\pi/3$.

8.257 $\tan 2x + 2 \sin x = 0$

▌ Using $\tan 2x = \dfrac{\sin 2x}{\cos 2x} = \dfrac{2 \sin x \cos x}{\cos 2x}$, we have $\dfrac{2 \sin x \cos x}{\cos 2x} + 2 \sin x = 2 \sin x \left(\dfrac{\cos x}{\cos 2x} + 1 \right) = 2 \sin x \left(\dfrac{\cos x + \cos 2x}{\cos 2x} \right) = 0$.

From $\sin x = 0$, $x = 0, \pi$; from $\cos x + \cos 2x = \cos x + 2 \cos^2 x - 1 = (2 \cos x - 1)(\cos x + 1) = 0$, $x = \pi/3, 5\pi/3$, and π. The required solutions are $x = 0, \pi/3, \pi, 5\pi/3$.

8.258 $\sin 2x = \cos 2x$

▌ First solution: Let $2x = \theta$; then we are to solve $\sin \theta = \cos \theta$ for $0 \le \theta < 4\pi$. Then $\theta = \pi/4, 5\pi/4, 9\pi/4, 13\pi/4$, and $x = \theta/2 = \pi/8, 5\pi/8, 9\pi/8, 13\pi/8$ are the solutions.

Second solution: Dividing by $\cos 2x$, we see the equation becomes $\tan 2x = 1$ for which $2x = \pi/4$, $5\pi/4, 9\pi/4, 13\pi/4$, as in the first solution.

8.259 $\sin 2x = \cos 4x$

▌ Since $\cos 4x = \cos 2(2x) = 1 - 2 \sin^2 2x$, the equation becomes

$$2 \sin^2 2x + \sin 2x - 1 = (2 \sin 2x - 1)(\sin 2x + 1) = 0$$

From $2 \sin 2x - 1 = 0$ or $\sin 2x = \frac{1}{2}$, $2x = \pi/6$, $5\pi/6$, $13\pi/6$, $17\pi/6$ and $x = \pi/12$, $5\pi/12$, $13\pi/12$, $17\pi/12$; from $\sin 2x + 1 = 0$ or $\sin 2x = -1$, $2x = 3\pi/2$, $7\pi/2$ and $x = 3\pi/4$, $7\pi/4$. All these values are solutions.

8.260 $\sin 3x = \cos 2x$

▌ To avoid the substitution for $\sin 3x$, we use one of the procedures below.
First solution: Since $\cos 2x = \sin(\frac{1}{2}\pi - 2x)$ and $\cos 2x = \sin(\frac{1}{2}\pi + 2x)$:

(a) $\sin 3x = \sin(\frac{1}{2}\pi - 2x)$, so $3x = \pi/2 - 2x$, $5\pi/2 - 2x$, $9\pi/2 - 2x$, ...
(b) $\sin 3x = \sin(\frac{1}{2}\pi + 2x)$, so $3x = \pi/2 + 2x$, $5\pi/2 + 2x$, $9\pi/2 + 2x$, ...

From **(a)**, $5x = \pi/2$, $5\pi/2$, $9\pi/2$, $13\pi/2$, $17\pi/2$ (since $5x < 10\pi$); and from **(b)**, $x = \pi/2$. The required solutions are $x = \pi/10$, $\pi/2$, $9\pi/10$, $13\pi/10$, $17\pi/10$.
Second solution: Since $\sin 3x = \cos(\frac{1}{2}\pi - 3x)$ and $\cos 2x = \cos(-2x)$:

(a) $\cos 2x = \cos(\frac{1}{2}\pi - 3x)$, so $5x = \pi/2$, $5\pi/2$, $9\pi/2$, $13\pi/2$, $17\pi/2$,
(b) $\cos(-2x) = \cos(\frac{1}{2}\pi - 3x)$, so $x = \pi/2$, as before.

8.261 $\tan 4x = \cot 6x$

▌ Since $\cot 6x = \tan(\frac{1}{2}\pi - 6x)$, we consider the equation $\tan 4x = \tan(\frac{1}{2}\pi - 6x)$. Then $4x = \pi/2 - 6x$, $3\pi/2 - 6x$, $5\pi/2 - 6x$, ..., the function $\tan\theta$ being of period π. Thus, $10x = \pi/2$, $3\pi/2$, $5\pi/2$, $7\pi/2$, $9\pi/2$, ..., $39\pi/2$, and the required solutions are $x = \pi/20$, $3\pi/20$, $\pi/4$, $7\pi/20$, ..., $39\pi/20$.

8.262 $\sin 5x - \sin 3x - \sin x = 0$

▌ Replacing $\sin 5x - \sin 3x$ by $2 \cos 4x \sin x$, we see that the given equation becomes

$$2 \cos 4x \sin x - \sin x = \sin x (2 \cos 4x - 1) = 0$$

From $\sin x = 0$, $x = 0$, π; from $2 \cos 4x - 1 = 0$ or $\cos 4x = \frac{1}{2}$, $4x = \pi/3$, $5\pi/3$, $7\pi/3$, $11\pi/3$, $13\pi/3$, $17\pi/3$, $19\pi/3$, $23\pi/3$ and $x = \pi/12$, $5\pi/12$, $7\pi/12$, $11\pi/12$, $13\pi/12$, $17\pi/12$, $19\pi/12$, $23\pi/12$. Each of the values obtained is a solution.

8.263 Solve the system: $\begin{cases} (1) & r \sin\theta = 2 \\ (2) & r \cos\theta = 3 \end{cases}$ $r > 0$, $0 \le \theta < 2\pi$

▌ Squaring the two equations and adding, we get $r^2 \sin^2\theta + r^2 \cos^2\theta = r^2 = 13$ and $r = \sqrt{13} = 3.606$. When $r > 0$, $\sin\theta$ and $\cos\theta$ are both greater than 0 and θ is acute. Dividing (1) by (2) gives $\tan\theta = \frac{2}{3} = 0.6667$ and $\theta = 33°41'$.

8.264 Solve the system: $\begin{cases} (1) & r \sin\theta = 3 \\ (2) & r = 4(1 + \sin\theta) \end{cases}$ $r > 0$, $0 \le \theta \le 2\pi$

▌ Dividing (2) by (1) gives $\dfrac{1}{\sin\theta} = \dfrac{4(1 + \sin\theta)}{3}$ or $4 \sin^2\theta + 4 \sin\theta - 3 = 0$ and

$$(2 \sin\theta + 3)(2 \sin\theta - 1) = 0$$

From $2 \sin\theta - 1 = 0$, $\sin\theta = \frac{1}{2}$, $\theta = \pi/6$ and $5\pi/6$; using (1), we see that $r(\frac{1}{2}) = 3$ and $r = 6$. Note that $2 \sin\theta + 3 = 0$ is excluded since when $r > 0$, $\sin\theta > 0$ by (1). The required solutions are $\theta = \pi/6$, $r = 6$ and $\theta = 5\pi/6$, $r = 6$.

8.265 Solve the system: $\begin{cases} (1) & \sin x + \sin y = 1.2 \\ (2) & \cos x + \cos y = 1.5 \end{cases}$ $0 \le x, y < 2\pi$

▌ Since each sum on the left is greater than 1, each of the four functions is positive and both x and y are acute. Using the appropriate formulas of Chapters 7 and 8, we obtain

(1') $\qquad\qquad\qquad\qquad 2 \sin\frac{1}{2}(x + y) \cos\frac{1}{2}(x - y) = 1.2$

(2') $\qquad\qquad\qquad\qquad 2 \cos\frac{1}{2}(x + y) \cos\frac{1}{2}(x - y) = 1.5$

Dividing (1′) by (2′) gives

$$\frac{\sin\frac{1}{2}(x+y)}{\cos\frac{1}{2}(x+y)} = \tan\frac{1}{2}(x+y) = \frac{1.2}{1.5} = 0.8000$$

and $\frac{1}{2}(x+y) = 38°40'$ since $\frac{1}{2}(x+y)$ is also acute. Substituting for $\sin\frac{1}{2}(x+y) = 0.6248$ in (1′), we have $\cos\frac{1}{2}(x-y) = 0.6/0.6248 = 0.9603$ and $\frac{1}{2}(x-y) = 16°12'$. Then $x = \frac{1}{2}(x+y) + \frac{1}{2}(x-y) = 54°52'$, and $y = \frac{1}{2}(x+y) - \frac{1}{2}(x-y) = 22°28'$.

8.266 Solve: Arccos $2x$ = Arcsin x

▮ If x is positive, α = Arccos $2x$ and β = Arcsin x terminate in quadrant I; if x is negative, α terminates in quadrant II and β terminates in quadrant IV. Thus, x must be positive.
For x positive, $\sin\beta = x$ and $\cos\beta = \sqrt{1-x^2}$. Taking the cosine of both members of the given equation, we have

$$\cos(\text{Arccos } 2x) = \cos(\text{Arcsin } x) = \cos\beta \qquad \text{or} \qquad 2x = \sqrt{1-x^2}$$

Squaring we get $4x^2 = 1 - x^2$, $5x^2 = 1$, and $x = \sqrt{5}/5 = 0.4472$.
Check: Arccos $2x$ = Arccos $0.8944 = 26°30'$ = Arcsin 0.4472, approximating the angle to the nearest 10′.

8.267 Solve: Arccos $(2x^2 - 1) = 2$ Arccos $\frac{1}{2}$

▮ Let α = Arccos $(2x^2 - 1)$ and β = Arccos $\frac{1}{2}$; then $\cos\alpha = 2x^2 - 1$ and $\cos\beta = \frac{1}{2}$. Taking the cosine of both members of the given equation, we get

$$\cos\alpha = 2x^2 - 1 = \cos 2\beta = 2\cos^2\beta - 1 = 2(\tfrac{1}{2})^2 - 1 = -\tfrac{1}{2}$$

Then $2x^2 = \frac{1}{2}$ and $x = \pm\frac{1}{2}$.
Check: For $x = \pm\frac{1}{2}$, Arccos $(-\frac{1}{2}) = 2$ Arccos $\frac{1}{2}$ or $120° = 2(60°)$.

8.268 Solve: Arccos $2x$ − Arccos $x = \pi/3$

▮ If x is positive, $0 < $ Arccos $2x < $ Arccos x; if x is negative, Arccos $2x > $ Arccos $x > 0$. Thus, x must be negative.
Let α = Arccos $2x$ and β = Arccos x; then $\cos\alpha = 2x$, $\sin\alpha = \sqrt{1-4x^2}$, $\cos\beta = x$, and $\sin\beta = \sqrt{1-x^2}$ since both α and β terminate in quadrant II. Taking the cosine of both members of the given equation, we get $\cos(\alpha-\beta) = \cos\alpha\cos\beta + \sin\alpha\sin\beta = 2x^2 + \sqrt{1-4x^2}\sqrt{1-x^2} = \cos(\pi/3) = \frac{1}{2}$ or $\sqrt{1-4x^2}\sqrt{1-x^2} = \frac{1}{2} - 2x^2$. Squaring gives $1 - 5x^2 + 4x^4 = \frac{1}{4} - 2x^2 + 4x^4$, $3x^2 = \frac{3}{4}$, and $x = -\frac{1}{2}$.
Check: Arccos (-1) − Arccos $(-\frac{1}{2}) = \pi - 2\pi/3 = \pi/3$.

8.269 Solve: Arcsin $2x = \frac{1}{4}\pi - $ Arcsin x

▮ Let α = Arcsin $2x$ and β = Arcsin x; then $\sin\alpha = 2x$ and $\sin\beta = x$. If x is negative, α and β terminate in quadrant IV; thus, x must be positive and β acute. Taking the sine of both members of the given equation, we get

$$\sin\alpha = \sin(\tfrac{1}{4}\pi - \beta) = \sin\tfrac{1}{4}\pi\cos\beta - \cos\tfrac{1}{4}\pi\sin\beta$$

or $\qquad 2x = \frac{1}{2}\sqrt{2}\sqrt{1-x^2} - \frac{1}{2}\sqrt{2}x \qquad$ and $\qquad (2\sqrt{2}+1)x = \sqrt{1-x^2}$

Squaring gives $(8 + 4\sqrt{2} + 1)x^2 = 1 - x^2$, $x^2 = 1/(10 + 4\sqrt{2})$, and $x = 0.2527$.
Check: Arcsin $0.5054 = 30°22'$, Arcsin $0.2527 = 14°38'$, and $\frac{1}{4}\pi = 14°38' + 30°22'$.

8.270 Solve: Arctan x + Arctan $(1-x)$ = Arctan $\frac{4}{3}$

▮ Let α = Arctan x and β = Arctan $(1-x)$; then $\tan\alpha = x$ and $\tan\beta = 1 - x$. Taking the tangent of both members of the given equation gives

$$\tan(\alpha + \beta) = \frac{\tan\alpha + \tan\beta}{1 - \tan\alpha\tan\beta} = \frac{x + (1 - x)}{1 - x(1 - x)} = \frac{1}{1 - x + x^2} = \tan(\text{Arctan}\,\tfrac{4}{3}) = \tfrac{4}{3}$$

Then $3 = 4 - 4x + 4x^2$, $4x^2 - 4x + 1 = (2x - 1)^2 = 0$, and $x = \tfrac{1}{2}$.

Check: $\text{Arctan}\,\tfrac{1}{2} + \text{Arctan}\,(1 - \tfrac{1}{2}) = 2\,\text{Arctan}\,0.5000 = 53°8'$ and $\text{Arctan}\,\tfrac{4}{3} = \text{Arctan}\,1.3333 = 53°8'$.

For Probs. 8.271 to 8.279, solve for the indicated variable over the given interval.

8.271 $2\sin^2 x = 3\sin x - 1$, $[0, 2\pi)$

▮ $2\sin^2 x - 3\sin x + 1 = 0$, so $(2\sin x - 1)(\sin x - 1) = 0$. If $2\sin x = 1$, then $\sin x = \tfrac{1}{2}$ and $x = \pi/6$ or $5\pi/6$. If $\sin x = 1$, then $x = \pi/2$.

8.272 $2\cos^2 x + \cos x = 1$, $[0, 2\pi)$

▮ $2\cos^2 x + \cos x - 1 = 0$, so $(2\cos x - 1)(\cos x + 1) = 0$. If $2\cos x = 1$, then $\cos x = \tfrac{1}{2}$, and $x = \pi/3$ or $5\pi/3$. If $\cos x = -1$, then $x = \pi$.

8.273 $2\cos 2x = 1$, $[0, 2\pi)$

▮ If $2\cos 2x = 1$, then $2(2\cos^2 x - 1) = 1$, $4\cos^2 x - 2 = 1$, $4\cos^2 x = 3$, $\cos^2 x = \tfrac{3}{4}$, or $\cos x = \pm\sqrt{3}/2$. If $\cos x = \sqrt{3}/2$, then $x = \pi/6$ or $11\pi/6$. If $\cos x = -\sqrt{3}/2$, then $x = 5\pi/6$ or $7\pi/6$.

8.274 $\sqrt{3}\sin x - \cos x = 0$, $(-\infty, \infty)$

▮ $\sqrt{3}\sin x = \cos x$, or $\tan x = \sqrt{3}/3$ $(= 1/\sqrt{3})$. Then x has a reference angle of $\pi/6$. The solution is $x = \pi/6 + K\pi\ \forall K \in \mathscr{Z}$.

8.275 $4\cos^2 2x - 4\cos 2x + 1 = 0$, $[0, \pi]$

▮ $(2\cos 2x - 1)(2\cos 2x - 1) = 0$. If $2\cos 2x - 1 = 0$, then $\cos 2x = \tfrac{1}{2}$. Thus, $x = \pi/6$ or $5\pi/6$ (note the domain); no other angles in $[0, \pi]$ have the property that the cosine of their double is $\tfrac{1}{2}$.

8.276 $\sin^2 \theta + 2\cos\theta = -2$, $[0°, 360°)$

▮ $\sin^2 \theta = 1 - \cos^2 \theta$. Then $1 - \cos^2 \theta + 2\cos\theta + 2 = 0$, $\cos^2 \theta - 2\cos\theta - 3 = 0$, or $(\cos\theta - 3)(\cos\theta + 1) = 0$. If $\cos\theta = 3$, there is no solution. If $\cos\theta = -1$, then $\theta = \pi = 180°$. Thus, the solution is $\theta = 180°$.

8.277 $2\sin^2 x + 3\cos x = 3$, $(-\infty, \infty)$

▮ $2(1 - \cos^2 x) + 3\cos x = 3$, $2 - 2\cos^2 x + 3\cos x = 3$, $2\cos^2 x - 3\cos x + 1 = 0$, or $(2\cos x - 1)(\cos x - 1) = 0$. If $2\cos x = 1$, then $\cos x = \tfrac{1}{2}$, and $x = \pi/3 + K(2\pi)\ \forall K \in \mathscr{Z}$. If $\cos x = 1$, then $x = K(2\pi)\ \forall K \in \mathscr{Z}$.

8.278 $\sec x + \tan x = 1$, $[0, 2\pi)$

▮ The idea here is to put the equation in a form so that the $\sec^2 \theta = 1 + \tan^2 \theta$ identity can be used. Here is one way: $\sec x = 1 - \tan x$. Squaring both sides of the equation, we get $\sec^2 x = (1 - \tan x)^2 = 1 - 2\tan x + \tan^2 x$, or $\tan^2 x + 1 = 1 - 2\tan x + \tan^2 x$. Then $2\tan x = 0$, $\tan x = 0$, and $x = 0$ or π. It is crucial to check these answers since we squared and may have picked up extraneous solutions. If $x = 0$ then $\sec 0 + \tan 0 = 1 + 0$, and $1 = 1$. If $x = \pi$, then $\sec\pi + \tan\pi = -1 + 0$, and $-1 \neq 1$. Thus, the only solution is $x = 0$.

8.279 $\sin 3x + \sin x = 0$, $[0, \pi)$

▮ $\sin 3x + \sin x = 2\sin\dfrac{3x + x}{2}\cos\dfrac{3x - x}{2} = 0$. (This is the formula for a sum! See Sec. 8.4.) Then

$2 \sin 2x \cos x = 0$. If $\sin 2x = 0$, then $2x = 0$, π, 2π, and $x = 0$, $\pi/2$, π. If $\cos x = 0$, then $x = \pi/2$. Thus, the solutions are $x = 0$, $\pi/2$, π.

For Probs. 8.280 to 8.285, solve for all real x, using a calculator. Give answers to four significant digits.

8.280 $\sin x = 0.2977$

▌ Recall that if $\sin x = r$ ($r \in [-1, 1]$), then $x = 2k\pi + \text{Sin}^{-1} r$ or $x = 2k + (\pi - \text{Sin}^{-1} r) = (2k + 1)\pi - \text{Sin}^{-1} r$ where $k \in \mathscr{L}$. Here, using a calculator, we find $\text{Sin}^{-1} 0.2977 \approx 0.3023$. Thus, $x = 2k\pi + 0.3023$ or $x = (2k + 1)\pi - 0.3023$ $\forall k \in \mathscr{L}$.

8.281 $\cos x = -0.8861$

▌ See Prob. 8.280. If $\cos x = r$, then $r = 2k\pi + \text{Cos}^{-1} r$ or $2k\pi - \text{Cos}^{-1} r$ $\forall k \in \mathscr{L}$, where $r \in (-1, 1)$. Hence, using a calculator, we get $\text{Cos}^{-1}(-0.8861) \approx 2.660$ and $x = 2k\pi + 2.660$ or $2k\pi - 2.660$ $\forall k \in \mathscr{L}$.

8.282 $\tan x = 13.08$

▌ See Probs. 8.280 and 8.281. If $\tan x = r$, $r \in \mathscr{R}$, then $x = k\pi + \text{Arctan } r$, $\forall k \in \mathscr{L}$. Here, Arctan $13.08 \approx 1.494$, and $x = k\pi + 1.494$ $\forall k \in \mathscr{L}$.

8.283 $\sec x = 8.613$

▌ $\cos x \approx -0.1161$ and $\text{Cos}^{-1}(-0.1161) \approx 1.687$. Then $x = 2k\pi + 1.687$ or $2k\pi - 1.687$ $\forall k \in \mathscr{L}$.

8.284 $\cot x = -3.478$

▌ $\tan x \approx -0.2875$ and $\text{Tan}^{-1}(-0.2875) = -2.800$. Then $x = k\pi + (-2.800)$ $\forall k \in \mathscr{L}$.

8.285 $\csc x = 42.29$

▌ $\sin x = 0.0236$ and $\text{Sin}^{-1} 0.0236 = 0.0236$. Then $x = 2k\pi + 0.0236$ or $(2k + 1)\pi - 0.0236$ $\forall k \in \mathscr{L}$.

CHAPTER 9
Additional Topics in Trigonometry

9.1 RIGHT TRIANGLES

For Probs. 9.1 to 9.9, find the remaining parts of $\triangle ABC$, where $\angle C = 90°$. Round all angles to the nearest minute and all lengths to the nearest hundredth.

9.1 $b = 12$, $c = 13$

▌ See Fig. 9.1. Then $12^2 + a^2 = 13^2$ (by the pythagorean theorem) and $a^2 = 169 - 144 = 25$, or $a = 5$. Using a calculator or a table, we get $\sin A = \text{opposite/hypotenuse} = \frac{5}{13}$, $\angle A = \text{Sin}^{-1} \frac{5}{13} \approx 22°37'$; and from a calculator or a table, $\sin B = \frac{12}{13}$, $\angle B = \text{Sin}^{-1} \frac{12}{13} \approx 67°23'$.

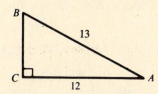

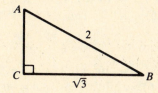

Fig. 9.1   Fig. 9.2

9.2 $a = \sqrt{3}$, $c = 2$

▌ See Fig. 9.2. Then $b^2 + (\sqrt{3})^2 = 4$, $b^2 + 3 = 4$, $b^2 = 1$, or $b = 1$. Then $\cos B = \sqrt{3}/2$. $B = \text{Cos}^{-1} \sqrt{3}/2 = 30°$. Thus, $\angle A = 60°$ ($90° = 60° + 30°$).

9.3 $a = 2$, $b = 2$

▌ See Fig. 9.3. If $a = b = 2$, $\triangle ABC$ is an isosceles right triangle and $\angle A = \angle B = 45°$. Then $c^2 = a^2 + b^2 = 2^2 + 2^2 = 8$, $c^2 = 8$, or $c = 2\sqrt{2}$.

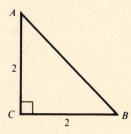

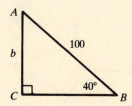

Fig. 9.3   Fig. 9.4

9.4 $c = 100$, $B = 40°$

▌ See Fig. 9.4. $\sin B = b/100$, $b = 100 \sin B \approx 64.28$ (use a calculator or table). Since $\angle B = 40°$. $\angle A = 50°$. Then $\sin 50° = a/100$, and $a = 100 \sin 50° \approx 76.60$.

9.5 $b = 14.2$, $c = 23.7$

▌ See Fig. 9.5. $a^2 + (14.2)^2 = (23.7)^2$, or $a \approx 18.98$. Then $\cos A = 14.2/23.7$, $A = \text{Cos}^{-1}(14.2/23.7) \approx 53°11'$. Thus. $B \approx 90° - (53°11') = 36°49'$.

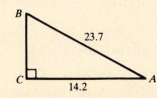

Fig. 9.5

9.6 $B = 17°50'$, $c = 3.45$

❚ See Fig. 9.6. $\angle A = 90° - 17°50' = 72°10'$. Then $\sin B = \sin 17°50' = b/3.45$, or $b = 3.45 \sin 17°50' \approx 1.06$. Thus, $\sin A = a/3.45$, $a = 3.45 \sin A \approx 3.28$.

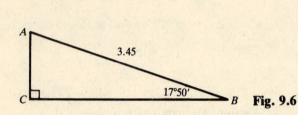

Fig. 9.6   Fig. 9.7

9.7 $a = 6$, $b = 8.46$

❚ See Fig. 9.7. $c^2 = (8.46)^2 + 6^2 = 107.57$, so $c \approx 10.37$. Then $\tan B = 8.46/6$, $\angle B = \text{Tan}^{-1}(8.46/6) \approx 54°40'$. Thus, $\angle A = 35°20'$.

9.8 $b = 10$, $c = 12.6$

❚ $a^2 + b^2 = c^2$, $a^2 + 10^2 = (12.6)^2$, $a^2 = 58.76$, or $a \approx 7.67$. Then, $\sin B = 10/12.6$, so $\angle B = \text{Sin}^{-1}(10/12.6) \approx 52°30'$. Thus, $\angle A = 90° - \angle B = 37°30'$. Finally, $\tan B = 10/a$, $a \tan B = 10$, $a = 10/\tan B = 10/\tan 52°30'$, or $a \approx 7.67$ (using a calculator).

9.9 $a = 2.42$, $c = 3.22$

❚ $a^2 + b^2 = c^2$, $b^2 = c^2 - a^2 = (3.22)^2 - (2.42)^2 \approx 16.22$, or $b \approx 4.03$. Then $\cos B = 2.42/3.22$, $\angle B = \text{Cos}^{-1}(2.42/3.22) \approx 41°20'$. Thus, $\angle A = 90° - 41°20' = 48°40'$.

For Probs. 9.10 to 9.12, find the perimeter of the described regular polygon.

9.10 A hexagon inscribed in a circle of radius 5 m

❚ See Fig. 9.8. Then $\angle AOB = 360°/6 = 60°$, so $\angle AOD = 60°/2 = 30°$. Since $\sin 30° = \sin(\angle AOD) = AD/5$, $AD = 5 \sin 30° = 5 \cdot \frac{1}{2} = 2.5$ m. Thus, $AB = 5$ m, and the perimeter $= P = 6 \cdot 5 = 30$ m.

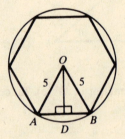

Fig. 9.8

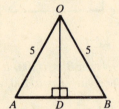

Fig. 9.9

9.11 An octagon inscribed in a circle of radius 5 m

❚ See Fig. 9.9 and Prob. 9.10. Here, the central angle, $\angle AOB$, $= 360°/8 = 45°$. Then $\frac{1}{2}$(central angle) $= 45°/2 = 22.5°$, and $\angle AOD = 22.5°$. Since $\sin 22.5° = AD/5$, $AD = 5 \sin 22.5° \approx 1.91$ m. Thus, $AB \approx 3.82$ m, and $P = 30.56$ m.

9.12 A pentagon circumscribed about a circle of radius 5 m

❚ See Fig. 9.10. $\angle AOG = \frac{1}{2} \cdot (360°/5) = 36°$. Then $\angle AGO = 90°$ (it is formed by a tangent and radius), and $AO = 5$ m (radius of the circle). Thus, $\sin 36° = AG/5$, and $AG = 5 \sin 36° \approx 2.94$ m, $AE \approx 5.88$ m, $P \approx 29.4$ m.

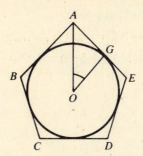

Fig. 9.10

9.13 If a train climbs at a constant angle of 1°23′, how many vertical feet has it climbed after going 1 mi?

▌ See Fig. 9.11. Let x = number of feet the train climbs. Then $\tan 1°23′ = x/5280$, and $x = 5280 \tan 1°23′ \approx 127.5$ ft.

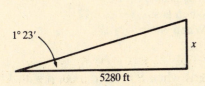

Fig. 9.11

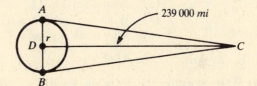

Fig. 9.12

9.14 Find the diameter of the moon (to the nearest mile) if at 239,000 mi from earth it subtends an angle of 32′ relative to an observer on the earth.

▌ See Fig. 9.12, where $\angle ACB = 32′$ and $\angle ACD = 16′$. The diameter of the moon = $2r$. Then $\tan 16′ = r/239,000$, and $r = (\tan 16′) \cdot 239,000$. Thus, the moon's diameter = $2 \cdot (\tan 16′) \cdot 239,000 \approx 2225$ mi.

9.15 An object 4 ft tall casts a 3 ft shadow when the angle of elevation of the sun is $\theta°$. Find θ to the nearest degree.

▌ See Fig. 9.13. Then $\tan \theta = \frac{4}{3}$, and $\theta = \text{Arctan } \frac{4}{3} \approx 59°$.

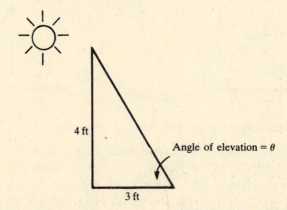

Angle of elevation = θ

Fig. 9.13

9.16 Find the angle formed by the intersection of a diagonal of the face of a cube with a diagonal of the cube drawn from the same vertex.

▌ See Fig. 9.14a. Let a = the edge of the cube, b = face diagonal, and d = cube's diagonal. Then $a^2 + a^2 = b^2$, $b^2 = 2a^2$, or $b = \sqrt{2}a$. Also, $a^2 + b^2 = d^2$, $a^2 + (\sqrt{2}a)^2 = d^2$, $a^2 + 2a^2 = d^2$, $d^2 = 3a^2$, or $d = \sqrt{3}a$. See Fig. 9.14b. Then $\sin \theta = a/d = a/(\sqrt{3}a) = 1/\sqrt{3}$, and $\theta = \text{Sin } (1/\sqrt{3}) \approx 35°$ (to the nearest degree).

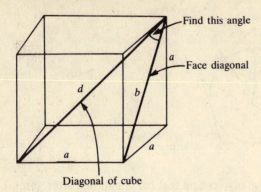

Find this angle

a — Face diagonal

d

b

a

a

Diagonal of cube

Fig. 9.14a

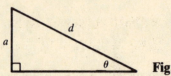

d

a

θ

Fig. 9.14b

9.17 If A is an acute angle:

(**a**) Why is $\sin A < 1$?	(**d**) Why is $\sin A < \tan A$?
(**b**) When is $\sin A = \cos A$?	(**e**) When is $\sin A < \cos A$?
(**c**) Why is $\sin A < \csc A$?	(**f**) When is $\tan A > 1$?

❚ In any right triangle ABC:

(**a**) Side $a <$ side c; therefore $\sin A = a/c < 1$. (**b**) $\sin A = \cos A$ when $a/c = b/c$; then $a = b$, $A = B$, and $A = 45°$. (**c**) $\sin A < 1$ (above) and $\csc A = 1/\sin A > 1$. (**d**) $\sin A = a/c$, $\tan A = a/b$, and $b < c$; therefore $a/c < a/b$ or $\sin A < \tan A$. (**e**) $\sin A < \cos A$ when $a < b$; then $A < B$ or $A < 90° - A$, and $A < 45°$. (**f**) $\tan A = a/b > 1$ when $a > b$; then $A > B$ and $A > 45°$.

9.18 Find the values of the trigonometric functions of 45°.

❚ See Fig. 9.15. In any isosceles right triangle ABC, $A = B = 45°$ and $a = b$. Let $a = b = 1$; then $c = \sqrt{1 + 1} = \sqrt{2}$ and

$$\sin 45° = 1/\sqrt{2} = \tfrac{1}{2}\sqrt{2} \qquad \cot 45° = 1$$
$$\cos 45° = 1/\sqrt{2} = \tfrac{1}{2}\sqrt{2} \qquad \sec 45° = \sqrt{2}$$
$$\tan 45° = 1/1 = 1 \qquad \csc 45° = \sqrt{2}$$

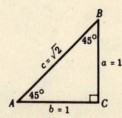

B

45°

$c = \sqrt{2}$

$a = 1$

45°

A

$b = 1$

C

Fig. 9.15

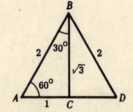

B

2   30°   2

$\sqrt{3}$

60°

A   1   C   D

Fig. 9.16

9.19 Find the values of the trigonometric functions of 30° and 60°.

❚ See Fig. 9.16. In any equilateral triangle ABD, each angle is 60°. The bisector of any angle, as B, is the perpendicular bisector of the opposite side. Let the sides of the equilateral triangle be 2 units long. Then in the right triangle ABC, $AB = 2$, $AC = 1$, and $BC = \sqrt{2^2 - 1^2} = \sqrt{3}$.

$$\sin 30° = 1/2 = \cos 60° \qquad \cot 30° = \sqrt{3} = \tan 60°$$
$$\cos 30° = \sqrt{3}/2 = \sin 60° \qquad \sec 30° = 2/\sqrt{3} = 2\sqrt{3}/3 = \csc 60°$$
$$\tan 30° = 1/\sqrt{3} = \sqrt{3}/3 = \cot 60° \qquad \csc 30° = 2 = \sec 60°$$

9.20 When the sun is 20° above the horizon, how long is the shadow cast by a building 150 ft high?

❚ In Fig. 9.17, $A = 20°$ and $CB = 150$. Then $\cot A = AC/CB$ and $AC = CB \cot A = 150 \cot 20° = 150(2.7) = 405$ ft.

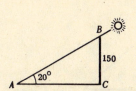

Fig. 9.17

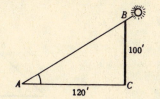

Fig. 9.18

9.21 A tree 100 ft tall casts a shadow 120 ft long. Find the angle of elevation of the sun.

▮ In Fig. 9.18, $CB = 100$ and $AC = 120$. Then $\tan A = CB/AC = 100/120 = 0.83$ and $A = 40°$.

9.22 A ladder leans against the side of a building with its foot 12 ft from the building. How far from the ground is the top of the ladder, and how long is the ladder if it makes an angle of 70° with the ground?

▮ From Fig. 9.19, $\tan A = CB/AC$; then $CB = AC \tan A = 12 \tan 70° = 12(2.7) = 32.4$. The top of the ladder is 32 ft above the ground. And $\sec A = AB/AC$; then $AB = AC \sec A = 12 \sec 70° = 12(2.9) = 34.8$. The ladder is 35 ft long.

Fig. 9.19

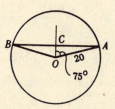

Fig. 9.20

9.23 Find the length of the chord of a circle of radius 20 m subtended by a central angle of 150°.

▮ In Fig. 9.20, OC bisects $\angle AOB$. Then $BC = AC$ and OAC is a right triangle. In $\triangle OAC$, $\sin \angle COA = AC/OA$ and $AC = OA \sin \angle COA = 20 \sin 75° = 20(0.97) = 19.4$. Then $BA = 38.8$, and the length of the chord is 39 m.

9.24 Find the height of a tree if the angle of elevation of its top changes from 20° to 40° as the observer advances 75 ft toward its base. See Fig. 9.21.

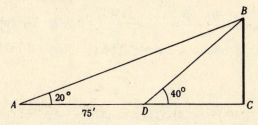

Fig. 9.21

▮ In right triangle ABC, $\cot A = AC/CB$; then $AC = CB \cot A$ or $DC + 75 = CB \cot 20°$. In right triangle DBC, $\cot D = DC/CB$; then $DC = CB \cot 40°$. Then $DC = CB \cot 20° - 75 = CB \cot 40°$, $CB(\cot 20° - \cot 40°) = 75$, $CB(2.7 - 1.2) = 75$, and $CB = 75/1.5 = 50$ ft.

9.2 LAW OF SINES

For Probs. 9.25 to 9.36, find the measure of the indicated angle or the length of the indicated side. Refer to Fig. 9.22, and use a calculator.

9.25 $\beta = 12°40'$, $\gamma = 100°$, $b = 13.1$; find a.

▮ $\alpha = 180° - 12°40' - 100° = 67°20'$. The law of sines is $\dfrac{\sin \alpha}{a} = \dfrac{\sin \beta}{b}$. Then
$$a = \frac{b \sin \alpha}{\sin \beta} = \frac{13.1 \sin 67°20'}{\sin 12°40'} = 55.1.$$

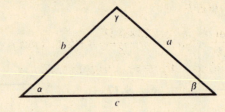

Fig. 9.22

9.26 Find c in Prob. 9.25.

∎ $\dfrac{\sin \gamma}{c} = \dfrac{\sin \beta}{b}$, so $c = \dfrac{b \sin \gamma}{\sin \beta} = \dfrac{13.1 \sin 100°}{\sin 12°40'} = 58.8$.

9.27 $\beta = 27°30'$, $\gamma = 54°30'$, $a = 9.27$; find b.

∎ $\alpha = 180° - 54°30' - 27°30' = 98°$. Then from $\dfrac{\sin \beta}{b} = \dfrac{\sin \alpha}{a}$, $b = \dfrac{a \sin \beta}{\sin \alpha} = \dfrac{9.27 \sin 27°30'}{\sin 98°} = 4.32$.

9.28 Find c in Prob. 9.27.

∎ $\dfrac{\sin \gamma}{c} = \dfrac{\sin \alpha}{a}$, so $c = \dfrac{a \sin \gamma}{\sin \alpha} = \dfrac{9.27 \sin 54°30'}{\sin 98°} = 7.62$.

9.29 $\alpha = 25°50'$, $a = 65$, $b = 105$, β obtuse; find β.

∎ $\dfrac{\sin \beta}{b} = \dfrac{\sin \alpha}{a}$, so $\sin \beta = \dfrac{b \sin \alpha}{a} = \dfrac{105 \sin 25°50'}{65} = 0.7039$. We choose $180 - \text{Sin}^{-1} 0.7039$, not $\text{Sin}^{-1} 0.7039$, since β is obtuse. Then $180 - \text{Sin}^{-1} 0.7039 = 135°15'$ ($= 135.25°$).

9.30 Find c in Prob. 9.29.

∎ $\gamma = 180° - (25°50' + 135°15') = 18°55'$. Then from $\dfrac{\sin \alpha}{a} = \dfrac{\sin \gamma}{c}$, $c = \dfrac{a \sin \gamma}{\sin \alpha} = \dfrac{65 \sin 18°55'}{\sin 25°50'} = 48.36$.

9.31 $\beta = 31°40'$, $a = 12$, $b = 8$, α acute; find α.

∎ $\dfrac{\sin \alpha}{a} = \dfrac{\sin \beta}{b}$, so $\sin \alpha = \dfrac{a \sin \beta}{b} = \dfrac{12 \sin 31°40'}{8} = 0.7875$. We choose $\alpha = \text{Sin}^{-1} 0.7875$, *not* $\alpha = 180 - \text{Sin}^{-1} 0.7875$ since α is acute. Then $\alpha = \text{Sin}^{-1} 0.7875 = 52°$.

9.32 Find c in Prob. 9.31.

∎ $\gamma = 180° - (52° + 31°40') = 96°20'$. Then from $\dfrac{\sin \beta}{b} = \dfrac{\sin \gamma}{c}$, $c = \dfrac{b \sin \gamma}{\sin \beta} = \dfrac{8 \sin 96°20'}{\sin 31°40'} = 15.1$.

9.33 $\alpha = 50$, $c = 40$, $\gamma = 30°$; find α.

∎ Here α can be obtuse or acute. Then from $\sin \alpha / a = \sin \gamma / c$, $\sin \alpha = a \sin \gamma / c = 50 \sin 30° / 40 = 0.6250$. Thus, (1) $\alpha = \text{Sin}^{-1} 0.6250 = 39°$, or (2) $\alpha = 180° - \text{Sin}^{-1} 0.6250 = 141°$.

9.34 Find b in Prob. 9.33 if $\alpha < 90°$.

∎ $\beta = 180° - 39° - 30° = 111°$. Then from $\dfrac{\sin \beta}{b} = \dfrac{\sin \gamma}{c}$, $b = \dfrac{40 \sin 111°}{\sin 30°} = 75$.

9.35 Find b in Prob. 9.33 if $\alpha > 90°$.

∎ $\beta = 180° - 141° - 30° = 9°$. Then from $\dfrac{\sin \beta}{b} = \dfrac{\sin \gamma}{c}$, $b = \dfrac{40 \sin 9°}{\sin 30°} = 13$.

9.36 $a = 14$, $b = 23$, $\alpha = 41°$; find β.

▮ Then from $\dfrac{\sin \alpha}{a} = \dfrac{\sin \beta}{b}$, $\sin \beta = \dfrac{23 \sin 41°}{14} = 1.078$. Thus $\beta = \text{Sin}^{-1} 1.078$; there is no solution. Draw the triangle; do you see why there is no solution?

9.3 LAW OF COSINES

For Probs. 9.37 to 9.48, find the indicated piece of information concerning the triangle in Fig. 9.22. Use a calculator.

9.37 $\alpha = 50°40'$, $b = 7.03$, $c = 7.00$; find a.

▮ From the law of cosines $a^2 = b^2 + c^2 - 2bc \cos \alpha$, we have $a^2 = (7.03)^2 + (7)^2 - 2(7.03)(7) \cos 50°40' = 36.03925 \cdots$, or $a = 6.00$.

9.38 Find β in Prob. 9.37.

▮ From the law of sines $a/\sin \alpha = b/\sin \beta$, $\sin \beta = 7.03 \sin 50°40'/6 = 0.9063$. Then $\beta = \text{Sin}^{-1} 0.9063 = 65°0'$. (Do *not* choose $180° - \text{Sin}^{-1} 0.9063$ since there cannot be two obtuse angles in a triangle.)

9.39 $\gamma = 120°20'$, $a = 5.73$, $b = 10.2$; find c.

▮ $c^2 = a^2 + b^2 - 2ab \cos \gamma = (5.73)^2 + (10.2)^2 - 2(5.73)(10.2) \cos (120°20') = 195.90686 \cdots$. Thus, $c = 14.0$.

9.40 Find β in Prob. 9.39.

▮ From $\sin \beta/b = \sin \gamma/c$, $\sin \beta = b \sin \gamma/c$. Then $\beta = \text{Sin}^{-1} 0.6288 = 39°0'$. (Do *not* use $180° - \text{Sin}^{-1} 0.6288$ since γ is obtuse.)

9.41 Find α in Prob. 9.39.

▮ We do not need the law of sines or cosines here; $\alpha = 180° - (\beta + \gamma) = 180° - 39°0' - 120°20' = 20°40'$.

9.42 $a = 4.00$, $b = 10.0$, $c = 9.00$; find α.

▮ $a^2 = b^2 + c^2 - 2bc \cos \alpha$. Then $\cos \alpha = [(10.0)^2 + (9.00)^2 - (4.00)^2]/180 = 0.9167$. Thus, $\alpha = \text{Cos}^{-1} 0.9167 = 23°30'$. (Note that, given the three lengths we have, α and γ are both less than 90°.)

9.43 Find γ in Prob. 9.42.

▮ From $\sin \gamma/c = \sin \alpha/a$, $\sin \gamma = 9.00 \sin 23°30'/4.00 = 0.8972$. Thus, $\gamma = \text{Sin}^{-1} 0.8972$ ($\gamma < 90°$) $= 63°50'$.

9.44 Find γ in Prob. 9.42, using the law of cosines.

▮ $\cos \gamma = [(4.00)^2 + (10.0)^2 - (9.00)^2]/[2(4)(10)] = 0.4375$. Thus $\gamma = \text{Cos}^{-1} 0.4375 = 64°0'$. (Do not be concerned by the small difference in results; rounding causes the discrepancy.)

9.45 $a = 10.5$, $b = 20.7$, $c = 12.2$; find α.

▮ $a^2 = b^2 + c^2 - 2bc \cos \alpha$. Then $\cos \alpha = \dfrac{b^2 + c^2 - a^2}{2bc} = \dfrac{(20.7)^2 + (12.2)^2 - (10.5)^2}{2(20.7)(12.2)} = 0.9248$. Thus, $\alpha = \text{Cos}^{-1} 0.9248 = 22°20'$. (Here, again, α and γ are acute. Draw a picture of the triangle whose sides are the lengths given in this problem.)

9.46 Find γ in Prob. 9.45.

9.61 (3, 0)

▌ $\tan \theta = \frac{0}{3} = 0$; $\theta = 0°$.

For Probs. 9.62 to 9.68, let $\mathbf{a} = (1, 0)$, $\mathbf{b} = (3, 0)$, $\mathbf{c} = (4, 6)$, $\mathbf{d} = (4, 9)$, $\mathbf{e} = (1, 6)$. Find each of the following.

9.62 2**a**

▌ Recall that $k(x, y) = (kx, ky)$. Thus $2\mathbf{a} = 2(1, 0) = (2, 0)$.

9.63 $\frac{2}{3}\mathbf{b}$

▌ $\frac{2}{3}\mathbf{b} = \frac{2}{3}(3, 0) = (\frac{2}{3} \cdot 3, \frac{2}{3} \cdot 0) = (2, 0) = 2\mathbf{a}$ (see Prob. 9.62).

9.64 **a** + **c**

▌ Recall that $(x_1, y_1) + (x_2, y_2) = (x_1 + x_2, y_1 + y_2)$. Thus, $\mathbf{a} + \mathbf{c} = (1, 0) + (4, 6) = (1 + 4, 0 + 6) = (5, 6)$.

9.65 **a** − **d**

▌ Recall that $\mathbf{a} - \mathbf{b} = \mathbf{a} + (-\mathbf{b})$. Here, $\mathbf{d} = (4, 9)$, so $-\mathbf{d} = -1\mathbf{d} = (-4, -9)$ and $\mathbf{a} - \mathbf{d} = \mathbf{a} + (-\mathbf{d}) = (1, 0) + (-4, -9) = (1 - 4, 0 - 9) = (-3, -9)$.

9.66 **e** − 2**c**

▌ $\mathbf{e} - 2\mathbf{c} = (1, 6) + (-2)(4, 6) = (1, 6) + (-8, -12) = (-7, -6)$.

9.67 −6(**b** + 2**d**)

▌ $-6(\mathbf{b} + 2\mathbf{d}) = -6\mathbf{b} - 12\mathbf{d}$ (distributive law) $= -6(3, 0) - 12(4, 9) = (-18, 0) + (-48, -108) = (-66, -108)$.

9.68 **a** · **c**

▌ If $\mathbf{x} = (x_1, y_1)$ and $\mathbf{y} = (x_2, y_2)$, then $\mathbf{x} \cdot \mathbf{y} = x_1 x_2 + y_1 y_2 =$ the dot product. Here, $\mathbf{a} \cdot \mathbf{c} = (1, 0) \cdot (4, 6) = 1 \cdot 4 + 0 \cdot 6 = 4 + 0 = 4$.

For Probs. 9.69 to 9.74, prove the given statement.

9.69 $\mathbf{u} \cdot \mathbf{u} = |\mathbf{u}|^2$

▌ Let $\mathbf{u} = (a, b)$. Then $\mathbf{u} \cdot \mathbf{u} = (a, b) \cdot (a, b) = a \cdot a + b \cdot b = a^2 + b^2$ or $|\mathbf{u}|^2 = (\sqrt{a^2 + b^2})^2 = a^2 + b^2$, and $\mathbf{u} \cdot \mathbf{u} = |\mathbf{u}|^2$.

9.70 $\mathbf{u} \cdot \mathbf{v} = \mathbf{v} \cdot \mathbf{u}$

▌ Let $\mathbf{u} = (a, b)$, $\mathbf{v} = (c, d)$. Then $\mathbf{u} \cdot \mathbf{v} = ac + bd = ca + db$ (multiplication is commutative) $= (c, d) \cdot (a, b) = \mathbf{v} \cdot \mathbf{u}$. Thus, $\mathbf{u} \cdot \mathbf{v} = \mathbf{v} \cdot \mathbf{u}$.

9.71 $\mathbf{u} \cdot (\mathbf{v} + \mathbf{w}) = \mathbf{u} \cdot \mathbf{v} + \mathbf{u} \cdot \mathbf{w}$

▌ Let $\mathbf{u} = (a, b)$, $\mathbf{v} = (c, d)$, $\mathbf{w} = (e, f)$. Then $\mathbf{u} \cdot (\mathbf{v} + \mathbf{w}) = (a, b) \cdot (c + e, d + f) = a(c + e) + b(d + f) = ac + ae + bd + bf = ac + bd + ae + bf = \mathbf{u} \cdot \mathbf{v} + \mathbf{u} \cdot \mathbf{w}$.

9.72 $(k\mathbf{u}) \cdot \mathbf{v} = k(\mathbf{u} \cdot \mathbf{v})$

▌ $k\mathbf{u} = (ka, kb)$ where $\mathbf{u} = (a, b)$. Let $\mathbf{v} = (c, d)$. Then $(k\mathbf{u}) \cdot \mathbf{v} = (ka, kb) \cdot (c, d) = (ka)c + (kb)d$. Since $\mathbf{u} \cdot \mathbf{v} = ac + bd$, we have $k(\mathbf{u} \cdot \mathbf{v}) = k(ac + bd) = k(ac) + k(bd) = (ka)c + (kb)d$ (associative law for multiplication), and $(k\mathbf{u}) \cdot \mathbf{v} = k(\mathbf{u} \cdot \mathbf{v})$.

9.73 If $\mathbf{u} \perp \mathbf{v}$, then $\mathbf{u} \cdot \mathbf{v} = 0$ ($\mathbf{u}, \mathbf{v} \neq \mathbf{0}$).

▮ If $\mathbf{u} \perp \mathbf{v}$, then $\mathbf{u} \cdot \mathbf{v} = |\mathbf{u}|\,|\mathbf{v}| \cos \theta$ (θ = angle between $\mathbf{u}, \mathbf{v}$) = $|\mathbf{u}|\,|\mathbf{v}| \cos (\pi/2)(\mathbf{u} \perp \mathbf{v}) = |\mathbf{u}|\,|\mathbf{v}| \cdot 0 = 0$.

9.74 If $\mathbf{u} \cdot \mathbf{v} = 0$, then $\mathbf{u} \perp \mathbf{v}$ ($\mathbf{u}, \mathbf{v} \neq \mathbf{0}$).

▮ If $\mathbf{u} \cdot \mathbf{v} = 0$, then $|\mathbf{u}|\,|\mathbf{v}| \cos \theta = 0$, $\cos \theta = 0$ ($\mathbf{u}, \mathbf{v} \neq \mathbf{0}$), $\theta = \pi/2$, and $\mathbf{u} \perp \mathbf{v}$.

For Probs. 9.75 to 9.78, prove the given statement.

9.75 $\mathbf{a} + \mathbf{c} = \mathbf{c} + \mathbf{a}$

▮ Let $\mathbf{a} = (x_1, y_1)$ and $\mathbf{c} = (x_2, y_2)$. Then $\mathbf{a} + \mathbf{c} = (x_1, y_1) + (x_2, y_2) = (x_1 + x_2, y_1 + y_2) = (x_2 + x_1, y_2 + y_1)$ (by commutative law for addition) = $\mathbf{c} + \mathbf{a}$.

9.76 $\mathbf{a} - \mathbf{c} = -(\mathbf{c} - \mathbf{a})$

▮ Let $\mathbf{a} = (x_1, y_1)$ and $\mathbf{c} = (x_2, y_2)$. Then $\mathbf{a} - \mathbf{c} = (x_1 - x_2, y_1 - y_2)$ and $-(\mathbf{c} - \mathbf{a}) = -1(x_2 - x_1, y_2 - y_1) = (x_1 - x_2, y_1 - y_2) = \mathbf{a} - \mathbf{c}$.

9.77 $\mathbf{u} + \mathbf{0} = \mathbf{u}$

▮ Let $\mathbf{u} = (x, y)$. Then $\mathbf{u} + \mathbf{0} = (x, y) + (0, 0) = (x + 0, y + 0) = (x, y) = \mathbf{u}$.

9.78 $\mathbf{u} + (-\mathbf{u}) = \mathbf{0}$

▮ Let $\mathbf{u} = (a, b)$. Then $\mathbf{u} + (-\mathbf{u}) = (a, b) + (-a, -b) = (a - a, b - b) = (0, 0) = \mathbf{0}$.

For Probs. 9.79 and 9.80, rewrite the given vector in terms of $\mathbf{i}$ and $\mathbf{j}$.

9.79 $(4, 7)$

▮ $\mathbf{i} = (1, 0)$ and $\mathbf{j} = (0, 1)$. Then $(4, 7) = 4(1, 0) + 7(0, 1) = 4\mathbf{i} + 7\mathbf{j}$.

9.80 5, S30°E

▮ See Fig. 9.30. The length of line $AB = |AB| = 5$. Then B has coordinates $(\frac{5}{2}, -\frac{5}{2}\sqrt{3})$, and so the vector is $(\frac{5}{2}, -\frac{5}{2}\sqrt{3}) = \frac{5}{2}(1, 0) - \frac{5}{2}\sqrt{3}(0, 1) = \frac{5}{2}\mathbf{i} - \frac{5}{2}\sqrt{3}\mathbf{j}$.

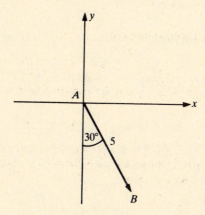

Fig. 9.30

For Probs. 9.81 and 9.82, find the unit vector in the direction of the given vector.

9.81 $(6, 4)$

▮ $|(6, 4)| = \sqrt{36 + 16} = \sqrt{52}$. Then $|\mathbf{A}| = 1$, where $\mathbf{A} = (6/\sqrt{52}, 4/\sqrt{52})$, and the direction of $\mathbf{A}$ = the direction of $(6, 4)$.

9.82 (e, f)

▮ $|(e, f)| = \sqrt{e^2 + f^2}$. Let $\mathbf{A} = \left(\dfrac{e}{\sqrt{e^2+f^2}}, \dfrac{f}{\sqrt{e^2+f^2}}\right)$. Then $|\mathbf{A}| = 1$, and the direction of $\mathbf{A}$ = the direction of (e, f).

9.83 Find the angle between the vectors $(1, 0)$ and $(\sqrt{2}, \sqrt{2})$.

▮ $(1, 0) \cdot (\sqrt{2}, \sqrt{2}) = 1 \cdot \sqrt{2} + 0 \cdot \sqrt{2} = \sqrt{2}$. Also $(1, 0) \cdot (\sqrt{2}, \sqrt{2}) = |(1, 0)| \, |(\sqrt{2}, \sqrt{2})| \cos \theta$. Thus $\sqrt{2} = 1 \cdot 2 \cos \theta$, $\cos \theta = \sqrt{2}/2$, and $\theta = \pi/4$.

9.84 An automobile weighing 2000 lb is standing on a smooth driveway that is inclined 5.0° with the horizontal. Find the force parallel to the driveway necessary to keep the car from rolling down the hill. Neglect all friction.

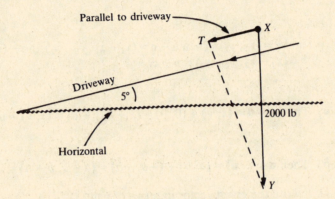

Fig. 9.31

▮ See Fig. 9.31. We need the component of ray XY parallel to the driveway; i.e., we need $|TX|$. Since $|TX|/2000 = \sin 5°$, $|TX| = 2000 \sin 5° = 174.5$ lb. See Prob. 9.98.

For Probs. 9.85 to 9.87, answer true or false, and justify your answer.

9.85 If $\mathbf{u} = \mathbf{v}$, then $|\mathbf{u}| = |\mathbf{v}|$.

▮ True. Two vectors are equal if and only if their magnitude and direction are the same. Thus, if $\mathbf{u} = \mathbf{v}$, $|\mathbf{u}|$ must equal $|\mathbf{v}|$.

9.86 Vector $(1, 1)$ = ray $\mathbf{AB}$ where the coordinates of $A = (1, 0)$ and the coordinates of $B = (2, 1)$.

▮ $|AB| = \sqrt{(2 - 1)^2 + (1 - 0)^2} = \sqrt{1 + 1} = \sqrt{2}$. Thus, $|AB| = |(1, 1)|$. Are they in the same direction? Since the slope of the line connecting the points $(0, 0)$ and $(1, 1)$ = the slope of line AB, the directions are the same, and the vector $(1, 1) = \mathbf{AB}$. True.

9.87 If $\mathbf{a} \cdot \mathbf{b} = \mathbf{c} \cdot \mathbf{d}$, then $\mathbf{a} = \mathbf{c}$ and $\mathbf{b} = \mathbf{d}$ or $\mathbf{a} = \mathbf{d}$ and $\mathbf{b} = \mathbf{c}$.

▮ False. Let $\mathbf{a} = (1, 0)$, $\mathbf{b} = (2, 0)$. Then $\mathbf{a} \cdot \mathbf{b} = 2 + 0 = 2$. Let $\mathbf{c} = (1, 1) = \mathbf{d}$. Then $\mathbf{c} \cdot \mathbf{d} = (1, 1) \cdot (1, 1) = 1 + 1 = 2$. But $\mathbf{a} \neq \mathbf{d}$ and $\mathbf{a} \neq \mathbf{c}$.

9.88 A motorboat moves in the direction N40°E for 3 h at 20 mi/h. How far north and how far east does it travel? See Fig. 9.32.

▮ Suppose the boat leaves A. Using the north-south line through A, draw the half-line AD so that the bearing of D from A is N40°E. On AD locate B such that $AB = 3(20) = 60$ mi. Through B pass a line perpendicular to the line NAS, meeting it in C. In right triangle ABC,

$$AC = AB \cos A = 60 \cos 40° = 60(0.7660) = 45.96$$

and

$$CB = AB \sin A = 60 \sin 40° = 60(0.6428) = 38.57$$

The boat travels 46 mi north and 39 mi east.

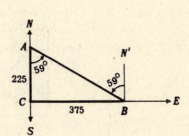

Fig. 9.32   **Fig. 9.33**

9.89 Three ships are situated as follows: A is 225 mi due north of C, and B is 375 mi due east of C. What is the bearing (*a*) of B from A, (*b*) of A from B?

▌ In right triangle ABC in Fig. 9.33, $\tan\angle CAB = 375/225 = 1.6667$ and $\angle CAB = 59°0'$.
(*a*) The bearing of B from A (angle SAB) is S59°0′E.
(*b*) The bearing of A from B (angle $N'BA$) is N59°0′W.

9.90 Three ships are situated as follows: A is 225 mi west of C while B, due south of C, bears S25°10′E from A. (*a*) How far is B from A? (*b*) How far is B from C? (*c*) What is the bearing of A from B?

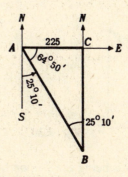

Fig. 9.34

▌ From Fig. 9.34, $\angle SAB = 25°10'$ and $\angle BAC = 64°50'$. Then

$$AB = AC \sec \angle BAC = 225 \sec 64°50' = 225(2.3515) = 529.1$$

or

$$AB = \frac{AC}{\cos \angle BAC} = \frac{225}{\cos 64°50'} = \frac{225}{0.4253} = 529.0$$

and

$$CB = AC \tan \angle BAC = 225 \tan 64°50' = 225(2.1283) = 478.9$$

(*a*) B is 529 mi from A. (*b*) B is 479 mi from C. (*c*) Since $\angle CBA = 25°10'$, the bearing of A from B is N25°10′W.

9.91 From a boat sailing due north at 16.5 mi/h, a wrecked ship K and an observation tower T are observed in a line due east. One hour later the wrecked ship and the tower have bearings S34°40′E and S65°10′E. Find the distance between the wrecked ship and the tower.

▌ In Fig. 9.35, C, K, and T represent, respectively, the boat, the wrecked ship, and the tower when in a line. One hour later the boat is at A, 16.5 mi due north of C. In right triangle ACK,

$$CK = 16.5 \tan 34°40' = 16.5(0.6916)$$

In right triangle ACT,

$$CT = 16.5 \tan 65°10' = 16.5(2.1609)$$

Then $KT = CT - CK = 16.5(2.1609 - 0.6916) = 24.2$ mi.

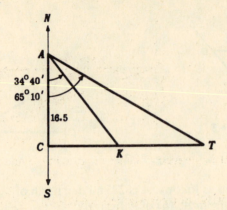

Fig. 9.35

9.92 A ship is sailing due east when a light is observed bearing N62°10′E. After the ship has traveled 2250 ft, the light bears N48°25′E. If the course is continued, how close will the ship approach the light?

▌ In Fig. 9.36, L is the position of the light, A is the first position of the ship, B is the second position, and C is the position when nearest L. In right triangle ACL, $AC = CL \cot \angle CAL = CL \cot 27°50′ = 1.8940CL$. In right triangle BCL, $BC = CL \cot \angle CBL = CL \cot 41°35′ = 1.1270CL$. Since $AC = BC + 2250$, $1.8940CL = 1.1270CL + 2250$, and $CL = 2250/(1.8940 - 1.1270) = 2934$ ft.

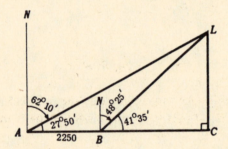

Fig. 9.36

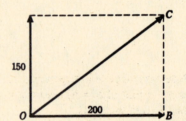

Fig. 9.37

9.93 Refer to Fig. 9.37. A body at O is being acted upon by two forces, one of 150 lb due north and the other of 200 lb due east. Find the magnitude and direction of the resultant.

▌ In right triangle OBC, $OC = \sqrt{(OB)^2 + (BC)^2} = \sqrt{(200)^2 + (150)^2} = 250$ lb, $\tan \angle BOC = \frac{150}{200} = 0.7500$ and $\angle BOC = 36°50′$. The magnitude of the resultant force is 250 lb, and its direction is N53°10′E.

9.94 An airplane is moving horizontally at 240 mi/h when a bullet is shot from the plane with speed 2750 ft/s at right angles to the path of the airplane. Find the resultant speed and direction of the bullet.

▌ The speed of the airplane is 240 mi/h = 240(5280)/[60(60)] ft/s = 352 ft/s. In Fig. 9.38, vector AB represents the velocity of the airplane, vector AC represents the initial velocity of the bullet, and vector AD represents the resultant velocity of the bullet. In right triangle ACD, $AD = \sqrt{(352)^2 + (2750)^2} = 2770$ ft/s, $\tan \angle CAD = \frac{352}{2750} = 0.1280$, and $\angle CAD = 7°20′$. Thus, the bullet travels at 2770 ft/s along a path making an angle of 82°40′ with the path of the airplane.

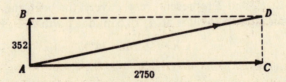

Fig. 9.38

9.95 A river flows due south at 125 ft/min. A motorboat, moving at 475 ft/min in still water, is headed due east across the river. (*a*) Find the direction in which the boat moves and its speed. (*b*) In what

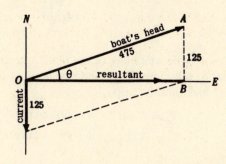

Fig. 9.39

direction must the boat be headed in order for it to move due east, and what is its speed in that direction?

▌ (**a**) Refer to Fig. 9.39. In right triangle OAB, $OB = \sqrt{(475)^2 + (125)^2} = 491$, $\tan \theta = \frac{125}{475} = 0.2632$, and $\theta = 14°40'$. Thus the boat moves at 491 ft/min in the direction S75°20′E.

(**b**) Refer to Fig. 9.40. In right triangle OAB, $\sin \theta = \frac{125}{475} = 0.2632$ and $\theta = 15°20'$. Thus the boat must be headed N74°40′E, and its speed in that direction is $OB = \sqrt{(475)^2 - (125)^2} = 458$ ft/min.

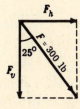

Fig. 9.40

9.96 A telegraph pole is kept vertical by a guy wire which makes an angle of 25° with the pole and which exerts a pull of $\mathbf{F} = 300$ lb on the top. Find the horizontal and vertical components $\mathbf{F}_h$ and $\mathbf{F}_v$ of the pull $\mathbf{F}$. See Fig. 9.41.

▌ $F_h = 300 \sin 25° = 300(0.4226) = 127$ lb
$F_v = 300 \cos 25° = 300(0.9063) = 272$ lb

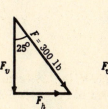

Fig. 9.41

9.97 A woman pulls a rope attached to a sled with a force of 100 lb. The rope makes an angle of 27° with the ground. (**a**) Find the effective pull tending to move the sled along the ground and the effective pull tending to lift the sled vertically. (**b**) Find the force which the woman must exert in order for the effective force tending to move the sled along the ground to be 100 lb.

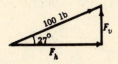
Fig. 9.42

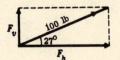
Fig. 9.43

▌ (**a**) In Figs. 9.42 and 9.43, the 100-lb pull in the rope is resolved into horizontal and vertical components $\mathbf{F}_h$ and $\mathbf{F}_v$, respectively. Then $\mathbf{F}_h$ is the force tending to move the sled along the ground, and $\mathbf{F}_v$ is the force tending to lift the sled.

$$F_h = 100 \cos 27° = 100(0.8910) = 89 \text{ lb} \qquad F_v = 100 \sin 27° = 100(0.4540) = 45 \text{ lb}$$

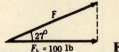

Fig. 9.44

(b) In Fig. 9.44, the horizontal component of the required force **F** is $F_h = 100$ lb. Then $F = 100/\cos 27° = 100/0.8910 = 112$ lb.

9.98 A block weighing **W** = 500 lb rests upon a ramp inclined 29° with the horizontal. Find the force tending to move the block down the ramp and the force of the block on the ramp.

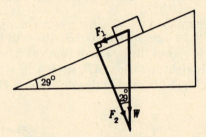

Fig. 9.45

▌ Refer to Fig. 9.45. Resolve the weight **W** of the block into components **F₁** and **F₂**, respectively, parallel and perpendicular to the ramp. **F₁** is the force tending to move the block down the ramp, and **F₂** is the force of the block on the ramp.

$$F_1 = W \sin 29° = 500(0.4848) = 242 \text{ lb}$$
$$F_2 = W \cos 29° = 500(0.8746) = 437 \text{ lb}$$

CHAPTER 10
Conic Sections

10.1 THE CIRCLE

For Probs. 10.1 to 10.12, find the center and radius of the given circle.

10.1 $x^2 + y^2 = 100$

▌ If $(x-h)^2 + (y-k)^2 = r^2$, then (h, k) is the center and r is the radius. Here, $C(0, 0)$, $r = \sqrt{100} = 10$.

10.2 $(x+2)^2 + (y+4)^2 = 10$

▌ $C(-2, -4)$, $r = \sqrt{10}$.

10.3 $(x-1)^2 + y^2 = 16$.

▌ $C(1, 0)$, $r = \sqrt{16} = 4$.

10.4 $x^2 + y^2 = 3$

▌ There are no real numbers satisfying this equation.

10.5 $(x-1)^2 + (y-5)^2 = 0$

▌ This is just the point $(1, 5)$; the only way $(x-1)^2 + (y-5)^2 = 0$ is if $(x-1)^2 = 0$ and $(y-5)^2 = 0$.

10.6 $x^2 - 8x + y^2 + 10y = 12$

▌ $(x^2 - 8x) + (y^2 + 10y) = 12$, $(x^2 - 8x + 16) + (y^2 + 10y + 25) = 12 + 16 + 25$, and $(x-4)^2 + (y+5)^2 = 53$. Here $C(4, -5)$, $r = \sqrt{53}$.

10.7 $3x^2 + 3y^2 - 4x + 2y + 6 = 0$

▌ $3(x^2 - \frac{4}{3}x + \frac{16}{36}) + 3(y^2 + \frac{2}{3}y + \frac{4}{36}) = -6 + \frac{48}{36} + \frac{12}{36}$, and $3(x - \frac{2}{3})^2 + 3(y + \frac{1}{3})^2 = -6 + \frac{60}{36}$. There are no real x and y satisfying this equation, since $r^2 < 0$.

10.8 $7x^2 + 7y^2 + 14x - 56y - 25 = 0$

▌ $7(x^2 + 2x + 1) + 7(y^2 - 8y + 16) = 25 + 7 + 112 = 144$, $(x+1)^2 + (y-4)^2 = 144/7$, and $C(-1, 4)$, $r = 12/\sqrt{7}$.

10.9 $x^2 + y^2 - 4x - 6y - \frac{10}{3} = 0$

▌ $x^2 - 4x + 4 + y^2 - 6y + 9 = \frac{10}{3} + 4 + 9$, $(x-2)^2 + (y-3)^2 = 16\frac{1}{3} = \frac{49}{3}$, and $C(2, 3)$, $r = 7/\sqrt{3} = 7\sqrt{3}/3$.

10.10 $x^2 + y^2 - 6x + 8y - 11 = 0$

▌ $x^2 - 6x + 9 + y^2 + 8y + 16 = 11 + 9 + 16$, $(x-3)^2 + (y+4)^2 = 36$, and $C(3, -4)$, $r = 6$.

10.11 $2x^2 + 2y^2 - x = 0$

▌ $2(x^2 - x/2 + \frac{1}{16}) + 2y^2 = \frac{2}{16} = \frac{1}{8}$, $(x - \frac{1}{4})^2 + (y - 0)^2 = \frac{1}{16}$, and $C(\frac{1}{4}, 0)$, $r = \frac{1}{4}$.

10.12 $x^2 + y^2 - 8x - 7y = 0$

▮ $x^2 - 8x + 16 + y^2 - 7y + \frac{49}{4} = 16 + \frac{49}{4}$, $(x - 4)^2 + (y - \frac{7}{2})^2 = \frac{113}{4}$, and $C(4, \frac{7}{2})$, $r = \sqrt{113}/2$.

For Probs. 10.13 to 10.24, write an equation of the circle satisfying the given conditions.

10.13 Center $(1, 0)$, radius 2

▮ $(x - 1)^2 + (y - 0)^2 = 2^2$, or $(x - 1)^2 + y^2 = 4$.

10.14 Center $(0, -5)$, radius 1

▮ $(x - 0)^2 + [y - (-5)]^2 = 1^2$, or $x^2 + (y + 5)^2 = 1$.

10.15 Center $(-2, -3)$, radius $\sqrt{7}$

▮ $(x + 2)^2 + (y + 3)^2 = (\sqrt{7})^2$, or $(x + 2)^2 + (y + 3)^2 = 7$.

10.16 $C(1, 2)$, passing through $(0, 0)$

▮ The line segment connecting $(1, 2)$ to $(0, 0)$ is a radius: $r = \sqrt{(1 - 0)^2 + (2 - 0)^2} = \sqrt{1 + 4} = \sqrt{5}$. Then $(x - 1)^2 + (y - 2)^2 = (\sqrt{5})^2 = 5$.

10.17 $C(1, -3)$, passing through $(2, 6)$

▮ $r = \sqrt{(2 - 1)^2 + [6 - (-3)]^2} = \sqrt{1^2 + 9^2} = \sqrt{82}$, so the equation is $(x - 1)^2 + (y + 3)^2 = 82$.

10.18 Center $(0, 6)$; diameter has $(0, -1)$ and $(0, 13)$ as endpoints.

▮ $r = \sqrt{(0 - 0)^2 + (13 + 1)^2}/2 = 14/2 = 7$, so the equation is $(x - 0)^2 + (y - 6)^2 = 49$.

10.19 Center $(0, 6)$; diameter has $(0, 1)$ as an endpoint.

▮ $r = \sqrt{(0 - 0)^2 + (1 - 6)^2} = 5$, so the equation is $x^2 + (y - 6)^2 = 25$.

10.20 Center at $(1, 2)$, diameter $= 6$

▮ If $d = 6$, then $r = d/2 = 3$, and the equation is $(x - 1)^2 + (y - 2)^2 = 9$.

10.21 Center $(-4, 3)$, tangent to y axis

▮ See Fig. 10.1. If the circle is tangent to the y axis, it must intersect it at P, where $AP \parallel x$ axis. Thus, P has coordinates $(0, 3)$, and $r = \sqrt{(4 + 0)^2 + (3 - 3)^2} = 4$. Then the equation is $(x + 4)^2 + (y - 3)^2 = 16$.

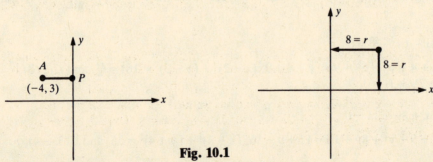

Fig. 10.1   Fig. 10.2

10.22 Circle is tangent to both axes, center is in quadrant I, $r = 8$.

▮ See Fig. 10.2. The center must be at $(8, 8)$, so the equation is $(x - 8)^2 + (y - 8)^2 = 64$.

10.23 Center is at the origin; circle crosses the x axis at 6.

▮ See Fig. 10.3. Then $r = 6$, and the equation is $(x - 0)^2 + (y - 0)^2 = 36$, or $x^2 + y^2 = 36$.

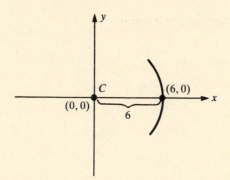

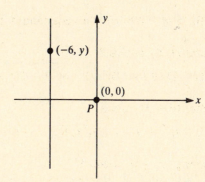

Fig. 10.3   **Fig. 10.4**

10.24 Passes through the origin; $r = 10$; abscissa of center is -6.

▮ See Fig. 10.4. The center is on the line $x = -6$. We want the distance from $(-6, y)$ to $(0, 0)$ (which is r) to be 10. Then $\sqrt{36 + y^2} = 10$, $36 + y^2 = 100$, $y^2 = 64$, or $y = 8, -8$. Thus, $C = (-6, 8)$ or $C = (-6, -8)$, so the equation is either $(x + 6)^2 + (y - 8)^2 = 100$ or $(x + 6)^2 + (y + 8)^2 = 100$.

For Probs. 10.25 to 10.27, find the equation of the circle passing through the three given points.

10.25 $(0, 0), (1, 1), (1, 2)$

▮ If $x^2 + y^2 + Ax + By + C = 0$ is the circle, then each of the abscissas and ordinates given must satisfy that equation. Then using $(0, 0)$: $0 + C = 0$, or $C = 0$. Using $(1, 1)$: $1^2 + 1^2 + A + B + C = 0$, but $C = 0$, so $2 + A + B = 0$. Using $(1, 2)$: $1^2 + 2^2 + A + 2B + C = 0$, but again since $C = 0$, $5 + A + 2B = 0$. Then $5 + (-B - 2) + 2B = 0$, $B + 3 = 0$, and $B = -3$. Using the equation obtained from point $(1, 1)$ gives $2 + A - 3 = 0$ and $A = 1$. Finally, substituting $A = 1$, $B = -3$, and $C = 0$ into $x^2 + y^2 + Ax + By + C = 0$, we get $x^2 + y^2 + x - 3y = 0$.

10.26 $(0, 1), (1, 0), (0, -1)$

▮ $x^2 + y^2 + Ax + By + C = 0$. Using $(0, 1)$: $0 + 1 + 0 + B + C = 0$. Using $(1, 0)$: $1 + 0 + A + 0 + C = 0$. Using $(0, -1)$: $0 + 1 + 0 - B + C = 0$. Then $2 + 2C = 0$, so $C = -1$; $1 + A - 1 = 0$, so $A = 0$; and $1 - B - 1 = 0$, so $B = 0$. Thus, $x^2 + y^2 - 1 = 0$, or $x^2 + y^2 = 1$.

10.27 $(0, 0), (1, -1), (2, 0)$

▮ $x^2 + y^2 + Ax + By + C = 0$. Using $(0, 0)$: $0 + 0 + 0 + 0 + C = 0$, so $C = 0$. Using $(1, -1)$: $1 + 1 + A - B = 0$. Using $(2, 0)$: $4 + 0 + 2A + 0 + 0 = 0$, $2A = -4$, and $A = -2$. Since $2 + A - B = 0$, $2 - 2 - B = 0$, and $B = 0$. Then the equation is $x^2 + y^2 - 2x = 0$ ($B = C = 0$).

For Probs. 10.28 to 10.30, find the center and radius of the circle passing through the given points.

10.28 $(0, 0), (1, 1), (1, 2)$

▮ $(x - h)^2 + (y - k)^2 = r^2$. Using $(0, 0)$: $(-h)^2 + (-k)^2 = r^2$, and $h^2 + k^2 = r^2$. Using $(1, 1)$: $(1 - h)^2 + (1 - k)^2 = r^2$, $1 - 2h + h^2 + 1 - 2k + k^2 = r^2$, and $2 + h^2 - 2h - 2k + k^2 = r^2$. Using $(1, 2)$: $(1 - h)^2 + (2 - k)^2 = r^2$, $1 - 2h + h^2 + 4 - 4k + k^2 = r^2$, and $5 + h^2 - 2h - 4k + k^2 = r^2$. From these equations we have $-2 + 2h + 2k = 0$ and $-3 + 2k = 0$, so $k = \frac{3}{2}$. Thus, $-2 + 2h + 3 = 0$, $2h = 1$, or $h = \frac{1}{2}$. Then, $\frac{1}{4} + \frac{1}{4} = r^2$, $r^2 = \frac{1}{2}$, $r = 1/\sqrt{2} = \sqrt{2}/2$. Thus, $C = (\frac{1}{2}, \frac{1}{2})$, $r = \sqrt{2}/2$.

10.29 $(0, 1), (1, 0), (0, 1)$

▮ See Prob. 10.26. If $x^2 + y^2 = 1$, then $C = (0, 0)$, $r = 1$. The technique in Prob. 10.28 will also work.

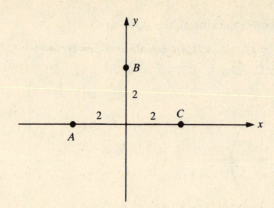

Fig. 10.5

10.30 $(0, 2), (2, 0), (-2, 0)$

▮ See Fig. 10.5. For the circle to pass through A, B, and C, it must have $C = (0, 0)$, $r = 2$.

For Probs. 10.31 to 10.33, find the equation of the circle passing through the given points, with center on the given line.

10.31 Through $(0, 1)$ and $(1, 0)$, center on $x = 0$

▮ Let $(x - h)^2 + (y - k)^2 = r^2$ be the circle. Then, (h, k) must be the same distance from $(0, 1)$ and $(1, 0)$: $\sqrt{(h - 0)^2 + (k - 1)^2} = \sqrt{(h - 1)^2 + (k - 0)^2}$, $\sqrt{h^2 + (k - 1)^2} = \sqrt{(h - 1)^2 + k^2}$, $h^2 + k^2 - 2k + 1 = h^2 - 2h + 1 + k^2$, so $-2k + 2h = 0$. If the center (h, k) lies on $x = 0$, then $h = 0$; since $-k + h = 0$, $k = 0$. Thus, $r = \sqrt{(0 - 0)^2 + (1 + 0)^2} = 1$, and the equation is $x^2 + y^2 = 1$.

10.32 Through $(0, 1)$ and $(1, 0)$, center on $y = -1$

▮ $\sqrt{(h - 0)^2 + (k - 1)^2} = \sqrt{(h - 1)^2 + (k - 0)^2}$, $h^2 + k^2 - 2k + 1 = h^2 - 2h + 1 + k^2$, $-2k + 1 = -2h + 1$, and $k = h$. If the center lies on $y = -1$, then $k = -1$ and $h = -1$. Thus $(h, k) = (-1, -1)$. Therefore, $r = \sqrt{(-1 - 0)^2 + (-1 - 1)^2} = \sqrt{1 + 4} = \sqrt{5}$.

10.33 Through $(0, 0)$ and $(1, 0)$, center on $y = 0$

▮ $\sqrt{(h - 0)^2 + (y - 0)^2} = \sqrt{(h - 1)^2 + y^2}$, $h^2 + y^2 = h^2 - 2h + 1 + y^2$, $-2h = -1$, so $h = \frac{1}{2}$. If the center is on $y = 0$, then $k = 0$. Also, $r = \sqrt{(\frac{1}{2} - 0)^2 + 0} = \sqrt{\frac{1}{4}} = \frac{1}{2}$, so $(x - \frac{1}{2})^2 + y^2 = \frac{1}{4}$.

For Probs. 10.34 to 10.38, find the intersection(s) (if any) of the graphs of the given equations. Use algebraic techniques.

10.34 $x^2 + y^2 = 1$ and $x = 1$

▮ $x^2 + y^2 = 1$ and $x = 1$. Then $1^2 + y^2 = 1$, $y^2 = 0$, and $y = 0$. But $x = 1$, so the intersection point is $(1, 0)$.

10.35 $x^2 + y^2 = 1$ and $(x - 1)^2 + y^2 = 1$

▮ $y^2 = 1 - x^2$, so $(x - 1)^2 + (1 - x^2) = 1$. Then $x^2 - 2x + 1 + 1 - x^2 = 1$, $-2x + 2 = 1$, $-2x = -1$, or $x = \frac{1}{2}$. Since $y^2 = 1 - x^2$, then $y^2 = 1 - (\frac{1}{2})^2 = \frac{3}{4}$. Thus, $y = \pm\sqrt{3}/2$, and $(x, y) = (\frac{1}{2}, \sqrt{3}/2)$ or $(x, y) = (\frac{1}{2}, -\sqrt{3}/2)$.

10.36 $x^2 + y^2 = 1$ and $x^2 + y^2 = 2$

▮ $x^2 = 1 - y^2$, $x = \sqrt{1 - y^2}$, or $(1 - y^2) + y^2 = 2$; no solution. These two circles do not intersect.

10.37 $x + y = 2$ and $(x - 1)^2 + (y - 2)^2 = 3$

▌ $(x - 1)^2 + (2 - x - 2)^2 = 3$, $x^2 - 2x + 1 + x^2 = 3$, $2x^2 - 2x - 2 = 0$, $x^2 - x - 1 = 0$, or $x = \dfrac{\sqrt{1 \pm (-1)^2 - 4(1)(-1)}}{2} = \dfrac{1 \pm \sqrt{1 + 4}}{2} = \dfrac{1 \pm \sqrt{5}}{2}$. If $x = \dfrac{1 + \sqrt{5}}{2}$, $y = 2 - \dfrac{1 + \sqrt{5}}{2}$. If $x = \dfrac{1 - \sqrt{5}}{2}$, $y = 2 - \dfrac{1 - \sqrt{5}}{2}$.

10.38 $x^2 + y^2 = y$ and $y = 3$

▌ $x^2 + 9 = 4$, or $x^2 = -5$. There is no solution, so there is no intersection.

For Probs. 10.39 to 10.41, tell whether the statement is true or false and why.

10.39 Every triple of points determines a circle.

▌ False; if the three points are collinear, they do not determine a circle.

10.40 If a circle and line intersect, they intersect twice.

▌ False; if the line is tangent to the circle, it intersects it only once.

10.41 Two circles either intersect in two places or do not intersect.

▌ False; if they are tangent, they intersect at one point.

10.42 Write the equation of the locus of a point, the sum of the squares of whose distances from $(-2, -5)$ and $(3, 4)$ is equal to 70.

▌ Let (x, y) be the point. Then $d[(x, y), (-2, -5)] = \sqrt{(x + 2)^2 + (y + 5)^2}$ and $d[(x, y), (3, 4)] = \sqrt{(x - 3)^2 + (y - 4)^2}$. We are told that $[(x + 2)^2 + (y + 5)^2] + [(x - 3)^2 + (y - 4)^2] = 70$. Then $x^2 + 4x + 4 + y^2 + 10y + 25 + x^2 - 6x + 9 + y^2 - 8y + 16 = 70$. Thus, $x^2 + y^2 - x + y - 8 = 0$. This is a circle with center at $(\frac{1}{2}, -\frac{1}{2})$ and radius $\sqrt{\frac{17}{2}}$.

10.43 Find the length of the tangent from $(3, -1)$ to $(x - 1)^2 + y^2 = 1$.

▌ See Fig. 10.6 and apply the pythagorean theorem. Then $d^2 = [\sqrt{(3 - 1)^2 + (-1 - 0)^2}]^2 - r^2 = (\sqrt{4 + 1})^2 - 1 = (\sqrt{5})^2 - 1 = 5 - 1 = 4$, and $d = 2$.

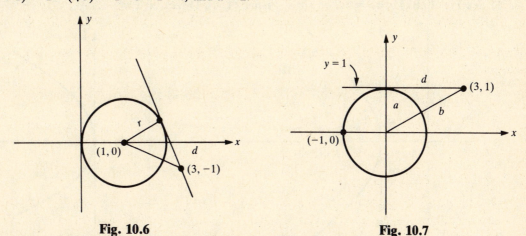

Fig. 10.6   **Fig. 10.7**

10.44 Find the length of the tangent from $(3, 1)$ to $x^2 + y^2 = 1$.

▌ See Fig. 10.7 and Prob. 10.43. Notice that by the pythagorean theorem $d^2 = b^2 - a^2$. Then $b = \sqrt{3^2 + 1^2} = \sqrt{10}$ and $a = 1$. Thus, $d^2 = (\sqrt{10})^2 - 1 = 9$ and $d = 3$. Does this result look correct?

10.2 THE PARABOLA

For Probs. 10.45 to 10.52, find the focus and directrix for the given parabola.

10.45 $y^2 = 7x$

▋ If $y^2 = 4ax$, then the focus is at $(a, 0)$, and the directrix is $x = -a = -\frac{7}{4}$. In this case, $y^2 = 4 \cdot \frac{7}{4} \cdot x$.

10.46 $y^2 = 40x$

▋ $y^2 = 40x$ means $y^2 = 4ax$ where $a = 10$. Focus: $(10, 0)$. Directrix: $x = -10$.

10.47 $y^2 = 3x$

▋ $3x = 4ax$, $3 = 4a$, or $a = \frac{4}{3}$. Focus: $(\frac{3}{4}, 0)$. Directrix: $x = -\frac{3}{4}$.

10.48 $y^2 = -x$

▋ $4ax = -x$, $4a = -1$, or $a = -\frac{1}{4}$. Focus: $(-\frac{1}{4}, 0)$. Directrix: $x = \frac{1}{4}$.

10.49 $y^2 = -5x$

▋ $-5x = 4ax$, $-5 = 4a$, or $a = -\frac{5}{4}$. Focus: $(-\frac{5}{4}, 0)$. Directrix: $x = \frac{5}{4}$. (Careful! $x = -a$.)

10.50 $x^2 = 8y$

▋ If $x^2 = 4ay$, then the focus is $(0, a)$ and the directrix is $y = -a$. In this case, $8 = 4a$, or $a = 2$. Focus: $(0, 2)$. Directrix: $y = -2$.

10.51 $x^2 = -16y$

▋ $4a = -16$, or $a = -4$. Focus: $(0, -4)$. Directrix: $y = 4$.

10.52 $x^2 = y$

▋ Then $4ay = y$, $4a = 1$, or $a = \frac{1}{4}$. Focus: $(0, \frac{1}{4})$. Directrix: $y = -\frac{1}{4}$.

For Probs. 10.53 to 10.62, sketch the graph of the given parabola.

10.53 $y^2 = 4x$

▋ See Fig. 10.8. Then $4 = 4a$, or $a = 1$. Focus: $(1, 0)$. Directrix: $x = -1$.

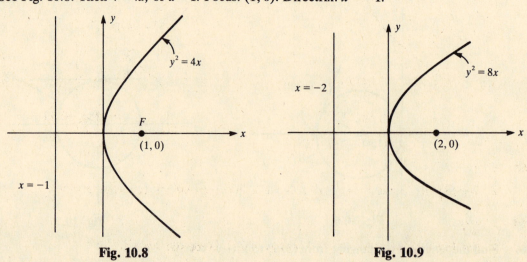

Fig. 10.8   Fig. 10.9

10.54 $y^2 = 8x$

▋ See Fig. 10.9. $8 = 4a$, or $a = 2$. Focus: $(2, 0)$. Directrix: $x = -2$.

10.55 $y^2 = x$

  ▌ See Fig. 10.10. Focus: $(\frac{1}{4}, 0)$. Directrix: $x = -\frac{1}{4}$.

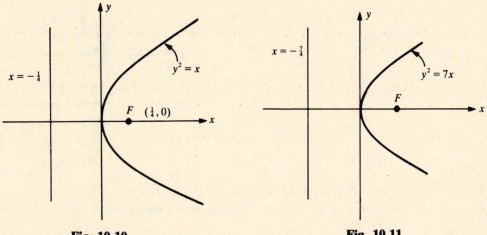

Fig. 10.10   Fig. 10.11

10.56 $y^2 = 7x$

  ▌ See Fig. 10.11. Then $4a = 7$, or $a = \frac{7}{4}$. Focus: $(\frac{7}{4}, 0)$. Directrix: $x = -\frac{7}{4}$.

10.57 $y^2 = -4x$

  ▌ See Fig. 10.12. Then $4a = -4$, or $a = -1$. Focus: $(-1, 0)$. Directrix: $x = 1$.

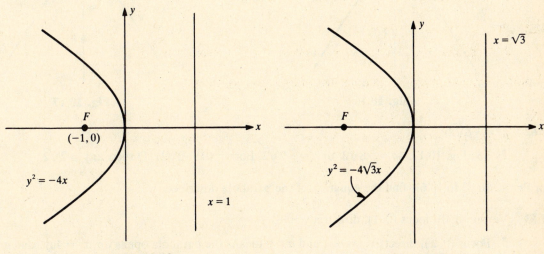

Fig. 10.12   Fig. 10.13

10.58 $y^2 = -4\sqrt{3}x$

  ▌ See Fig. 10.13. $4a = -4\sqrt{3}$, or $a = -\sqrt{3}$. Focus: $(-\sqrt{3}, 0)$. Directrix: $x = \sqrt{3}$.

10.59 $x^2 = 2y$

  ▌ See Fig. 10.14. $2 = 4a$, or $a = \frac{1}{2}$. Focus: $(0, \frac{1}{2})$. Directrix: $y = -\frac{1}{2}$.

10.60 $x^2 = 10y$

  ▌ See Fig. 10.15. $10 = 4a$, or $a = \frac{5}{2}$. Focus: $(0, \frac{5}{2})$. Directrix: $y = -\frac{5}{2}$.

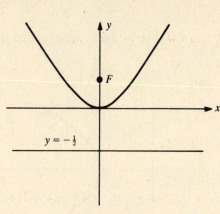

Fig. 10.14

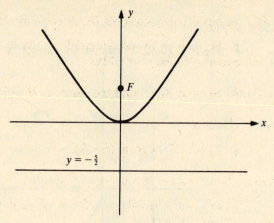

Fig. 10.15

10.61 $x^2 = -10y$

▮ See Fig. 10.16. $4a = -10$, or $a = -\frac{5}{2}$. Focus: $(0, -\frac{5}{2})$. Directrix: $y = \frac{5}{2}$.

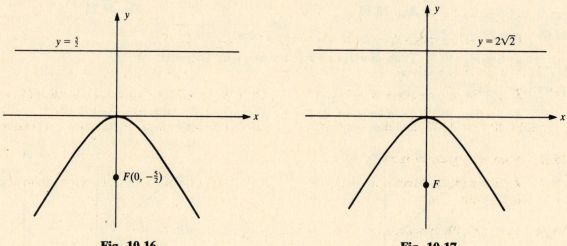

Fig. 10.16

Fig. 10.17

10.62 $x^2 = -8\sqrt{2}y$

▮ See Fig. 10.17. $4a = -8\sqrt{2}$, or $a = -2\sqrt{2}$. Focus: $(0, -2\sqrt{2})$. Directrix: $y = 2\sqrt{2}$.

For Probs. 10.63 to 10.67, find the equation of the parabola described.

10.63 Vertex $(0, 0)$, focus $(0, 1)$, directrix $y = -1$

▮ Focus $(0, a)$, directrix $y = -a$, and $a > 0$ means the parabola opens up. $x^2 = 4ay$, and $a = 1$; so $x^2 = 4y$.

10.64 Vertex $(0, 0)$, focus $(1, 0)$, directrix $x = -1$

▮ Focus $(a, 0)$, directrix $x = -a$, and $a > 0$ means the parabola opens to the right. $y^2 = 4ax$, and $a = 1$; so $y^2 = 4x$.

10.65 Vertex $(0, 0)$, focus $(0, 2)$, directrix $y = -2$

▮ $x^2 = 4ay$, and $a = 2$; so $x^2 = 8y$ (opens up).

10.66 Vertex $(0, 0)$, focus $(0, -2)$, directrix $y = 2$

▮ $x^2 = 4ay$, and $a = -2$; $x^2 = -8y$.

10.67 Vertex $(0, 0)$, focus $(-3, 0)$, directrix $x = 3$

▌ Focus $(-3, 0)$ and directrix $x = 3$ means $y^2 = 4ax$, $a < 0$. The parabola opens to the left. $y^2 = 4ax$, and $a = -3$; so $y^2 = -12x$.

For Probs. 10.68 to 10.72, find the vertex of the given parabola.

10.68 $x^2 = -10y$

▌ This is of the form $(x - h)^2 = -10(y - k)$, where $(h, k) = (0, 0)$. Vertex: $(0, 0)$.

10.69 $x^2 - 6x + 8y + 25 = 0$

▌ $(x^2 - 6x + 9) + 8y = -25 + 9$, $(x - 3)^2 = -8y - 16$, $(x - 3)^2 = -8(y + 2)$, so $h = 3$, $k = -2$. Vertex: $(3, -2)$.

10.70 $y^2 - 16x + 2y + 49 = 0$

▌ $y^2 + 2y + 1 = 16x - 49 + 1 = 16x - 48$, $(y + 1)^2 = 16(x - 3)$, so $k = -1$, $h = 3$. Vertex: $(3, -1)$.

10.71 $x^2 - 2x - 6y - 53 = 0$

▌ $x^2 - 2x + 1 = 6y + 53 + 1$, $(x - 1)^2 = 6(y + 9)$. Vertex: $(1, -9)$.

10.72 $y^2 + 20x + 4y - 60 = 0$

▌ $y^2 + 4y + 4 = -20x + 60 + 4$, $(y + 2)^2 = -20x + 64$, $(y + 2)^2 = -20(x - \frac{16}{5})$. Vertex: $(\frac{16}{5}, -2)$.

For Probs. 10.73 to 10.78, find the equation of the parabola satisfying the given conditions, if possible.

10.73 Focus $(3, 0)$, directrix $x + 3 = 0$

▌ Focus $(3, 0)$, directrix $x = -3$ means the parabola is of the form $y^2 = 4ax$, $a = 3$. So the equation is $y^2 = 12x$.

10.74 Vertex $(0, 0)$; axis along the x axis; passing through $(-3, -6)$.

▌ See Fig. 10.18. The axis passes through the vertex, to the directrix. Then the directrix must be $x = 3$, and the focus must be $(-3, 0)$. Then $y^2 = 4ax$, $a < 0$. When $a = -3$, $y^2 = -12x$. See Prob. 10.73.

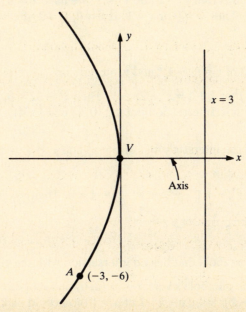

Fig. 10.18

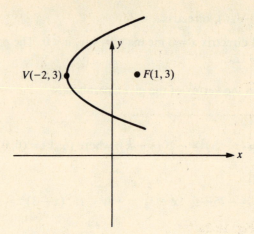

Fig. 10.19

10.75 Vertex $(-2, 3)$, focus $(1, 3)$

❚ See Fig. 10.19. Here $a = 3$. The equation must be of the form $(y - k)^2 = 4a(x - h)$. Then $(y - 3)^2 = 4 \cdot 3(x + 2) = 12(x + 2)$, or $y^2 - 6y - 12x - 15 = 0$.

10.76 Vertex $(1, 6)$, focus $(2, 6)$

❚ Then $a = 6$, $(y - k)^2 = 4a(x - h)$, and $(y - 6)^2 = 24(x - 1)$.

10.77 Vertex $(2, 5)$, focus $(2, 8)$

❚ Then $a = 2$, and $(x - h)^2 = 4a(y - k)$, and $(x - 2)^2 = 8(y - 5)$.

10.78 Axis parallel to x axis, passing through $(3, 3)$, $(6, 5)$, $(6, -3)$

❚ If the axis is parallel to the x axis, then $y^2 + Ax + Ey + F = 0$. Using $(3, 3)$: $9 + 3A + 3E + F = 0$. Using $(6, 5)$: $25 + 6A + 5E + F = 0$. Using $(6, -3)$: $9 + 6A - 3E + F = 0$. Then, using these equations, we get $-3A + 6E = 0$ and $16 + 8E = 0$, so $E = -2$. Then $-3A - 12 = 0$, or $A = -4$; and $9 - 12 - 6 + F = 0$, or $F = 9$. Thus, the equation is $y^2 - 2y - 4x + 9 = 0$.

10.79 Find the intersection(s) of $y^2 = x$ and $y = x^2$.

❚ If $y = x^2$, then $y^2 = x^4$. From $y^2 = x$ and $y^2 = x^4$ we have $x = x^4$. Then $x^4 - x = 0$, $x(x^3 - 1) = 0$, and $x = 0$ or $x^3 - 1 = 0$. Thus, $x = 0$ or $x = 1$. If $x = 0$, $y = 0$; if $x = 1$, $y = 1$. Intersection points: $(0, 0)$, $(1, 1)$.

10.80 Find the intersection(s) of $y = (x - 1)^2$ and $y = x + 1$.

❚ Then $x + 1 = (x - 1)^2$ or $x + 1 = x^2 - 2x + 1$. Then $x^2 - 3x = 0$, $x(x - 3) = 0$, and $x = 0$ or $x = 3$. If $x = 0$, $y = 1$; if $x = 3$, $y = 4$. Intersection points: $(0, 1)$ and $(3, 4)$.

10.81 Find the ordinate(s) of the intersection(s) of $y = x^2$ and $x^2 + y^2 = 1$.

❚ $y = x^2$ and $x^2 = 1 - y^2$; $y = 1 - y^2$, or $y^2 + y - 1 = 0$. Then $y = \dfrac{-1 \pm \sqrt{1^2 - 4(1)(-1)}}{2} = \dfrac{-1 \pm \sqrt{1 + 4}}{2}$, so $y = \dfrac{-1 + \sqrt{5}}{2}$, or $y = \dfrac{-1 - \sqrt{5}}{2}$.

10.82 Give an example of a parabola that has no intersection with the circle $x^2 + y^2 = 1$.

❚ See Fig. 10.20. If the parabola's vertex is at $(2, 0)$ and it opens to the right, there will be no intersection. Then $(y - k)^2 = 4a(x - h)$. Let $a = 1$ (you could pick others). Then $y^2 = 4(x - 2)$.

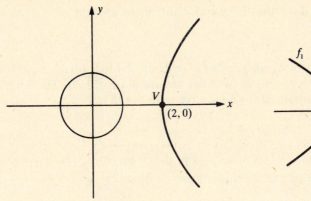

Fig. 10.20

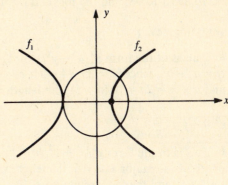

Fig. 10.21

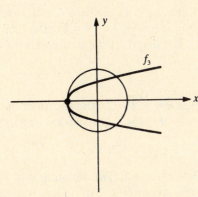

Fig. 10.22

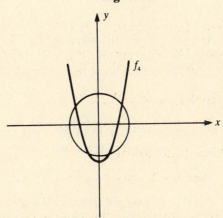

Fig. 10.23

10.83 What are the different possibilities for the intersection of a circle and a parabola?

▮ The possibilities are no intersections (see Fig. 10.20), one intersection (f_1) or two intersections (f_2) (see Fig. 10.21), three intersections (f_3) (see Fig. 10.22), and four intersections (f_4) (see Fig. 10.23).

10.3 THE ELLIPSE

For Probs. 10.84 to 10.88, put the given equation in the standard form for an ellipse.

10.84 $x^2 + 8y^2 = 8$

▮ Divide by 8: $x^2/8 + 8y^2/8 = 1$. Then $x^2/8 + y^2 = 1$.

10.85 $x^2 + 2y^2 = 3$

▮ Divide by 3: $x^2/3 + 2y^2/3 = 1$, $x^2/3 + y^2/\frac{3}{2} = 1$.

10.86 $2x^2 + 5y^2 = 6$

▮ Divide by 6: $2x^2/6 + 5y^2/6 = 1$, $x^2/3 + y^2/\frac{6}{5} = 1$.

10.87 $-x^2 - 3y^2 = -1$

▮ Divide by -1: $x^2 + 3y^2 = 1$, $x^2 + y^2/\frac{1}{3} = 1$.

10.88 $(x - 1)^2 + 2(y + 3)^2 = 4$

▮ Divide by 4: $(x - 1)^2/4 + (y + 3)^2/2 = 1$.

For Probs. 10.89 to 10.98, find the foci and lengths of the major and minor axes for each given ellipse.

10.89 $x^2/25 + y^4/4 = 1$

▮ This is of the form $x^2/a^2 + y^2/b^2 = 1$, where $a > b > 0$. Thus, the foci are at $(\pm c, 0)$, where $c^2 = a^2 - b^2$. Since $c^2 = 25 - 4 = 21$, $c = \pm\sqrt{21}$, and the foci are at $(\pm\sqrt{21}, 0)$. The length of the major axis $= 2a = 10$ ($a = 5$); the length of the minor axis $= 2b = 4$ ($b = 2$).

10.90 $x^2/9 + y^2/4 = 1$

▮ This is of the form $x^2/a^2 + y^2/b^2 = 1$, $a > b > 0$ (see Prob. 10.89). Thus, $a = 3$ and $b = 2$. The length of the major axis $= 6$; the length of the minor axis $= 4$. Since $c^2 = a^2 - b^2 = 9 - 4 = 5$, $c = \pm\sqrt{5}$, and the foci are at $(\pm\sqrt{5}, 0)$.

10.91 $x^2/4 + y^2/25 = 1$

▮ Here, $x^2/b^2 + y^2/a^2 = 1$, $a > b > 0$. Then the foci are at $(0, \pm c)$, where $c^2 = a^2 - b^2$. The major axis length $= 2a$; the minor axis length $= 2b$. Since $c^2 = a^2 - b^2 = 25 - 4 = 21$, the foci are at $(0, \pm\sqrt{21})$. The length of the major axis $= 10$ ($a = 5$); the length of the minor axis $= 4$ ($b = 2$).

10.92 $x^2/4 + y^2/9 = 1$

▮ See Prob. 10.91. $x^2/b^2 + y^2/a^2 = 1$, $a > b > 0$. Then $a = 3$, $b = 2$. The length of the major axis $= 6$; the length of the minor axis $= 4$; and since $c^2 = 9 - 4 = 5$, the foci are at $(0, \pm\sqrt{5})$.

10.93 $x^2 + 9y^2 = 9$

▮ Put this in standard form: $x^2/9 + y^2/1 = 1$ or $x^2/3^2 + y^2/1^2 = 1$. The equation is of the form $x^2/a^2 + y^2/b^2 = 1$ ($a > b$). Thus, the length of the major axis $= 6$ ($2 \cdot 3$); the length of the minor axis $= 2$ ($2 \cdot 1$); and since $c^2 = 9 - 1 = 8$, $c = \pm 2\sqrt{2}$ and the foci are at $(\pm 2\sqrt{2}, 0)$.

10.94 $4x^2 + y^2 = 4$

▮ Divide by 4: $x^2/1^2 + y^2/2^2 = 1$. Then $b^2 = 1^2$, $a^2 = 2^2$; $b = 1$, $a = 2$. Since $c^2 = 2^2 - 1^2 = 3$, $c = \pm\sqrt{3}$, and the foci are at $(0, \pm\sqrt{3})$. The length of the major axis $= 4$; the length of the minor axis $= 2$.

10.95 $2x^2 + y^2 = 12$

▮ Divide by 12: $x^2/6 + y^2/12 = 1$. Then $b = \sqrt{6}$ and $a = 2\sqrt{3}$. Since $c^2 = 12 - 6 = 6$, $c = \pm\sqrt{6}$ and the foci are at $(0, \pm\sqrt{6})$. The length of the major axis $= 2a = 4\sqrt{3}$; the length of the minor axis $= 2b = 2\sqrt{6}$.

10.96 $4x^2 + 3y^2 = 24$

▮ $x^2/6 + y^2/8 = 1$. Then $b^2 = 6$, $a^2 = 8$; thus $c^2 = 2$ and the foci are at $(0, \pm\sqrt{2})$. The length of the major axis $= 2a = 2\sqrt{8} = 4\sqrt{2}$; the length of the minor axis $= 2b = 2\sqrt{6}$.

10.97 $4x^2 + 7y^2 = 28$

▮ $x^2/7 + y^2/4 = 1$. Then $a^2 = 7$, $b^2 = 4$; thus $c^2 = 3$. The foci are at $(\pm\sqrt{3}, 0)$; the length of the major axis $= 2\sqrt{7}$; the length of the minor axis $= 4$.

10.98 $3x^2 + 2y^2 = 24$

▮ $x^2/8 + y^2/12 = 1$. Then $b^2 = 8$, $a^2 = 12$; thus $c^2 = 4$. The foci are at $(0, \pm 2)$; the length of the major axis $= 2b = 2\sqrt{8} = 4\sqrt{2}$; the length of the minor axis $= 2a = 2\sqrt{12} = 4\sqrt{3}$.

For Probs. 10.99 to 10.108, sketch the graph of the given ellipse.

10.99 $x^2/25 + y^2/4 = 1$

▮ See Fig. 10.24 and Prob. 10.89. $a^2 = 25$ and $b^2 = 4$, where $x^2/a^2 + y^2/b^2 = 1$. Thus, the x intercepts $= \pm a$ and the y intercepts $= \pm b$. F_1 and F_2 are the foci.

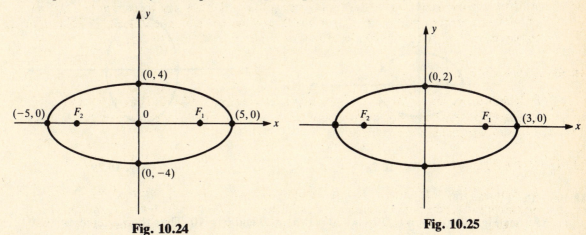

Fig. 10.24   Fig. 10.25

10.100 $x^2/9 + y^2/4 = 1$

▮ See Fig. 10.25 and Prob. 10.90. $a^2 = 9$ and $b^2 = 4$. Then the x intercepts $= \pm 3$, and the y intercepts $= \pm 2$.

10.101 $x^2/4 + y^2/25 = 1$

▮ See Fig. 10.26 and Prob. 10.91. $a^2 = 25$ and $b^2 = 4$. Then the x intercepts $= \pm 2$, and the y intercepts $= \pm 5$.

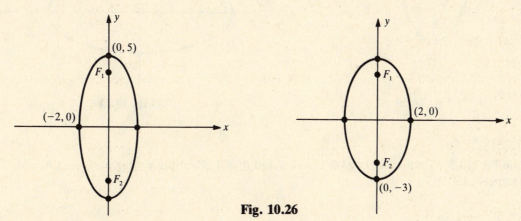

Fig. 10.26   Fig. 10.27

10.102 $x^2/4 + y^2/9 = 1$

▮ See Fig. 10.27 and Prob. 10.92. $a^2 = 9$ and $b^2 = 4$. Then the x intercepts $= \pm 2$, and the y intercepts $= \pm 3$.

10.103 $x^2 + 2y^2 = 2$

▮ See Fig. 10.28. Then $x^2/2 + y^2 = 1$. $a^2 = 2$ and $b^2 = 1$. Then the x intercepts $= \pm\sqrt{2}$, and the y intercepts $= \pm 1$.

10.104 $5x^2 + 2y^2 = 2$

▮ See Fig. 10.29. Then $5x^2/2 + y^2 = 1$, or $x^2/\frac{2}{5} + y^2 = 1$. $a^2 = 1$ and $b^2 = \frac{2}{5}$. Then the x intercepts $= \pm\sqrt{\frac{2}{5}}$, and the y intercepts $= \pm 1$.

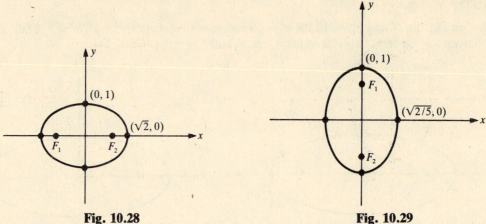

Fig. 10.28   Fig. 10.29

10.105 $10x^2 + 5y^2 = 100$

▎ See Fig. 10.30. Then $x^2/10 + y^2/20 = 1$. $a^2 = 20$ and $b^2 = 10$. Then the x intercepts $= \pm\sqrt{10}$, and the y intercepts $= \pm 2\sqrt{5}$.

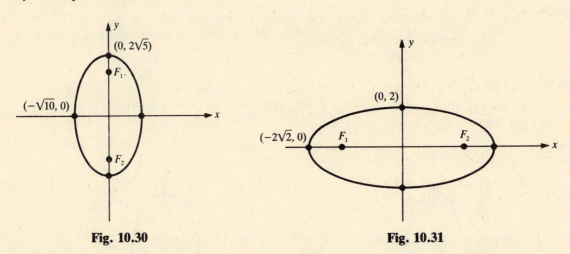

Fig. 10.30   Fig. 10.31

10.106 $-x^2 - 2y^2 = -8$.

▎ See Fig. 10.31. Then $x^2/8 + y^2/4 = 1$. $a^2 = 8$ and $b^2 = 4$. Then the x intercepts $= \pm\sqrt{8}$, and the y intercepts $= \pm 2$.

10.107 $x^2/\sqrt{2} + y^2/\sqrt{3} = 1$

▎ See Fig. 10.32. $\sqrt{3} > \sqrt{2}$. Thus, $a^2 = \sqrt{3}$, $b^2 = \sqrt{2}$, and the x intercepts $= \pm 2^{1/4}$ and the y intercepts $= \pm 3^{1/4}$.

10.108 $\dfrac{2x^2}{0.1} + \dfrac{3y^2}{0.2} = 0.7$

▎ See Fig. 10.33. Then $1 = \dfrac{2x^2}{(0.1)(0.7)} + \dfrac{3y^2}{(0.2)(0.7)} = \dfrac{2x^2}{0.07} + \dfrac{3y^2}{0.14} = \dfrac{2x^2}{\frac{7}{100}} + \dfrac{3y^2}{\frac{14}{100}} = \dfrac{x^2}{\frac{7}{200}} + \dfrac{y^2}{\frac{14}{300}}$. Since $\frac{14}{300} > \frac{7}{200}$, $a^2 = \frac{14}{300}$ and $b^2 = \frac{7}{200}$. Then the x intercepts $= \pm\sqrt{\frac{7}{200}}$, and the y intercepts $= \pm\sqrt{\frac{14}{300}}$.

For Probs. 10.109 to 10.116, find an equation of the ellipse with center $(0, 0)$ satisfying the given information. Put the equation in standard form for the ellipse.

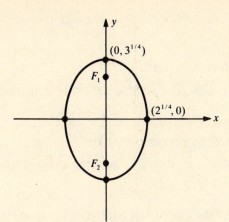

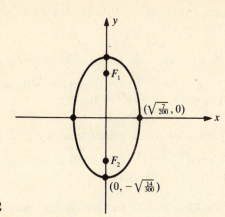

Fig. 10.32   Fig. 10.33

10.109 Major axis on x axis, major axis length $= 8$, minor axis length $= 6$

▮ If the major axis is on the x axis, then the ellipse is of the form $x^2/a^2 + y^2/b^2 = 1$, $a > b > 0$. Major axis length 8 means $2a = 8$, $a = 4$. Minor axis length 6 means $2b = 6$, $b = 3$. Thus, the equation is $x^2/16 + y^2/9 = 1$.

10.110 Major axis on x axis, major axis length 14, minor axis length 10

▮ If the major axis is on the x axis, then $x^2/a^2 + y^2/b^2 = 1$, $a > b > 0$. Then, $2a = 14$, $a = 7$; and $2b = 10$, $b = 5$. Thus, the equation is $x^2/49 + y^2/25 = 1$.

10.111 Major axis on y axis, major axis length 22, minor axis length 16

▮ If the major axis is on the y axis, then $x^2/b^2 + y^2/a^2 = 1$, $a > b > 0$. Then $2a = 22$, $a = 11$; and $2b = 16$, $b = 8$. Thus, the equation is $x^2/64 + y^2/121 = 1$.

10.112 Major axis on y axis, major axis length 24, minor axis length 18

▮ Since the major axis is on the y axis, $x^2/b^2 + y^2/a^2 = 1$. Then $2a = 24$, $a = 12$; and $2b = 18$, $b = 9$. Thus, the equation is $x^2/81 + y^2/144 = 1$.

10.113 Major axis on x axis, major axis length 16, distance of foci from center $= 6$

▮ See Fig. 10.34. $x^2/a^2 + y^2/b^2 = 1$ (since the major axis is on the x axis). Then $2a = 16$, $a = 8$. From the given information $d((0, 0), F) = 6$. We know that $b^2 = a^2 - c^2$, and that the distance from the focus to $(0, b)$ is the same as half the length of the major axis (a). Thus, $b^2 = a^2 - c^2 = 64 - 36 = 28$, and $b = \pm\sqrt{28} = \pm 2\sqrt{7}$. Then $x^2/64 + y^2/28 = 1$.

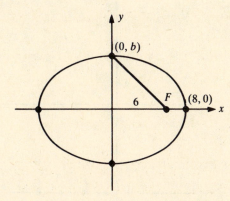

Fig. 10.34

10.114 Major axis on x axis, major axis length 24, distance of foci from center $= 10$

▮ See Prob. 10.113. $x^2/a^2 + y^2/b^2 = 1$ $(a > b > 0)$. Then $2a = 24$, $a = 12$. From the given information $d((0, 0), F) = 10$. Since $b^2 = a^2 - c^2$, $b^2 = 144 - 100 = 44$. Thus, the equation is $x^2/100 + y^2/44 = 1$.

10.115 Major axis on y axis, minor axis length 20, distance of foci from center $= \sqrt{70}$

▮ Then $x^2/b^2 + y^2/a^2 = 1$, and $2b = 20$, or $b = 10$. Since $a^2 = b^2 + c^2$, $a^2 = 100 + 70 = 170$. Thus, the equation is $x^2/100 + y^2/170 = 1$.

10.116 Major axis on y axis, minor axis length 14; distance of foci from center $= \sqrt{200}$

▮ Then $x^2/b^2 + y^2/a^2 = 1$, and $2b = 14$, or $b = 7$. Since $a^2 = b^2 + c^2$, $a^2 = 49 + 200 = 249$. Thus, the equation is $x^2/49 + y^2/249 = 1$.

For Probs. 10.117 to 10.120, find the coordinates of the center of the ellipse and put the equation in standard form for an ellipse.

10.117 $4x^2 + 4x + y^2 = 10$

▮ $4(x^2 + x + \frac{1}{4}) + y^2 = 10 + 1$, $4(x + \frac{1}{2})^2 + y^2 = 11$, and $(x + \frac{1}{2})^2/\frac{11}{4} + y^2/11 = 1$. Center: $(-\frac{1}{2}, 0)$, since this is of the form $(x - h)^2/b^2 + (y - k)^2/a^2 = 1$.

10.118 $2x^2 + y^2 + 2y = 15$

▮ $2x^2 + (y^2 + 2y + 1) = 15 + 1$, $2x^2 + (y + 1)^2 = 16$, $2x^2/16 + (y + 1)^2/16 = 1$, and $x^2/8 + (y + 1)^2/16 = 1$. Center: $(0, -1)$.

10.119 $x^2 + 4y^2 - 6x + 32y + 69 = 0$

▮ $(x^2 - 6x + 9) + 4(y^2 + 8y + 16) = -69 + 9 + 64$, $(x - 3)^2 + 4(y + 4)^2 = 4$, and $(x - 3)^2/4 + (y + 4)^2 = 1$. Center: $(3, -4)$.

10.120 $16x^2 + 9y^2 + 32x - 36y - 92 = 0$

▮ $16(x^2 + 2x + 1) + 9(y^2 - 4y + 4) = 92 + 16 + 36$, $16(x + 1)^2 + 9(y - 2)^2 = 144$, and $(x + 1)^2/9 + (y - 2)^2/16 = 1$. Center: $(-1, 2)$.

For Probs. 10.121 and 10.122, find the vertices and foci for the given ellipse.

10.121 $x^2 + 4y^2 - 6x + 32y + 69 = 0$

▮ See Fig. 10.35 and Prob. 10.119. $(x - 3)^2/4 + (y + 4)^2/1 = 1$. Then $a^2 = 4$, $a = 2$; and $b^2 = 1$, $b = 1$. Thus the vertices must be at a distance of $a\ (= 2)$ from the center. The vertices are also along

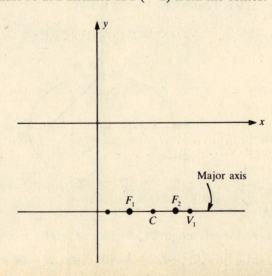

Fig. 10.35

the major axis (here, the major axis is $y = -4$ since a^2 is the denominator of the x term). Vertices: $V_1(5, -4)$ and $V_2(1, -4)$. Also $c = \sqrt{a^2 - b^2} = \sqrt{4 - 1} = \sqrt{3}$. The foci are along the major axis at a distance c (here $c = \sqrt{3}$) from the center. Foci: $(3 \pm \sqrt{3}, -4)$.

10.122 $16x^2 + 9y^2 + 32x - 36y - 92 = 0$

▎ See Prob. 10.120. $(x + 1)^2/9 + (y - 2)^2/16 = 1$. The center is at $(-1, 2)$. Then $a^2 = 16$, $b^2 = 9$, and $c^2 = \sqrt{a^2 - b^2} = \sqrt{16 - 9} = \sqrt{7}$. The vertices are along the major axis. Vertices: $V(-1, 6)$ and $(-1, -2)$. Foci: $(-1, 2 \pm \sqrt{7})$.

For Probs. 10.123 and 10.124, find the equation of the ellipse satisfying the given conditions.

10.123 One vertex at $(0, 10)$, one focus at $(0, -2)$, center at $(0, 0)$

▎ The major axis must be along the y axis. Since the center is $(0, 0)$, $a = 10$, $c = 2$, and $b^2 = a^2 - c^2 = 100 - 4 = 96$. Thus, $x^2/96 + y^2/100 = 1$.

10.124 Vertices $(7, 2)$ and $(-3, 2)$; one focus at $(6, 2)$

▎ The center is the midpoint of the segment joining the two vertices. Center: $\left(\dfrac{7 - 3}{2}, 2\right) = (2, 2)$. Then $a = d(C, V) = 7 - 2 = 5$, $c = d(C, F) = 6 - 2 = 4$, and $b^2 = a^2 - c^2 = 25 - 16 = 9$. The major axis is parallel to the x axis, so $(x - 2)^2/25 + (y - 2)^2/9 = 1$.

For Probs. 10.125 to 10.127, find the intersection(s) of the graphs of the given relations.

10.125 $x^2/1 + y^2/4 = 1$ and $y = x$

▎ If $y = x$, then $x^2 + x^2/4 = 1$ where they meet. Then $5x^2/4 = 1$, $x^2 = \frac{4}{5}$, or $x = \pm 2/\sqrt{5}$. If $x = 2/\sqrt{5}$, then $y = 2/\sqrt{5}$; if $x = -2/\sqrt{5}$, then $y = -2/\sqrt{5}$. Intersection points: $(2/\sqrt{5}, 2\sqrt{5})$ and $(-2/\sqrt{5}, -2/\sqrt{5})$.

10.126 $x^2/1 + y^2/4 = 1$ and $x^2 + y^2 = 1$

▎ $x^2 = 1 - y^2/4$. Thus, where they meet, $1 - y^2/4 + y^2 = 1$. Then $3y^2/4 = 0$, $y = 0$. If $y = 0$, then $x^2 + 0 = 1$, or $x = \pm 1$. Intersection points: $(1, 0)$ and $(-1, 0)$.

10.127 $x^2/1 + y^2/4 = 1$ and $y = x^2$

▎ If $y = x^2$, then where they meet, $y + y^2/4 = 1$. So $4y + y^2 = 4$, $y^2 + 4y - 4 = 0$, and
$$y = \frac{-4 \pm \sqrt{4^2 - 4(1)(-4)}}{2(1)} = \frac{-4 \pm \sqrt{32}}{2} = \frac{-4 \pm 4\sqrt{2}}{2} = -2 \pm 2\sqrt{2}.$$ Then if $y = x^2$, $x = \pm\sqrt{y}$.
If $y = -2 + 2\sqrt{2}$, $x = \sqrt{-2 + 2\sqrt{2}}$ or $x = -\sqrt{-2 + 2\sqrt{2}}$. If $y = -2 - 2\sqrt{2}$, then there is no x corresponding since $-2 - 2\sqrt{2} < 0$. Intersection points: $(\pm\sqrt{-2 + 2\sqrt{2}}, -2 + 2\sqrt{2})$.

10.4 THE HYPERBOLA

For Probs. 10.128 to 10.132, put the equation in standard form for the hyperbola.

10.128 $x^2 - 5y^2 = 10$

▎ The standard form for the hyperbola with center at $(0, 0)$ is $x^2/a^2 - y^2/b^2 = 1$, or $y^2/a^2 - x^2/b^2 = 1$. Dividing by 10 above, we get $x^2/10 - 5y^2/10 = 1$, or $x^2/10 - y^2/2 = 1$.

10.129 $-6y^2 + 7x^2 = 15$

▎ Divide by 15: $-6y^2/15 + 7x^2/15 = 1$, or $x^2/\frac{15}{7} - y^2/\frac{15}{6} = 1$.

10.130 $x^2 + 7y^2 = 14$

▎ This is an ellipse, not a hyperbola. It *cannot* be made to fit either form in Prob. 10.128.

10.131 $-x^2 + \sqrt{2}y^2 = 17$

❚ $1 = \dfrac{-x^2}{17} + \dfrac{\sqrt{2}}{17}y^2 = \dfrac{\sqrt{2}}{17}y^2 - \dfrac{x^2}{17} = \dfrac{y^2}{17/\sqrt{2}} - \dfrac{x^2}{17}$.

10.132 $x^2 - 2y^2 = \dfrac{1}{\sqrt{3}}$

❚ $1 = \sqrt{3}x^2 - 2\sqrt{3}y^2 = \dfrac{x^2}{1/\sqrt{3}} - \dfrac{y^2}{1/2\sqrt{3}}$.

For Probs. 10.133 to 10.144, find the foci and lengths of the transverse and conjugate axes.

10.133 $x^2/9 - y^2/4 = 1$

❚ If $x^2/a^2 - y^2/b^2 = 1$, then the foci are $(\pm c, 0)$, where $c^2 = a^2 + b^2$, and the lengths of the transverse and conjugate axes are $2a$ and $2b$, respectively. Here $a^2 = 9$, $b^2 = 4$, and $c^2 = 9 + 4 = 13$. Thus, the foci are $(\pm\sqrt{13}, 0)$. The length of the transverse axis $= 6$ ($= 2a$), and the length of the conjugate axis $= 4$ ($= 2b$).

10.134 $x^2/9 - y^2/25 = 1$

❚ See Prob. 10.133. $x^2/a^2 - y^2/b^2 = 1$; $a^2 = 9$, $b^2 = 25$; thus, $c^2 = 9 + 25 = 34$. The foci are at $(\pm\sqrt{34}, 0)$; the length of the transverse axis $= 6$; the length of the conjugate axis $= 10$.

10.135 $y^2/4 - x^2/9 = 1$

❚ Here, $y^2/a^2 - x^2/b^2 = 1$. If $c^2 = a^2 + b^2$, then the foci are at $(0, \pm c)$; the transverse axis length is $2a$, the conjugate axis length is $2b$. Then $c^2 = 4 + 9 = 13$. The foci are at $(0, \pm\sqrt{13})$; the length of the transverse axis $= 4$; and the length of the conjugate axis $= 6$.

10.136 $y^2/25 - x^2/9 = 1$

❚ See Prob. 10.135. $c^2 = a^2 + b^2 = 25 + 9 = 34$. The foci are at $(0, \pm\sqrt{34})$; the length of the transverse axis $= 10$; the length of the conjugate axis $= 6$.

10.137 $4x^2 - y^2 = 16$

❚ Divide by 16: $x^2/4 - y^2/16 = 1$. Then $c^2 = a^2 + b^2 = 20$, and $c = \pm\sqrt{20} = \pm2\sqrt{5}$. The foci are at $(\pm2\sqrt{5}, 0)$; the length of the transverse axis $= 4$; and the length of the conjugate axis $= 8$.

10.138 $x^2 - 9y^2 = 9$

❚ $x^2/9 - y^2 = 1$. Then $c^2 = a^2 + b^2 = 9 + 1 = 10$. The foci at $(\pm\sqrt{10}, 0)$; the length of the transverse axis $= 6$; and the length of the conjugate axis $= 2$.

10.139 $9y^2 - 16x^2 = 144$

❚ $y^2/16 - x^2/9 = 1$. Then $c^2 = a^2 + b^2 = 16 + 9 = 25$, and $c = \pm5$. The foci are at $(0, \pm5)$; the length of the transverse axis $= 8$; and the length of the conjugate axis $= 6$.

10.140 $4y^2 - 25x^2 = 100$

❚ $y^2/25 - x^2/4 = 1$. Then $c^2 = a^2 + b^2 = 25 + 4 = 29$. The foci are at $(0, \pm\sqrt{29})$; the length of the transverse axis $= 10$; and the length of the conjugate axis $= 4$.

10.141 $3x^2 - 2y^2 = 12$

❚ $3x^2/12 - 2y^2/12 = 1$; $x^2/4 - y^2/6 = 1$. Then $c^2 = 4 + 6 = 10$. The foci are at $(\pm\sqrt{10}, 0)$; the length of the transverse axis $= 4$; and the length of the conjugate axis $= 2\sqrt{6}$.

10.142 $3x^2 - 4y^2 = 24$

▌ $3x^2/24 - 4y^2/24 = 1$; $x^2/8 - y^2/6 = 1$. Then $c^2 = a^2 + b^2 = 8 + 6 = 14$. The foci are at $(\pm\sqrt{14}, 0)$; the length of the transverse axis $= 2a = 2(2\sqrt{2}) = 4\sqrt{2}$; and the length of the conjugate axis $= 2b = 2\sqrt{6}$.

10.143 $7y^2 - 4x^2 = 28$

▌ $y^2/4 - x^2/7 = 1$. Then $c^2 = a^2 + b^2 = 4 + 7 = 11$. The foci are at $(0, \pm\sqrt{11})$; the length of the transverse axis $= 4$; and the length of the conjugate axis $= 2\sqrt{7}$.

10.144 $3y^2 - 2x^2 = 24$

▌ $y^2/8 - x^2/12 = 1$. Then $c^2 = 8 + 12 = 20$, $c = 2\sqrt{5}$. The foci are at $(0, \pm2\sqrt{5})$; the length of the transverse axis $= 2(2\sqrt{2}) = 4\sqrt{2}$; and the length of the conjugate axis $= 2(2\sqrt{3}) = 4\sqrt{3}$.

For Probs. 10.145 to 10.156, sketch the given hyperbola. *Sketching hint*: Sketch first the rectangle formed by the two axes, then the diagonals to get the asymptotes, then the vertices and the hyperbola.

10.145 $x^2/9 - y^2/4 = 1$

▌ See Fig. 10.36 and Prob. 10.133. Vertices are V_1, V_2 $(\pm3, 0)$ (when $y = 0$). The center is at $C(0, 0)$; the length of the transverse axis $= 6$; and the length of the conjugate axis $= 4$.

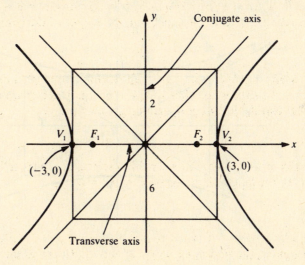

Fig. 10.36

10.146 $x^2/9 - y^2/25 = 1$

▌ See Fig. 10.37 and Prob. 10.134.

10.147 $y^2/4 - x^2/9 = 1$

▌ See Fig. 10.38 and Prob. 10.135. Vertices at $(0, \pm2)$ (when $x = 0$).

10.148 $y^2/25 - x^2/9 = 1$

▌ See Fig. 10.39 and Prob. 10.136.

10.149 $4x^2 - y^2 = 16$

▌ See Fig. 10.40 and Prob. 10.137. Then $x^2/4 - y^2/16 = 1$.

10.150 $x^2 - 9y^2 = 9$

▌ See Fig. 10.41 and Prob. 10.138. Then $x^2/9 - y^2 = 1$. Ask yourself, What significance does the y axis have?

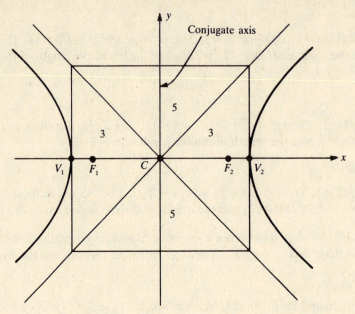

Fig. 10.37

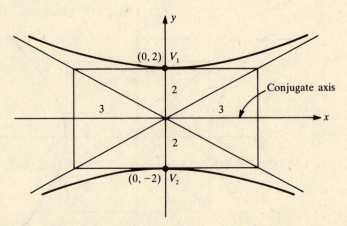

Fig. 10.38

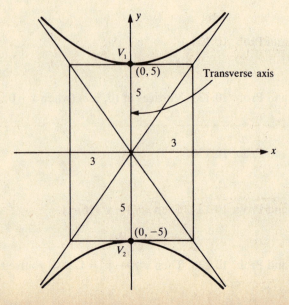

Fig. 10.39

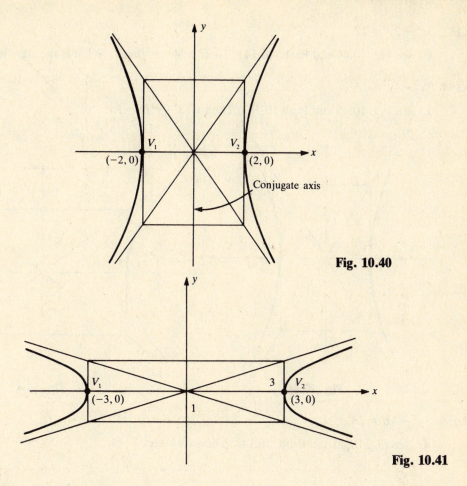

Fig. 10.40

Fig. 10.41

10.151 $9y^2 - 16x^2 = 144$

▮ See Fig. 10.42 and Prob. 10.139. Then $y^2/16 - x^2/9 = 1$. $\overline{V_1V_2}$ is the transverse axis.

10.152 $4y^2 - 25x^2 = 100$

▮ See Fig. 10.43 and Prob. 10.140. Then $y^2/25 - x^2/4 = 1$.

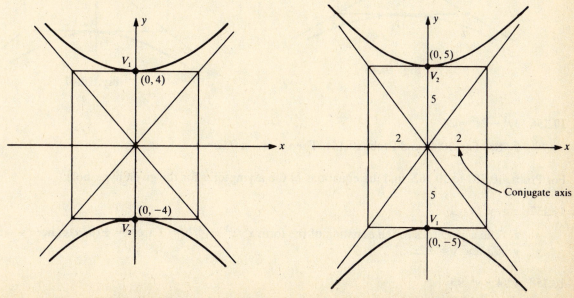

Fig. 10.42   Fig. 10.43

10.153 $3x^2 - 2y^2 = 12$

▎ See Fig. 10.44 and Prob. 10.141. Then $x^2/4 - y^2/6 = 1$. What is the significance of the x axis?

10.154 $3x^2 - 4y^2 = 24$

▎ See Fig. 10.45 and Prob 10.142. Then $x^2/8 - y^2/6 = 1$.

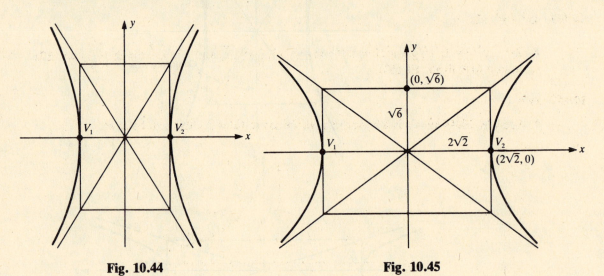

Fig. 10.44   Fig. 10.45

10.155 $7y^2 - 4x^2 = 28$

▎ See Fig. 10.46 and Prob. 10.143. Then $y^2/4 - x^2/7 = 1$.

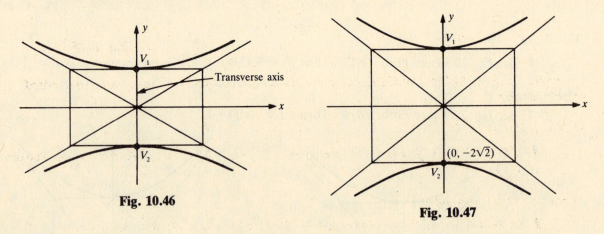

Fig. 10.46   Fig. 10.47

10.156 $3y^2 - 2x^2 = 24$

▎ See Fig. 10.47 and Prob. 10.144. Then $y^2/8 - x^2/12 = 1$.

For Probs. 10.157 to 10.162, find the equations of the asymptotes for the given hyperbola.

10.157 $x^2 - y^2/4 = 1$

▎ The asymptotes for an equation of the form $x^2/a^2 - y^2/b^2 = 1$ are $y = \pm(b/a)x$; here $y = \pm\frac{2}{1}x$ or $y = 2x$ and $y = -2x$.

10.158 $x^2/4 - y^2/9 = 1$

▎ Here, $a = 2$, $b = 3$; $y = \pm\frac{3}{2}x$, or $y = \frac{3}{2}x$ and $y = -\frac{3}{2}x$.

10.159 $y^2/3 - x^2/2 = 1$

▌ This hyperbola is of the form $y^2/a^2 - x^2/b^2 = 1$. Here, $y = \pm(\sqrt{3}/\sqrt{2})x$, or $y = \pm\sqrt{\frac{3}{2}}x$.

10.160 $y^2 - 4x^2 = 1$

▌ This is $y^2 - x^2/\frac{1}{4} = 1$ in standard form. Then $a = 1$, $b = \frac{1}{2}$, and $y = \pm(a/b)x = \pm(1/\frac{1}{2})x = \pm2x$.

10.161 $4x^2 - y^2 = 2$

▌ $2x^2 - y^2/2 = 1$, $x^2/\frac{1}{2} - y^2/2 = 1$. Then $a = \sqrt{\frac{1}{2}}$, $b = \sqrt{2}$, and $y = \pm\dfrac{b}{a}x = \pm\dfrac{\sqrt{2}}{1/\sqrt{2}}x = \pm2x$ (compare with Prob. 10.160).

10.162 $xy = 1$

▌ See Fig. 10.48 $xy = a$ is a hyperbola whose asymptotes are the x and y axes.

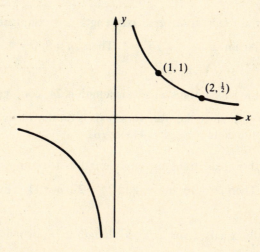

Fig. 10.48

For Probs. 10.163 to 10.165, find the center of the given hyperbola and put the equation in standard hyperbola form.

10.163 $16y^2 - 9x^2 = 144$

▌ $y^2/9 - x^2/16 = 1$. This is of the form $(y - k)^2/a^2 - (x - h)^2/b^2 = 1$, where $h = k = 0$. The center is at $(0, 0)$.

10.164 $x^2 - 4y^2 + 6x + 16y - 11 = 0$

▌ $(x^2 + 6x + 9) - 4(y^2 - 4y + 4) = 11 + 9 - 16$, $(x + 3)^2 - 4(y - 2)^2 = 4$, and $(x + 3)^2/4 - (y - 2)^2/1 = 1$. The center is at $(-3, 2)$ since $h = -3$, $k = 2$.

10.165 $144x^2 - 25y^2 - 576x + 200y + 3776 = 0$

▌ $144(x^2 - 4x + 4) - 25(y^2 - 8y + 16) = -3776 + 576 - 400$, $144(x - 2)^2 - 25(y - 4)^2 = -3600$, $\dfrac{25(y - 4)^2}{3600} - \dfrac{144(x - 2)^2}{3600} = 1$, $\dfrac{(y - 4)^2}{\frac{3600}{25}} - \dfrac{(x - 2)^2}{\frac{3600}{144}} = 1$, $\dfrac{(y - 4)^2}{144} - \dfrac{(x - 2)^2}{25} = 1$. The center is at $(2, 4)$.

For Probs. 10.166 and 10.167, find the vertices, foci, and asymptotes.

10.166 $x^2 - 4y^2 + 6x + 16y - 11 = 0$

▌ See Prob. 10.164. Then $(x + 3)^2/4 - (y - 2)^2/1 = 1$. Thus, $a = 2$, $b = 1$, and $c = \sqrt{a^2 + b^2} = \sqrt{5}$. The vertices are on the transverse axis at a distance a from the center $(-3, 2)$. Thus, the vertices are

$(-1, 2)$ and $(-5, 2)$. The foci are at a distance c from the center $(-3, 2)$. Thus, the foci are $(-3 \pm \sqrt{5}, 2)$.

10.167 $144x^2 - 25y^2 - 576x + 200y + 3776 = 0$

▮ See Prob. 10.165. Then $(y - 4)^2/144 - (x - 2)^2/25 = 1$. Thus, $a = 12$, $b = 5$, and $c = \sqrt{a^2 + b^2} = \sqrt{144 + 25} = 13$. Then the vertices are $V_1 = (2, 4 + 12) = (2, 16)$ and $V_2 = (2, 4 - 12) = (2, -8)$. The foci are $F_1 = (2, 4 + 13) = (2, 17)$ and $F_2 = (2, 4 - 13) = (2, -9)$.

For Probs. 10.168 to 10.173, find the equation of the hyperbola with center $(0, 0)$ satisfying the given conditions.

10.168 Transverse axis is on the x axis; transverse axis length is 14; conjugate axis length is 10.

▮ Since the transverse axis is on the x axis, $x^2/a^2 - y^2/b^2 = 1$. Then $2a = 14$, $a = 7$; and $2b = 10$, $b = 5$. Thus, the equation is $x^2/49 - y^2/25 = 1$.

10.169 Transverse axis is on the x axis; transverse axis length is 8; conjugate axis length is 6.

▮ See Prob. 10.168. Again, $x^2/a^2 - y^2/b^2 = 1$. Then $2a = 8$, $a = 4$; and $2b = 6$, $b = 3$. Thus, the equation is $x^2/16 - y^2/9 = 1$.

10.170 Transverse axis is on the y axis; transverse axis length is 24; conjugate axis length is 18.

▮ Here $y^2/a^2 - x^2/b^2 = 1$, since the transverse axis is on the y axis. Then $2a = 24$, $a = 12$; and $2b = 18$, $b = 9$. Thus, the equation is $y^2/144 - x^2/81 = 1$.

10.171 Transverse axis is on the y axis; transverse axis length is 16; conjugate axis length is 22.

▮ $y^2/a^2 - x^2/b^2 = 1$. Then $2a = 16$, $a = 8$; and $2b = 22$, $b = 11$. Thus, the equation is $y^2/64 - x^2/121 = 1$.

10.172 Transverse axis is on the x axis; transverse axis length is 18; distance from foci to center is 11.

▮ $x^2/a^2 - y^2/b^2 = 1$. Then $2a = 18$, $a = 9$; and $c = 11 =$ distance from foci to center. Since $a^2 + b^2 = c^2$, $b^2 = c^2 - a^2 = 121 - 81 = 40$. Thus, the equation is $x^2/81 - y^2/40 = 1$.

10.173 Conjugate axis is on the x axis; conjugate axis length is 14; distance from foci to center is $\sqrt{200}$.

▮ The transverse axis is on the y axis, so $y^2/a^2 - x^2/b^2 = 1$. Then $2b = 14$, $b = 7$; and $a^2 = c^2 - b^2 = 200 - 49 = 151$. Thus, $y^2/151 - x^2/49 = 1$.

For Probs. 10.174 to 10.176, find the equation of the hyperbola fitting the given conditions.

10.174 Vertices $(\pm 5, 0)$; focus $(13, 0)$

▮ Here the transverse axis is on the x axis since $(5, 0)$ and $(-5, 0)$ are vertices. Also $c =$ distance from center to focus $= 13$, since the center is at $(0, 0)$ [midway between $(5, 0)$ and $(-5, 0)$]. Thus, $a = 5$, $c = 13$, and $b^2 = c^2 - a^2 = 169 - 25 = 144$. Thus, $x^2/25 - y^2/144 = 1$.

10.175 Center $(0, 0)$; $a = 5$; focus $(0, 6)$

▮ If the focus is at $(0, 6)$, then the transverse axis is on the y axis and $c = 6$. Then $b^2 = c^2 - a^2 = 36 - 25 = 11$. Thus, $y^2/25 - x^2/11 = 1$.

10.176 Center $(2, -3)$; vertex $(7, -3)$; asymptote $3x - 5y - 21 = 0$

▮ If the center is at $(2, -3)$ and one vertex is $(7, -3)$, the other must be at $(-3, -3)$ so that c is the midpoint of $\overline{V_1 V_2}$. If $3x - 5y - 21 = 0$, then $5y = 3x - 21$ and $y = \frac{3}{5}x - \frac{21}{5}$. The slope of $\frac{3}{5}$ of the asymptotes is b/a. But $CV = a = 5$; thus $\frac{3}{5} = b/5$, $b = 3$. $(x - 2)^2/25 - (y + 3)^2/9 = 1$.

For Probs. 10.177 to 10.180, find the intersection(s) of the given curves.

10.177 $x^2 - y^2 = 1$ and $y = x$

    ▮ Then $x^2 - x^2 = 1$, $0 = 1$; no intersection.

10.178 $x^2 - 2y^2 = 1$ and $x^2 + y^2 = 4$

    ▮ $x^2 = 4 - y^2$. Thus, $4 - y^2 - 2y^2 = 1$, $-3y^2 = -3$, $y^2 = 1$, or $y = \pm 1$. If $y = 1$, $x^2 = 4 - 1$, or $x = \pm\sqrt{3}$. If $y = -1$, $x^2 = 4 - 1$, or $x = \pm\sqrt{3}$. Intersections: $(\sqrt{3}, 1)$, $(-\sqrt{3}, 1)$, $(\sqrt{3}, -1)$, $(-\sqrt{3}, -1)$.

10.179 $x^2 - 2y^2 = 1$ and $x^2 + 2y^2 = 1$

    ▮ Adding, we get $2x^2 = 2$, $x^2 = 1$, or $x = \pm 1$. If $x = 1$, $1 + 2y^2 = 1$, or $y = 0$. If $x = -1$, $1 + 2y^2 = 1$, or $y = 0$. Intersections: $(1, 0)$, $(-1, 0)$.

10.180 $x^2 - 2y^2 = 1$ and $y = 2x^2$

    ▮ $x^2 - 2(2x^2)^2 = 1$, $x^2 - 8x^4 = 1$, and $8x^4 - x^2 + 1 = 0$. Let $u = x^2$. Then $8u^2 - u + 1 = 0$ and $u = \dfrac{1 \pm \sqrt{1 - 4(8)(1)}}{16}$. There is no solution (negative discriminant), and, therefore, there are no intersections.

10.5 MISCELLANEOUS PROBLEMS

For Probs. 10.181 to 10.186, write the equation of the circle satisfying the given conditions.

10.181 $C(0, 0)$, $r = 5$

    ▮ Using $(x - h)^2 + (y - k)^2 = r^2$, we see that the equation is $(x - 0)^2 + (y - 0)^2 = 25$ or $x^2 + y^2 = 25$.

10.182 $C(4, -2)$, $r = 8$

    ▮ Using $(x - h)^2 + (y - k)^2 = r^2$, we get $(x - 4)^2 + (y + 2)^2 = 64$.

10.183 $C(-4, -2)$ and passing through $P(1, 3)$

    ▮ Since the center is at $C(-4, -2)$, the equation has the form $(x + 4)^2 + (y + 2)^2 = r^2$. The condition that $P(1, 3)$ lie on this circle is $(1 + 4)^2 + (3 + 2)^2 = r^2 = 50$. Hence, the required equation is $(x + 4)^2 + (y + 2)^2 = 50$.

10.184 $C(-5, 6)$ and tangent to x axis

    ▮ The tangent to a circle is perpendicular to the radius drawn to the point of tangency; hence, $r = 6$. The equation of the circle is $(x + 5)^2 + (y - 6)^2 = 36$. See Fig. 10.49.

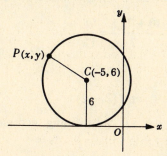

Fig. 10.49

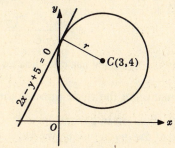

Fig. 10.50

10.185 $C(3, 4)$ and tangent to $2x - y + 5 = 0$

    ▮ The radius is the undirected distance of the point $C(3, 4)$ from the line $2x - y + 5 = 0$; thus, $r = \dfrac{|2 \cdot 3 - 4 + 5|}{|-\sqrt{5}|} = \dfrac{7}{\sqrt{5}}$. The equation of the circle is $(x - 3)^2 + (y - 4)^2 = \frac{49}{5}$. See Fig. 10.50.

10.186 Center on $y = x$, tangent to both axes, $r = 4$

> **▌** Since the center (h, k) lies on the line $x = y$, $h = k$; since the circle is tangent to both axes, $|h| = |k| = r$. Thus, there are two circles satisfying the conditions, one with center $(4, 4)$ and equation $(x - 4)^2 + (y - 4)^2 = 16$, the other with center $(-4, -4)$ and equation $(x + 4)^2 + (y + 4)^2 = 16$. See Fig. 10.51.

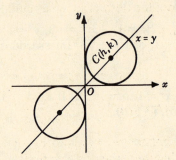

Fig. 10.51

For Probs. 10.187 through 10.190, describe the locus represented by the given equation.

10.187 $x^2 + y^2 - 10x + 8y + 5 = 0$

> **▌** From the standard form $(x - 5)^2 + (y + 4)^2 = 36$, the locus is a circle with center at $C(5, -4)$ and radius 6.

10.188 $x^2 + y^2 - 6x - 8y + 25 = 0$

> **▌** From the standard form $(x - 3)^2 + (y - 4)^2 = 0$, the locus is a point circle or the point $(3, 4)$.

10.189 $x^2 + y^2 + 4x - 6y + 24 = 0$

> **▌** Here we have $(x + 2)^2 + (y - 3)^2 = -11$; the locus is imaginary.

10.190 $4x^2 + 4y^2 + 80x + 12y + 265 = 0$

> **▌** Dividing by 4, we have $x^2 + y^2 + 20x + 3y + \frac{265}{4} = 0$. From $(x + 10)^2 + (y + \frac{3}{2})^2 = 36$, the locus is a circle with center at $C(-10, -\frac{3}{2})$ and radius 6.

10.191 Show that the circles $x^2 + y^2 - 16x - 20y + 115 = 0$ and $x^2 + y^2 + 8x - 10y + 5 = 0$ are tangent. Find the point of tangency.

> **▌** The first circle has center $C_1(8, 10)$ and radius 7; the second has center $C_2(-4, 5)$ and radius 6. The two circles are tangent externally since the distance between their centers $C_1 C_2 = \sqrt{144 + 25} = 13$ is equal to the *sum* of the radii. The point of tangency $P(x, y)$ divides the segment $C_2 C_1$ in the ratio $6:7$. Then
>
> $$x = \frac{6 \cdot 8 + 7(-4)}{6 + 7} = \frac{20}{13} \qquad y = \frac{6 \cdot 10 + 7 \cdot 5}{6 + 7} = \frac{95}{13}$$

and the point of tangency has coordinates $(\frac{20}{13}, \frac{95}{13})$.

10.192 Find the equation of the circle through the points $(5, 1)$, $(4, 6)$, and $(2, -2)$.

> **▌** Take the equation in general form $x^2 + y^2 + 2Dx + 2Ey + F = 0$. Substituting successively the coordinates of the given points, we have
>
> $$\begin{array}{ll} 25 + 1 + 10D + 2E + F = 0 & \qquad 10D + 2E + F = -26 \\ 16 + 36 + 8D + 12E + F = 0 \qquad \text{or} & \qquad 8D + 12E + F = -52 \\ 4 + 4 + 4D - 4E + F = 0 & \qquad 4D - 4E + F = -8 \end{array}$$

with solution $D = -\frac{1}{3}$, $E = -\frac{8}{3}$, $F = -\frac{52}{3}$. Thus the required equation is

$$x^2 + y^2 - \frac{2}{3}x - \frac{16}{3}y - \frac{52}{3} = 0 \qquad \text{or} \qquad 3x^2 + 3y^2 - 2x - 16y - 52 = 0$$

10.193 Write the equations of the circles having radius $\sqrt{13}$ and tangent to the line $2x - 3y + 1 = 0$ at $(1, 1)$.

▌ Let the equation of the circle be $(x - h)^2 + (y - k)^2 = 13$. Since the coordinates $(1, 1)$ satisfy this equation, we have

(1)
$$(1 - h)^2 + (1 - k)^2 = 13$$

The undirected distance from the tangent to the center of the circle is equal to the radius, that is,

(2)
$$\left| \frac{2h - 3k + 1}{-\sqrt{13}} \right| = \sqrt{13} \quad \text{and} \quad \frac{2h - 3k + 1}{\sqrt{13}} = \pm\sqrt{13}$$

Finally, the radius through $(1, 1)$ is perpendicular to the tangent there, that is,

(3)
$$\text{Slope of radius through } (1, 1) = -\frac{1}{\text{slope of given line}} \quad \text{or} \quad \frac{k - 1}{h - 1} = -\frac{3}{2}$$

Since there are only two unknowns, we may solve simultaneously any two of the three equations. Using (2) and (3), noting that there are two equations in (2), we find $h = 3$, $k = -2$ and $h = -1$, $k = 4$. The equations of the circles are $(x - 3)^2 + (y + 2)^2 = 13$ and $(x + 1)^2 + (y - 4)^2 = 13$.

10.194 Write the equation of the circle through $(2, 3)$ and $(-1, 6)$, with center on $2x + 5y + 1 = 0$.

▌ Take the equation of the circle in standard form $(x - h)^2 + (y - k)^2 = r^2$. We obtain the following system of equations:
(1) $2h + 5k + 1 = 0$ center (h, k) on $2x + 5y + 1 = 0$
(2) $(2 - h)^2 + (3 - k)^2 = r^2$ point $(2, 3)$ on the circle
(3) $(-1 - h)^2 + (6 - k)^2 = r^2$ point $(-1, 6)$ on the circle

The elimination of r between (2) and (3) yields $h - k + 4 = 0$, and when this is solved simultaneously with (1), we obtain $h = -3$, $k = 1$. By (2), $r^2 = (2 + 3)^2 + (3 - 1)^2 = 29$; the equation of the circle is $(x + 3)^2 + (y - 1)^2 = 29$.

For Probs. 10.195 through 10.198, for the given parabola, sketch the curve, find the coordinates of the vertex and focus, and give the equations of the axis and directrix.

10.195 $y^2 = 16x$

▌ The parabola opens to the right $(p > 0)$ with vertex at $V(0, 0)$. The equation of its axis is $y = 0$. Moving from V to the right along the axis a distance $|p| = 4$, we locate the focus at $F(4, 0)$. Moving from V to the left along the axis a distance $|p| = 4$, we locate the point $D(-4, 0)$. Since the directrix passes through D perpendicular to the axis, its equation is $x + 4 = 0$. See Fig. 10.52.

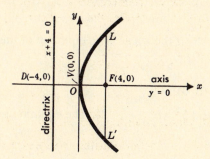

Fig. 10.52

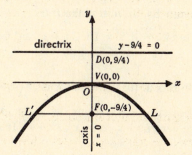

Fig. 10.53

10.196 $x^2 = -9y$

▌ The parabola opens downward $(p < 0)$ with vertex at $V(0, 0)$. The equation of its axis is $x = 0$. Moving from V downward along the axis a distance $|p| = \frac{9}{4}$, we locate the focus at $F(0, -\frac{9}{4})$. Moving from V upward along the axis a distance $|p| = \frac{9}{4}$, we locate the point $D(0, \frac{9}{4})$; the equation of the directrix is $4y - 9 = 0$. See Fig. 10.53.

10.197 $x^2 - 2x - 12y + 25 = 0$

▌ Here $(x - 1)^2 = 12(y - 2)$. The parabola opens upward ($p > 0$) with vertex at $V(1, 2)$. The equation of its axis is $x - 1 = 0$. Since $|p| = 3$, the focus is at $F(1, 5)$, and the equation of the directrix is $y + 1 = 0$. See Fig. 10.54.

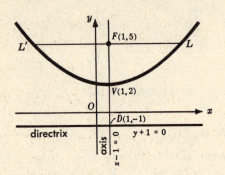

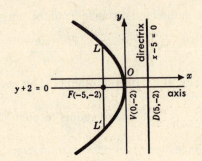

Fig. 10.54   Fig. 10.55

10.198 $y^2 + 4y + 20x + 4 = 0$

▌ Here $(y + 2)^2 = -20x$. The parabola opens to the left ($p < 0$) with vertex at $V(0, -2)$, and the equation of the directrix is $x - 5 = 0$. See Fig. 10.55.

For Probs. 10.199 through 10.201, find the equation of the parabola.

10.199 $V(0, 0)$; $F(0, -4)$

▌ Since the directed distance $p = VF = -4$, the parabola opens downward. Its equation is $x^2 = -16y$.

10.200 $V(0, 0)$; directrix: $x = -5$

▌ The parabola opens to the right (away from the directrix). Since $p = DV = 5$, the equation is $y^2 = 20x$.

10.201 $V(0, 0)$; axis: $y = 0$; passing through $(4, 5)$

▌ The equation of this parabola is of the form $y^2 = 4px$. If $(4, 5)$ is a point on it, then $5^2 = 4p(4)$, $4p = \frac{25}{4}$, and the equation is $y^2 = \frac{25}{4}x$.

For Probs. 10.202 through 10.207, find the equation of the ellipse.

10.202 Vertices $(\pm 8, 0)$, minor axis $= 6$

▌ Here $2a = V'V = 16$, $2b = 6$, and the major axis is along the x axis. The equation of the ellipse is $x^2/a^2 + y^2/b^2 = x^2/64 + y^2/9 = 1$.

10.203 One vertex at $(0, 13)$, one focus at $(0, -12)$, center at $(0, 0)$

▌ The major axis is along the y axis, $a = 13$, $c = 12$, and $b^2 = a^2 - c^2 = 25$. The equation of the ellipse is $x^2/b^2 + y^2/a^2 = x^2/25 + y^2/169 = 1$.

10.204 Foci $(\pm 10, 0)$, eccentricity $= \frac{5}{6}$

▌ Here the major axis is along the x axis and $c = 10$. Since $e = c/a = 10/a = \frac{5}{6}$, $a = 12$ and $b^2 = a^2 - c^2 = 44$. The equation of the ellipse is $x^2/144 + y^2/44 = 1$.

10.205 Vertices $(8, 3)$ and $(-4, 3)$, one focus at $(6, 3)$

▌ The center is at the midpoint of $V'V$, that is, at $C(2, 3)$. Then $a = CV = 6$, $c = CF = 4$, and

$b^2 = a^2 - c^2 = 20$. Since the major axis is parallel to the x axis, the equation of the ellipse is $\dfrac{(x-2)^2}{36} + \dfrac{(y-3)^2}{20} = 1$.

10.206 Vertices $(3, -10)$ and $(3, 2)$, length of latus rectum $= 4$

▌ The center is at $C(3, -4)$ and $a = 6$. Since $2b^2/a = 2b^2/6 = 4$, $b^2 = 12$. The major axis is parallel to the y axis, and the equation of the ellipse is $\dfrac{(x-3)^2}{12} + \dfrac{(y+4)^2}{36} = 1$.

10.207 Directrices $4y - 33 = 0$, $4y + 17 = 0$; major axis on $x + 1 = 0$; eccentricity $= \frac{4}{5}$

▌ The major axis intersects the directrices in $D(-1, \frac{33}{4})$ and $D'(-1, -\frac{17}{4})$. The center of the ellipse bisects $D'D$ and hence is at $C(-1, 2)$. Since $CD = a/e = \frac{25}{4}$ and $e = \frac{4}{5}$, $a = 5$ and $c = ae = 4$. Then $b^2 = a^2 - c^2 = 9$. Since the major axis is parallel to the y axis, the equation of the ellipse is $\dfrac{(x+1)^2}{9} + \dfrac{(y-2)^2}{25} = 1$.

10.208 Find the equation of the locus of the midpoints of a system of parallel chords of slope m of the ellipse $x^2/a^2 + y^2/b^2 = 1$.

▌ Let $P(x, y)$ be any point on the locus, and let $Q(p, q)$ and $R(r, s)$ be the extremities of the chord of which P is the midpoint. Then

(1) $\qquad\qquad b^2p^2 + a^2q^2 = a^2b^2 \qquad$ since $Q(p, q)$ is on ellipse

(2) $\qquad\qquad b^2r^2 + a^2s^2 = a^2b^2 \qquad$ since $R(r, s)$ is on ellipse

(3) $\qquad\qquad \dfrac{q-s}{p-r} = m \qquad$ since m is slope of chord

(4) $\qquad\qquad x = \frac{1}{2}(p+r) \qquad y = \frac{1}{2}(q+s) \qquad$ since P is midpoint of QR

Equating the left members of (1) and (2), we have

$$b^2p^2 + a^2q^2 = b^2r^2 + a^2s^2$$

Then, using (3) and (4), we get

$$b^2(p^2 - r^2) = a^2(s^2 - q^2) \qquad \text{and} \qquad \frac{b^2}{a^2} = \frac{s^2 - q^2}{p^2 - r^2} = -\frac{q-s}{p-r} \cdot \frac{q+s}{p+r} = -m\,\frac{2y}{2x} = -m\,\frac{y}{x}$$

Thus the desired equation is $y = -\dfrac{b^2}{a^2m}\,x$. The locus passes through the center $C(0, 0)$ and is called a *diameter* of the ellipse. See Fig. 10.56.

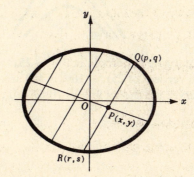

Fig. 10.56

For Probs. 10.209 through 10.214, find the equation of the hyperbola.

10.209 Center $(0, 0)$, vertex $(4, 0)$, focus $(5, 0)$

┃ Here $a = CV = 4$, $c = CF = 5$, and $b^2 = c^2 - a^2 = 25 - 16 = 9$. The transverse axis is along the x axis, and the equation of the hyperbola is $x^2/16 - y^2/9 = 1$.

10.210 Center $(0, 0)$, focus $(0, -4)$, eccentricity $= 2$

┃ Since $c = F'C = 4$ and $e = c/a = 2$, $a = 2$ and $b^2 = c^2 - a^2 = 12$. The transverse axis is along the y axis, and the equation of the hyperbola is $y^2/4 - x^2/12 = 1$.

10.211 Center $(0, 0)$, vertex $(5, 0)$, one asymptote $5y + 3x = 0$

┃ The slope of the asymptote is $-b/a = -\frac{3}{5}$, and since $a = CV = 5$, $b = 3$. The transverse axis is along the x axis, and the equation of the hyperbola is $x^2/25 - y^2/9 = 1$.

10.212 Center $(-5, 4)$, vertex $(-11, 4)$, eccentricity $= \frac{5}{3}$

┃ Here $a = V'C = 6$, and $e = c/a = \frac{5}{3} = \frac{10}{6}$; then $c = 10$ and $b^2 = c^2 - a^2 = 64$. The transverse axis is parallel to the x axis, and the equation of the hyperbola is $(x + 5)^2/36 - (y - 4)^2/64 = 1$.

10.213 Vertices $(-11, 1)$ and $(5, 1)$, one asymptote $x - 4y + 7 = 0$

┃ The center is at $(-3, 1)$, the midpoint of VV'. The slope of the asymptote is $b/a = \frac{1}{4}$, and since $a = CV = 8$, $b = 2$. The transverse axis is parallel to the x axis, and the equation of the hyperbola is $(x + 3)^2/64 - (y - 1)^2/4 = 1$.

10.214 Transverse axis parallel to the x axis, asymptotes $3x + y - 7 = 0$ and $3x - y - 5 = 0$, passes through $(4, 4)$

┃ The asymptotes intersect in the center $C(2, 1)$. Since the slope of the asymptote $3x - y - 5 = 0$ is $b/a = 3/1$, we take $a = m$ and $b = 3m$. The equation of the hyperbola may be written as $(x - 2)^2/m^2 - (y - 1)^2/(9m^2) = 1$. In order that the hyperbola pass through $(4, 4)$, $4/m^2 - 9/(9m^2) = 1$ and $m = \sqrt{3}$. Then $a = m = \sqrt{3}$, $b = 3m = 3\sqrt{3}$, and the required equation is $(x - 2)^2/3 - (y - 1)^2/27 = 1$.

10.215 Write the equation of the conjugate of the hyperbola $25x^2 - 16y^2 = 400$, and sketch both curves.

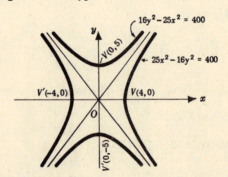

Fig. 10.57

┃ The equation of the conjugate hyperbola is $16y^2 - 25x^2 = 400$. The common asymptotes have equations $y = \pm 5x/4$. The vertices of $25x^2 - 16y^2 = 400$ are at $(\pm 4, 0)$. The vertices of $16y^2 - 25x^2 = 400$ are at $(0, \pm 5)$. The curves are shown in Fig. 10.57.

CHAPTER 11
The Complex Numbers

11.1 POLAR FORM

For Probs. 11.1 to 11.5, plot the given complex number.

11.1 $2 + i$

▌ Remember that, given $a + bi$ where $a, b \in \mathscr{R}$, $a + bi$ corresponds to the point (a, b) in the cartesian plane. Thus, $2 + i$ corresponds to $(2, 1)$. See Fig. 11.1.

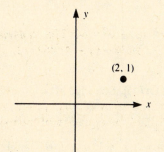

Fig. 11.1

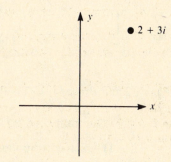

Fig. 11.2

11.2 $2 + 3i$

▌ See Prob. 11.1. $2 + 3i$ corresponds to the point $(2, 3)$. See Fig. 11.2.

11.3 $2 - i$

▌ $2 - i$ corresponds to $(2, -1)$. See Fig. 11.3.

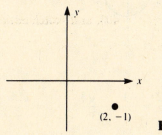

Fig. 11.3

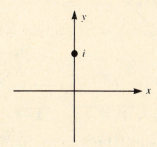

Fig. 11.4

11.4 i

▌ Since $i = 0 + 1i$, i corresponds to $(0, 1)$. See Fig. 11.4.

11.5 6

▌ $6 = 6 + 0i$ and corresponds to the point $(6, 0)$. See Fig. 11.5.

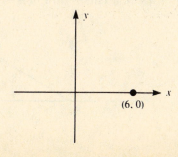

Fig. 11.5

For Probs. 11.6 to 11.26, give the polar (or trigonometric) form for the given complex number.

11.6 $4 + 0i$

▌ $d(A, P) = 4$ and $\theta = 0$ (θ is the angle between AP and the positive x axis), so $4 + 0i = 4(\cos 0 + i \sin 0)$. See Fig. 11.6.

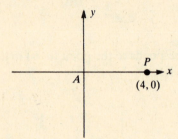

Fig. 11.6

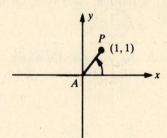

Fig. 11.7

11.7 $1 + i$

▌ $d(A, P) = \sqrt{2}$ and $\theta = \pi/4$, so $1 + i = \sqrt{2}[\cos(\pi/4) + i \sin(\pi/4)]$. See Fig. 11.7.

11.8 $-1 - i$

▌ Let $P = (-1, -1)$ which corresponds to $-1 - i$. $d(A, P) = r = \sqrt{1^2 + 1^2} = \sqrt{2}$. θ = angle from the positive x axis to the distance segment $= 225° = 5\pi/4$. $-1 - i = \sqrt{2}[\cos(5\pi/4) + i \sin(5\pi/4)]$. See Fig. 11.8.

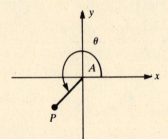

Fig. 11.8

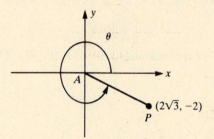

Fig. 11.9

11.9 $2\sqrt{3} - 2i$

▌ $d(A, P) = \sqrt{(2\sqrt{3})^2 + 4} = \sqrt{12 + 4} = 4 = r$. $\tan \theta = \dfrac{b}{a} = \dfrac{-2}{2\sqrt{3}} = \dfrac{-1}{\sqrt{3}} = \dfrac{-\sqrt{3}}{3}$, so $\theta = \dfrac{11\pi}{6}$.
$2\sqrt{3} - 2i = 4[\cos(11\pi/6) + i \sin(11\pi/6)]$. See Fig. 11.9.

11.10 $\sqrt{3} + i$

▌ $d(A, P) = \sqrt{(\sqrt{3})^2 + 1^2} = \sqrt{3 + 1} = 2$. To find θ: $\tan \theta_1 = b/a$ for $a + bi$. In this case, $\tan \theta = 1/\sqrt{3} = \sqrt{3}/3$. Thus, $\theta = \text{Tan}^{-1}(\sqrt{3}/3) = \pi/6$. $\sqrt{3} + i = 2[\cos(\pi/6) + i \sin(\pi/6)]$. *Note*: The formulas we are using in these problems can be summarized as follows. For the number $a + bi$, with corresponding point (a, b), $r^2 = a^2 + b^2$, $\tan \theta = b/a$, $a = r \cos \theta$, and $a + bi = r(\cos \theta + i \sin \theta)$. See Fig. 11.10.

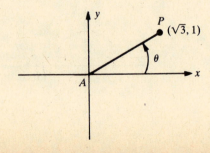

Fig. 11.10

11.11 $0 - 3i$

▌ $r = \sqrt{a^2 + b^2} = \sqrt{0 + 9} = 3$. $\tan \theta = b/a = -3/0$ which is undefined, so $\theta = 3\pi/2$.
$0 - 3i = 3[\cos (3\pi/2) + i \sin (3\pi/2)]$. See Fig. 11.11.

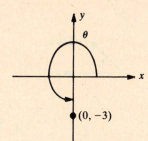

Fig. 11.11

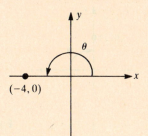

Fig. 11.12

11.12 $-4 + 0i$

▌ $r = \sqrt{16 + 0} = 4$. $\tan \theta = 0/-4 = 0$, so $\theta = \pi$. $-4 + 0i = 4(\cos \pi + i \sin \pi)$. See Fig. 11.12.

11.13 $1 - i\sqrt{3}$

▌ $r = \sqrt{a^2 + b^2} = \sqrt{1^2 + (-\sqrt{3})^2} = \sqrt{1 + 3} = 2$. $\tan \theta = b/a = -\sqrt{3}/1 = -\sqrt{3}$, so $\theta = 5\pi/3$.
$1 - i\sqrt{3} = 2[\cos (5\pi/3) + i \sin (5\pi/3)]$. See Fig. 11.13.

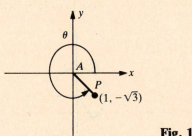

Fig. 11.13

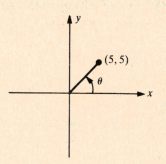

Fig. 11.14

11.14 $5 + 5i$

▌ $r = \sqrt{5^2 + 5^2} = \sqrt{50} = 5\sqrt{2}$. $\tan \theta = 5/5 = 1$, so $\theta = \pi/4$. $5 + 5i = 5\sqrt{2}[\cos (\pi/4) + i \sin (\pi/4)]$. See Fig. 11.14.

11.15 $7 + 7i$

▌ $r = \sqrt{7^2 + 7^2} = \sqrt{98} = 7\sqrt{2}$. $\tan \theta = 7/7 = 1$, so $\theta = \pi/4$. $7 + 7i = 7\sqrt{2}[\cos (\pi/4) + i \sin (\pi/4)]$.
Compare this to Prob. 11.14. Do you see a pattern developing?

11.16 $6 - 6i\sqrt{3}$

▌ $r = \sqrt{6^2 + (6\sqrt{3})^2} = \sqrt{36 + 108} = 12$. $\tan \theta = -6\sqrt{3}/6 = -\sqrt{3}$, so $\theta = 5\pi/3$.
$6 - 6i\sqrt{3} = 12[\cos (5\pi/3) + i \sin (\pi/3)]$. See Fig. 11.15.

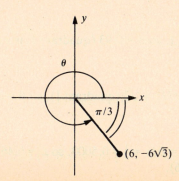

Fig. 11.15

11.38 $4[\cos{(11\pi/6)} + i\sin{(\pi/6)}]$

▮ $4\cos{(11\pi/6)} = 4 \cdot (\sqrt{3}/2) = 2\sqrt{3}$, and $i \cdot 4\sin{(11\pi/6)} = -2i$. So the answer is $2\sqrt{3} - 2i$. See Prob. 11.9.

11.39 $2[\cos{(5\pi/3)} + i\sin{(5\pi/3)}]$

▮ $2\cos{(5\pi/3)} = 1$, and $i \cdot 2\sin{(5\pi/3)} = i \cdot -2(\sqrt{3}/2) = i\sqrt{3}$. So the answer is $1 - i\sqrt{3}$. See Prob. 11.13.

11.40 $3[\cos{(3\pi/2)} + i\sin{(3\pi/2)}]$

▮ $\cos{(3\pi/2)} = 0$, and $\sin{(3\pi/2)} = -1$. Then $3\cos{(3\pi/2)} = 0$ and $i \cdot 3\sin{(3\pi/2)} = -3i$. So the answer is $0 - 3i$. See Prob. 11.11.

11.41 $4(\cos{\pi} + i\sin{\pi})$

▮ $\cos{\pi} = -1$, and $\sin{\pi} = 0$. Then $4\cos{\pi} = -4$, and $i \cdot 4\sin{\pi} = 0i$. So the answer is $-4 + 0i$. See Prob. 11.12.

11.42 $2[\cos{(\pi/4)} + i\sin{(\pi/4)}]$

▮ $\cos{(\pi/4)} = \sin{(\pi/4)} = \sqrt{2}/2$. $2\cos{(\pi/4)} = 2 \cdot \sqrt{2}/2 = \sqrt{2}$, and $i \cdot 2\sin{(\pi/4)} = i \cdot 2\sqrt{2}/2 = \sqrt{2}i$. So the answer is $\sqrt{2} + i\sqrt{2}$. See Prob. 11.18.

For Probs. 11.43 to 11.47, perform the operation indicated, and write the result in $a + bi$ form.

11.43 $2[\cos{(\pi/6)} + i\sin{(\pi/6)}] \cdot 3[\cos{(\pi/3)} + i\sin{(\pi/3)}]$

▮ The product is $6[\cos{(4\pi/6)} + i\sin{(4\pi/6)}] = 6(-\frac{1}{2}) + 6(i \cdot \sqrt{3}/2) = -3 + 3i\sqrt{3}$.

11.44 $3(\cos{25°} + i\sin{25°}) \cdot 8(\cos{200°} + i\sin{200°})$

▮ The product is $24(\cos{225°} + i\sin{225°}) = -12\sqrt{2} - 12\sqrt{2}i$.

11.45 $4(\cos{50°} + i\sin{50°}) \cdot 2(\cos{100°} + i\sin{100°})$

▮ The product is $8(\cos{150°} + i\sin{150°}) = -4\sqrt{3} + 4i$.

11.46 $\dfrac{4(\cos{190°} + i\sin{190°})}{2(\cos{70°} + i\sin{70°})}$

▮ The quotient is $2(\cos{120°} + i\sin{120°}) = -1 + i\sqrt{3}$.

11.47 $\dfrac{12(\cos{200°} + i\sin{200°})}{3(\cos{350°} + i\sin{350°})}$

▮ The quotient is $4[\cos{(-150°)} + i\sin{(-150°)}] = -2\sqrt{3} - 2i$.

For Probs. 11.48 to 11.51, use polar form to perform the given calculation.

11.48 $(1 + i)(\sqrt{2} - i\sqrt{2})$

▮ $1 + i = \sqrt{2}[\cos{(\pi/4)} + i\sin{(\pi/4)}]$, and $\sqrt{2} - i\sqrt{2} = 2[\cos{(7\pi/4)} + i\sin{(7\pi/4)}]$. The product is $2\sqrt{2}(\cos{2\pi} + i\sin{2\pi}) = 2\sqrt{2} \cdot 1 = 2\sqrt{2}$.

11.49 $\dfrac{1 + i}{1 + i}$

▮ $1 + i = \sqrt{2}[\cos{(\pi/4)} + i\sin{(\pi/4)}]$. Then $\dfrac{1 + i}{1 + i} = (\sqrt{2}/\sqrt{2})(\cos{0} + i\sin{0}) = (\sqrt{2}/\sqrt{2}) \cdot 1 = 1$ (but you knew that prior to doing the calculation!).

11.50 $(1 + i)^2$

❙ $(1 + i)^2 = (1 + i)(1 + i) = \sqrt{2}[\cos(\pi/4) + i \sin(\pi/4)]\sqrt{2}[\cos(\pi/4) + i \sin(\pi/4)] = 2[\cos(\pi/2) + i \sin(\pi/2)] = 2(0 + i) = 2i$.

11.51 $(1 + i)^3$

❙ $(1 + i)^3 = (1 + i)^2(1 + i) = 2[\cos(\pi/2) + i \sin(\pi/2)]\sqrt{2}[\cos(\pi/4) + i \sin(\pi/4)]$ (see Prob. 11.50) $= 2\sqrt{2}[\cos(3\pi/4) + i \sin(3\pi/4)] = 2\sqrt{2}\cos(3\pi/4) + i \cdot 2\sqrt{2}\sin(3\pi/4) = 2\sqrt{2} \cdot (-\sqrt{2}/2) + i \cdot 2\sqrt{2} \cdot (\sqrt{2}/2) = -2 + 2i$.

11.2 ROOTS AND DE MOIVRE'S THEOREM

11.52 State De Moivre's theorem.

❙ Let $n \in \mathscr{Z}$. Then $[r(\cos\theta + i \sin\theta)]^n = r^n(\cos n\theta + i \sin n\theta)$.

For Probs. 11.53 to 11.56, find the indicated powers.

11.53 $(\cos 15° + i \sin 15°)^6$

❙ Since $1^6 = 1$ and $15° \cdot 6 = 90°$, the answer is $\cos 90° + i \sin 90°$. (Do you notice that this is just 1? That is a surprising result of the original problem!)

11.54 $[2(\cos 35° + i \sin 35°)]^2$

❙ Since $2^2 = 4$, and $2 \cdot 35° = 70°$, the answer is $4(\cos 70° + i \sin 70°)$.

11.55 $[3(\cos 10° + i \sin 10°)]^3$

❙ Since $3^3 = 27$, and $10° \cdot 3 = 30°$, the answer is $27(\cos 30° + i \sin 30°)$.

11.56 $[4(\cos 310° + i \sin 310°)]^3$

❙ $64(\cos 930° + i \sin 930°)$.

For Probs. 11.57 to 11.63, use De Moivre's theorem to rewrite each in standard form.

11.57 $[\cos(\pi/3) + i \sin(\pi/3)]^5$

❙ $[\cos(\pi/3) + i \sin(\pi/3)]^5 = \cos(5\pi/3) + i \sin(5\pi/3) = \dfrac{1}{2} - \dfrac{\sqrt{3}}{2}i$.

11.58 $(\cos 40° + i \sin 40°)^3$

❙ $(\cos 40 + i \sin 40°)^3 = \cos 120° + i \sin 120° = \dfrac{-1}{2} + \dfrac{\sqrt{3}}{2}i$.

11.59 $8(1 - i)^6$

❙ $1 - i = \sqrt{2}[\cos(7\pi/4) + i \sin(7\pi/4)]$. Then $8(1 - i)^6 = 8\{\sqrt{2}[\cos(7\pi/4) + i \sin(7\pi/4)]\}^6 = 8 \cdot (\sqrt{2})^6[\cos(42\pi/4) + i \sin(42\pi/4)] = 64(0 + i) = 64i$. See Fig. 11.20.

11.60 $(\sqrt{3} + i)^{12}$

❙ $\tan\theta = b/a = 1/\sqrt{3}$. $\sqrt{3} + i = 2[\cos(\pi/6) + i \sin(\pi/6)]$. Then $(\sqrt{3} + i)^{12} = (2)^{12}(\cos 2\pi + i \sin 2\pi) = 4096 + 0i$. See Fig. 11.21.

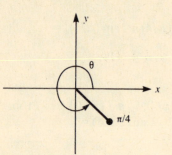

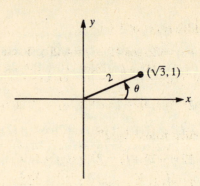

Fig. 11.20   **Fig. 11.21**

11.61 $(1+i)^{16}(1-i)^{10}$

▐ $(1+i)^{16} = \{\sqrt{2}[\cos{(\pi/4)} + i\sin{(\pi/4)}]\}^{16} = (\sqrt{2})^{16}(\cos 4\pi + i\sin 4\pi) = 256(1+0) = 256$, and
$(1-i)^{10} = (\sqrt{2})^{10}[\cos{(70\pi/4)} + i\sin{(70\pi/4)}] = 32(0 - i) = -32i$. Therefore, the answer is
$(32i)(256) = -8192i$.

11.62 $(1+i)^{-6}$

▐ $1+i = \sqrt{2}[\cos{(\pi/4)} + i\sin{(\pi/4)}]$. Then $(1+i)^{-6} = (\sqrt{2})^{-6}[\cos{(-6\pi/4)} + i\sin{(-6\pi/4)}] =$
$[1/(\sqrt{2})^6](0 + i) = \frac{1}{8}i$.

11.63 $\dfrac{(1-i)^6}{(1+i)^{10}}$

▐ $(1-i)^6 = 8i$. (See Prob. 11.59). $(1+i)^{10} = (\sqrt{2})^{10}[\cos(10\pi/4) + i\sin{(10\pi/4)}] = 32(0 + i) = 32i$.
Therefore, the answer is $8i/(32i) = \frac{1}{4}$.

For Probs. 11.64 to 11.68, find the indicated roots.

11.64   The square roots of i

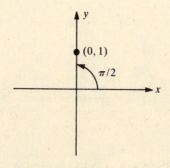

Fig. 11.22

▐   See Fig. 11.22. Recall that if $z(\neq 0) = r(\cos\theta + i\sin\theta)$, then z has n, nth roots given by the
formula

$$r^{1/n}\left(\cos\frac{\theta + 2\pi k}{n} + i\sin\frac{\theta + 2\pi k}{n}\right) \qquad \text{for} \qquad k = 0, 1, \ldots, n-1$$

Since $i = 0 + i = 1[\cos{(\pi/2)} + i\sin{(\pi/2)}]$, the square roots of i are $1^{1/2}\left(\cos\dfrac{\theta + 2\pi k}{n} + i\sin\dfrac{\theta + 2\pi k}{n}\right)$
for $k = 0, 1$ where $\theta = \pi/2$, $n = 2$. The two roots are

$$w_1 = \cos\frac{\pi/2}{2} + i\sin\frac{\pi/2}{2} = \cos\frac{\pi}{4} + i\sin\frac{\pi}{4} = \frac{\sqrt{2}}{2} + i\frac{\sqrt{2}}{2}$$

and

$$w_2 = \cos\frac{\pi/2 + 2\pi}{2} + i\sin\frac{\pi/2 + 2\pi}{2} = \frac{-\sqrt{2}}{2} - i\frac{\sqrt{2}}{2}$$

11.65 The fifth roots of $1 + i$

$\blacksquare$ $1 + i = \sqrt{2}(\cos 45° + i \sin 45°)$. Then the fifth roots of $1 + i$ are $(\sqrt{2})^{1/5}\left(\cos\dfrac{45° + k \cdot 360°}{5} + i \sin\dfrac{45° + k \cdot 360°}{5}\right)$ where $k = 0, 1, 2, 3, 4$. Then $w_1 = 2^{1/10}(\cos 9° + i \sin 9°)$, $w_2 = 2^{1/10}(\cos 81° + i \sin 81°)$, etc.

11.66 The cube roots of -27

$\blacksquare$ $-27 = -27 + 0i = 27(\cos \pi + i \sin \pi)$. Then the cube roots are $w_1 = 27^{1/3}\left(\cos\dfrac{\pi}{3} + i \sin\dfrac{\pi}{3}\right) = \dfrac{3}{2} + \dfrac{3\sqrt{3}}{2}i$, $w_2 = 27^{1/3}\left(\cos\dfrac{2\pi + \pi}{3} + i \sin\dfrac{2\pi + \pi}{3}\right) = -3$, and $w_3 = 27^{1/3}\left(\cos\dfrac{4\pi + \pi}{3} + i \sin\dfrac{4\pi + \pi}{3}\right) = \dfrac{3}{2} - \dfrac{3\sqrt{3}}{2}i$.

11.67 The cube roots of -8

$\blacksquare$ See Fig. 11.23. $-8 = -8 + 0i = 8(\cos \pi + i \sin \pi)$. The cube roots are $w_1 = 8^{1/3}\left(\cos\dfrac{\pi}{3} + i \sin\dfrac{\pi}{3}\right) = 1 + \sqrt{3}i$, $w_2 = 8^{1/3}\left(\cos\dfrac{2\pi + \pi}{3} + i \sin\dfrac{2\pi + \pi}{3}\right) = -2$, and $w_3 = 8^{1/3}\left(\cos\dfrac{4\pi + \pi}{3} + i \sin\dfrac{4\pi + \pi}{3}\right) = 1 - \sqrt{3}i$.

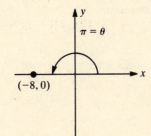

Fig. 11.23

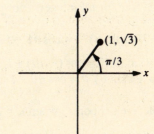

Fig. 11.24

11.68 The square roots of $1 + i\sqrt{3}$

$\blacksquare$ See Fig. 11.24. $1 + i\sqrt{3} = 2\left(\cos\dfrac{\pi}{3} + i \sin\dfrac{\pi}{3}\right)$. Then $w_1 = 2^{1/2}\left(\cos\dfrac{\pi/3}{2} + i \sin\dfrac{\pi/3}{2}\right) = \dfrac{\sqrt{6}}{2} + \dfrac{i\sqrt{2}}{2}$ and $w_2 = 2^{1/2}\left(\cos\dfrac{2\pi + \pi/3}{2} + i \sin\dfrac{2\pi + \pi/3}{2}\right) = \dfrac{\sqrt{6}}{2} - \dfrac{i\sqrt{2}}{2}$.

For Probs. 11.69 to 11.73, solve the given equation.

11.69 $x^3 + 1 = 0$

$\blacksquare$ $x^3 + 1 = 0$; then $x^3 = -1$, and x can be any one of the three cube roots of -1. $-1 = -1 + 0i = 1(\cos \pi + i \sin \pi)$. Then $w_1 = -1$, $w_2 = \dfrac{1 + \sqrt{3}i}{2}$, and $w_3 = \dfrac{1 - \sqrt{3}i}{2}$.

11.70 $x^2 - i = 0$

$\blacksquare$ If $x^2 - i = 0$, then $x^2 = i$. The two square roots of i are $w_1 = \sqrt{2}/2 + i\sqrt{2}/2$, and $w_2 = -\sqrt{2}/2 - i\sqrt{2}/2$.

11.71 $x^2 - (1 + i\sqrt{3}) = 0$

$\blacksquare$ If $x^2 - (1 + i\sqrt{3}) = 0$, $x = $ square roots of $1 + i\sqrt{3}$. Then $w_1 = \sqrt{6}/2 + i\sqrt{2}/2$, and $w_2 = -\sqrt{6}/2 - i\sqrt{2}/2$. (See Prob. 11.68.)

11.72 $x^3 + 8 = 0$

▮ If $x^3 + 8 = 0$, then $x^3 = -8$. We need to find the three cube roots of -8. Then $w_1 = 1 + \sqrt{3}i$, $w_2 = -2$, and $w_3 = 1 - \sqrt{3}i$. (See Prob. 11.67.)

11.73 $x^5 - (1 + i) = 0$

▮ See Prob. 11.65 for the fifth roots of $1 + i$. If $x^5 - (1 + i) = 0$, then $x^5 = 1 + i$ and x is the 5 fifth roots of $1 + i$.

For Probs. 11.74 to 11.77, tell whether the given statement is true or false and why.

11.74 Every complex number $a + bi$ has n distinct nth roots.

▮ True; this is the major theorem concerning roots of complex numbers.

11.75 Only one of the fifth roots of 32 is real.

▮ True; $2^5 = 32$, and no other real number has that property.

11.76 The modulus of the product of two complex numbers is the sum of their moduli.

▮ False; if $z_1 = r_1(\cos\theta + i\sin\theta)$ and $z_2 = r_2(\cos\theta_2 - i\sin\theta_2)$, then modulus $(z_1, z_2) = r_1 \cdot r_2$.

11.77 De Moivre's theorem is true for $n = -7$.

▮ True; it is true for all integers, not just the natural or whole numbers.

11.78 Prove: $(\cos\theta + i\sin\theta)^{-n} = \cos n\theta - i\sin n\theta$

▮ $(\cos\theta + i\sin\theta)^{-n} = \cos(-n\theta) + i\sin(-n\theta)$, but $\cos(-n\theta) = \cos n\theta$ and $\sin(-n\theta) = -\sin n\theta$. Thus, $(\cos\theta + i\sin\theta)^{-n} = \cos n\theta - i\sin n\theta$.

11.79 Prove De Moivre's theorem when $n = 2$.

▮ Let $Z = r(\cos\theta + i\sin\theta)$. Then

$$[r(\cos\theta + i\sin\theta)]^2 = r(\cos\theta + i\sin\theta) \cdot r(\cos\theta + i\sin\theta)$$
$$= r^2(\cos\theta + i\sin\theta)(\cos\theta + i\sin\theta)$$
$$= r^2[\cos^2\theta + i^2\sin^2\theta + i(\sin\theta\cos\theta + \sin\theta\cos\theta)]$$
$$= r^2[\cos^2\theta - \sin^2\theta + i(\sin 2\theta)]$$
$$= r^2(\cos 2\theta + i\sin 2\theta)$$

11.80 Find the reciprocal of i, using De Moivre's theorem.

▮ $1/i = i^{-1}$, and $i = 0 + 1i = 1[\cos(\pi/2) + i\sin(\pi/2)]$. Thus, $i^{-1} = 1^{-1}[\cos(-\pi/2) + i\sin(-\pi/2)] = \cos(-\pi/2) + i\sin(-\pi/2) = \cos(\pi/2) - i\sin(\pi/2) = 0 - i = -i$.

11.81 Prove that your result in Prob. 11.80 is correct.

▮ $i \cdot (-i) = -(i^2) = -(-1) = 1$. Thus, $-i$ is the reciprocal of i.

11.3 MISCELLANEOUS PROBLEMS

For Probs. 11.82 to 11.86, find the distance between the two given complex numbers.

11.82 $z_1 = 1 + i$, $z_2 = 2 + 2i$

▮ If $z = a + bi$ and $w = c + di$, then $d(z, w) = \sqrt{(a-c)^2 + (b-d)^2}$. Notice that this formula comes from the ordinary distance formula in the cartesian plane. Thus, $d(z_1, z_2) = \sqrt{(1-2)^2 + (1-2)^2} = \sqrt{2}$.

11.83 $z_1 = 3i$, $z_2 = 5 + 6i$

$\blacksquare$ $d(z_1, z_2) = \sqrt{(0 - 5)^2 + (3 - 6)^2} = \sqrt{25 + 9} = \sqrt{34}$.

11.84 $z_1 = 4$, $z_2 = 6 - zi$

$\blacksquare$ $d(z_1, z_2) = \sqrt{(4 - 6)^2 + (0 + 2)^2} = \sqrt{4 + 4} = 2\sqrt{2}$.

11.85 $z_1 = -7$, $z_2 = 7i$

$\blacksquare$ $d(z_1, z_2) = \sqrt{(-7 - 0)^2 + (0 - 7)^2} = \sqrt{98} = 7\sqrt{2}$.

11.86 $z_1 = a - bi$, $z_2 = b - ai$

$\blacksquare$ $d(z_1, z_2) = \sqrt{(a - b)^2 + (-b + a)^2} = \sqrt{2(a - b)^2} = |a - b|\sqrt{2}$.

For Probs. 11.87 to 11.90, perform the given operations graphically.

11.87 $(1 + i) + (2 + 3i)$

$\blacksquare$ See Fig. 11.25. We represent $1 + i$ and $2 + 3i$ as vectors in the plane and find the diagonal of the resulting parallelogram. The point $A = 3 + 4i$ represents the sum.

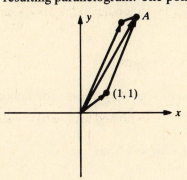

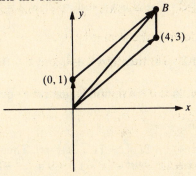

Fig. 11.25   Fig. 11.26

11.88 $i + (4 + 3i)$

$\blacksquare$ See Fig. 11.26. We proceed as in Prob. 11.87. The point $B = 4 + 4i$ represents the sum.

11.89 $(2 + i) - (4 - i)$

$\blacksquare$ See Fig. 11.27. $(2 + i) - (4 - i) = (2 + i) + [-(4 - i)] = (2 + i) + (-4 + i)$. The point $C = -2 + 2i$ represents the difference.

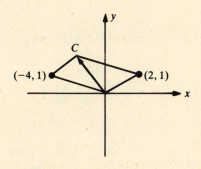

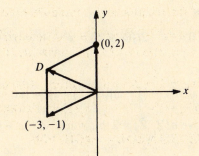

Fig. 11.27   Fig. 11.28

11.90 $2i - (3 + i)$

$\blacksquare$ See Fig. 11.28. We find $2i + (-3 - i)$ graphically. The point $D = (-3 + i)$ represents the difference.

For Probs. 11.91 and 11.92, solve the given equation.

11.91 $ix + 3i = 5i$

▌ Let $x = a + bi$; then $5i = i(a + bi) + 3i = ai + 3i + bi^2 = i(a + 3) - b$. Thus, $a + 3 = 5$, and $-b = 0$, so $a = 2$, and $b = 0$. Thus, $x = 2 + 0i$. In this problem, you could see that $x = 2$ without going through this procedure by looking carefully at the equation.

11.92 $(1 - i)x + 3i = 2i$

▌ Let $x = a + bi$; then $(1 - i)(a + bi) + 3i = 2i$. Then $a - ai + bi - bi^2 + 3i = 2i$, and $i(-a + b + 3) + (b + a) = 0 + 2i$. If $b + a = 0$, then $a = -b$. If $-a + b + 3 = 2$, then $-a + (-a) + 3 = 2$, so $-2a = -1$, or $a = \frac{1}{2}$. Then $b = -\frac{1}{2}$. Thus, $x = \frac{1}{2} - \frac{1}{2}i$.

For Probs. 11.93 to 11.95, the given number is a solution of a quadratic equation with real coefficients. Find the other solution.

11.93 $1 + 2i$

▌ If $1 + 2i = Z$ is a solution of a quadratic equation with real coefficients, then so is $\bar{Z} = 1 - 2i$.

11.94 $2 - 5i$

▌ If $Z = 2 - 5i$ is a solution, so is $\bar{Z} = 2 + 5i$.

11.95 $4i$

▌ If $Z = 4i$, then $Z = 0 + 4i$ and $\bar{Z} = 0 - 4i = -4i$.

For Probs. 11.96 to 11.100, find a quadratic equation with real coefficients having the given solution(s).

11.96 $1 \pm 3i$

▌ $[x - (1 + 3i)][x - (1 - 3i)] = 0$. Then $[(x - 1) - 3i][(x - 1) + 3i] = 0$. Thus, $0 = (x - 1)^2 - 9i^2 = (x - 1)^2 + 9 = (x^2 - 2x + 1) + 9 = x^2 - 2x + 10$.

11.97 $1 + i$

▌ $1 + i$ and $1 - i$ must be solutions. Then $0 = [x - (1 + i)][x - (1 - i)] = [(x - 1) - i][(x - 1) + i] = (x - 1)^2 - i^2 = x^2 - 2x + 1 + 1 = x^2 - 2x + 2$.

11.98 i

▌ i and $-i$ are solutions. Then $0 = (x - i)(x + i) = x^2 - ix + ix - i^2 = x^2 + 1$.

11.99 $-2i$

▌ $0 = (x - 2i)(x + 2i) = x^2 - 4i^2 = x^2 + 4$.

11.100 $\frac{1}{2} - i$

▌ $0 = [x - (\frac{1}{2} - i)][x - (\frac{1}{2} + i)] = [(x - \frac{1}{2}) + i][(x - \frac{1}{2}) - i] = (x - \frac{1}{2})^2 - i^2 = x^2 - x + \frac{1}{4} + 1 = x^2 - x + \frac{5}{4}$.

11.101 Prove $d(Z, \bar{Z}) = 2|b|$, where $Z = a + bi$.

▌ If $Z = a + bi$, then $\bar{Z} = a - bi$. $d(Z, \bar{Z}) = \sqrt{(a - a)^2 + [b - (-b)]^2} = \sqrt{(2b)^2} = \sqrt{4b^2} = 2|b|$.

11.102 Prove $|Z| = |\bar{Z}|$ for all Z.

▌ Let $Z = a + bi$; then $\bar{Z} = a - bi$. $|Z| = d(Z, (0, 0)) = \sqrt{a^2 + b^2}$, and $|\bar{Z}| = d(\bar{Z}, (0, 0)) = \sqrt{a^2 + (-b)^2} = \sqrt{a^2 + b^2}$. Thus, $|Z| = |\bar{Z}|$.

11.103 Prove $\overline{(\bar{Z})} = Z$ for all Z.

▌ Let $Z = a + bi$. Then, $\bar{Z} = a - bi = a + (-bi)$, and $\overline{(\bar{Z})} = a + [-(-bi)] = a + bi = Z$. (Question: What is Z's fifteenth conjugate? Sixteenth?)

11.104 Find x and y such that $2x - yi = 4 + 3i$.

▌ Here $2x = 4$ and $-y = 3$; then $x = 2$ and $y = -3$.

11.105 Show that the conjugate complex numbers $2 + i$ and $2 - i$ are roots of the equation $x^2 - 4x + 5 = 0$.

▌ For $x = 2 + i$: $(2 + i)^2 - 4(2 + i) + 5 = 4 + 4i + i^2 - 8 - 4i + 5 = 0$. For $x = 2 - i$: $(2 - i)^2 - 4(2 - i) + 5 = 4 - 4i + i^2 - 8 + 4i + 5 = 0$. Since each number satisfies the equation, it is a root of the equation.

For Probs. 11.106 through 11.109, perform the given operations graphically.

11.106 $(3 + 4i) + (2 + 5i)$

▌ See Fig. 11.29.

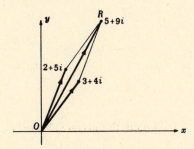

Fig. 11.29

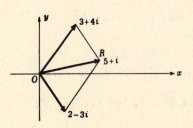

Fig. 11.30

For Probs. 11.106 and 11.107, draw the two vectors as in Figs. 11.29 and 11.30 and apply the parallelogram law.

11.107 $(3 + 4i) + (2 - 3i)$

▌ See Fig. 11.30.

11.108 $(4 + 3i) - (2 + i)$

▌ See Fig. 11.31.

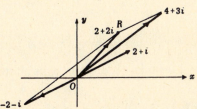

Fig. 11.31

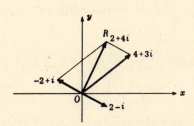

Fig. 11.32

For Prob. 11.108, draw the vectors representing $4 + 3i$ and $-2 - i$ and apply the parallelogram law as in Fig. 11.31.

11.109 $(4 + 3i) - (2 - i)$

▌ See Fig. 11.32.

For Prob. 11.109 draw the vectors representing $4 + 3i$ and $-2 + i$ and apply the parallelogram law as in Fig. 11.32.

CHAPTER 12
Sequences, Series, and Probability ▱

12.1 SEQUENCES

For Probs. 12.1 to 12.8, write the first four terms of the sequence whose general term is given.

12.1 $n - 3$

▮ Let $n = 1, 2, 3$, and 4 to generate the first four terms: $1 - 3, 2 - 3, 3 - 3, 4 - 3$; or $-2, -1, 0, 1$.

12.2 $n^2 + 2$

▮ $1^2 + 2, 2^2 + 2, 3^2 + 2, 4^2 + 2$, or $3, 6, 11, 18$.

12.3 $1/n$

▮ $1/1, 1/2, 1/3, 1/4$; or $1, \frac{1}{2}, \frac{1}{3}, \frac{1}{4}$.

12.4 $1/3^n$

▮ $1/3^1, 1/3^2, 1/3^3, 1/3^4$, or $\frac{1}{3}, \frac{1}{9}, \frac{1}{27}, \frac{1}{81}$.

12.5 $\dfrac{1}{4^{n-2} + 1}$

▮ $\dfrac{1}{4^{1-2} + 1}, \dfrac{1}{4^{2-2} + 1}, \dfrac{1}{4^{3-2} + 1}, \dfrac{1}{4^{4-2} + 1}$; or $\frac{4}{5}, 1, \frac{1}{5}, \frac{1}{17}$.

12.6 $\dfrac{n^2 + 1}{n^3 + 1}$

▮ $\dfrac{1^2 + 1}{1^3 + 1}, \dfrac{2^2 + 1}{2^3 + 1}, \dfrac{3^2 + 1}{3^3 + 1}, \dfrac{4^2 + 1}{4^3 + 1}$; or $1, \frac{5}{9}, \frac{10}{28}, \frac{17}{65}$.

12.7 $(1 + 1/n)^n$

▮ $(1 + 1/1)^2, (1 + \frac{1}{2})^2, (1 + \frac{1}{3})^3, (1 + \frac{1}{4})^4$; or $2, \frac{9}{4}, \frac{64}{27}, \frac{625}{256}$.

12.8 $\dfrac{(-1)^n}{n + 2}$

▮ When $n = 1$, $(-1)^n = -1^1 = -1$. Then the sign of the $n + 1$ term is -1, and the signs alternate $-1, +1, -1$, etc. Thus, the sequence is $-\frac{1}{3}, \frac{1}{4}, -\frac{1}{5}, \frac{1}{6}$.

For Probs. 12.9 to 12.18, find the general term of the given sequence.

12.9 $2, 3, 4, 5, \ldots$

▮ Notice that the terms of this sequence begin at 2 (which is $n + 1$ when $n = 1$) and progress consecutively. The general term, then, is $n + 1$. (Check this by writing out the terms.)

12.10 $-1, -2, -3, -4, \ldots$

▮ When $n = 1$, $-n = -1$. The sequence is $-n$. (Check this by writing out the terms.)

12.11 $-4, -3, -2, -1, 0, 1, \ldots$

▌ When $n = 1$, $-4 = n - 5$. The general term of this sequence is $n - 5$. (Check this by writing out the terms.)

12.12 $2, 4, 6, 8, 10, \ldots$

▌ When $n = 1$, $2 = 2n$. The general term is $2n$. Notice that we could not use $n + 1$ since we must generate only even numbers.

12.13 $2, -4, 6, -8, 10, \ldots$

▌ This is similar to Prob. 12.12, but we want the terms to alternate in sign beginning with a positive. Using $(-1)^{n+1}$ to alternate signs (-1^{n+1} is $+1$ when $n = 1$), we see that the general term is $(-1)^{n+1} \cdot 2n$.

12.14 $-2, 4, -6, 8, -10, \ldots$

▌ See Prob. 12.13. This is very similar, except that we want the sign of the $n = 1$ term to be negative. The sequence is $(-1)^n \cdot 2n$.

12.15 $x, x^2/2, x^3/3, x^4/4, \ldots$

▌ Each term is of the form x^n/n (beginning with $n = 1$).

12.16 $\frac{1}{2}, \frac{3}{4}, \frac{5}{6}, \frac{7}{8}, \ldots$

▌ Notice that each term's numerator is 1 less than its denominator. Also the numerator of the first term is 1. We try $n/(n + 1)$. Checking, we see that this is not the general term! Check it. However, when we notice that the numerators are all odd and the denominators even, we find the general term: $(2n - 1)/(2n)$. Check this; it is correct.

12.17 $x, -x^3, x^5, -x^7, \ldots$

▌ Alternate signs beginning with the positive; moreover, we need to generate odd powers of x: $(-1)^{n+1} \cdot x^{2n-1}$.

12.18 $-x, x^3/2, -x^5/3, x^7/4, \ldots$

▌ See Prob. 12.17. We reverse the alternation in sign and divide each term by n: $(-1)^n (x^{2n-1}/n)$.

For Probs. 12.19 to 12.23, find the first four terms of the sequence defined recursively by the given equations.

12.19 $A_1 = 1$, $A_{n+1} = A_n + 1$

▌ $A_1 = 1$; $A_2 = A_{1+1} = A_1 + 1 = 1 + 1 = 2$; $A_3 = A_{2+1} = A_2 + 1 = 2 + 1 = 3$; etc. The first four terms are 1, 2, 3, 4.

12.20 $A_1 = 2$, $A_{n+1} = 2A_n + 1$

▌ $A_2 = A_{1+1} = 2A_1 + 1 = 2(2) + 1 = 5$; $A_3 = A_{2+1} = 2A_2 + 1 = 11$; $A_4 = A_{3+1} = 2A_3 + 1 = 23$. The terms are 2, 5, 11, 23.

12.21 $A_1 = 1$, $A_{n+1} = -A_n$

▌ $A_1 = 1$, $A_2 = -A_1$, $A_3 = -A_2$, etc. The terms are 1, -1, 1, -1.

12.22 $A_1 = 1$, $A_2 = 1$, $A_{n+2} = A_n + A_{n+1}$

▌ This is the famous *Fibonacci sequence*. $A_1 = 1$; $A_2 = 1$; $A_3 = A_2 + A_1 = 1 + 1 = 2$; $A_4 = A_3 + A_2 = 2 + 1 = 3$. The terms are 1, 1, 2, 3.

12.23 $A_1 = 5$, $A_2 = 1$, $A_{n+2} = A_{n+1}/A_n$

▮ $A_3 = A_{1+2} = A_2/A_1 = \frac{1}{5}$; $A_4 = A_{2+2} = A_3/A_2 = \frac{1}{5}/1 = \frac{1}{5}$. The terms are 5, 1, $\frac{1}{5}$, $\frac{1}{5}$.

For Probs. 12.24 and 12.25, define the given sequence by using a recursive formula.

12.24 $1, 2, 3, 5, 8, \ldots$

▮ $A_1 = 1$, $A_2 = 2$, $A_{n+2} = A_{n+1} + A_n$. Do you see how this resembles the Fibonacci sequence?

12.25 $1, 1, 5, 17, 61, \ldots$

▮ How can we get 5 from 1 and 1? How about $3(1) + 2(1)$? Note that $17 = 3(5) + 2(1)$, etc. $A_1 = 1$, $A_2 = 1$, $A_{n+2} = 3A_{n+1} + 2A_n$.

12.2 SERIES

For Probs. 12.26 to 12.36, write the series without summation (sigma) notation.

12.26 $\displaystyle\sum_{i=1}^{5} i$

▮ Remember that to expand a given series with sigma notation, we replace the variable (in this case i) in the series expression one number at a time and add successively. In this case, $1 + 2 + 3 + 4 + 5$.

12.27 $\displaystyle\sum_{i=1}^{4} 2i$

▮ $2(1) + 2(2) + 2(3) + 2(4) = 2 + 4 + 6 + 8$.

12.28 $\displaystyle\sum_{i=2}^{5} i + 3$

▮ $(2 + 3) + (3 + 3) + (4 + 3) + (5 + 3) = 5 + 6 + 7 + 8$.

12.29 $\displaystyle\sum_{i=0}^{5} i^2$

▮ $0^2 + 1^2 + 2^2 + 3^2 + 4^2 + 5^2 = 0 + 1 + 4 + 9 + 16 + 25$.

12.30 $\displaystyle\sum_{i=1}^{4} i^i$

▮ $1^1 + 2^2 + 3^3 + 4^4 = 1 + 4 + 27 + 256$.

12.31 $\displaystyle\sum_{i=1}^{4} i(i - 1)$

▮ $1(1 - 1) + 2(2 - 1) + 3(3 - 1) + 4(4 - 1) = 0 + 2 + 6 + 12$.

12.32 $\displaystyle\sum_{i=1}^{4} (-1)^i i(i - 1)$

▮ This is the same as Prob. 12.31, but with an alternating sign beginning with a negative. Thus, we have, $-0 + 2 - 6 + 12$.

12.33 $\displaystyle\sum_{i=1}^{3} (-1)^{i+1}(x^i - 1)$

▮ When $i = 1$, $(-1)^2(x^1 - 1) = x - 1$. When $i = 2$, $(-1)^3(x^2 - 1) = -(x^2 - 1) = 1 - x^2$. When $i = 3$, $(-1)^4(x^3 - 1) = x^3 - 1$. Thus, we have $(x - 1) + (1 - x^2) + (x^3 - 1)$.

12.34 $\displaystyle\sum_{i=1}^{\infty} \frac{1}{i}$

▮ $1 + \frac{1}{2} + \frac{1}{3} + \frac{1}{4} + \cdots + 1/n + \cdots$. (This is the harmonic series. It is an important series, as you will see later.)

12.35 $\displaystyle\sum_{i=1}^{\infty} (-1)^i \frac{1}{i+1}$

▮ When $i = 1$, we get $-\frac{1}{2}$; after that, the signs alternate, and the denominators increase consecutively: $-\frac{1}{2} + \frac{1}{3} - \frac{1}{4} + \frac{1}{5} - \cdots + \cdots (-1)^n[1/(n+1)] + \cdots$.

12.36 $\displaystyle\sum_{i=1}^{\infty} (-1)^{i-1} \frac{1}{i^2}$

▮ When $i = 1$, $(-1)^{i-1}(1/i^2) = 1/1 = 1$. When $i = 2$, $(-1)^{i-1}(1/i^2) = -\frac{1}{4}$; and then the signs alternate. $1 - \frac{1}{4} + \frac{1}{9} - \frac{1}{16} + \cdots - \cdots + (-1)^{n-1}(1/n^2) + \cdots$.

For Probs. 12.37 to 12.47, write the given series, using summation (sigma) notation.

12.37 $1 + 2 + 3 + 4 + 5$

▮ The terms of this series are increasing consecutively, beginning at $i = 1$: $\displaystyle\sum_{i=1}^{5} i$.

12.38 $2 + 3 + 4 + 5 + 6$

▮ This is the same as the series in Prob. 12.37, except i's range is from 2 to 6: $\displaystyle\sum_{i=2}^{6} i$. See Prob. 12.39.

12.39 $2 + 3 + 4 + 5 + 6$

▮ We could have written this as $\displaystyle\sum_{i=1}^{5} (i + 1)$. Notice that you can change the general term if you want (or need) to if you make an appropriate change in the range of variable i.

12.40 $2 + 4 + 6 + 8 + 10 + 12$

▮ These are the consecutive even numbers $2i$ beginning with $i = 1$: $\displaystyle\sum_{i=1}^{6} 2i$.

12.41 $1 - 3 + 5 - 7 + 9 - 11$

▮ These are the consecutive odd numbers $2i - 1$ beginning with $i = 1$, alternating with a positive lead term: $\displaystyle\sum_{i=1}^{6} (-1)^{i+1}(2i - 1)$.

12.42 $1 - 3 + 5 - 7 + 9 - 11 + \cdots - \cdots$

▮ This is the series in Prob. 12.41, but the infinite version of it: $\displaystyle\sum_{i=1}^{\infty} (-1)^{i+1}(2i - 1)$.

12.43 $1 - \frac{1}{2} + \frac{1}{4} - \frac{1}{8}$

▮ The denominators are the powers of 2, beginning with 2, which is 2^{i-1} when $i = 1$, so $\displaystyle\sum_{i=1}^{4} (-1)^{i+1} \frac{1}{2^{i-1}}$.

12.44 $-1 + \frac{1}{4} - \frac{1}{9} + \frac{1}{16} + \cdots$

▎ The denominators are the squares of the consecutive integers beginning at 1: $\sum\limits_{i=1}^{\infty} (-1)^i \frac{1}{i^2}$.

12.45 $1 - 1 + 1 - 1 + 1 - 1 + \cdots$

▎ This is the infinite series of alternating signs of 1: $\sum\limits_{i=1}^{\infty} (-1)^{i+1}$.

12.46 $1 + 2 + \frac{3}{2} + \frac{4}{6} + \frac{5}{24} + \cdots$

▎ The denominators are the $i!$ beginning with 0! (which is 1): $\sum\limits_{i=0}^{\infty} \frac{i+1}{i!}$.

12.47 $1 + x + x^2/2 + x^3/3! + \cdots$

▎ Remember that $0! = 1! = 1$: $\sum\limits_{i=0}^{\infty} x^i/i!$.

12.3 ARITHMETIC AND GEOMETRIC SEQUENCES

For Probs. 12.48 to 12.54, tell whether the given sequence is arithmetic or geometric or neither.

12.48 $1, 3, 5, 7, 9, \ldots$

▎ Arithmetic; the common difference is 2.

12.49 $2, 6, 10, 11, 15, 19, \ldots$

▎ Neither; there is neither a common difference nor a common ratio.

12.50 $1, \frac{1}{2}, \frac{1}{3}, \frac{1}{4}, \ldots$

▎ Neither; there is neither a common difference nor a common ratio.

12.51 $1, \frac{1}{2}, \frac{1}{4}, \frac{1}{8}, \ldots$

▎ Geometric; ratio $= \frac{1}{2}$.

12.52 $1, -\frac{1}{2}, \frac{1}{4}, -\frac{1}{8}, \frac{1}{16}, \ldots$

▎ Geometric; common ratio $= -\frac{1}{2}$.

12.53 $10, 6, 2, -2, -6, \ldots$

▎ Arithmetic; difference $= -4$.

12.54 $10, 10x, 10x^2, 10x^3, \ldots$

▎ Geometric; ratio $= x$.

For Probs. 12.55 to 12.64, find all indicated quantities, where $a_1, a_2, \ldots, a_n, \ldots$ is an arithmetic sequence, $d =$ common difference, and $S =$ sum of first n terms of the sequence.

12.55 $a_1 = -5$, $d = 4$, $a_2 = ?$, $a_3 = ?$, $a_4 = ?$

▎ $a_2 = a_1 + d = -5 + 4 = -1$; $a_3 = a_2 + d = 3$; $a_4 = a_3 + d = 7$.

12.56 $a_1 = -3$, $d = 5$, $a_{15} = ?$, $S_{11} = ?$

▮ $a_n = a_1 + (n - 1)d$ for every $n > 1$. Thus, $a_{15} = -3 + 14(5) = 67$. Also $S_n = (n/2)(a_1 + a_n) = (n/2)[2a_1 + (n - 1)d]$, so $S_{11} = \frac{11}{2}[2(-3) + 10(5)] = \frac{11}{2}(44) = 242$.

12.57 $a_1 = 1$, $a_2 = 5$, $S_{21} = ?$

▮ $S_n = (n/2)[2a_1 + (n - 1)d]$, so $S_{21} = \frac{21}{2}[2(1) + (20)(4)]$ (since $a_2 - a_1 = d = 4$) $= \frac{21}{2}(82) = 861$.

12.58 $a_1 = 7$, $a_2 = 5$, $a_{15} = ?$

▮ $a_n = a_1 + (n - 1)d$, for $n \geq 1$; $d = a_2 - a_1 = 5 - 7 = -2$; $a_{15} = 7 + 14(-2) = -21$.

12.59 $a_1 = 12$, $a_{40} = 22$, $S_{40} = ?$

▮ $S_{40} = \frac{40}{2}(a_1 + a_{40}) = 20(12 + 22) = 20(34) = 680$.

12.60 $a_1 = 24$, $a_{24} = -28$, $S_{24} = ?$

▮ $S_{24} = \frac{24}{2}[24 + (-28)] = 12(-4) = -48$.

12.61 $a_1 = \frac{1}{3}$, $a_2 = \frac{1}{2}$, $a_{11} = ?$, $S_{11} = ?$

▮ $d = a_2 - a_1 = \frac{1}{2} - \frac{1}{3} = \frac{1}{6}$; $a_{11} = a_1 + (n - 1)d = \frac{1}{3} + 10(\frac{1}{6}) = \frac{1}{3} + \frac{10}{6} = 2$; $S_{11} = \frac{11}{2}(\frac{1}{3} + 2) = \frac{11}{2} \cdot \frac{7}{3} = \frac{77}{6}$.

12.62 $a_3 = 13$, $a_{10} = 55$, $a_1 = ?$

▮ $a_{10} - a_3 = 42 = 7d$; $d = \frac{42}{7} = 6$; $a_2 = 13 - 6 = 7$; $a_1 = 7 - 6 = 1$.

12.63 $a_9 = -12$, $a_{13} = 4$, $a_1 = ?$

▮ $a_{13} - a_9 = 4 - (-12) = 16 = 4d$; $d = 4$; $a_9 = a_1 + 8d = -12 = a_1 + 8(4)$; $a_1 = -12 - 32 = -44$.

12.64 $a_3 = 13$, $a_{10} = 55$, $a_1 = ?$

▮ $7d = a_{10} - a_3 = 42$; $d = 6$; $a_{10} = a_1 + 9d = a_1 + 54 = 55$; $a_1 = 1$.

For Probs. 12.65 to 12.69, the sequence is geometric, and r stands for the common ratio.

12.65 Find d for the sequence $1, \frac{1}{2}, \frac{1}{4}, \frac{1}{8}, \ldots$

▮ $d = a_{n+1}/a_n = \frac{1}{8}/\frac{1}{4} = \frac{1}{2}$. (If the sequence is geometric, any two consecutively numbered terms can be used.)

12.66 Find the sum of the first eight terms of the sequence $1, \frac{1}{3}, \frac{1}{9}, \frac{1}{27}, \ldots$

▮ $S_8 = \frac{a_1(1 - r^8)}{1 - r} = \frac{1[1 - (\frac{1}{3})^8]}{1 - \frac{1}{3}} = \frac{1 - \frac{1}{6561}}{\frac{2}{3}} = \frac{6560}{6561} \cdot \frac{3}{2} = \frac{19,680}{13,122}$.

12.67 Find the sum of the first five terms of a geometric sequence where $a_2 = 3$, $r = \frac{1}{2}$.

▮ $S_5 = \frac{a_1(1 - r^5)}{1 - \frac{1}{2}} = \frac{6[1 - (\frac{1}{2})^5]}{1 - \frac{1}{2}} = \frac{6(\frac{31}{32})}{\frac{1}{2}} = 12\left(\frac{31}{32}\right) = \frac{372}{32}$.

12.68 If $a_1 = 2$ and $r = x$ ($x \neq 1$), find S_4.

▮ $S_4 = \frac{a_1(1 - r^4)}{1 - r} = \frac{2(1 - x^4)}{1 - x} = \frac{2(1 + x^2)(1 - x)(1 + x)}{1 - x} = 2(1 + x^2)(1 + x)$.

12.69 If $r = 1/x$ and $S_5 = 10x$, find a_1.

▮ $S_5 = \dfrac{a_1[1 - (1/x)^5]}{1 - 1/x}$; $10x = \dfrac{a_1(1 - 1/x^5)}{1 - 1/x}$; $a_1 = \dfrac{10x(1 - 1/x)}{1 - 1/x^5}$.

12.70 The sum of three numbers in an arithmetic progression is 24. If the first number is decreased by 1 and the second decreased by 2, the three numbers are in a geometric progression. Find the three numbers.

▮ Let x = 1st number. Then $x + d$ = 2d number and $x + 2d$ = 3d number (since they are in arithmetic progression). Furthermore, the sum of these numbers is 24: $x + (x + d) + (x + 2d) = 24$, or $x + d = 8$. Also $x - 1$, $(x + d) - 2$, $x + 2d$ are in geometric progression, so $\dfrac{x + d - 2}{x - 1} = \dfrac{x + 2d}{x + d - 2}$, or $(x + d - 2)^2 = (x + 2d)(x - 1)$. Simplifying gives $d^2 - 2d = 3x - 4$. But $x + d = 8$, and solving simultaneously, we find $d^2 + d - 20 = 0$, or $d = 4, -5$. If $d = 4$, $x = 4$. If $d = -5$, $x = 13$. Therefore, the solutions are 4, 8, 12; *or* 13, 8, 3.

12.4 GEOMETRIC SERIES

For Probs. 12.71 to 12.75, tell whether the given series is geometric.

12.71 $1 + \frac{1}{2} + \frac{1}{4} + \frac{1}{8} + \cdots$

▮ We notice that $a_{n+1}/a_n = \frac{1}{2}$ for all successive pairs. The series is geometric.

12.72 $1 + 2 + 4 + 8 + 16 + \cdots$

▮ $a_{n+1}/a_n = 2$ for all successive pairs. The series is geometric.

12.73 $1 - 2 + 4 - 8 + 16 - \cdots$

▮ $a_{n+1}/a_n = -2$ for all successive pairs. The series is geometric.

12.74 $1 + \frac{1}{4} + \frac{1}{9} + \frac{1}{16} + \cdots$

▮ Notice that $a_2/a_1 \neq a_3/a_2$. Thus, the series is nongeometric.

12.75 $1 + \frac{1}{2} - \frac{1}{4} - \frac{1}{8} + \frac{1}{16} + \frac{1}{25} - \cdots$

▮ $a_2/a_1 \neq a_3/a_2$. The series is nongeometric.

For Probs. 12.76 to 12.80, write the given geometric series, using sigma (summation) notation.

12.76 $2 + 1 + \frac{1}{2} + \frac{1}{4} + \frac{1}{8} + \cdots$

▮ These terms are each $\frac{1}{2}$ of the previous term, beginning with 2: $\sum\limits_{i=1}^{\infty} 2(\frac{1}{2})^{i-1}$. When $i = 1$, this is $(\frac{1}{2})^0 = 1$; when $i = 2$, this is $\frac{1}{2}$; etc. See Prob. 12.77.

12.77 $2 + 1 + \frac{1}{2} + \frac{1}{4} + \frac{1}{8} + \cdots$

▮ Each term is a power of $\frac{1}{2}$. The first is $(\frac{1}{2})^{-1}$, the second is $(\frac{1}{2})^0$, the third is $(\frac{1}{2})^1$, etc. And $\sum\limits_{i=1}^{\infty} (\frac{1}{2})^{i-2}$ represents this. Note that this is the same series as in Prob. 12.76.

12.78 $x + x^2 + x^3 + \cdots$

▮ $\sum\limits_{i=1}^{\infty} x^i$

12.79 $x - x^2 + x^3 + \cdots$

▮ $\sum\limits_{i=1}^{\infty} x^i (-1)^{i+1}$. This will alternate signs, beginning with a positive.

12.80 $\frac{1}{6} - (\frac{1}{6})^2 + (\frac{1}{6})^3 + \cdots$

▮ $\sum\limits_{i=1}^{\infty} (-1)^{i+1}(\frac{1}{6})^i$.

For Probs. 12.81 to 12.88, find the sum or show that the series has no sum.

12.81 $1 + \frac{1}{2} + \frac{1}{4} + \frac{1}{8} + \cdots$

▮ $S = \dfrac{a}{1-r}$ where $|r| < 1$. In this case $r = \frac{1}{2}$ and $a = 1$, so $S = \dfrac{1}{1-\frac{1}{2}} = \dfrac{1}{\frac{1}{2}} = 2$.

12.82 $2 + \frac{2}{3} + \frac{2}{9} + \frac{2}{27} + \cdots$

▮ $r = \frac{1}{3}$, so $|r| < 1$. Thus, $S = \dfrac{a}{1-r} = \dfrac{2}{1-\frac{1}{3}} = \dfrac{2}{\frac{2}{3}} = 3$.

12.83 $1 + \frac{1}{9} + \frac{1}{81} + \cdots$

▮ $r = \frac{1}{9}$, so $|r| < 1$. Thus, $S = \dfrac{a}{1-r} = \dfrac{1}{1-\frac{1}{9}} = \dfrac{1}{\frac{8}{9}} = \dfrac{9}{8}$.

12.84 $3 + 6 + 12 + 24 + \cdots$

▮ Here, $r = 2$; thus, $|r| \geq 1$, and S does not exist.

12.85 $3 - \frac{3}{2} + \frac{3}{4} - \frac{3}{8} + \cdots$

▮ $r = -\frac{1}{2}$, so $|r| < 1$. $s = \dfrac{a}{1-r} = \dfrac{3}{1-(-\frac{1}{2})} = \dfrac{3}{\frac{3}{2}} = 2$.

12.86 $3 + \frac{3}{2} + \frac{3}{4} + \frac{3}{8} + \cdots$

▮ $r = \frac{1}{2}$, so $|r| < 1$. Thus, $S = \dfrac{a}{1-r} = \dfrac{3}{1-\frac{1}{2}} = \dfrac{3}{\frac{1}{2}} = 6$. Compare this to Prob. 12.85.

12.87 $\frac{1}{2} + (\frac{1}{2})^2 + (\frac{1}{2})^3 + \cdots$

▮ $r = \frac{1}{2}$, so $|r| < 1$. Thus, $S = \dfrac{a}{1-r} = \dfrac{\frac{1}{2}}{1-\frac{1}{2}} = \dfrac{\frac{1}{2}}{\frac{1}{2}} = 1$.

12.88 $-\frac{1}{2} + (\frac{1}{2})^2 - (\frac{1}{2})^3 + \cdots$

▮ This can be rewritten as $-\frac{1}{2} + \frac{1}{4} - \frac{1}{8} + \cdots$, which is geometric, $a = -\frac{1}{2}$, $r = -\frac{1}{2}$. Then $|r| < 1$, and
$S = \dfrac{a}{1-r} = \dfrac{-\frac{1}{2}}{1-(-\frac{1}{2})} = \dfrac{-\frac{1}{2}}{\frac{3}{2}} = -\dfrac{1}{3}$.

12.5 BINOMIAL THEOREM

For Probs. 12.89 to 12.105, evaluate the given expression.

12.89 $7!$

▮ $7! = 7 \cdot 6 \cdot 5 \cdots 1 = 5040$.

12.90 $\dfrac{7!}{6!}$

▮ $\dfrac{7!}{6!} = \dfrac{7 \cdot 6 \cdots 2 \cdot 1}{6 \cdot 5 \cdots 2 \cdot 1} = 7.$

12.91 $\dfrac{7!}{4!}$

▮ $\dfrac{7!}{4!} = \dfrac{7 \cdot 6 \cdot 5 \cdot 4 \cdots 2 \cdot 1}{4 \cdot 3 \cdot 2 \cdot 1} = 7 \cdot 6 \cdot 5 = 210.$

12.92 $\dfrac{7!}{4! \, 3!}$

▮ $\dfrac{7!}{4! \, 3!} = \dfrac{7 \cdot 6 \cdot 5 \cdot 4 \cdot 3 \cdot 2 \cdot 1}{\underbrace{3 \cdot 2}_{3!} \cdot \underbrace{4 \cdot 3 \cdot 2 \cdot 1}_{4!}} = \dfrac{7 \cdot 6 \cdot 5}{6} = 35.$

12.93 $\dfrac{8!}{6! \, 2!}$

▮ $\dfrac{8!}{6! \, 2!} = \dfrac{8 \cdot 7 \cdot 6 \cdot 5 \cdot 4 \cdot 3 \cdot 2}{2 \cdot 6 \cdot 5 \cdot 4 \cdot 3 \cdot 2} = \dfrac{56}{2} = 28.$

12.94 $\dfrac{14!}{12! \, 2!}$

▮ $\dfrac{14!}{12! \, 2!} = \dfrac{14 \cdot 13 \cdot 12 \cdot 11 \cdot 10 \cdot 9 \cdot 8 \cdot 7 \cdot 6 \cdot 5 \cdot 4 \cdot 3 \cdot 2}{2 \cdot 12 \cdot 11 \cdot 10 \cdot 9 \cdot 8 \cdot 7 \cdot 6 \cdot 5 \cdot 4 \cdot 3 \cdot 2} = \dfrac{182}{2} = 91.$

12.95 $\dfrac{102!}{2! \, 100!}$

▮ $\dfrac{102!}{2! \, 100!} = \dfrac{102 \cdot 101 \cdot 100!}{2 \cdot 100!} = 51 \cdot 101 = 5151.$

12.96 $\dfrac{6!}{2! \, 2! \, 2!}$

▮ $\dfrac{6!}{2! \, 2! \, 2!} = \dfrac{6 \cdot 5 \cdot 4 \cdot 3 \cdot 2}{2 \cdot 2 \cdot 2} = 3 \cdot 5 \cdot 2 \cdot 3 = 90.$

12.97 $\dfrac{5!}{2! \, (5-2)!}$

▮ $\dfrac{5!}{2! \, (5-2)!} = \dfrac{5!}{2! \, 3!} = \dfrac{5 \cdot 4 \cdot 3 \cdot 2}{2 \cdot 3 \cdot 2} = \dfrac{20}{2} = 10.$

12.98 $\dfrac{15!}{0! \, (15-0)!}$

▮ $\dfrac{15!}{0! \, (15-0)!} = \dfrac{15!}{0! \, 15!} = \dfrac{15!}{15!} = 1.$ (Remember: $0! = 1! = 1$.)

12.99 $\dbinom{2}{1}$

▮ $\dbinom{2}{1} = \dfrac{2!}{1! \, (2-1)!} = \dfrac{2!}{1} = 2$, since $\dbinom{n}{r} = \dfrac{n!}{r! \, (n-r)!}$.

12.100 $\binom{4}{2}$

■ $\binom{4}{2} = \dfrac{4!}{2!\,(4-2)!} = \dfrac{4!}{2!\,2!} = \dfrac{24}{4} = 6.$

12.101 $\binom{5}{2}$

■ $\binom{5}{2} = \dfrac{5!}{2!\,(5-2)!} = \dfrac{5!}{2!\,3!} = \dfrac{5\cdot 4\cdot 3\cdot 2}{2\cdot 3\cdot 2} = \dfrac{20}{2} = 10.$

12.102 $\binom{6}{2}$

■ $\binom{6}{2} = \dfrac{6!}{2!\,4!} = \dfrac{6\cdot 5\cdot 4\cdot 3\cdot 2}{2\cdot 4\cdot 3\cdot 2} = \dfrac{30}{2} = 15.$

12.103 $\binom{7}{0}$

■ $\binom{7}{0} = \dfrac{7!}{0!\,(7-0)!} = \dfrac{7!}{7!} = 1.$

12.104 $\binom{150}{0}$

■ $\binom{150}{0} = \dfrac{150!}{0!\,150!} = 1.$

12.105 $\binom{80}{80}$

■ $\binom{80}{80} = \dfrac{80!}{80!\,(80-80)!} = \dfrac{80!}{80!} = 1.$

For Probs. 12.106 to 12.113, expand the given expression, using the binomial theorem.

12.106 $(x+y)^2$

■ Remember that $(a+b)^n = \sum\limits_{k=0}^{n} \binom{n}{k} a^{n-k} b^k,\ n \ge 1.$ In this case, $(x+y)^2 =$

$\sum\limits_{k=0}^{2} \binom{2}{k} x^{2-k} y^k = \binom{2}{0} x^2 + \binom{2}{1} xy + \binom{2}{2} y^2$ $(k=0,\ 1,\ \text{and } 2,\ \text{respectively}) = x^2 + 2xy + y^2.$ All problems in which we use the binomial theorem to expand an exponential form of a binomial are done in this way.

12.107 $(x+y)^3$

■ $(x+y)^3 = \binom{3}{0} x^3 + \binom{3}{1} x^2 y + \binom{3}{2} xy^2 + \binom{3}{3} y^3 = x^3 + 3x^2 y + 3xy^2 + y^3.$

$\left[\text{Here } (x+y)^3 = \sum\limits_{k=0}^{3} \binom{3}{k} x^{3-k} y^k.\right]$

12.108 $(p+q)^4$

■ $(p+q)^4 = \binom{4}{0} p^4 + \binom{4}{1} p^3 q + \binom{4}{2} p^2 q^2 + \binom{4}{3} pq^3 + \binom{4}{4} q^4 = p^4 + 4p^3 q + 6p^2 q^2 + 4pq^3 + q^4.$

12.109 $(a+3)^3$

$$(a+3)^3 = \binom{3}{0}a^3 + \binom{3}{1}a^2 \cdot 3 + \binom{3}{2}a \cdot 3^2 + \binom{3}{3}3^3 = a^3 + 3a^2 \cdot 3 + 3a \cdot 9 + 27 = a^3 + 9a^2 + 27a + 27.$$

12.110 $(a-3)^3$

$$(a-3)^3 = [a+(-3)]^3 = \binom{3}{0}a^3 + \binom{3}{1}a^2 \cdot (-3) + \binom{3}{2}a \cdot (-3)^2 + \binom{3}{3}(-3)^3 =$$
$$a^3 + 3a^2(-3) + 3a(9) + (-27) = a^3 - 9a^2 + 27a - 27.$$

12.111 $(2a+3)^5$

$$(2a+3)^5 = \binom{5}{0}(2a)^5 + \binom{5}{1}(2a)^4(3) + \binom{5}{2}(2a)^3 3^2 + \binom{5}{3}(2a)^2 3^3 + \binom{5}{4}(2a)3^4 + \binom{5}{5}3^5 =$$
$$(2a)^5 + 5(3)(2a)^4 + 10(2a)^3(9) + 10(2a)^2(27) + 5(2a)81 + 243 = 32a^5 + 30a^4 + 180a^3 + 540a^2 + 810a + 243.$$

12.112 $(2x-4)^3$

$$(2x-4)^3 = \binom{3}{0}(2x)^3 + \binom{3}{1}(2x)^2(-4) + \binom{3}{2}(2x)(-4)^2 + \binom{3}{3}(-4)^3 =$$
$$8x^3 + (3)(4x^2)(-4) + 3(2x)(16) - 64 = 8x^3 - 48x^2 + 96x - 64.$$

12.113 $(4-2u)^4$

$$(4-2u)^4 = \binom{4}{0}4^4 + \binom{4}{1}4^3(-2u) + \binom{4}{2}4^2(-2u)^2 + \binom{4}{3}4(-2u)^3 + \binom{4}{4}(-2u)^4 =$$
$$256 + 4 \cdot 4^3(-2u) + 6 \cdot 4^2(-2u)^2 + 4 \cdot 4(-2u)^3 + (-2u)^4 = 256 - 512u + 384u^2 - 128u^3 + 16u^4.$$

12.114 Find the coefficient of x^2y^5 in the expansion of $(x-y)^7$.

Recall that $(x-y)^7 = \sum_{k=0}^{7}\binom{7}{k}x^{7-k}(-y)^k$. Thus, the coefficient of x^2y^5 occurs when $7-k=2$, or $k=5$. The coefficient $\binom{7}{k}$, then, is $\binom{7}{5}$ with a negative sign since $(-y)^k < 0$ when k is odd. Then $-\binom{7}{5} = -\frac{7\cdot6\cdot5\cdot4\cdot3\cdot2}{2\cdot5\cdot4\cdot3\cdot2} = -21$.

12.115 Find the seventh term in the expansion of $(u+v)^{15}$.

The seventh term is $\binom{15}{6}u^9v^6 = 5005u^9v^6$.

12.6 PERMUTATIONS, COMBINATIONS, AND PROBABILITY

For Probs. 12.116 to 12.125, calculate the given permutation or combination.

12.116 $P(7, 1)$

(a) $P(n, r) = n(n-1)(n-2)\cdots(n-r+1)$. Thus, $P(7, 1) = 7(6)(5)\cdots(7-1+1) = 7$.

(b) $P(n, r) = \frac{n!}{(n-r)!} = \frac{7!}{(7-1)!} = \frac{7!}{6!} = 7$.

12.117 $P(7, 3)$

$$P(7, 3) = \frac{7!}{(7-3)!} = \frac{7!}{4!} = 7\cdot6\cdot5 = 210.$$

12.118 $P(7, 4)$

▮ $P(7, 4) = \dfrac{7!}{(7-4)!} = \dfrac{7!}{3!} = 7 \cdot 6 \cdot 5 \cdot 4 = 840.$

12.119 $P(7, 7)$

▮ $P(7, 7) = \dfrac{7!}{(7-7)!} = \dfrac{7!}{0!} = 7! = 5040.$

12.120 $P(4, 1) + P(4, 2) + P(4, 3) + P(4, 4)$

▮ $P(4, 1) + P(4, 2) + P(4, 3) + P(4, 4) = \dfrac{4!}{3!} + \dfrac{4!}{2!} + \dfrac{4!}{1!} + \dfrac{4!}{0!} = 4 + 12 + 24 + 24 = 64.$

12.121 $C(31, 2)$

▮ $C(n, r) = \dfrac{n!}{r!\,(n-r)!}$. Thus, $C(31, 2) = \dfrac{31!}{2!\,29!} = \dfrac{31 \cdot 30}{2!} = 465.$

12.122 $C(26, 3)$

▮ $C(26, 3) = \dfrac{26!}{3!\,23!} = \dfrac{26 \cdot 25 \cdot 24}{6} = 2600.$

12.123 $C(24, 4)$

▮ $C(24, 4) = \dfrac{24!}{4!\,20!} = \dfrac{24 \cdot 23 \cdot 22 \cdot 21}{24} = 10,626.$

12.124 $C(13, 5)$

▮ $C(13, 5) = \dfrac{13!}{5!\,8!} = \dfrac{13 \cdot 12 \cdots 8 \cdot 7 \cdots 2 \cdot 1}{5! \quad 8 \cdot 7 \cdots 2 \cdot 1} = \dfrac{13 \cdot 12 \cdot 11 \cdot 10 \cdot 9}{5!} = 1287.$

12.125 $C(16, 4)$

▮ $C(16, 4) = \dfrac{16!}{4!\,12!} = \dfrac{16 \cdot 15 \cdot 14 \cdot 13 \cdot 12!}{4! \cdot 12!} = \dfrac{16 \cdot 15 \cdot 14 \cdot 13}{4!} = 1820.$

12.126 How many two-digit numbers can be made with the digits 2, 4, 6, 8 if repetitions are allowed?

▮ There are 4 choices (2, 4, 6, 8) for each digit: $4 \cdot 4 = 16$.

12.127 Do as in Prob. 12.126, but no repetitions are allowed.

▮ In this case there are 4 choices for the first digit and 3 choices for the second digit: $4 \cdot 3 = 12$.

12.128 How many three-digit numbers can be formed with the digits 1, 2, 3, 4, 5 if repetitions are allowed?

▮ There are 5 choices for each digit: $5 \cdot 5 \cdot 5 = 125$.

12.129 Do as in Prob. 12.128, but no repetitions are allowed.

▮ There are 5 choices for the first digit, 4 for the second, and 3 for the third: $5 \cdot 4 \cdot 3 = 60$.

12.130 How many three-digit numbers (with 0 not the first digit) can be formed with digits 0, 1, 2, 3, 4 if repetitions are allowed?

▮ There are 4 choices for the first digit (1, 2, 3, 4) and 5 choices for the others: $4 \cdot 5 \cdot 5 = 100$.

12.131 Do as in Prob. 12.130 but with no repetitions.

▌ There are 4 choices for the first, 4 for the second (since 0 is added as a possibility), and 3 for the third: $4 \cdot 4 \cdot 3 = 48$.

12.132 How many four-digit numbers can be formed, each less than 5000, with the digits 1, 2, 4, 6, 8 if repetitions are allowed?

▌ The first digit can be 1, 2, or 4. The second, third, and fourth digits can each be any one of 1, 2, 4, 6, 8: $3 \cdot 5 \cdot 5 \cdot 5 = 375$.

12.133 Do as in Prob. 12.132, with no repetitions.

▌ The first digit can be 1, 2, or 4. The second digit can be 1, 2, 4, 6, or 8 but not the digit in the first slot: $3 \cdot 4 \cdot 3 \cdot 2 = 72$.

12.134 How many committees, consisting of 1 first-year student, 1 sophomore, and 1 junior can be selected from 40 first-year students, 30 sophomores, and 25 juniors?

▌ $40 \cdot 30 \cdot 25 = 30,000$.

12.135 Five boys are in a room which has 4 doors. In how many ways can they leave the room?

▌ Each boy has 4 choices. $4^5 = 4 \cdot 4 \cdot 4 \cdot 4 \cdot 4 = 1024$.

For Probs. 12.136 to 12.138, find the number of permutations, each with 7 letters, which can be made with the letters given.

12.136 WYOMING

▌ $P(7, 7)$ [The first 7 indicates that we have 7 letters to choose from; the second 7 indicates that we want to choose 7 letters from those available.] $= \dfrac{7!}{(7-7)!} = \dfrac{7!}{0!} = 7! = 5040$.

12.137 KENTUCKY

▌ $P(8, 7) = \dfrac{8!}{(8-7)!} = \dfrac{8!}{1!} = 8! = 40,320$.

12.138 WASHINGTON

▌ $P(10, 7) = \dfrac{10!}{(10-7)!} = \dfrac{10!}{6} = 604,800$.

12.139 How many even numbers of three digits can be made with the digits 1, 2, 3, 4, 5, 6, 7 if no digit is repeated?

▌ The units digit must be 2, 4, or 6. The other digits are any one of 1, 2, 3, 4, 5, 6, 7 with no repetitions: $3 \cdot 6 \cdot 5 = 90$.

12.140 Seven songs are to be given in a program. In how many different orders could they be rendered?

▌ $7! = 5040$.

For Probs. 12.141 to 12.143, find the number of permutations which can be formed by using all the given letters.

12.141 ALABAMA

▌ $\dfrac{7!}{4!} = 7 \cdot 6 \cdot 5 = 210$ (7 objects, 4 objects are alike).

12.142 ARKANSAS

▌ $\dfrac{8!}{3!\,2!} = 3360$ (8 letters, 3 A's, 2 S's).

12.143 MISSISSIPPI

▌ $\dfrac{11!}{4!\,2!\,4!} = 34{,}650$ (4 S's, 2 P's, 4 I's).

12.144 Find the number of ways 2 half dollars, 4 quarters, and 6 dimes can be distributed among 12 girls if each gets a coin.

▌ $\dfrac{12!}{2!\,4!\,6!} = 13{,}860$.

12.145 A line is formed by 5 girls and 5 boys, with boys and girls alternating. Find the number of ways of making the line.

▌ There are 10 spots. The first is filled by a boy or girl. There are 10 possibilities. Then there are 5 boys or girls left, and 4 left of the other gender, leaving 9 altogether: $10 \cdot 5 \cdot 4 \cdot 4 \cdot 3 \cdot 3 \cdot 2 \cdot 2 \cdot 1 \cdot 1 = 28{,}800$.

12.146 Do as in Prob. 12.145, but with a circular arrangement.

▌ $\dfrac{28{,}000}{10} = 2880$. In a circle, there is no first or last object.

12.147 Show that $P(5, 4) = P(5, 5)$.

▌ $P(5, 4) = \dfrac{5!}{(5-4)!} = \dfrac{5!}{1!} = \dfrac{5!}{0!} = P(5, 5)$ since $0! = 1!$.

12.148 Show that $P(n, n - 1) = P(n, n)$.

▌ $P(n, n) = \dfrac{n!}{(n-n)!} = \dfrac{n!}{0!}$; $P(n, n-1) = \dfrac{n!}{[n-(n-1)]!} = \dfrac{n!}{1!} = \dfrac{n!}{0!}$ since $0! = 1!$.

12.149 In how many different ways can a tennis team of 4 be chosen from 17 players?

▌ Here we are looking for the number of combinations of 17 objects taken 4 at a time. The concern here is not the order of selection or the arrangement of the players. Compare this to, for example, Prob. 12.140 which is a permutation:

$$C(17, 4) = \dfrac{17!}{4!\,13!} = 2380.$$

12.150 Nine points, no three of which are on a straight line, are marked on a chalkboard. How many lines, each through two of the points, can be drawn?

▌ See Fig. 12.1. Point A can be connected to any of the other points.

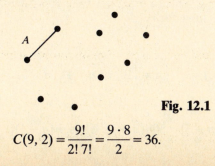

Fig. 12.1

$$C(9, 2) = \dfrac{9!}{2!\,7!} = \dfrac{9 \cdot 8}{2} = 36.$$

12.151 In Prob. 12.150, how many triangles are determined?

▮ $C(9, 3) = \dfrac{9!}{3!\,6!} = 84$.

12.152 Seven different coins are tossed simultaneously. In how many ways can 3 heads and 4 tails come up?

▮ $C(7, 3) \cdot C(4, 4) = 35$. [Choose 3 heads, and then the other 4 must all be tails; $C(4, 4) = 1$.]

12.153 In how many ways can a court of 9 judges make a 5-to-4 decision?

▮ $C(9, 5) \cdot C(4, 4) = \dfrac{9!}{4!\,5!} \cdot 1 = 126$.

For Probs. 12.154 to 12.156, 4 delegates are to chosen from 8 members of a club.

12.154 How many choices are possible?

▮ $C(8, 4) = \dfrac{8!}{4!\,4!} = 70$.

12.155 How many choices contain member A?

▮ This is the same as choosing 3 from a group of 7, since we are fixing A as a chosen member: $7!/3!\,4! = 35$.

12.156 How many contain A and B?

▮ Here 2 members are fixed: $C(6, 2) = \dfrac{6!}{4!\,2!} = 15$.

The alphabet consists of 21 consonants and 5 vowels. Refer to this information for Probs. 12.157 to 12.160.

12.157 In how many ways can 5 consonants and 3 vowels be selected?

▮ $C(21, 5) \cdot C(5, 3) = 20,349 \cdot 40 = 813,960$.

12.158 How many words consisting of 5 consonants and 3 vowels can be formed?

▮ 8! words can be found from each combination in Prob. 12.157 by taking all permutations of 8 letters: $C(21, 5) \cdot C(5, 3) \cdot 8!$.

12.159 How many of the words in Prob. 12.158 begin with X?

▮ X's position is fixed. We select 4 other consonants and the 3 vowels. There are 7 choices altogether since X is fixed: $7! \cdot C(20, 4) \cdot C(5, 3)$.

12.160 How many of the words in Prob. 12.159 contain A?

▮ A's position is not fixed. We select 4 other consonants and 2 other vowels: $C(20, 4) \cdot C(4, 2) \cdot 7!$

For Probs. 12.161 to 12.163, solve for n.

12.161 $P(n, 2) = 110$

▮ $P(n, 2) = n(n - 1) = n^2 - n = 110$, or $n = 11$ ($n > 0$).

12.162 $P(n, 4) = 30P(n, 2)$

▮ $n(n - 1)(n - 2)(n - 3) = 30n(n - 1)$. Thus, $n(n - 1)(n - 8)(n + 3) = 0$ (check the algebra). But $n \geq 4$, so $n = 8$.

12.163 $C(n, 3) = 84$

▮ $\dfrac{n!}{3!\,(n-3)!} = 84$; $\dfrac{n(n-2)(n-1)}{6} = 84$. $n(n-2)(n-1) = 504$, $n^3 - 3n^2 + 2n - 504 = 0$, $(n-9)(n^2+6n+56) = 0$, or $n = 9$.

For Probs. 12.164 to 12.166, a ball is chosen from a bag containing 2 white, 1 black, and 2 red balls.

12.164 Find the probability that a white ball is chosen.

▮ $P(W) = \frac{2}{5}$. Since there are 2 whites (we are looking for the probability of drawing white) and there are 5 balls altogether, $P(E) = S/n$, where E is an event, $S =$ number of ways the event can occur, and $n =$ total number of possibilities.

12.165 Find the probability that a black ball is chosen.

▮ $P(B) = \frac{1}{5}$. (There is only 1 black ball; there are 5 balls.)

12.166 Find the probability of choosing a green ball.

▮ $P(G) = 0/5 = 0$ since there are 0 green balls.

For Probs. 12.167 to 12.170, a single die is rolled.

12.167 Find the probability that the die comes up 6.

▮ $P(6) = \frac{1}{6}$ since only one side has a 6 and there are six sides altogether.

12.168 Find the probability that it lands on an even number.

▮ $P(\text{even}) = P(2, 4, \text{or } 6) = \frac{3}{6} = \frac{1}{2}$.

12.169 Find the probability that it lands on 2 or an odd number.

▮ $P(2 \text{ or odd}) = P(1, 2, 3, 5) = \frac{4}{6} = \frac{2}{3}$.

12.170 Find the probability it lands on a number less than 10.

▮ $P(<10) = 6/6$ since all possibilities are less than 10.

For Probs. 12.171 to 12.173, three pennies are tossed at the same time.

12.171 Find the probability that all 3 land on heads (H).

▮ $P(H, H, H) = \frac{1}{8}$ since there are $2^3 = 8$ possibilities. Altogether each coin has 2 possibilities, and only one of these is H, H, H.

12.172 Find the probability that 2 are H and 1 is T (tails).

▮ There are $2^3 = 8$ possibilities altogether. How many are H, H, T? There are $C(3, 1)$ ways of one coin being T: $C(3, 1) = \dfrac{3!}{1!\,2!} = 3$. Thus, $P(H, H, T) = \frac{3}{8}$. Note: Since $C(3, 1) = C(3, 2)$, we could have used $C(3, 2)$ instead. The number of ways of 1 coin out of 3 being T is the same as 2 coins out of 3 being H.

12.173 Find the probability that 1 is H and 2 are T.

▮ $P(T, T, H) = \dfrac{C(3, 2)}{8} = \dfrac{3}{8}$.

For Probs. 12.174 to 12.177, 3 dice are rolled.

12.174 Find the probability that their sum is 18.

    ▌ $P(\text{sum} = 18) = P(6, 6, 6) = 1/6^3$ (since that is the only way to get 18) $= \frac{1}{216}$.

12.175 Find the probability of the sum being less than 5.

    ▌ To be less than 5, the sum must be 3 or 4. To be 3, the sum must be 1, 1, 1. To be 4, it could be 2, 1, 1 or 1, 2, 1 or 1, 1, 2 [i.e., there are $C(3, 1)$, or 3, ways]. Altogether there are 3 ways to get 4; altogether, there is 1 way to get 3.

$$P(<5) = P(3 \text{ or } 4) = \frac{3+1}{6^2} = \frac{4}{36} = \frac{1}{9}.$$

12.176 Find the probability that the sum is 6.

    ▌ $P(6) = P(1, 1, 4) + P(1, 4, 1) + P(4, 1, 1) + P(2, 2, 2) + P(2, 3, 1) + P(2, 1, 3) + P(3, 1, 2) + P(3, 2, 1) + P(1, 2, 3) + P(1, 3, 2) = \frac{10}{36}$.

12.177 Find the probability that the sum is even.

    ▌ $P(E) = \frac{8}{16} = \frac{1}{2}$ since there are 16 possibilities (3 through 18) and 8 of those are even (4, 6, 8, 10, 12, 14, 16, 18).

For Probs. 12.178 to 12.180, 3 balls are drawn from a bag containing 6 red and 5 black balls.

12.178 Find the probability that all are red.

    ▌ $P(\text{all red}) = P(R, R, R)$. There are $C(11, 3)$ possibilities altogether. $C(11, 3) = \frac{11!}{3! \, 8!} = 165$. Since 6 balls are red, there are $C(6, 3)$ ways to pick R, R, R. $C(6, 3) = 20$. $P(R, R, R) = \frac{20}{165} = \frac{4}{33}$.

12.179 Find the probability that all are black.

    ▌ $P(B, B, B) = \frac{C(5, 3)}{C(11, 3)} = \frac{10}{165} = \frac{2}{33}$.

12.180 Find the probability of 2 being red and 1 being black.

    ▌ There are $C(6, 2)$ ways to pick 2 red balls and $C(5, 1)$ ways to pick 1 black ball. Therefore,

$$P(2R, 1B) = \frac{C(6, 2)C(5, 1)}{C(11, 3)} = \frac{5}{11}.$$

For Probs. 12.181 to 12.184, 1 card is picked from an ordinary 52-card deck.

12.181 Find the probability that a red card is chosen.

    ▌ $P(R) = \frac{26}{52} = \frac{1}{2}$ since 26 cards are red and 26 are black.

12.182 Find the probability that a queen is chosen.

    ▌ $P(Q) = \frac{4}{52} = \frac{1}{13}$ since there are 4 queens (1 per suit).

12.183 Find the probability that a red 8 is chosen.

    ▌ $P(\text{red } 8) = \frac{2}{52} = \frac{1}{26}$ since only the 8 of hearts and 8 of diamonds are red.

12.184 Find the probability that the card chosen is less than 5 (counting the ace as high).

    ▌ $P(<5) = P(2, 3, \text{ or } 4)$. For each of 2, 3, 4 there are 4 suits. $P(2, 3, \text{ or } 4) = \frac{12}{52} = \frac{3}{13}$.

For Probs. 12.185 to 12.188, 2 cards are chosen simultaneously from an ordinary deck.

12.185 Find the probability both cards chosen are 5s.

▌ There are $C(4, 2)$ ways to choose two 5s from four 5s, and $C(52, 2)$ ways to choose 2 cards from 52 cards. $P(5, 5) = \dfrac{C(4, 2)}{C(52, 2)} = \dfrac{6}{1326}$.

12.186 Find the probability both cards chosen are aces.

▌ See Prob. 12.185. This is the exact same situation. $P(A, A) = \frac{6}{1326}$.

12.187 Find the probability that 1 ace and 1 king are chosen.

▌ There are $C(4, 1)$ ways to choose 1 ace and 1 king. Therefore,
$$P(A, K) = \frac{C(4, 1)C(4, 1)}{C(52, 2)} = \frac{4 \cdot 4}{1275} = \frac{16}{1326}.$$

12.188 Find the probability of choosing 2 black cards.

▌ $P(B, B) = \dfrac{C(26, 2)}{C(52, 2)} = \dfrac{325}{1326}$.

For Probs. 12.189 to 12.193, the probability that a brush salesperson will make a sale at one house is $\frac{2}{3}$ and at a second house is $\frac{1}{2}$.

12.189 Find the probability the salesperson will make both sales.

▌ $P(\text{both sales}) = \frac{2}{3} \cdot \frac{1}{2} = \frac{2}{6} = \frac{1}{3}$.

12.190 Find the probability he or she will make neither sale.

▌ $P(\text{no sales}) = (1 - \frac{2}{3})(1 - \frac{1}{2}) = \frac{1}{3} \cdot \frac{1}{2} = \frac{1}{6}$.

12.191 Find the probability the salesperson will make a sale at the first but not the second house.

▌ $P(\text{sale, no sale}) = \frac{2}{3}(1 - \frac{1}{2}) = \frac{2}{3} \cdot \frac{1}{2} = \frac{1}{3}$.

12.192 Find the probability she or he will make at least 1 sale.

▌ $P(\geq 1) = P(1 \text{ sale}) + P(2 \text{ sales}) = P(\text{sale, no sale}) + P(\text{no sale, sale}) + P(\text{sale, sale}) = \frac{1}{3} + \frac{1}{3} + \frac{1}{6} = \frac{5}{6}$. [Alternatively, $P(\geq 1) = 1 - P(\text{no sale}) = 1 - \frac{1}{6} = \frac{5}{6}$.] (See Prob. 12.190.)

12.193 Find the probability the salesperson will make *exactly* 1 sale.

▌ $P(1 \text{ sale}) = P(\text{sale, no sale}) + P(\text{no sale, sale}) = \frac{1}{3} + \frac{1}{3} = \frac{2}{3}$.

For Probs. 12.194 to 12.198, 4 cards are drawn from a box containing 10 cards numbered $1, 2, \ldots, 10$. Find the given probability.

12.194 All are even if the cards are all drawn together.

▌ $P(4 \text{ evens}) = C(5, 4)/C(10, 4)$ (draw 4 evens from 5 evens) $= 5/210 = 1/42$.

12.195 All are even if each card is replaced before the next is drawn.

▌ $P(\text{1st is even}) = \frac{1}{2}$; $P(\text{2d even given 1st even replaced}) = \frac{1}{2}$, etc.; $(\frac{1}{2})^4 = \frac{1}{16}$.

12.196 Exactly 3 cards are even if all the cards are drawn together.

▌ $P(3 \text{ even}) = P(3 \text{ even, 1 odd}) = C(5, 3)C(5, 1)/C(10, 4) = \frac{50}{210} = \frac{5}{21}$.

12.197 Exactly 3 cards are even if the cards are replaced after being drawn.

▌ $P(\text{1st even}) = \frac{1}{2}$, $P(\text{2d even}) = \frac{1}{2}$, $P(\text{3d even}) = \frac{1}{2}$, and there are $C(4, 3)$ ways to do this. $P(\text{3 even}) = P(\text{3 even, 1 odd}) = (\frac{1}{2})^3 \cdot (\frac{1}{2})^1 \cdot 4 = \frac{1}{8} \cdot \frac{1}{2} \cdot 4 = \frac{1}{4}$.

12.198 At least 3 cards are even if all cards are drawn at the same time.

▌ $P(\geq 3 \text{ even}) = P(\text{3 even}) + P(\text{4 even}) = \frac{5}{21} + \frac{1}{42}$ (see Probs. 12.194 and 12.196).

For Probs. 12.199 and 12.200, bag 1 contains 2 white balls and 1 red ball; bag 2 contains 3 white balls and 1 red ball. A bag is chosen at random, and a ball is picked at random. What is the given probability?

12.199 The ball chosen is red.

▌ The probability of choosing bag 1 is $\frac{1}{2}$. The probability of choosing a red ball is then $\frac{1}{3}$. Similarly, the probability of choosing 2 is $\frac{1}{2}$, but here the probability of choosing a red ball is $\frac{1}{4}$. Therefore, $P(\text{red}) = \frac{1}{2} \cdot \frac{1}{3} + \frac{1}{2} \cdot \frac{1}{4} = \frac{1}{6} + \frac{1}{8} = \frac{7}{24}$.

12.200 The ball chosen is white.

▌ $P(\text{white}) = \frac{1}{2} \cdot \frac{2}{3} + \frac{1}{2} \cdot \frac{3}{4} = \frac{17}{24}$. [Alternatively, $P(\text{W}) = 1 - P(\text{R}) = 1 - \frac{7}{24} = \frac{17}{24}$.]

APPENDIX

TABLE A.1 The Exponential Function

x	e^x	e^{-x}	x	e^x	e^{-x}
.00	1.00000	1.00000	.60	1.82212	.54881
.01	1.01005	.99005	.61	1.84043	.54335
.02	1.02020	.98020	.62	1.85893	.53794
.03	1.03045	.97045	.63	1.87761	.53259
.04	1.04081	.96079	.64	1.89648	.52729
.05	1.05127	.95123	.65	1.91554	.52205
.06	1.06184	.94176	.66	1.93479	.51685
.07	1.07251	.93239	.67	1.95424	.51171
.08	1.08329	.92312	.68	1.97388	.50662
.09	1.09417	.91393	.69	1.99372	.50158
.10	1.10517	.90484	.70	2.01375	.49659
.11	1.11628	.89583	.71	2.03399	.49164
.12	1.12750	.88692	.72	2.05443	.48675
.13	1.13883	.87810	.73	2.07508	.48191
.14	1.15027	.86936	.74	2.09594	.47711
.15	1.16183	.86071	.75	2.11700	.47237
.16	1.17351	.85214	.76	2.13828	.46767
.17	1.18530	.84366	.77	2.15977	.46301
.18	1.19722	.83527	.78	2.18147	.45841
.19	1.20925	.82696	.79	2.20340	.45384
.20	1.22140	.81873	.80	2.22554	.44933
.21	1.23368	.81058	.81	2.24791	.44486
.22	1.24608	.80252	.82	2.27050	.44043
.23	1.25860	.79453	.83	2.29332	.43605
.24	1.27125	.78663	.84	2.31637	.43171
.25	1.28403	.77880	.85	2.33965	.42741
.26	1.29693	.77105	.86	2.36316	.42316
.27	1.30996	.76338	.87	2.38691	.41895
.28	1.32313	.75578	.88	2.41090	.41478
.29	1.33643	.74826	.89	2.43513	.41066
.30	1.34986	.74082	.90	2.45960	.40657
.31	1.36343	.73345	.91	2.48432	.40252
.32	1.37713	.72615	.92	2.50929	.39852
.33	1.39097	.71892	.93	2.53451	.39455
.34	1.40495	.71177	.94	2.55998	.39063
.35	1.41907	.70469	.95	2.58571	.38674
.36	1.43333	.69768	.96	2.61170	.38289
.37	1.44773	.69073	.97	2.63794	.37908
.38	1.46228	.68386	.98	2.66446	.37531
.39	1.47698	.67706	.99	2.69123	.37158
.40	1.49182	.67032	1.00	2.71828	.36788
.41	1.50682	.66365	1.01	2.74560	.36422
.42	1.52196	.65705	1.02	2.77319	.36059
.43	1.53726	.65051	1.03	2.80107	.35701
.44	1.55271	.64404	1.04	2.82922	.35345
.45	1.56831	.63763	1.05	2.85765	.34994
.46	1.58407	.63128	1.06	2.88637	.34646
.47	1.59999	.62500	1.07	2.91538	.34301
.48	1.61607	.61878	1.08	2.94468	.33960
.49	1.63232	.61263	1.09	2.97427	.33622
.50	1.64872	.60653	1.10	3.00417	.33287
.51	1.66529	.60050	1.11	3.03436	.32956
.52	1.68203	.59452	1.12	3.06485	.32628
.53	1.69893	.58860	1.13	3.09566	.32303
.54	1.71601	.58275	1.14	3.12677	.31982
.55	1.73325	.57695	1.15	3.15819	.31664
.56	1.75067	.57121	1.16	3.18993	.31349
.57	1.76827	.56553	1.17	3.22199	.31037
.58	1.78604	.55990	1.18	3.25437	.30728
.59	1.80399	.55433	1.19	3.28708	.30422

From Koshy: *College Algebra and Trigonometry with Applications*, 1986. McGraw-Hill. Reprinted with permission.

TABLE A.1 The Exponential Function (*Continued*)

x	e^x	e^{-x}	x	e^x	e^{-x}
1.20	3.32012	.30119	1.80	6.04965	.16530
1.21	3.35348	.29820	1.81	6.11045	.16365
1.22	3.38719	.29523	1.82	6.17186	.16203
1.23	3.42123	.29229	1.83	6.23389	.16041
1.24	3.45561	.28938	1.84	6.29654	.15882
1.25	3.49034	.28650	1.85	6.35982	.15724
1.26	3.52542	.28365	1.86	6.42374	.15567
1.27	3.56085	.28083	1.87	6.48830	.15412
1.28	3.59664	.27804	1.88	6.55350	.15259
1.29	3.63279	.27527	1.89	6.61937	.15107
1.30	3.66930	.27253	1.90	6.68589	.14957
1.31	3.70617	.26982	1.91	6.75309	.14808
1.32	3.74342	.26714	1.92	6.82096	.14661
1.33	3.78104	.26448	1.93	6.88951	.14515
1.34	3.81904	.26185	1.94	6.95875	.14370
1.35	3.85743	.25924	1.95	7.02869	.14227
1.36	3.89619	.25666	1.96	7.09933	.14086
1.37	3.93535	.25411	1.97	7.17068	.13946
1.38	3.97490	.25158	1.98	7.24274	.13807
1.39	4.01485	.24908	1.99	7.31553	.13670
1.40	4.05520	.24660	2.00	7.38906	.13534
1.41	4.09596	.24414	2.01	7.46332	.13399
1.42	4.13712	.24171	2.02	7.53832	.13266
1.43	4.17870	.23931	2.03	7.61409	.13134
1.44	4.22070	.23693	2.04	7.69061	.13003
1.45	4.26311	.23457	2.05	7.76790	.12873
1.46	4.30596	.23224	2.06	7.84597	.12745
1.47	4.34924	.22993	2.07	7.92482	.12619
1.48	4.39295	.22764	2.08	8.00447	.12493
1.49	4.43710	.22537	2.09	8.08491	.12369
1.50	4.48169	.22313	2.10	8.16617	.12246
1.51	4.52673	.22091	2.11	8.24824	.12124
1.52	4.57223	.21871	2.12	8.33114	.12003
1.53	4.61818	.21654	2.13	8.41487	.11884
1.54	4.66459	.21438	2.14	8.49944	.11765
1.55	4.71147	.21225	2.15	8.58486	.11648
1.56	4.75882	.21014	2.16	8.67114	.11533
1.57	4.80665	.20805	2.17	8.75828	.11418
1.58	4.85496	.20598	2.18	8.84631	.11304
1.59	4.90375	.20393	2.19	8.93521	.11192
1.60	4.95303	.20190	2.20	9.02501	.11080
1.61	5.00281	.19989	2.21	9.11572	.10970
1.62	5.05309	.19790	2.22	9.20733	.10861
1.63	5.10387	.19593	2.23	9.29987	.10753
1.64	5.15517	.19398	2.24	9.39333	.10646
1.65	5.20698	.19205	2.25	9.48774	.10540
1.66	5.25931	.19014	2.26	9.58309	.10435
1.67	5.31217	.18825	2.27	9.67940	.10331
1.68	5.36556	.18637	2.28	9.77668	.10228
1.69	5.41948	.18452	2.29	9.87494	.10127
1.70	5.47395	.18268	2.30	9.97418	.10026
1.71	5.52896	.18087	2.31	10.07442	.09926
1.72	5.58453	.17907	2.32	10.17567	.09827
1.73	5.64065	.17728	2.33	10.27794	.09730
1.74	5.69734	.17552	2.34	10.38124	.09633
1.75	5.75460	.17377	2.35	10.48557	.09537
1.76	5.81244	.17204	2.36	10.59095	.09442
1.77	5.87085	.17033	2.37	10.69739	.09348
1.78	5.92986	.16864	2.38	10.80490	.09255
1.79	5.98945	.16696	2.39	10.91349	.09163

TABLE A.1 The Exponential Function (*Continued*)

x	e^x	e^{-x}	x	e^x	e^{-x}
2.40	11.02318	.09072	**3.00**	20.08554	.04979
2.41	11.13396	.08982	3.01	20.28740	.04929
2.42	11.24586	.08892	3.02	20.49129	.04880
2.43	11.35888	.08804	3.03	20.69723	.04832
2.44	11.47304	.08716	3.04	20.90524	.04783
2.45	11.58835	.08629	3.05	21.11534	.04736
2.46	11.70481	.08543	3.06	21.32756	.04689
2.47	11.82245	.08458	3.07	21.54190	.04642
2.48	11.94126	.08374	3.08	21.75840	.04596
2.49	12.06128	.08291	3.09	21.97708	.04550
2.50	12.18249	.08208	**3.10**	22.19795	.04505
2.51	12.30493	.08127	3.11	22.42104	.04460
2.52	12.42860	.08046	3.12	22.64638	.04416
2.53	12.55351	.07966	3.13	22.87398	.04372
2.54	12.67967	.07887	3.14	23.10387	.04328
2.55	12.80710	.07808	3.15	23.33606	.04285
2.56	12.93582	.07730	3.16	23.57060	.04243
2.57	13.06582	.07654	3.17	23.80748	.04200
2.58	13.19714	.07577	3.18	24.04675	.04159
2.59	13.32977	.07502	3.19	24.28843	.04117
2.60	13.46374	.07427	**3.20**	24.53253	.04076
2.61	13.59905	.07353	3.21	24.77909	.04036
2.62	13.73572	.07280	3.22	25.02812	.03996
2.63	13.87377	.07208	3.23	25.27966	.03956
2.64	14.01320	.07136	3.24	25.53372	.03916
2.65	14.15404	.07065	3.25	25.79034	.03877
2.66	14.29629	.06995	3.26	26.04954	.03839
2.67	14.43997	.06925	3.27	26.31134	.03801
2.68	14.58509	.06856	3.28	26.57577	.03763
2.69	14.73168	.06788	3.29	26.84286	.03725
2.70	14.87973	.06721	**3.30**	27.11264	.03688
2.71	15.02928	.06654	3.31	27.38512	.03652
2.72	15.18032	.06587	3.32	27.66035	.03615
2.73	15.33289	.06522	3.33	27.93834	.03579
2.74	15.48698	.06457	3.34	28.21913	.03544
2.75	15.64263	.06393	3.35	28.50273	.03508
2.76	15.79984	.06329	3.36	28.78919	.03474
2.77	15.95863	.06266	3.37	29.07853	.03439
2.78	16.11902	.06204	3.38	29.37077	.03405
2.79	16.28102	.06142	3.39	29.66595	.03371
2.80	16.44465	.06081	**3.40**	29.96410	.03337
2.81	16.60992	.06020	3.41	30.26524	.03304
2.82	16.77685	.05961	3.42	30.56941	.03271
2.83	16.94546	.05901	3.43	30.87664	.03239
2.84	17.11577	.05843	3.44	31.18696	.03206
2.85	17.28778	.05784	3.45	31.50039	.03175
2.86	17.46153	.05727	3.46	31.81698	.03143
2.87	17.63702	.05670	3.47	32.13674	.03112
2.88	17.81427	.05613	3.48	32.45972	.03081
2.89	17.99331	.05558	3.49	32.78595	.03050
2.90	18.17414	.05502	**3.50**	33.11545	.03020
2.91	18.35680	.05448	3.51	33.44827	.02990
2.92	18.54129	.05393	3.52	33.78443	.02960
2.93	18.72763	.05340	3.53	34.12397	.02930
2.94	18.91585	.05287	3.54	34.46692	.02901
2.95	19.10595	.05234	3.55	34.81332	.02872
2.96	19.29797	.05182	3.56	35.16320	.02844
2.97	19.49192	.05130	3.57	35.51659	.02816
2.98	19.68782	.05079	3.58	35.87354	.02788
2.99	19.88568	.05029	3.59	36.23408	.02760

TABLE A.1 The Exponential Function (*Continued*)

x	e^x	e^{-x}	x	e^x	e^{-x}
3.60	36.59823	.02732	4.20	66.68633	.01500
3.61	36.96605	.02705	4.21	67.35654	.01485
3.62	37.33757	.02678	4.22	68.03348	.01470
3.63	37.71282	.02652	4.23	68.71723	.01455
3.64	38.09184	.02625	4.24	69.40785	.01441
3.65	38.47467	.02599	4.25	70.10541	.01426
3.66	38.86134	.02573	4.26	70.80998	.01412
3.67	39.25191	.02548	4.27	71.52163	.01398
3.68	39.64639	.02522	4.28	72.24044	.01384
3.69	40.04485	.02497	4.29	72.96647	.01370
3.70	40.44730	.02472	4.30	73.69979	.01357
3.71	40.85381	.02448	4.31	74.44049	.01343
3.72	41.26439	.02423	4.32	75.18863	.01330
3.73	41.67911	.02399	4.33	75.94429	.01317
3.74	42.09799	.02375	4.34	76.70754	.01304
3.75	42.52108	.02352	4.35	77.47846	.01291
3.76	42.94843	.02328	4.36	78.25713	.01278
3.77	43.38006	.02305	4.37	79.04363	.01265
3.78	43.81604	.02282	4.38	79.83803	.01253
3.79	44.25640	.02260	4.39	80.64042	.01240
3.80	44.70118	.02237	4.40	81.45087	.01228
3.81	45.15044	.02215	4.41	82.26946	.01216
3.82	45.60421	.02193	4.42	83.09628	.01203
3.83	46.06254	.02171	4.43	83.93141	.01191
3.84	46.52547	.02149	4.44	84.77494	.01180
3.85	46.99306	.02128	4.45	85.62694	.01168
3.86	47.46535	.02107	4.46	86.48751	.01156
3.87	47.94238	.02086	4.47	87.35672	.01145
3.88	48.42421	.02065	4.48	88.23467	.01133
3.89	48.91089	.02045	4.49	89.12144	.01122
3.90	49.40245	.02024	4.50	90.01713	.01111
3.91	49.89895	.02004	4.51	90.92182	.01100
3.92	50.40044	.01984	4.52	91.83560	.01089
3.93	50.90698	.01964	4.53	92.75856	.01078
3.94	51.41860	.01945	4.54	93.69080	.01067
3.95	51.93537	.01925	4.55	94.63240	.01057
3.96	52.45732	.01906	4.56	95.58347	.01046
3.97	52.98453	.01887	4.57	96.54411	.01036
3.98	53.51703	.01869	4.58	97.51439	.01025
3.99	54.05489	.01850	4.59	98.49443	.01015
4.00	54.59815	.01832	4.60	99.48431	.01005
4.01	55.14687	.01813	4.61	100.48415	.00995
4.02	55.70110	.01795	4.62	101.49403	.00985
4.03	56.26091	.01777	4.63	102.51406	.00975
4.04	56.82634	.01760	4.64	103.54435	.00966
4.05	57.39745	.01742	4.65	104.58498	.00956
4.06	57.97431	.01725	4.66	105.63608	.00947
4.07	58.55696	.01708	4.67	106.69774	.00937
4.08	59.14547	.01691	4.68	107.77007	.00928
4.09	59.73989	.01674	4.69	108.85318	.00919
4.10	60.34029	.01657	4.70	109.94717	.00910
4.11	60.94671	.01641	4.71	111.05216	.00900
4.12	61.55924	.01624	4.72	112.16825	.00892
4.13	62.17792	.01608	4.73	113.29556	.00883
4.14	62.80282	.01592	4.74	114.43420	.00874
4.15	63.43400	.01576	4.75	115.58428	.00865
4.16	64.07152	.01561	4.76	116.74592	.00857
4.17	64.71545	.01545	4.77	117.91924	.00848
4.18	65.36585	.01530	4.78	119.10435	.00840
4.19	66.02279	.01515	4.79	120.30136	.00831

TABLE A.1 The Exponential Function (*Continued*)

x	e^x	e^{-x}	x	e^x	e^{-x}
4.80	121.51041	.00823	**8.00**	2980.95779	.00034
4.81	122.73161	.00815	8.10	3294.46777	.00030
4.82	123.96509	.00807	8.20	3640.95004	.00027
4.83	125.21096	.00799	8.30	4023.87219	.00025
4.84	126.46935	.00791	8.40	4447.06665	.00022
4.85	127.74039	.00783	8.50	4914.76886	.00020
4.86	129.02420	.00775	8.60	5431.65906	.00018
4.87	130.32091	.00767	8.70	6002.91180	.00017
4.88	131.63066	.00760	8.80	6634.24371	.00015
4.89	132.95357	.00752	8.90	7331.97339	.00014
4.90	134.28978	.00745	**9.00**	8103.08295	.00012
4.91	135.63941	.00737	9.10	8955.29187	.00011
4.92	137.00261	.00730	9.20	9897.12830	.00010
4.93	138.37951	.00723	9.30	10938.01868	.00009
4.94	139.77024	.00715	9.40	12088.38049	.00008
4.95	141.17496	.00708	9.50	13359.72522	.00007
4.96	142.59379	.00701	9.60	14764.78015	.00007
4.97	144.02688	.00694	9.70	16317.60608	.00006
4.98	145.47438	.00687	9.80	18033.74414	.00006
4.99	146.93642	.00681	9.90	19930.36987	.00005
5.00	148.41316	.00674	**10.00**	22026.46313	.00005
5.10	164.02190	.00610	10.10	24343.00708	.00004
5.20	181.27224	.00552	10.20	26903.18408	.00004
5.30	200.33680	.00499	10.30	29732.61743	.00003
5.40	221.40641	.00452	10.40	32859.62500	.00003
5.50	244.69192	.00409	10.50	36315.49854	.00003
5.60	270.42640	.00370	10.60	40134.83350	.00002
5.70	298.86740	.00335	10.70	44355.85205	.00002
5.80	330.29955	.00303	10.80	49020.79883	.00002
5.90	365.03746	.00274	10.90	54176.36230	.00002
6.00	403.42877	.00248	**11.00**	59874.13477	.00002
6.10	445.85775	.00224	11.10	66171.15430	.00002
6.20	492.74903	.00203	11.20	73130.43652	.00001
6.30	544.57188	.00184	11.30	80821.63379	.00001
6.40	601.84502	.00166	11.40	89321.72168	.00001
6.50	665.14159	.00150	11.50	98715.75879	.00001
6.60	735.09516	.00136	11.60	109097.78906	.00001
6.70	812.40582	.00123	11.70	120571.70605	.00001
6.80	897.84725	.00111	11.80	133252.34570	.00001
6.90	992.27469	.00101	11.90	147266.62109	.0000!
7.00	1096.63309	.00091	—	—	—
7.10	1211.96703	.00083	—	—	—
7.20	1339.43076	.00075	—	—	—
7.30	1480.29985	.00068	—	—	—
7.40	1635.98439	.00061	—	—	—
7.50	1808.04231	.00055	—	—	—
7.60	1998.19582	.00050	—	—	—
7.70	2208.34796	.00045	—	—	—
7.80	2440.60187	.00041	—	—	—
7.90	2697.28226	.00037	—	—	—

TABLE A.2 Common Logarithms

N	0	1	2	3	4	5	6	7	8	9
1.0	.0000	.0043	.0086	.0128	.0170	.0212	.0253	.0294	.0334	.0374
1.1	.0414	.0453	.0492	.0531	.0569	.0607	.0645	.0682	.0719	.0755
1.2	.0792	.0828	.0864	.0899	.0934	.0969	.1004	.1038	.1072	.1106
1.3	.1139	.1173	.1206	.1239	.1271	.1303	.1335	.1367	.1399	.1430
1.4	.1461	.1492	.1523	.1553	.1584	.1614	.1644	.1673	.1703	.1732
1.5	.1761	.1790	.1818	.1847	.1875	.1903	.1931	.1959	.1987	.2014
1.6	.2041	.2068	.2095	.2122	.2148	.2175	.2201	.2227	.2253	.2279
1.7	.2304	.2330	.2355	.2380	.2405	.2430	.2455	.2480	.2504	.2529
1.8	.2553	.2577	.2601	.2625	.2648	.2672	.2695	.2718	.2742	.2765
1.9	.2788	.2810	.2833	.2856	.2878	.2900	.2923	.2945	.2967	.2989
2.0	.3010	.3032	.3054	.3075	.3096	.3118	.3139	.3160	.3181	.3201
2.1	.3222	.3243	.3263	.3284	.3304	.3324	.3345	.3365	.3385	.3404
2.2	.3424	.3444	.3464	.3483	.3502	.3522	.3541	.3560	.3579	.3598
2.3	.3617	.3636	.3655	.3674	.3692	.3711	.3729	.3747	.3766	.3784
2.4	.3802	.3820	.3838	.3856	.3874	.3892	.3909	.3927	.3945	.3962
2.5	.3979	.3997	.4014	.4031	.4048	.4065	.4082	.4099	.4116	.4133
2.6	.4150	.4166	.4183	.4200	.4216	.4232	.4249	.4265	.4281	.4298
2.7	.4314	.4330	.4346	.4362	.4378	.4393	.4409	.4425	.4440	.4456
2.8	.4472	.4487	.4502	.4518	.4533	.4548	.4564	.4579	.4594	.4609
2.9	.4624	.4639	.4654	.4669	.4683	.4698	.4713	.4728	.4742	.4757
3.0	.4771	.4786	.4800	.4814	.4829	.4843	.4857	.4871	.4886	.4900
3.1	.4914	.4928	.4942	.4955	.4969	.4983	.4997	.5011	.5024	.5038
3.2	.5051	.5065	.5079	.5092	.5105	.5119	.5132	.5145	.5159	.5172
3.3	.5185	.5198	.5211	.5224	.5237	.5250	.5263	.5276	.5289	.5302
3.4	.5315	.5328	.5340	.5353	.5366	.5378	.5391	.5403	.5416	.5428
3.5	.5441	.5453	.5465	.5478	.5490	.5502	.5514	.5527	.5539	.5551
3.6	.5563	.5575	.5587	.5599	.5611	.5623	.5635	.5647	.5658	.5670
3.7	.5682	.5694	.5705	.5717	.5729	.5740	.5752	.5763	.5775	.5786
3.8	.5798	.5809	.5821	.5832	.5843	.5855	.5866	.5877	.5888	.5899
3.9	.5911	.5922	.5933	.5944	.5955	.5966	.5977	.5988	.5999	.6010
4.0	.6021	.6031	.6042	.6053	.6064	.6075	.6085	.6096	.6107	.6117
4.1	.6128	.6138	.6149	.6160	.6170	.6180	.6191	.6201	.6212	.6222
4.2	.6232	.6243	.6253	.6263	.6274	.6284	.6294	.6304	.6314	.6325
4.3	.6335	.6345	.6355	.6365	.6375	.6385	.6395	.6405	.6415	.6425
4.4	.6435	.6444	.6454	.6464	.6474	.6484	.6493	.6503	.6513	.6522
4.5	.6532	.6542	.6551	.6561	.6571	.6580	.6590	.6599	.6609	.6618
4.6	.6628	.6637	.6646	.6656	.6665	.6675	.6684	.6693	.6702	.6712
4.7	.6721	.6730	.6739	.6749	.6758	.6767	.6776	.6785	.6794	.6803
4.8	.6812	.6821	.6830	.6839	.6848	.6857	.6866	.6875	.6884	.6893
4.9	.6902	.6911	.6920	.6928	.6937	.6946	.6955	.6964	.6972	.6981
5.0	.6990	.6998	.7007	.7016	.7024	.7033	.7042	.7050	.7059	.7067
5.1	.7076	.7084	.7093	.7101	.7110	.7118	.7126	.7135	.7143	.7152
5.2	.7160	.7168	.7177	.7185	.7193	.7202	.7210	.7218	.7226	.7235
5.3	.7243	.7251	.7259	.7267	.7275	.7284	.7292	.7300	.7308	.7316
5.4	.7324	.7332	.7340	.7348	.7356	.7364	.7372	.7380	.7388	.7396

From Koshy: *College Algebra and Trigonometry with Applications*, 1986. McGraw-Hill. Reprinted with permission.

TABLE A.2 Common Logarithms (*Continued*)

N	0	1	2	3	4	5	6	7	8	9
5.5	.7404	.7412	.7419	.7427	.7435	.7443	.7451	.7459	.7466	.7474
5.6	.7482	.7490	.7497	.7505	.7513	.7520	.7528	.7536	.7543	.7551
5.7	.7559	.7566	.7574	.7582	.7589	.7597	.7604	.7612	.7619	.7627
5.8	.7634	.7642	.7649	.7657	.7664	.7672	.7679	.7686	.7694	.7701
5.9	.7709	.7716	.7723	.7731	.7738	.7745	.7752	.7760	.7767	.7774
6.0	.7782	.7789	.7796	.7803	.7810	.7818	.7825	.7832	.7839	.7846
6.1	.7853	.7860	.7868	.7875	.7882	.7889	.7896	.7903	.7910	.7917
6.2	.7924	.7931	.7938	.7945	.7952	.7959	.7966	.7973	.7980	.7987
6.3	.7993	.8000	.8007	.8014	.8021	.8028	.8035	.8041	.8048	.8055
6.4	.8062	.8069	.8075	.8082	.8089	.8096	.8102	.8109	.8116	.8122
6.5	.8129	.8136	.8142	.8149	.8156	.8162	.8169	.8176	.8182	.8189
6.6	.8195	.8202	.8209	.8215	.8222	.8228	.8235	.8241	.8248	.8254
6.7	.8261	.8267	.8274	.8280	.8287	.8293	.8299	.8306	.8312	.8319
6.8	.8325	.8331	.8338	.8344	.8351	.8357	.8363	.8370	.8376	.8382
6.9	.8388	.8395	.8401	.8407	.8414	.8420	.8426	.8432	.8439	.8445
7.0	.8451	.8457	.8463	.8470	.8476	.8482	.8488	.8494	.8500	.8506
7.1	.8513	.8519	.8525	.8531	.8537	.8543	.8549	.8555	.8561	.8567
7.2	.8573	.8579	.8585	.8591	.8597	.8603	.8609	.8615	.8621	.8627
7.3	.8633	.8639	.8645	.8651	.8657	.8663	.8669	.8675	.8681	.8686
7.4	.8692	.8698	.8704	.8710	.8716	.8722	.8727	.8733	.8739	.8745
7.5	.8751	.8756	.8762	.8768	.8774	.8779	.8785	.8791	.8797	.8802
7.6	.8808	.8814	.8820	.8825	.8831	.8837	.8842	.8848	.8854	.8859
7.7	.8865	.8871	.8876	.8882	.8887	.8893	.8899	.8904	.8910	.8915
7.8	.8921	.8927	.8932	.8938	.8943	.8949	.8954	.8960	.8965	.8971
7.9	.8976	.8982	.8987	.8993	.8998	.9004	.9009	.9015	.9020	.9025
8.0	.9031	.9036	.9042	.9047	.9053	.9058	.9063	.9069	.9074	.9079
8.1	.9085	.9090	.9096	.9101	.9106	.9112	.9117	.9122	.9128	.9133
8.2	.9138	.9143	.9149	.9154	.9159	.9165	.9170	.9175	.9180	.9186
8.3	.9191	.9196	.9201	.9206	.9212	.9217	.9222	.9227	.9232	.9238
8.4	.9243	.9248	.9253	.9258	.9263	.9269	.9274	.9279	.9284	.9289
8.5	.9294	.9299	.9304	.9309	.9315	.9320	.9325	.9330	.9335	.9340
8.6	.9345	.9350	.9355	.9360	.9365	.9370	.9375	.9380	.9385	.9390
8.7	.9395	.9400	.9405	.9410	.9415	.9420	.9425	.9430	.9435	.9440
8.8	.9445	.9450	.9455	.9460	.9465	.9469	.9474	.9479	.9484	.9489
8.9	.9494	.9499	.9504	.9509	.9513	.9518	.9523	.9528	.9533	.9538
9.0	.9542	.9547	.9552	.9557	.9562	.9566	.9571	.9576	.9581	.9586
9.1	.9590	.9595	.9600	.9605	.9609	.9614	.9619	.9624	.9628	.9633
9.2	.9638	.9643	.9647	.9652	.9657	.9661	.9666	.9671	.9675	.9680
9.3	.9685	.9689	.9694	.9699	.9703	.9708	.9713	.9717	.9722	.9727
9.4	.9731	.9736	.9741	.9745	.9750	.9754	.9759	.9763	.9768	.9773
9.5	.9777	.9782	.9786	.9791	.9795	.9800	.9805	.9809	.9814	.9818
9.6	.9823	.9827	.9832	.9836	.9841	.9845	.9850	.9854	.9859	.9863
9.7	.9868	.9872	.9877	.9881	.9886	.9890	.9894	.9899	.9903	.9908
9.8	.9912	.9917	.9921	.9926	.9930	.9934	.9939	.9943	.9948	.9952
9.9	.9956	.9961	.9965	.9969	.9974	.9978	.9983	.9987	.9991	.9996

TABLE A.3 Trigonometric Functions of an Angle

t	t degrees	$\sin t$	$\cos t$	$\tan t$	$\cot t$	$\sec t$	$\csc t$		
.0000	0° 00′	.0000	1.0000	.0000		1.000		90° 00′	1.5708
.0029	10	.0029	1.0000	.0029	343.8	1.000	343.8	50	1.5679
.0058	20	.0058	1.0000	.0058	171.9	1.000	171.9	40	1.5650
.0087	30	.0087	1.0000	.0087	114.6	1.000	114.6	30	1.5621
.0116	40	.0116	.9999	.0116	85.94	1.000	85.95	20	1.5592
.0145	50	.0145	.9999	.0145	68.75	1.000	68.76	10	1.5563
.0175	1° 00′	.0175	.9998	.0175	57.29	1.000	57.30	89° 00′	1.5533
.0204	10	.0204	.9998	.0204	49.10	1.000	49.11	50	1.5504
.0233	20	.0233	.9997	.0233	42.96	1.000	42.98	40	1.5475
.0262	30	.0262	.9997	.0262	38.19	1.000	38.20	30	1.5446
.0291	40	.0291	.9996	.0291	34.37	1.000	34.38	20	1.5417
.0320	50	.0320	.9995	.0320	31.24	1.001	31.26	10	1.5388
.0349	2° 00′	.0349	.9994	.0349	28.64	1.001	28.65	88° 00′	1.5359
.0378	10	.0378	.9993	.0378	26.43	1.001	26.45	50	1.5330
.0407	20	.0407	.9992	.0407	24.54	1.001	24.56	40	1.5301
.0436	30	.0436	.9990	.0437	22.90	1.001	22.93	30	1.5272
.0465	40	.0465	.9989	.0466	21.47	1.001	21.49	20	1.5243
.0495	50	.0494	.9988	.0495	20.21	1.001	20.23	10	1.5213
.0524	3° 00′	.0523	.9986	.0524	19.08	1.001	19.11	87° 00′	1.5184
.0553	10	.0552	.9985	.0553	18.07	1.002	18.10	50	1.5155
.0582	20	.0581	.9983	.0582	17.17	1.002	17.20	40	1.5126
.0611	30	.0610	.9981	.0612	16.35	1.002	16.38	30	1.5097
.0640	40	.0640	.9980	.0641	15.60	1.002	15.64	20	1.5068
.0669	50	.0669	.9978	.0670	14.92	1.002	14.96	10	1.5039
.0698	4° 00′	.0698	.9976	.0699	14.30	1.002	14.34	86° 00′	1.5010
.0727	10	.0727	.9974	.0729	13.73	1.003	13.76	50	1.4981
.0756	20	.0756	.9971	.0758	13.20	1.003	13.23	40	1.4952
.0785	30	.0785	.9969	.0787	12.71	1.003	12.75	30	1.4923
.0814	40	.0814	.9967	.0816	12.25	1.003	12.29	20	1.4893
.0844	50	.0843	.9964	.0846	11.83	1.004	11.87	10	1.4864
.0873	5° 00′	.0872	.9962	.0875	11.43	1.004	11.47	85° 00′	1.4835
.0902	10	.0901	.9959	.0904	11.06	1.004	11.10	50	1.4806
.0931	20	.0929	.9957	.0934	10.71	1.004	10.76	40	1.4777
.0960	30	.0958	.9954	.0963	10.39	1.005	10.43	30	1.4748
.0989	40	.0987	.9951	.0992	10.08	1.005	10.13	20	1.4719
.1018	50	.1016	.9948	.1022	9.788	1.005	9.839	10	1.4690
.1047	6° 00′	.1045	.9945	.1051	9.514	1.006	9.567	84° 00′	1.4661
.1076	10	.1074	.9942	.1080	9.255	1.006	9.309	50	1.4632
.1105	20	.1103	.9939	.1110	9.010	1.006	9.065	40	1.4603
.1134	30	.1132	.9936	.1139	8.777	1.006	8.834	30	1.4573
.1164	40	.1161	.9932	.1169	8.556	1.007	8.614	20	1.4544
.1193	50	.1190	.9929	.1198	8.345	1.007	8.405	10	1.4515
.1222	7° 00′	.1219	.9925	.1228	8.144	1.008	8.206	83° 00′	1.4486
		$\cos t$	$\sin t$	$\cot t$	$\tan t$	$\csc t$	$\sec t$	t degrees	t

From Koshy: *College Algebra and Trigonometry with Applications*, 1986. McGraw-Hill. Reprinted with permission.

TABLE A.3 Trigonometric Functions of an Angle (*Continued*)

t	t degrees	sin t	cos t	tan t	cot t	sec t	csc t		
.1222	**7° 00'**	.1219	.9925	.1228	8.144	1.008	8.206	**83° 00'**	1.4486
.1251	10	.1248	.9922	.1257	7.953	1.008	8.016	50	1.4457
.1280	20	.1276	.9918	.1287	7.770	1.008	7.834	40	1.4428
.1309	30	.1305	.9914	.1317	7.596	1.009	7.661	30	1.4399
.1338	40	.1334	.9911	.1346	7.429	1.009	7.496	20	1.4370
.1367	50	.1363	.9907	.1376	7.269	1.009	7.337	10	1.4341
.1396	**8° 00'**	.1392	.9903	.1405	7.115	1.010	7.185	**82° 00'**	1.4312
.1425	10	.1421	.9899	.1435	6.968	1.010	7.040	50	1.4283
.1454	20	.1449	.9894	.1465	6.827	1.011	6.900	40	1.4254
.1484	30	.1478	.9890	.1495	6.691	1.011	6.765	30	1.4224
.1513	40	.1507	.9886	.1524	6.561	1.012	6.636	20	1.4195
.1542	50	.1536	.9881	.1554	6.435	1.012	6.512	10	1.4166
.1571	**9° 00'**	.1564	.9877	.1584	6.314	1.012	6.392	**81° 00'**	1.4137
.1600	10	.1593	.9872	.1614	6.197	1.013	6.277	50	1.4108
.1629	20	.1622	.9868	.1644	6.084	1.013	6.166	40	1.4079
.1658	30	.1650	.9863	.1673	5.976	1.014	6.059	30	1.4050
.1687	40	.1679	.9858	.1703	5.871	1.014	5.955	20	1.4021
.1716	50	.1708	.9853	.1733	5.769	1.015	5.855	10	1.3992
.1745	**10° 00'**	.1736	.9848	.1763	5.671	1.015	5.759	**80° 00'**	1.3963
.1774	10	.1765	.9843	.1793	5.576	1.016	5.665	50	1.3934
.1804	20	.1794	.9838	.1823	5.485	1.016	5.575	40	1.3904
.1833	30	.1822	.9833	.1853	5.396	1.017	5.487	30	1.3875
.1862	40	.1851	.9827	.1883	5.309	1.018	5.403	20	1.3846
.1891	50	.1880	.9822	.1914	5.226	1.018	5.320	10	1.3817
.1920	**11° 00'**	.1908	.9816	.1944	5.145	1.019	5.241	**79° 00'**	1.3788
.1949	10	.1937	.9811	.1974	5.066	1.019	5.164	50	1.3759
.1978	20	.1965	.9805	.2004	4.989	1.020	5.089	40	1.3730
.2007	30	.1994	.9799	.2035	4.915	1.020	5.016	30	1.3701
.2036	40	.2022	.9793	.2065	4.843	1.021	4.945	20	1.3672
.2065	50	.2051	.9787	.2095	4.773	1.022	4.876	10	1.3643
.2094	**12° 00'**	.2079	.9781	.2126	4.705	1.022	4.810	**78° 00'**	1.3614
.2123	10	.2108	.9775	.2156	4.638	1.023	4.745	50	1.3584
.2153	20	.2136	.9769	.2186	4.574	1.024	4.682	40	1.3555
.2182	30	.2164	.9763	.2217	4.511	1.024	4.620	30	1.3526
.2211	40	.2193	.9757	.2247	4.449	1.025	4.560	20	1.3497
.2240	50	.2221	.9750	.2278	4.390	1.026	4.502	10	1.3468
.2269	**13° 00'**	.2250	.9744	.2309	4.331	1.026	4.445	**77° 00'**	1.3439
.2298	10	.2278	.9737	.2339	4.275	1.027	4.390	50	1.3410
.2327	20	.2306	.9730	.2370	4.219	1.028	4.336	40	1.3381
.2356	30	.2334	.9724	.2401	4.165	1.028	4.284	30	1.3352
.2385	40	.2363	.9717	.2432	4.113	1.029	4.232	20	1.3323
.2414	50	.2391	.9710	.2462	4.061	1.030	4.182	10	1.3294
.2443	**14° 00'**	.2419	.9703	.2493	4.011	1.031	4.134	**76° 00'**	1.3265
		cos t	sin t	cot t	tan t	csc t	sec t	t degrees	t

TABLE A.3   Trigonometric Functions of an Angle (*Continued*)

t	t degrees	sin t	cos t	tan t	cot t	sec t	csc t		
.2443	14° 00′	.2419	.9703	.2493	4.011	1.031	4.134	76° 00′	1.3265
.2473	10	.2447	.9696	.2524	3.962	1.031	4.086	50	1.3235
.2502	20	.2476	.9689	.2555	3.914	1.032	4.039	40	1.3206
.2531	30	.2504	.9681	.2586	3.867	1.033	3.994	30	1.3177
.2560	40	.2532	.9674	.2617	3.821	1.034	3.950	20	1.3148
.2589	50	.2560	.9667	.2648	3.776	1.034	3.906	10	1.3119
.2618	15° 00′	.2588	.9659	.2679	3.732	1.035	3.864	75° 00′	1.3090
.2647	10	.2616	.9652	.2711	3.689	1.036	3.822	50	1.3061
.2676	20	.2644	.9644	.2742	3.647	1.037	3.782	40	1.3032
.2705	30	.2672	.9636	.2773	3.606	1.038	3.742	30	1.3003
.2734	40	.2700	.9628	.2805	3.566	1.039	3.703	20	1.2974
.2763	50	.2728	.9621	.2836	3.526	1.039	3.665	10	1.2945
.2793	16° 00′	.2756	.9613	.2867	3.487	1.040	3.628	74° 00′	1.2915
.2822	10	.2784	.9605	.2899	3.450	1.041	3.592	50	1.2886
.2851	20	.2812	.9596	.2931	3.412	1.042	3.556	40	1.2857
.2880	30	.2840	.9588	.2962	3.376	1.043	3.521	30	1.2828
.2909	40	.2868	.9580	.2994	3.340	1.044	3.487	20	1.2799
.2938	50	.2896	.9572	.3026	3.305	1.045	3.453	10	1.2770
.2967	17° 00′	.2924	.9563	.3057	3.271	1.046	3.420	73° 00′	1.2741
.2996	10	.2952	.9555	.3089	3.237	1.047	3.388	50	1.2712
.3025	20	.2979	.9546	.3121	3.204	1.048	3.356	40	1.2683
.3054	30	.3007	.9537	.3153	3.172	1.049	3.326	30	1.2654
.3083	40	.3035	.9528	.3185	3.140	1.049	3.295	20	1.2625
.3113	50	.3062	.9520	.3217	3.108	1.050	3.265	10	1.2595
.3142	18° 00′	.3090	.9511	.3249	3.078	1.051	3.236	72° 00′	1.2566
.3171	10	.3118	.9502	.3281	3.047	1.052	3.207	50	1.2537
.3200	20	.3145	.9492	.3314	3.018	1.053	3.179	40	1.2508
.3229	30	.3173	.9483	.3346	2.989	1.054	3.152	30	1.2479
.3258	40	.3201	.9474	.3378	2.960	1.056	3.124	20	1.2450
.3287	50	.3228	.9465	.3411	2.932	1.057	3.098	10	1.2421
.3316	19° 00′	.3256	.9455	.3443	2.904	1.058	3.072	71° 00′	1.2392
.3345	10	.3283	.9446	.3476	2.877	1.059	3.046	50	1.2363
.3374	20	.3311	.9436	.3508	2.850	1.060	3.021	40	1.2334
.3403	30	.3338	.9426	.3541	2.824	1.061	2.996	30	1.2305
.3432	40	.3365	.9417	.3574	2.798	1.062	2.971	20	1.2275
.3462	50	.3393	.9407	.3607	2.773	1.063	2.947	10	1.2246
.3491	20° 00′	.3420	.9397	.3640	2.747	1.064	2.924	70° 00′	1.2217
.3520	10	.3448	.9387	.3673	2.723	1.065	2.901	50	1.2188
.3549	20	.3475	.9377	.3706	2.699	1.066	2.878	40	1.2159
.3578	30	.3502	.9367	.3739	2.675	1.068	2.855	30	1.2130
.3607	40	.3529	.9356	.3772	2.651	1.069	2.833	20	1.2101
.3636	50	.3557	.9346	.3805	2.628	1.070	2.812	10	1.2072
.3665	21° 00′	.3584	.9336	.3839	2.605	1.071	2.790	69° 00′	1.2043
		cos t	sin t	cot t	tan t	csc t	sec t	t degrees	t

TABLE A.3 Trigonometric Functions of an Angle (*Continued*)

t	t degrees	$\sin t$	$\cos t$	$\tan t$	$\cot t$	$\sec t$	$\csc t$		
.3665	**21° 00′**	.3584	.9336	.3839	2.605	1.071	2.790	**69° 00′**	1.2043
.3694	10	.3611	.9325	.3872	2.583	1.072	2.769	50	1.2014
.3723	20	.3638	.9315	.3906	2.560	1.074	2.749	40	1.1985
.3752	30	.3665	.9304	.3939	2.539	1.075	2.729	30	1.1956
.3782	40	.3692	.9293	.3973	2.517	1.076	2.709	20	1.1926
.3811	50	.3719	.9283	.4006	2.496	1.077	2.689	10	1.1897
.3840	**22° 00′**	.3746	.9272	.4040	2.475	1.079	2.669	**68° 00′**	1.1868
.3869	10	.3773	.9261	.4074	2.455	1.080	2.650	50	1.1839
.3898	20	.3800	.9250	.4108	2.434	1.081	2.632	40	1.1810
.3927	30	.3827	.9239	.4142	2.414	1.082	2.613	30	1.1781
.3956	40	.3854	.9228	.4176	2.394	1.084	2.595	20	1.1752
.3985	50	.3881	.9216	.4210	2.375	1.085	2.577	10	1.1723
.4014	**23° 00′**	.3907	.9205	.4245	2.356	1.086	2.559	**67° 00′**	1.1694
.4043	10	.3934	.9194	.4279	2.337	1.088	2.542	50	1.1665
.4072	20	.3961	.9182	.4314	2.318	1.089	2.525	40	1.1636
.4102	30	.3987	.9171	.4348	2.300	1.090	2.508	30	1.1606
.4131	40	.4014	.9159	.4383	2.282	1.092	2.491	20	1.1577
.4160	50	.4041	.9147	.4417	2.264	1.093	2.475	10	1.1548
.4189	**24° 00′**	.4067	.9135	.4452	2.246	1.095	2.459	**66° 00′**	1.1519
.4218	10	.4094	.9124	.4487	2.229	1.096	2.443	50	1.1490
.4247	20	.4120	.9112	.4522	2.211	1.097	2.427	40	1.1461
.4276	30	.4147	.9100	.4557	2.194	1.099	2.411	30	1.1432
.4305	40	.4173	.9088	.4592	2.177	1.100	2.396	20	1.1403
.4334	50	.4200	.9075	.4628	2.161	1.102	2.381	10	1.1374
.4363	**25° 00′**	.4226	.9063	.4663	2.145	1.103	2.366	**65° 00′**	1.1345
.4392	10	.4253	.9051	.4699	2.128	1.105	2.352	50	1.1316
.4422	20	.4279	.9038	.4734	2.112	1.106	2.337	40	1.1286
.4451	30	.4305	.9026	.4770	2.097	1.108	2.323	30	1.1257
.4480	40	.4331	.9013	.4806	2.081	1.109	2.309	20	1.1228
.4509	50	.4358	.9001	.4841	2.066	1.111	2.295	10	1.1199
.4538	**26° 00′**	.4384	.8988	.4877	2.050	1.113	2.281	**64° 00′**	1.1170
.4567	10	.4410	.8975	.4913	2.035	1.114	2.268	50	1.1141
.4596	20	.4436	.8962	.4950	2.020	1.116	2.254	40	1.1112
.4625	30	.4462	.8949	.4986	2.006	1.117	2.241	30	1.1083
.4654	40	.4488	.8936	.5022	1.991	1.119	2.228	20	1.1054
.4683	50	.4514	.8923	.5059	1.977	1.121	2.215	10	1.1025
.4712	**27° 00′**	.4540	.8910	.5095	1.963	1.122	2.203	**63° 00′**	1.0996
.4741	10	.4566	.8897	.5132	1.949	1.124	2.190	50	1.0966
.4771	20	.4592	.8884	.5169	1.935	1.126	2.178	40	1.0937
.4800	30	.4617	.8870	.5206	1.921	1.127	2.166	30	1.0908
.4829	40	.4643	.8857	.5243	1.907	1.129	2.154	20	1.0879
.4858	50	.4669	.8843	.5280	1.894	1.131	2.142	10	1.0850
.4887	**28° 00′**	.4695	.8829	.5317	1.881	1.133	2.130	**62° 00′**	1.0821
		$\cos t$	$\sin t$	$\cot t$	$\tan t$	$\csc t$	$\sec t$	t degrees	t

TABLE A.3 Trigonometric Functions of an Angle (*Continued*)

t	t degrees	sin t	cos t	tan t	cot t	sec t	csc t		
.4887	28° 00′	.4695	.8829	.5317	1.881	1.133	2.130	62° 00′	1.0821
.4916	10	.4720	.8816	.5354	1.868	1.134	2.118	50	1.0792
.4945	20	.4746	.8802	.5392	1.855	1.136	2.107	40	1.0763
.4974	30	.4772	.8788	.5430	1.842	1.138	2.096	30	1.0734
.5003	40	.4797	.8774	.5467	1.829	1.140	2.085	20	1.0705
.5032	50	.4823	.8760	.5505	1.816	1.142	2.074	10	1.0676
.5061	29° 00′	.4848	.8746	.5543	1.804	1.143	2.063	61° 00′	1.0647
.5091	10	.4874	.8732	.5581	1.792	1.145	2.052	50	1.0617
.5120	20	.4899	.8718	.5619	1.780	1.147	2.041	40	1.0588
.5149	30	.4924	.8704	.5658	1.767	1.149	2.031	30	1.0559
.5178	40	.4950	.8689	.5696	1.756	1.151	2.020	20	1.0530
.5207	50	.4975	.8675	.5735	1.744	1.153	2.010	10	1.0501
.5236	30° 00′	.5000	.8660	.5774	1.732	1.155	2.000	60° 00′	1.0472
.5265	10	.5025	.8646	.5812	1.720	1.157	1.990	50	1.0443
.5294	20	.5050	.8631	.5851	1.709	1.159	1.980	40	1.0414
.5323	30	.5075	.8616	.5890	1.698	1.161	1.970	30	1.0385
.5352	40	.5100	.8601	.5930	1.686	1.163	1.961	20	1.0356
.5381	50	.5125	.8587	.5969	1.675	1.165	1.951	10	1.0327
.5411	31° 00′	.5150	.8572	.6009	1.664	1.167	1.942	59° 00′	1.0297
.5440	10	.5175	.8557	.6048	1.653	1.169	1.932	50	1.0268
.5469	20	.5200	.8542	.6088	1.643	1.171	1.923	40	1.0239
.5498	30	.5225	.8526	.6128	1.632	1.173	1.914	30	1.0210
.5527	40	.5250	.8511	.6168	1.621	1.175	1.905	20	1.0181
.5556	50	.5275	.8496	.6208	1.611	1.177	1.896	10	1.0152
.5585	32° 00′	.5299	.8480	.6249	1.600	1.179	1.887	58° 00′	1.0123
.5614	10	.5324	.8465	.6289	1.590	1.181	1.878	50	1.0094
.5643	20	.5348	.8450	.6330	1.580	1.184	1.870	40	1.0065
.5672	30	.5373	.8434	.6371	1.570	1.186	1.861	30	1.0036
.5701	40	.5398	.8418	.6412	1.560	1.188	1.853	20	1.0007
.5730	50	.5422	.8403	.6453	1.550	1.190	1.844	10	.9977
.5760	33° 00′	.5446	.8387	.6494	1.540	1.192	1.836	57° 00′	.9948
.5789	10	.5471	.8371	.6536	1.530	1.195	1.828	50	.9919
.5818	20	.5495	.8355	.6577	1.520	1.197	1.820	40	.9890
.5847	30	.5519	.8339	.6619	1.511	1.199	1.812	30	.9861
.5876	40	.5544	.8323	.6661	1.501	1.202	1.804	20	.9832
.5905	50	.5568	.8307	.6703	1.492	1.204	1.796	10	.9803
.5934	34° 00′	.5592	.8290	.6745	1.483	1.206	1.788	56° 00′	.9774
.5963	10	.5616	.8274	.6787	1.473	1.209	1.781	50	.9745
.5992	20	.5640	.8258	.6830	1.464	1.211	1.773	40	.9716
.6021	30	.5664	.8241	.6873	1.455	1.213	1.766	30	.9687
.6050	40	.5688	.8225	.6916	1.446	1.216	1.758	20	.9657
.6080	50	.5712	.8208	.6959	1.437	1.218	1.751	10	.9628
.6109	35° 00′	.5736	.8192	.7002	1.428	1.221	1.743	55° 00′	.9599
		cos t	sin t	cot t	tan t	csc t	sec t	t degrees	t

TABLE A.3 Trigonometric Functions of an Angle (*Continued*)

t	t degrees	sin t	cos t	tan t	cot t	sec t	csc t		
.6109	**35° 00′**	.5736	.8192	.7002	1.428	1.221	1.743	**55° 00′**	.9599
.6138	10	.5760	.8175	.7046	1.419	1.223	1.736	50	.9570
.6167	20	.5783	.8158	.7089	1.411	1.226	1.729	40	.9541
.6196	30	.5807	.8141	.7133	1.402	1.228	1.722	30	.9512
.6225	40	.5831	.8124	.7177	1.393	1.231	1.715	20	.9483
.6254	50	.5854	.8107	.7221	1.385	1.233	1.708	10	.9454
.6283	**36° 00′**	.5878	.8090	.7265	1.376	1.236	1.701	**54° 00′**	.9425
.6312	10	.5901	.8073	.7310	1.368	1.239	1.695	50	.9396
.6341	20	.5925	.8056	.7355	1.360	1.241	1.688	40	.9367
.6370	30	.5948	.8039	.7400	1.351	1.244	1.681	30	.9338
.6400	40	.5972	.8021	.7445	1.343	1.247	1.675	20	.9308
.6429	50	.5995	.8004	.7490	1.335	1.249	1.668	10	.9279
.6458	**37° 00′**	.6018	.7986	.7536	1.327	1.252	1.662	**53° 00′**	.9250
.6487	10	.6041	.7969	.7581	1.319	1.255	1.655	50	.9221
.6516	20	.6065	.7951	.7627	1.311	1.258	1.649	40	.9192
.6545	30	.6088	.7934	.7673	1.303	1.260	1.643	30	.9163
.6574	40	.6111	.7916	.7720	1.295	1.263	1.636	20	.9134
.6603	50	.6134	.7898	.7766	1.288	1.266	1.630	10	.9105
.6632	**38° 00′**	.6157	.7880	.7813	1.280	1.269	1.624	**52° 00′**	.9076
.6661	10	.6180	.7862	.7860	1.272	1.272	1.618	50	.9047
.6690	20	.6202	.7844	.7907	1.265	1.275	1.612	40	.0918
.6720	30	.6225	.7826	.7954	1.257	1.278	1.606	30	.8988
.6749	40	.6248	.7808	.8002	1.250	1.281	1.601	20	.8959
.6778	50	.6271	.7790	.8050	1.242	1.284	1.595	10	.8930
.6807	**39° 00′**	.6293	.7771	.8098	1.235	1.287	1.589	**51° 00′**	.8901
.6836	10	.6316	.7753	.8146	1.228	1.290	1.583	50	.8872
.6865	20	.6338	.7735	.8195	1.220	1.293	1.578	40	.8843
.6894	30	.6361	.7716	.8243	1.213	1.296	1.572	30	.8814
.6923	40	.6383	.7698	.8292	1.206	1.299	1.567	20	8785
.6952	50	.6406	.7679	.8342	1.199	1.302	1.561	10	.8756
.6981	**40° 00′**	.6428	.7660	.8391	1.192	1.305	1.556	**50° 00′**	.8727
.7010	10	.6450	.7642	.8441	1.185	1.309	1.550	50	.8698
.7039	20	.6472	.7623	.8491	1.178	1.312	1.545	40	.8668
.7069	30	.6494	.7604	.8541	1.171	1.315	1.540	30	.8639
.7098	40	.6517	.7585	.8591	1.164	1.318	1.535	20	.8610
.7127	50	.6539	.7566	.8642	1.157	1.322	1.529	10	.8581
.7156	**41° 00′**	.6561	.7547	.8693	1.150	1.325	1.524	**49° 00′**	.8552
.7185	10	.6583	.7528	.8744	1.144	1.328	1.519	50	.8523
.7214	20	.6604	.7509	.8796	1.137	1.332	1.514	40	.8494
.7243	30	.6626	.7490	.8847	1.130	1.335	1.509	30	.8465
.7272	40	.6648	.7470	.8899	1.124	1.339	1.504	20	.8436
.7301	50	.6670	.7451	.8952	1.117	1.342	1.499	10	.8407
.7330	**42° 00′**	.6691	.7431	.9004	1.111	1.346	1.494	**48° 00′**	.8378
		cos t	sin t	cot t	tan t	csc t	sec t	t degrees	t

TABLE A.3	Trigonometric Functions of an Angle (*Continued*)								
t	t degrees	sin t	cos t	tan t	cot t	sec t	csc t		
.7330	**42° 00′**	.6691	.7431	.9004	1.111	1.346	1.494	**48° 00**	.8378
.7359	10	.6713	.7412	.9057	1.104	1.349	1.490	50	.8348
.7389	20	.6734	.7392	.9110	1.098	1.353	1.485	40	.8319
.7418	30	.6756	.7373	.9163	1.091	1.356	1.480	30	.8290
.7447	40	.6777	.7353	.9217	1.085	1.360	1.476	20	.8261
.7476	50	.6799	.7333	.9271	1.079	1.364	1.471	10	.8232
.7505	**43° 00′**	.6820	.7314	.9325	1.072	1.367	1.466	**47° 00′**	.8203
.7534	10	.6841	.7294	.9380	1.066	1.371	1.462	50	.8174
.7563	20	.6862	.7274	.9435	1.060	1.375	1.457	40	.8145
.7592	30	.6884	.7254	.9490	1.054	1.379	1.453	30	.8116
.7621	40	.6905	.7234	.9545	1.048	1.382	1.448	20	.8087
.7650	50	.6926	.7214	.9601	1.042	1.386	1.444	10	.8058
.7679	**44° 00′**	.6947	.7193	.9657	1.036	1.390	1.440	**46° 00′**	.8029
.7709	10	.6967	.7173	.9713	1.030	1.394	1.435	50	.7999
.7738	20	.6988	.7153	.9770	1.024	1.398	1.431	40	.7970
.7767	30	.7009	.7133	.9827	1.018	1.402	1.427	30	.7941
.7796	40	.7030	.7112	.9884	1.012	1.406	1.423	20	.7912
.7825	50	.7050	.7092	.9942	1.006	1.410	1.418	10	.7883
.7854	**45° 00′**	.7071	.7071	1.0000	1.0000	1.414	1.414	**45° 00′**	.7854
		cos t	sin t	cot t	tan t	csc t	sec t	t degrees	t

TABLE A.4	Trigonometric Functions of a Real Number					
t	$\sin t$	$\cos t$	$\tan t$	$\cot t$	$\sec t$	$\csc t$
.00	.0000	1.0000	.0000		1.000	
.01	.0100	1.0000	.0100	99.997	1.000	100.00
.02	.0200	.9998	.0200	49.993	1.000	50.00
.03	.0300	.9996	.0300	33.323	1.000	33.34
.04	.0400	.9992	.0400	24.987	1.001	25.01
.05	.0500	.9988	.0500	19.983	1.001	20.01
.06	.0600	.9982	.0601	16.647	1.002	16.68
.07	.0699	.9976	.0701	14.262	1.002	14.30
.08	.0799	.9968	.0802	12.473	1.003	12.51
.09	.0899	.9960	.0902	11.081	1.004	11.13
.10	.0998	.9950	.1003	9.967	1.005	10.02
.11	.1098	.9940	.1104	9.054	1.006	9.109
.12	.1197	.9928	.1206	8.293	1.007	8.353
.13	.1296	.9916	.1307	7.649	1.009	7.714
.14	.1395	.9902	.1409	7.096	1.010	7.166
.15	.1494	.9888	.1511	6.617	1.011	6.692
.16	.1593	.9872	.1614	6.197	1.013	6.277
.17	.1692	.9856	.1717	5.826	1.015	5.911
.18	.1790	.9838	.1820	5.495	1.016	5.586
.19	.1889	.9820	.1923	5.200	1.018	5.295
.20	.1987	.9801	.2027	4.933	1.020	5.033
.21	.2085	.9780	.2131	4.692	1.022	4.797
.22	.2182	.9759	.2236	4.472	1.025	4.582
.23	.2280	.9737	.2341	4.271	1.027	4.386
.24	.2377	.9713	.2447	4.086	1.030	4.207
.25	.2474	.9689	.2553	3.916	1.032	4.042
.26	.2571	.9664	.2660	3.759	1.035	3.890
.27	.2667	.9638	.2768	3.613	1.038	3.749
.28	.2764	.9611	.2876	3.478	1.041	3.619
.29	.2860	.9582	.2984	3.351	1.044	3.497
.30	.2955	.9553	.3093	3.233	1.047	3.384
.31	.3051	.9523	.3203	3.122	1.050	3.278
.32	.3146	.9492	.3314	3.018	1.053	3.179
.33	.3240	.9460	.3425	2.920	1.057	3.086
.34	.3335	.9428	.3537	2.827	1.061	2.999
.35	.3429	.9394	.3650	2.740	1.065	2.916
.36	.3523	.9359	.3764	2.657	1.068	2.839
.37	.3616	.9323	.3879	2.578	1.073	2.765
.38	.3709	.9287	.3994	2.504	1.077	2.696
.39	.3802	.9249	.4111	2.433	1.081	2.630

From Koshy: *College Algebra and Trigonometry with Applications*, 1986. McGraw-Hill. Reprinted with permission.

TABLE A.4	Trigonometric Functions of a Real Number (*Continued*)					
t	sin *t*	cos *t*	tan *t*	cot *t*	sec *t*	csc *t*
.40	.3894	.9211	.4228	2.365	1.086	2.568
.41	.3986	.9171	.4346	2.301	1.090	2.509
.42	.4078	.9131	.4466	2.239	1.095	2.452
.43	.4169	.9090	.4586	2.180	1.100	2.399
.44	.4259	.9048	.4708	2.124	1.105	2.348
.45	.4350	.9004	.4831	2.070	1.111	2.299
.46	.4439	.8961	.4954	2.018	1.116	2.253
.47	.4529	.8916	.5080	1.969	1.122	2.208
.48	.4618	.8870	.5206	1.921	1.127	2.166
.49	.4706	.8823	.5334	1.875	1.133	2.125
.50	.4794	.8776	.5463	1.830	1.139	2.086
.51	.4882	.8727	.5594	1.788	1.146	2.048
.52	.4969	.8678	.5726	1.747	1.152	2.013
.53	.5055	.8628	.5859	1.707	1.159	1.978
.54	.5141	.8577	.5994	1.668	1.166	1.945
.55	.5227	.8525	.6131	1.631	1.173	1.913
.56	.5312	.8473	.6269	1.595	1.180	1.883
.57	.5396	.8419	.6410	1.560	1.188	1.853
.58	.5480	.8365	.6552	1.526	1.196	1.825
.59	.5564	.8309	.6696	1.494	1.203	1.797
.60	.5646	.8253	.6841	1.462	1.212	1.771
.61	.5729	.8196	.6989	1.431	1.220	1.746
.62	.5810	.8139	.7139	1.401	1.229	1.721
.63	.5891	.8080	.7291	1.372	1.238	1.697
.64	.5972	.8021	.7445	1.343	1.247	1.674
.65	.6052	.7961	.7602	1.315	1.256	1.652
.66	.6131	.7900	.7761	1.288	1.266	1.631
.67	.6210	.7838	.7923	1.262	1.276	1.610
.68	.6288	.7776	.8087	1.237	1.286	1.590
.69	.6365	.7712	.8253	1.212	1.297	1.571
.70	.6442	.7648	.8423	1.187	1.307	1.552
.71	.6518	.7584	.8595	1.163	1.319	1.534
.72	.6594	.7518	.8771	1.140	1.330	1.517
.73	.6669	.7452	.8949	1.117	1.342	1.500
.74	.6743	.7385	.9131	1.095	1.354	1.483
.75	.6816	.7317	.9316	1.073	1.367	1.467
.76	.6889	.7248	.9505	1.052	1.380	1.452
.77	.6961	.7179	.9697	1.031	1.393	1.437
.78	.7033	.7109	.9893	1.011	1.407	1.422
.79	.7104	.7038	1.009	.9908	1.421	1.408

TABLE A.4 Trigonometric Functions of a Real Number (*Continued*)

t	sin t	cos t	tan t	cot t	sec t	csc t
.80	.7174	.6967	1.030	.9712	1.435	1.394
.81	.7243	.6895	1.050	.9520	1.450	1.381
.82	.7311	.6822	1.072	.9331	1.466	1.368
.83	.7379	.6749	1.093	.9146	1.482	1.355
.84	.7446	.6675	1.116	.8964	1.498	1.343
.85	.7513	.6600	1.138	.8785	1.515	1.331
.86	.7578	.6524	1.162	.8609	1.533	1.320
.87	.7643	.6448	1.185	.8437	1.551	1.308
.88	.7707	.6372	1.210	.8267	1.569	1.297
.89	.7771	.6294	1.235	.8100	1.589	1.287
.90	.7833	.6216	1.260	.7936	1.609	1.277
.91	.7895	.6137	1.286	.7774	1.629	1.267
.92	.7956	.6058	1.313	.7615	1.651	1.257
.93	.8016	.5978	1.341	.7458	1.673	1.247
.94	.8076	.5898	1.369	.7303	1.696	1.238
.95	.8134	.5817	1.398	.7151	1.719	1.229
.96	.8192	.5735	1.428	.7001	1.744	1.221
.97	.8249	.5653	1.459	.6853	1.769	1.212
.98	.8305	.5570	1.491	.6707	1.795	1.204
.99	.8360	.5487	1.524	.6563	1.823	1.196
1.00	.8415	.5403	1.557	.6421	1.851	1.188
1.01	.8468	.5319	1.592	.6281	1.880	1.181
1.02	.8521	.5234	1.628	.6142	1.911	1.174
1.03	.8573	.5148	1.665	.6005	1.942	1.166
1.04	.8624	.5062	1.704	.5870	1.975	1.160
1.05	.8674	.4976	1.743	.5736	2.010	1.153
1.06	.8724	.4889	1.784	.5604	2.046	1.146
1.07	.8772	.4801	1.827	.5473	2.083	1.140
1.08	.8820	.4713	1.871	.5344	2.122	1.134
1.09	.8866	.4625	1.917	.5216	2.162	1.128
1.10	.8912	.4536	1.965	.5090	2.205	1.122
1.11	.8957	.4447	2.014	.4964	2.249	1.116
1.12	.9001	.4357	2.066	.4840	2.295	1.111
1.13	.9044	.4267	2.120	.4718	2.344	1.106
1.14	.9086	.4176	2.176	.4596	2.395	1.101
1.15	.9128	.4085	2.234	.4475	2.448	1.096
1.16	.9168	.3993	2.296	.4356	2.504	1.091
1.17	.9208	.3902	2.360	.4237	2.563	1.086
1.18	.9246	.3809	2.427	.4120	2.625	1.082
1.19	.9284	.3717	2.498	.4003	2.691	1.077

TABLE A.4	Trigonometric Functions of a Real Number (*Continued*)					
t	sin *t*	cos *t*	tan *t*	cot *t*	sec *t*	csc *t*
1.20	.9320	.3624	2.572	.3888	2.760	1.073
1.21	.9356	.3530	2.650	.3773	2.833	1.069
1.22	.9391	.3436	2.733	.3659	2.910	1.065
1.23	.9425	.3342	2.820	.3546	2.992	1.061
1.24	.9458	.3248	2.912	.3434	3.079	1.057
1.25	.9490	.3153	3.010	.3323	3.171	1.054
1.26	.9521	.3058	3.113	.3212	3.270	1.050
1.27	.9551	.2963	3.224	.3102	3.375	1.047
1.28	.9580	.2867	3.341	.2993	3.488	1.044
1.29	.9608	.2771	3.467	.2884	3.609	1.041
1.30	.9636	.2675	3.602	.2776	3.738	1.038
1.31	.9662	.2579	3.747	.2669	3.878	1.035
1.32	.9687	.2482	3.903	.2562	4.029	1.032
1.33	.9711	.2385	4.072	.2456	4.193	1.030
1.34	.9735	.2288	4.256	.2350	4.372	1.027
1.35	.9757	.2190	4.455	.2245	4.566	1.025
1.36	.9779	.2092	4.673	.2140	4.779	1.023
1.37	.9799	.1994	4.913	.2035	5.014	1.021
1.38	.9819	.1896	5.177	.1931	5.273	1.018
1.39	.9837	.1798	5.471	.1828	5.561	1.017
1.40	.9854	.1700	5.798	.1725	5.883	1.015
1.41	.9871	.1601	6.165	.1622	6.246	1.013
1.42	.9887	.1502	6.581	.1519	6.657	1.011
1.43	.9901	.1403	7.055	.1417	7.126	1.010
1.44	.9915	.1304	7.602	.1315	7.667	1.009
1.45	.9927	.1205	8.238	.1214	8.299	1.007
1.46	.9939	.1106	8.989	.1113	9.044	1.006
1.47	.9949	.1006	9.887	.1011	9.938	1.005
1.48	.9959	.0907	10.983	.0910	11.029	1.004
1.49	.9967	.0807	12.350	.0810	12.390	1.003
1.50	.9975	.0707	14.101	.0709	14.137	1.003
1.51	.9982	.0608	16.428	.0609	16.458	1.002
1.52	.9987	.0508	19.670	.0508	19.695	1.001
1.53	.9992	.0408	24.498	.0408	24.519	1.001
1.54	.9995	.0308	32.461	.0308	32.476	1.000
1.55	.9998	.0208	48.078	.0208	48.089	1.000
1.56	.9999	.0108	92.620	.0108	92.626	1.000
1.57	1.0000	.0008	1255.8	.0008	1255.8	1.000